Journal of *Neural* Transmission

Supplementum 41

K. F. Tipton, M. B. H. Youdim, C. J. Barwell,
B. A. Callingham, and G. A. Lyles (eds.)

Amine Oxidases: Function and Dysfunction

*Proceedings of the
5ᵗʰ International Amine Oxidase Workshop,
Galway, Ireland, August 22–25, 1992*

Springer-Verlag   Wien New York

Prof. Dr. K. F. Tipton
Biochemistry Department, Trinity College, Dublin, Ireland

Prof. Dr. M. B. H. Youdim
Department of Pharmacology, Technion, Haifa, Israel

Dr. C. J. Barwell
School of Pharmacy and Biomedical Sciences, University of Portsmouth, United Kingdom

Prof. Dr. B. A. Callingham
Department of Pharmacology, University of Cambridge, United Kingdom

Dr. G. A. Lyles
Department of Pharmacology and Clinical Pharmacology, University of Dundee,
United Kingdom

Product Liability: The publisher can give no guarantee for information about drug dosage and
application thereof contained in this book. In every individual case the respective user must
check its accuracy by consulting other pharmaceutical literature. The use of registered names,
trademarks, etc. in this publication does not imply, even in the absence of a specific statement,
that such names are exempt from the relevant protective laws and regulations and therefore free
for general use.

Typesetting: Best-set Typesetter Ltd, Hong Kong

Printed on acid-free and chlorine-free bleached paper

With 113 Figures

Library of Congress Cataloging-in-Publication Data: International Amine
Oxidases Workshop (5th: 1992: Galway, Ireland)
Amino oxidases: form and dysfunction: proceedings of the 5th International Amine Oxidases Workshop, Galway, Ireland, August 22–25, 1992
/ K. F. Tipton ... [et al.], eds. p. cm. — (Journal of neural transmission.
Supplementum; 41)
ISBN 0-387-82521-5 1. Monoamine oxidase — Congresses. 2. Amine oxidase — Congresses. 3. Monoamine oxidase — Inhibitors — Congresses.
I. Tipton, Keith F. II. Title. III. Series. QP603.M6I58 1992.
612.8'042 — dc20. 94-6714 CIP

ISSN 0303-6995
ISBN-13:978-3-211-82521-1   e-ISBN-13:978-3-7091-9324-2
DOI: 10.1007/978-3-7091-9324-2

Preface

Amine Oxidase workshops are held in alternate years, and such is the pace of advance in this area that there is always a great deal of new and exciting material to report. New data on the structures of the monoamine oxidases and their promoter regions are presented in these proceedings along with unexpected discoveries about their behaviour, function, and dysfunction. Renewed interest in the design and operation of selective monoamine oxidase inhibitors has stemmed from suggestions that inhibitors of monoamine oxidase-B may be neuroprotective whereas reversible monoamine oxidase-A inhibitors are free of the "cheese effects" associated with irreversible inhibitors of that enzyme at doses where they are effective antidepressants. All these aspects and many more are discussed in this volume. The only thing still lacking is the X-ray crystal structure of the monoamine oxidases; membrane-bound proteins are so beastly! However, advances in other areas suggest that problem will eventually be solved, perhaps in time for the next Amine Oxidase workshop which is to be held in Saskatoon, Canada, in 1994.

The semicarbazide-sensitive amino oxidases (SSAO) remain the Cinderella of this field, nevertheless, there were many Prince (and Princess) Charmings at this particular ball to flirt with her. Despite this there is still no clear indication of the physiological function(s) of this group of enzymes. Several possibilities are discussed here and the developments reported in the design of new and more specific inhibitors should eventually lead to a better understanding of these aspects.

We would like to express our gratitude to ASTA Medica AG, Britania Pharmaceuticals, Chinoin, Ciba-Geigy (Basel), Convention Bureau of Ireland, Farmitalia Carlo Erba, Galway Regional Tourist Board, Hoffman-La Roche (Basel), International Society for Neurochemistry, Marian Merrell Dow, Medlabs, Merk Sharp and Dohme Research Laboratories U.S.A., Millipore, Pharmoa, Sandoz Pharmaceutical, Sanofi Winthrop, Somerset and Synthelabó-L.E.R.S., whose generous support made the meeting possible. Although the workshop organisers were officially listed as Keith Tipton and Moussa Youdim, the acutal work, was of course done by others and we are particularly grateful to Mary Anderson, Melina Lawless, John McCrodden, Gemma Tipton, and Gill Tipton for making everything happen. Perhaps a special word of praise is also due to the Irish weather over the period of the Meeting which ensured that most of those attending remained at the sessions rather than experiencing the delights of Western Ireland.

The 5th Amine Oxidase Workshop and these proceedings are dedicated to Irv Kopin. An appreciation of his contributions is published elsewhere in this volume and it is probably sufficient to say that on behalf of so many working in this field, we are proud to honour a man who has contributed so much to our understanding of the functions and metabolism of the biogenic amines.

Dublin, April 1994

K. F. TIPTON
M. B. H. YOUDIM
C. J. BARWELL
B. A. CALLINGHAM
G. A. LYLES

Irwin J. Kopin, M.D.

Irwin Kopin ("Irv") is one of the major contributors to understanding the function and metabolism of catecholamines. His research spans the whole spectrum of the field, from biosynthesis, release, and metabolism of the biogenic amines, to detailed pharmacology of the drugs affecting the system, metabolism of the endogenous compounds in laboratory animals and man, and their role in cardiovascular and mental disease. It is particularly appropriate that he should be honoured at the Galway meeting on MAO and trace amines, since he has made basic contributions to both areas. In a series of classical papers in the 1960s, he described the role of MAO in the metabolism of endogenous noradrenaline, and also demonstrated the accumulation and release of octopamine following MAO inhibition. More recently, he was a major force behind the unravelling of the MPTP story at NIH, which spawned so much excitement and renewed interest in selective neurotoxicity, with particular application to the role of MAO-B.

Irv Kopin was born (1929) in New York, and acquired an early interest in applied chemistry working in his father's mirror silvering factory. When he finally succeeded in making a really clean mirror (he relates), his father said:

"Now you can go to college". Chemistry, however, is not the only of his strong basic sciences. His first of many awards, while at the College of the City of New York, was for excellence in pure and applied calculus. His mathematical ability is immediately evident in the logical approach he brings to a variety of scientific problems, and has enabled him to unravel complex problems of metabolite distribution in the body. He obtained both his B.Sc. (biochemistry) and M.D. degrees at McGill University, Montreal, and joined the National Institutes of Health (Laboratory of Clinical Science), Bethesda, in 1957. There, as is now well known, he joined forces with Julius Axelrod. The dingy 2D corridor, deep in the labyrinth of the Clinical Center, seems an inappropriate place for such a scintillating team as Kopin and Axelrod, but the results of this combination, as well as with many other associates, are now history.

Since joining NIH, he has left the Bethesda campus for only one year, to complete his medical residency at Columbia, New York. The medical background has played an important role in his ability to apply basic science to the study and treatment of human illness, and is also evident in the comprehensiveness of his physiological knowledge. The application of his work to mental illness earned him the Anna-Monika Award for Research on Depression (twice). He is unrelenting in his pursuit of basic research, despite the rigors of a senior administrative appointment in the US Government (currently, he is Director of Intramural Research, National Institute of Neurological Disorders). Despite his heavy administrative load, he is readily available to a large group of postdoctoral fellows and other research workers, and revels in bringing scientific order to a chaotic problem, such as understanding the meaning of a complex PET scan, or the intricacies of CNS dopamine release.

In spite of responsibilities to a variety of scientific committees, editorial boards, international congresses and currently serving as President of the American Society of Neuropsychopharmacology, he maintains his regular swimming practice (was intercollegiate champion). At his home in Bethesda he is happy to show a visitor his collection of stamps and postcards from all over the world. With his wife Rita, whom he met at McGill, he has two daughters and a son, Alan, who has started his career in medical research with important publications on the molecular biology of gastro-intestinal hormones. Rita runs a center for Hebrew teachers, and is particularly proud of her innovative resource department.

Few researchers in the life sciences possess such diverse skills in chemistry, mathematics, and physiology as Irv Kopin. Still fewer, who possess such skills, are able to put them to such valuable use in their chosen branch of science; but only a very small number are capable of instilling as much appreciation for science in those around him, a quality for which many of us are indebted.

J. P. M. FINBERG

Contents

X   Contents

Semicarbazide-sensitive amine oxidases

XII   Contents

Listed in Current Contents

Monoamine oxidase: structure

J Neural Transm (1994) [Suppl] 41: 3–15

Functional expression of C-terminally truncated human monoamine oxidase type A in *Saccharomyces cerevisiae*

W. Weyler

Molecular Biology Division, Veterans Affairs Medical Center, and Division of Toxicology, School of Pharmacy, University of California, San Francisco, California, U.S.A.

Summary. The deduced amino acid sequence of human liver monoamine oxidase type A was analyzed with secondary structure programs. These analyses and comparison to other flavoproteins identified a single potential transmembrane hydrophobic peptide at the C-terminus suggesting that this peptide is a membrane anchor and that the remainder of the protein constitutes a soluble domain. Truncation of the C-terminus by 24 amino acids which are inclusive of the putative transmembrane peptide, however, gave a protein which exhibited solubility properties substantially similar to the wild type enzyme. This result indicates that the hydrophobic behavior of monoamine oxidase type A is due to more complex features than a single transmembrane anchor. The mutant enzyme expressed in yeast appears to form a disulfide bond which reduces catalytic effciency by up to 90%. Full activity, however, can be recovered by incubation with dithiothreitol, suggesting that in the wild type enzyme the amino acid residues deleted in the mutant protein protect two cysteine residues (those involved in the formation of the disulfide bond in the mutant) from oxidation and that the deleted residues are in close proximity to the active site. The activation experiments indicated that the deleted amino acids do not contribute any catalytic residues to the active site.

Introduction

Monoamine oxidase is a flavoprotein (Kearney et al., 1971; Weyler, 1989) occuring in two forms, types A and B, in the outer membrane of mitochondria (Schnaitman et al., 1967) and is found in many organisms in the animal kingdom. The role of the enzyme is the degradation of biogenic and xenobiotic amines in the nervous system and other tissues. Knowledge on monoamine oxidase has recently been reviewed by us (Weyler et al., 1990a). The enzyme's mode of binding to the membrane is unknown. Common modes of protein binding to biological membranes are by a single transmembrane helical segment, or membrane anchor, either at the C- or N-terminus of a protein. In the nomenclature of Blobel (1980) these are referred to as class I and class II intrinsic membrane proteins, respectively. A large number of membrane proteins, particularly receptors that are components of transmembrane signalling systems, are bound to the membranes by multiple transmembrane helixes which often make heterosubunit contacts that are thought to be involved in transmission of the signal; these are referred to as class III proteins. The final type of integral membrane protein in this system of nomenclature, type IV, are ion channel forming proteins such as proton pumps and porins which form aqueous channels through a membrane by the assembly, in the membrane, of multiple identical or similar subunits. This classification system is relatively gross and within each class of proteins there is considerable diversity in structure. Direct evidence for these modes of binding is the exception rather than the rule as only a few membrane proteins have been crystallized, e.g., the reaction center of *Rhodobacter sphaeroides* (Allen et al., 1988), and only a few membrane proteins have been solubilized by truncation of their putative transmembrane sequences, e.g., cytochrome b5 (Spatz and Strittmatter, 1971). An entirely separate mechanism of association of proteins with their target membrane is through the posttranslational modification of the protein with a lipid, fatty acid, or isoprenoid anchor. To date there are five classes of these. One is myristoylation, and this occurs exclusively at the N-terminal glycine of a number of proteins (Gordon et al., 1991). A second is palmitoylation mediated via a cysteine residue thioester linkage, which appears not restricted to a specific region of the protein (Sefton and Buss, 1987). A third type of anchor is the posttranslational addition of a glycosyl-phosphatidylinositol anchor added to a C-terminal residue that is part of a weak consensus sequence. This modification is usually accompanied by the concomitant trimming of a C-terminal targeting sequence (Ferguson and Williams, 1988). The fourth and fifth types are the most recently discovered membrane anchors and consist of the modification of a cysteine residue that is part of a consensus amino acid tetrad at the extreme C-terminus of a number of proteins by a farnesyl or geranylgeranyl moiety (Ishibashi et al., 1984; Sakagami et al., 1981; Maltese, 1990). There is direct evidence for many of these anchors, as specific processing enzymes have been identified for all but the palmitoyl anchor. Judging from the vast number of membrane proteins only poorly, or not at all, characterized with respect to their mode of membrane attachment it is likely that additional classes of this type will be found.

We and others recently cloned MAO from human liver (Hsu et al., 1988; Bach et al., 1988) and have expressed MAO A in *Saccharomyces cerevisiae* (Weyler et al., 1990b; Urban et al., 1991). These advances have made possible the study of membrane interactions of MAO using molecular biology approaches. Here I report on observations of a monoamine oxidase type A mutant protein with the putative C-terminal membrane spanning peptide deleted.

Methods

General

All reagents and DNA modifying enzymes were from commercial sources. Oligonucleotides were synthesized with a Biosearch DNA synthesizer (New Brunswick Scientific Company). PCR methods were as recommended by the manufacturer of the PCR kit (Perkin-Elmer Cetus). Growth and selection conditions for bacteria and DNA manipulations were carried out by standard methodologies found in Maniatis et al. (1983) or Sambrook et al. (1989). For culturing bacteria 2XYT liquid or LB solid media was used. Large scale plasmid preparations from bacteria (75 ml or larger cultures) were performed with the aid of a Qiagen plasmid kit (Studio City, CA). Restriction enzyme screening of clones was performed with DNA obtained by the rapid boiling method in Maniatis et al. (1983). Competent *Escherichia coli*, strain DH5αF' (BRL), were used for transformations during construction of plasmids. The *S. cerevisiae* strain used for enzyme expression is designated RH218 (*Mata*, *trp1*, *gal2*, *cir⁰*; ATCC #44076). Protein measurements were made with the biuret reaction (Layne, 1957). Centrifugations were performed in Beckman J21 or L8 centrifuges operated with JA-20 or type 65 rotors, respectively; average values of **g** are given.

Construction of C-terminal deletion mutant (MA-CΔ24)

The mutant human liver MAO A from which the last 24 amino acids were deleted was constructed from plasmid *p*BSMA3'ut (Weyler et al., 1990b) by deleting the cDNA fragment from the unique internal *Bst* EII site to the stop codon and replacing it with a PCR fragment having codons for the last 24 amino acids deleted. The construction was verified by restriction analysis. The modified gene was excised from the construction vector with *Bam* HI and inserted into the yeast expression vector, *p*GPD(G)-2 (Bitter and Eagan, 1988), as previously done for the wild type enzyme (Weyler et al., 1990b). The correct orientation could be ascertained by comparison of a *Hind* III restriction analysis with a similar analysis for the wild type enzyme the orientation of which had been verified by DNA sequencing and expression of the enzyme (Weyler et al., 1990b). The expression vector carrying the truncated MAO A gene in the correct orientation was designated *p*MA-CΔ24.

Expression of mutant enzyme in yeast

*p*MA-CΔ24 was transformed into *S. cerevisiae* by the method found in Sherman et al. (1986). Selection and growth conditions were as previously described for the wild type enzyme (Weyler et al., 1990b).

Partial purification of mutant MAO A

The enzyme was characterized both in crude yeast cell extracts and as a partially purified preparation. Crude homogenates were prepared by homogenizing yeast cells with glass beads in 0.1 M triethanolamine HCL buffer, adjusted to pH 7.2 with NH_4OH, 0.5 mM PMSF, and depending on the experiment, either with or without, 5 mM dithiothreitol, 5 mM EDTA, and 1 mM amphetamine. The homogenate was centrifuged at 800 **g** and the supernatant was recentrifuged at 48k **g** for 30 min. Further fractionation was achieved by centrifugation of the 48k **g** supernatant at 250k **g** and suspending the pellet at 30 mg protein/ml in the same buffer and treating with phospholipase A and C as previously reported for the isolation of human placenta enzyme (Weyler and Salach, 1985). After the 1 h treatment, Triton X-100 was added to 0.5% from a 20% stock solution; the mixture was homogenized with a glass/Teflon homogenizer at 4°C for 5 min and chromatographed on DEAE Sepharose CL-6B as previously reported in the purification of human placenta MAO A (Weyler and Salach, 1985). The fractions showing MAO A activity were combined into a single pool, concentrated by ultra-filtration with an Amicon PM30 membrane, made 50% in glycerol, and stored at 1.25 mg protein/ml in liquid nitrogen until used. Both the mutant enzyme and wild type enzyme did not lose activity under these storage conditions.

Attempt to solubilize MA-CΔ24 protein without prior phospholipase treatment

Triton X-100, at a final concentration of 0.5%, was homogenized into the 250k **g** pellet material from above prior to phospholipase treatment and the homogenate was recentrifuged at 250k **g**.

Primary structure analysis

Hydropathy plots and overall hydrophobicity of the deduced sequence of human liver cDNA were calculated by the method of Kyte and Doolittle (1982) using the computer program SOAP. Hydrophobic transmembrane peptides were predicted by the method of Klein et al. (1985), also using SOAP, and by the methods of Eisenberg et al. (1984) with the program HELIXMEM. All three analysis programs are part of the software PCGENE from IntelliGenetics Inc., CA.

Results

Primary structure analysis

Figure 1 shows the Kyte-Doolittle hydropathy plot of human liver MAO A. In conjunction with the transmembrane helix analysis, using the algorithms of Klein et al. (1985) and Eisenberg et al. (1984), two potential transmembrane peptides were identified. The first at the N-terminus and the other at the C-terminus. Only the peptide at the C-terminus was considered a plausible membrane anchor (see Discussion). The overall calculated hydrophobicity, using Kyte and Doolittle (1982) parameters, of wild type monoamine oxidase was −2.79 which is in the range of a soluble protein rather than indicative of an intrinsic membrane protein. This result supported the notion that the putative transmembrane peptide at the C-terminus was the only domain interacting with the membrane and that it represented a membrane anchor. The possibility was tested by deletion of 24 amino acids from the C-terminus of the wild type enzyme.

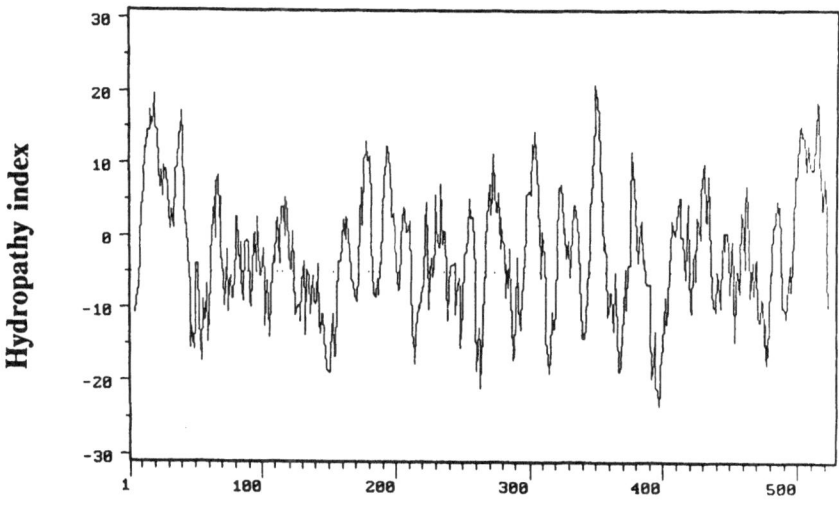

Residue number

Fig. 1. Hydropathy plot according to Kyte and Doolittle (1982) for human liver MAO A amino acid sequence deduced from cDNA. Sequence analyzed is that of Hsu et al. (1988). Window size was set at 9 residues for this analysis. Only peptides at the extreme N- and C-termini appear to be of sufficient hydrophobicity and length for classification as transmembrane sequences. The average hydrophobicity for the entire sequence has a GRAVY score of −2.79 (Kyte and Doolittle, 1982) suggesting that the protein is soluble. The methods of Klein et al. (1985) and Eisenberg et al. (1984) similarly predict the sequences at the termini as potential transmembrane spanning helices

Construction of the C-terminal deletion mutant

Construction of the mutant gene is described in "Methods". After substitution of the *Bst* Ell/*Afl* II fragment of *p*BSMA3′ut (Weyler et al., 1990b) with the PCR fragment lacking the codons for the last 24 amino acids, the mutant gene, contained in a *Bam* HI fragment, had the expected size as is indicated in Fig. 2a. This fragment was inserted into the *Bam* HI cloning site of the expression vector *p*GPD(G)-2 (Bitter and Eagan, 1988) to yield *p*MA-C△24.

Expression of the mutant enzyme in yeast

Yeast transformed with *p*MA-C△24 expressed MAO A activity at levels equal to about 10% of yeast transformed with wild type enzyme. The first evidence of expression was obtained by direct assay of yeast cells permeabilized to the MAO A substrate kynuramine (Weissbach et al., 1960) by 5 cycles of freezing at −70°C and thawing at 30°C, 2 and 1 min, respectively, as small pellets and immediately suspending treated cells in assay medium for 10 min at 30°C. The change in absorbance at 316 nm due

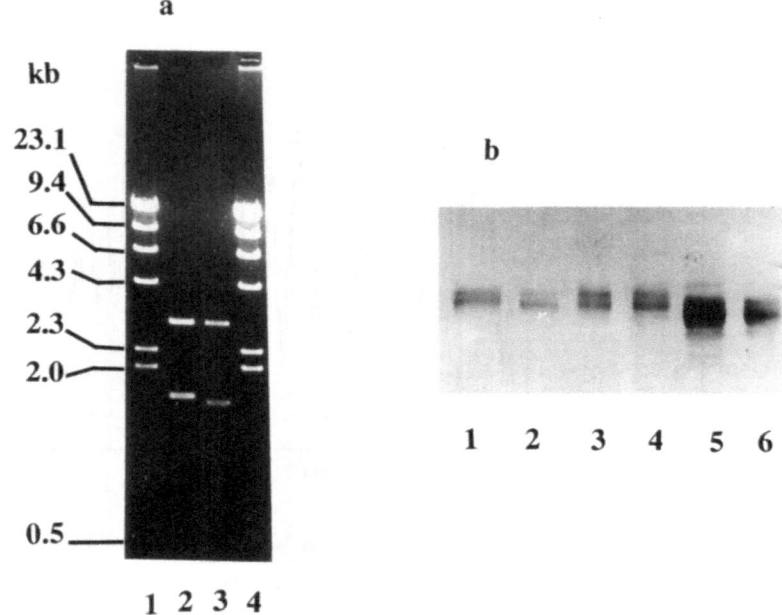

Fig. 2. DNA agarose gel and Western blot showing changes in size of human liver MAO A cDNA and protein due to truncation. The panels show: **a** DNA gel, 0.8% agarose. *1* and *4* 100 and 200 ng, respectively, of λDNA/Hind III fragments; size is indicated in kb on left margin. *2* and *3* upper bands are due to linearized Bluescript vector; lower bands are *Bam* HI fragments of full length MAO A gene in lane 2 and truncated gene in lane 3; **b** Western blot showing wild type and mutant MAO A expressed in yeast. Samples were electrophoresed in a 7.5% acrylamide SDS denaturing gel (Laemmli, 1970). Protein was transblotted onto nitrocellulose (Harlow and Lane, 1988) and developed with sheep anti-human placenta MAO A (Kirchgessner and Pintar, 1991) and anti-sheep IgG alkaline phosphatase-coupled antibodies. *1* and *2* purified human liver monoamine oxidase expressed in yeast, 600 and 300 ng, respectively; *3* and *4* crude homogenate of yeast transformed with wild type human liver monoamine oxidase, 15 and 7.5 μg total protein; *5* and *6* crude homogenate of yeast transformed with truncated monoamine oxidase, 15 and 7.5 μg total protein. Protein concentration for the purified enzyme was estimated from the flavin$^{\Delta ox\text{-}red}$ extinction coefficient of $10,800\,cm^{-1}M^{-1}$ at 450 nm (Weyler and Salach, 1985) and protein for the crude homogenates was determined with the biuret reaction (Layne, 1957)

to the product 4-hydroxyquinoline was then recorded. Enzyme was also detected by SDS-PAGE and Western blot assay (Fig. 2b). The Western blot indicated that although enzymatic activity was only 10% of the wild type the amount of protein expressed was at a level similar to wild type expression, suggesting that the catalytic efficiency of the mutant enzyme was substantially decreased.

Partial purification of the mutant protein

A significant difference of MA-CΔ24 protein was that centrifugation at 48k **g** was not sufficient to sediment the mutant protein as is observed for the wild type enzyme. Centrifugation at 250k g, however, gave a loose pellet containing most of the enzyme activity with less than 10 percent remaining in the supernatant. This indicated that the mutant enzyme is still membrane bound although probably not to the mitochondrial fraction. Treatment of this membrane fraction with 0.5% Triton X-100 alone did not liberate activity from the membrane. Both phospholipase A and C digestion and extraction with this detergent were required to solubilize activity. Thus, while the distribution of the enzyme in the membrane fractions differed from wild type enzyme the solubility behavior remained similar.

Size of the mutant enzyme

The size of the mutant enzyme was expected to be 2,400 Da smaller than the wild type enzyme. This was verified by analysis of crude homogenate and partially purified enzyme on SDS-PAGE and subsequent analysis by Western blot using sheep anti-human placenta monoamine oxidase A antibody (Kirchgessner and Pintar, 1991) and alkaline phosphatase-coupled rabbit anti-sheep lgG (Zymed Laboratories). The Western blot in Fig. 2b shows a comparison of wild type enzyme produced in yeast strain RH218, both in purified form and from a crude homogenate to the C-terminally deleted mutant protein from a crude cell homogenate. There is evidence of proteolytic modification in all samples indicated by the doublet structure of the immunoreactive bands. Since the purified wild type enzyme consisted of at least 95% catalytically active molecules (Weyler et al., 1990b) and extensive kinetic analysis has shown it to be virtually identical to human placenta monoamine oxidase A (Tan et al., 1991) we believe that any proteolytic modification is due to nicking of the native protein and that this has no significant effect on catalytic properties. It is important to note in Fig. 2b that both bands of the doublet structure of the mutant enzyme are shifted by similar amount when compared to the doublet structure of the wild type enzymes. The similar shifts of both bands in the wild type and mutant enzymes suggests that the proteolytic nick is near the N-terminus.

Activation of the mutant enzyme with dithiothreitol

During partial purification steps, that included dialysis against buffers containing low concentrations of DTT, the mutant enzyme exhibited small but consistent increases in catalytic activity. Consequently, aliquots of the enzyme were incubated with DTT to determine if the presence of DTT could account for the increased activity. Indeed, these incubations gave variable activation, the effect being greatest with the least pure enzyme preparations. The effect was seen in at least four independent enzyme purification experiments. An up to 14-fold increase in enzyme activity was observed with the 48k g supernatant, whereas, as is shown in Fig. 3, only a 3.6-fold increase in activity was observed after an overnight dialysis of

W. Weyler

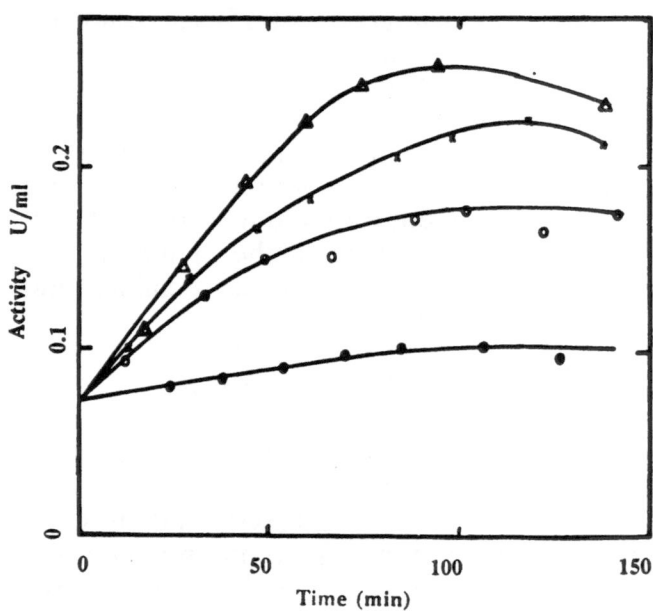

Fig. 3. Activation of mutant monoamine oxidase by dithiothreitol. Triton X-100 was
added to the 800 g supernatant described in "Methods" to a final concentration of 0.5%
and this was further centrifuged at 48k g for 30 min. The supernatant was then dialyzed
overnight against 100 volumes of 50 mM NaPO₄, pH 7.2, 20% glycerol, and 0.5 mM
PMSF. Reaction was initiated by the addition of DTT from a 1 M stock solution and
incubation at 30°C. ● no addition; ○ 5 mM; × 10 mM; △ 40 mM, DTT, respectively.
  Prior to dialysis the same sample gave a fourteen-fold activation with 40 mM DTT

this same fraction against buffer containing 20% glycerol, 5 mM DTT and
0.5 mM PMSF. As is seen in Fig. 3 extent of activation was dependent on
DTT concentration. The relative decrease in activation in the latter sample
could be attributed to an increase of basal activity occuring during the
dialysis, since the final activity was the same for both samples. Enzyme
purified with a DEAE-Sepharose column step could not be activated and
was very labile, losing activity with a half-life of 25 min at 30°C in the presence or absence of 0.5 mM PMSF or 40 mM DTT. Under similar conditions
the half-life of activity of enzyme from the 48k g supernatant was on the
order of 3 to 4 h, determined after maximal activity due to DTT activation
was recorded. In anaerobic spectrophotometric experiments (Weyler, 1987)
with partially purified enzyme the flavin spectrum was 50% reducible with
excess tyramine and the remainder could be reduced with dithionite. Based
on enzymatic assay with kynuramine and the substrate reducible fraction of
the flavin spectrum it was evident that the enzyme had a turnover number
2.5 to 3.1-fold lower than the wild type enzyme. Since this enzyme could
not be activated with DTT because of competing irreversible inactivation it
is not possible from these experiments to determine if the lower turnover
number is due to the putative disulfide bond or due to other factors.

Discussion

Monoamine oxidase is an intrinsic membrane protein which has been localized to the outer membrane of mitochondria (Schnaitman et al., 1967).
The mode of binding to the membrane is unknown. Analysis of the deduced
amino acid sequence by the method of Kyte and Doolittle (1982) indicates
that monoamine oxidase type A does not have multiple membrane spanning
sequences. This and the methods of Klein et al. (1985), and Eisenberg et
al. (1984), predict only two potential transmembrane spanning sequences;
one at the N-terminus and one at the C-terminus. The hydrophobic peptide
at the N-terminus could be ruled out as a transmembrane helix since this
sequence is involved in the binding of the adenine moiety of the FAD
prosthetic group as deduced from homology to an increasing number of
FAD proteins, many of which are soluble proteins (Weyler et al., 1987).
The overall hydrophobicity or GRAVY score (Kyte and Doolittle, 1982) of
the enzyme suggests it to be soluble, which is entirely contrary to observed
properties.

These observations suggested the possibility that monoamine oxidase
type A is a class I membrane protein (Bloble, 1980), anchored to the membrane with a single short hydrophobic C-terminial sequence. This prediction
was tested by deleting the C-terminal 24 animo acids which include all of
the hydrophobic residues in the predicted transmembrane peptide, and
expressing the truncated protein in yeast as previously reported for the
wild type enzyme (Weyler et al., 1990b). The solubility properties of the
resultant protein, however, do not support this model. Enzyme activity
remained associated with a membrane fraction through all centrifugation
steps and solubilization required treatment with phospholipase A and C and
detergent similar to the wild type enzyme, thereby suggesting that the
truncated protein was still membrane bound. These results clearly rule
out that monoamine oxidase is anchored to the outer membrane of the
mitochondria through a single C-terminal hydrophobic peptide. Other
modes of binding involving the C-terminus are also eliminated. Glycosyl-
phosphatidylinositol (Ferguson and Williams, 1988) and isoprenoid membrane anchors of the type discovered to date cannot be involved (Maltese,
1990). It is possible that monoamine oxidase is a multimeric structure where
subunit/subunit contacts mask charged faces and only a relatively small
fraction of the surface on each subunit interacts with the lipid bilayer as has
been suggested for porins (Kleffel et al., 1985). Support for a model which
has monoamine oxidase deeply buried in the membrane comes from very
recent protein import studies that have revealed the membrane bound
protein to be highly insensitive to proteinase K digestion (Zhuang et al.,
1992). A recent report has identified the C-terminal hydrophobic peptide to
be involved in targeting of monoamine oxidase type B to the mitochondrial
membrane (Mitoma and Ito, 1992). In this study, deletion of 28 amino acids
from the C-terminus prevented specific targeting of rat liver monoamine
oxidase type B, transiently overexpressed in COS cells, to the mitochondria.
In a sucrose gradient sedimentation experiment the CΔ28 mutant protein

co-sedimented with cytochrome P_{450} reductase, a microsomal marker protein. This observation could explain the higher **g**-force required to sediment the MA-CΔ24 mutant enzyme from the initial yeast homogenate in the present work and supports the notion that the truncated protein remains membrane bound.

Enzymatic activity of crude extract of the mutant protein was about 10% of wild type, but SDS-PAGE followed by Western blot analysis with antibody to human placenta MAO A indicated that the protein was expressed at similar levels as the wild type enzyme. This initially suggested that the truncated protein had only 10% of the wild type catalytic efficiency. During purification trials which included dialysis against DTT small but significant increases in activity were noted and subsequent experiments in which the enzyme was incubated with 5 to 40 mM DTT showed that activity could be increased 0 to 14-fold depending on the enzymes state of purity. Activation potential was inversely related to purity. The purest enzyme could not be activated and was labile under conditions of the experiment with a half-life for inactivation of 25 min. The activation behavior suggests that a fraction of the enzyme as isolated from the yeast cells has a pair of oxidized sulfhydryl groups which are reduced on incubation with DTT. This behavior is consistent with the C-terminus of the wild type enzyme protecting these sulfhydryl groups in the wild type enzyme against oxidation and that they are exposed in the mutant enzyme. These observations also suggest that the C-terminus may be in close proximity to the active site, but since the final activity of the crude activated enzyme is about the same as wild type enzyme it is clear that the deleted amino acids do not contribute any catalytic residues to the active site. In addition the C-terminus appears to make a major contribution to the enzymes thermal stability when it is removed from its native environment. Conversely, the membrane environment has a stabilizing effect on the mutant enzyme as is evident from the 7 to 10-fold greater half-life of enzyme activity in the crude homogenate compared to the purified enzyme under similar conditions of temperature and reducing agent.

The observation that the MA-CΔ24 mutant protein has virtually full activity suggests that the C-terminal peptide is part of the proteins envelope and not a protein core peptide. If the C-terminus were part of the core, the mutant enzyme would most likely not fold properly. This peptide is most probably in contact with the membrane in the native condition.

We previously reported that at least two cysteine residues of human placenta MAO A are modified by the sulfhydryl reagent 2-pyridyldisulfide (Weyler and Salach, 1985). Since modification of one of these residues led to only partial inactivation we concluded that this residue was not involved in catalysis. It is possible that the cysteine residues oxidized in the mutant protein are the same residues responsible for the inactivation in the chemical modification experiments. This hypothesis can be tested by site-directed mutagenesis experiments.

Results with this mutant protein have made significant contributions toward our understanding of how MAO interacts with its membrane environment and provides a basis for further investigations.

Conclusion

1. We demonstrated that MAO is not a class I membrane protein, i.e., it is not bound to the membrane by a simple C-terminal membrane anchor.
2. We deduced that the enzyme is not bound to the membrane by either glycosyl-phosphatidylinositol, farnesyl, or geranylgeranyl membrane anchors.
3. The high catalytic activity of this mutant enzyme (after DTT activation) suggests that the C-terminal 24 amino acids do not contribute any catalytic residues to the active site. Our data further suggests that this peptide is on the surface of the protein since proper folding would not be likely if the deleted peptide constituted a core peptide.
4. It is clear that the C-terminal peptide makes a large contribution to the stabilization of the purified wild type enzyme and conversely, the membrane environment makes a large contribution to the stability of the MA-CΔ24 mutant enzyme.

Acknowledgement

This work was supported by the Department of Veterans Affairs.

References

Allen JP, Fever G, Yeates TO, Komiya H, Reese CD (1988) Structure of the reaction center from *Rhodobacter sphaeroides* R-26: protein-cofactor (quinone and Fe^{2+}) interactions. Proc Natl Acad Sci USA 85: 8487–8491

Bach AWJ, Lan NC, Johnson DL, Abell CW, Bembenek ME, Kwan S-W, Seeburg PH, Shih JC (1988) cDNA cloning of human liver monoamine oxidase A and B: molecular basis of differences in enzymatic properties. Proc Natl Acad Sci USA 85: 4934–4938

Bitter GA, Eagan EM (1988) Expression of interferon-gamma from hybrid yeast GPD promoter containing upstream regulatory sequence from GAL1–GAL10 intronic region. Gene 69: 193–207

Blobel G (1980) Intracellular protein topogenesis. Proc Natl Acad Sci USA 77: 1496–1500

Eisenberg D, Schwarz E, Komaromy M, Wall R (1984) Description of the method used in HELIXMEM. J Mol Biol 179: 125–142

Ferguson MAJ, Williams AF (1988) Cell-surface anchoring of proteins via glycosyl-phosphatidylinositol structures. Ann Rev Biochem 57: 285–320

Gordon JI, Duronio RJ, Rudnick DA, Adams SP, Goke GW (1991) J Biol Chem 266: 8647–8650

Harlow E, Lane D (1988) Antibodies, a laboratory manual. Cold Spring Harbor Laboratory, Cold Spring Harbor, pp 761–510

Hsu Y-P, Weyler W, Chen S, Sims KB, Rinehart WB, Utterback M, Powell JF, Breakefield XO (1988) Structural features of human monoamine oxidase A elucidated from cDNA and peptide sequences. J Neurochem 51: 1321–1324

Ishibashi Y, Sakagami Y, Isogai A, Suzuki A (1984) Structure of tremerogens A-9291-I and A9291-VIII: peptidyl sex hormones of *Tremella brasiliensis*. Biochemistry 23: 1399–1404

Kearney EB, Salach JI, Walker WH, Seng RL, Kenney W, Zeszotek E, Singer TP (1971) The covalently-bound flavin of hepatic monoamine oxidase. Isolation and sequence of a flavin peptide and evidence for binding at the 8-alpha position. Eur J Biochem 24: 321–327

Kirchgessner AL, Pintar JE (1991) Guinea pig pancreatic ganglia: projection, transmitter content, and the type-specific localization of monoamine oxidase. J Comp Neurol 305: 613–631

Kleffel B, Garavito RM, Baumeister W, Rosenbusch JP (1985) Secondary structure of a channel-forming protein: porin from *E. coli* outer membrane. EMBO J 4: 1589–1592

Klein D, Kanehisa M, Delisi C (1985) The detection and classification of membrane-spanning proteins. Biochim Biophys Acta 815: 468–476

Kyte J, Doolittle RF (1982) A simple method for displaying the hydrophathic character of a protein. J Mol Biol 157: 105–132

Laemmli UK (1970) Cleavage of structural proteins during the assembly of the head of bacteriophage T4. Nature 227: 680–685

Layne E (1957) Spectrophotometric and turbimetric methods for measureing proteins. Meth Enzymol 3: 447–454

Maltese WA (1990) Posttranslational modification of proteins by isoprenoids in mammalian cells. FASEB J 4: 3319–3328

Maniatis T, Fritsch EF, Sambrook J (1983) Molecular cloning a laboratory manual. Cold Spring Harbor Laboratory, Cold Spring Harbor

Mitoma J, Ito, A (1992) Mitochondrial targeting signal of rat liver monoamine oxidase B is located at its carboxy terminus. J Biochem 111: 20–24

Sambrook J, Fritsch EF, Maniatis T (1989) Molecular cloning a laboratory manual, 2nd ed. Cold Spring Harbor Laboratory, Cold Spring Harbor

Sakagami Y, Yoshida M, Isogai A, Suzuki A (1981) Peptidyl sex hormones inducing conjugation tube formation in compatible mating type cells of *Tremella mesenterica*. Science 212: 1525–1527

Schnaitman C, Erwin V, Greenawalt JW (1967) The submitochondrial localization of monoamine oxidase. J Cell Biol 32: 719–735

Sefton BM, Buss JE (1987) The covalent modification of proteins with lipid. J Cell Biol 104: 1449–1453

Sherman F, Fink GR, Hicks JB (1986) Methods in yeast genetics. Cold Spring Harbor Laboratory, Cold Spring Harbor, p 117

Spatz L, Strittmatter P (1971) A form of cytochrome b_5 that contains an additional hydrophobic sequence of 40 amino acids. Proc Natl Acad Sci USA 68: 1042–1046

Tan AK, Weyler W, Salach JI, Singer TP (1991) Differences in substrate specificities of monoamine oxidase A from human liver and placenta. Biochem Biophys Res Commun 181: 1084–1088

Urban P, Andersen JK, Hsu H-PP, Pompon D (1991) Comparative membrane locations and activities of human monoamine oxidases expressed in yeast. FEBS Lett 286: 142–146

Weissbach H, Smith TE, Daly JW, Witkop B, Udenfriend S (1960) A rapid spectrophotometric assay of monoamine oxidase based on rate of disappearance of kynuramine. J Biol Chem 235: 1160–1163

Weyler W (1989) Monoamine oxidase A from human placenta and monoamine oxidase B from bovine liver both have one FAD per subunit. Biochem J 260: 726–729

Weyler W, Hsu Y-P P, Breakefield XO (1990a) Biochemistry and genetics of monoamine oxidase. Pharmacol Ther 47: 391–417

Weyler W, Titlow CC, Salach JI (1990b) Catalytically active monoamine oxidase type A from human liver expressed in *Saccharomyces cerevisiae* contains covalent FAD. Biochem Biophys Res Commun 173: 1205–1211

Weyler W, Salach JI (1985) Purification and properties of mitochondrial monoamine oxidase type A from human placenta. J Biol Chem 260: 13199–13207

Weyler W (1987) 2-Chloro-2-phenylethylamine as a mechanistic probe and active site directed inhibitor for monoamine oxidase from bovine liver mitochondria. Arch Biochem Biophys 255: 400–408

Weyler W, Powell J, Hsu Y-P, Chen S, Corey DP, Utterback M, Titlow C, Breakefield XO (1987) Comparison of partial amino acid sequences deduced from the nucleotide sequence of a bovine adrenal monoamine oxidase cDNA clone to amino acid sequences obtained from bovine liver monoamine oxidase type B. In: Flavins and flavoproteins. de Gruyter, New York, pp 725–728

Zhuang Z, Marks B, McCauley RB (1992) The insertion of monoamine oxidase A into the outer membrane of rat liver mitochondria. J Biol Chem 267: 591–596

Author's address: Dr. W. Weyler, Research and Development, Genencor Intñl, 180 Kimball Way, South San Francisco, CA 94080, U.S.A.

J Neural Transm (1994) [Suppl] 41: 17–26

Kinetic properties of cloned human liver monoamine oxidase A

R. R. Ramsay[1,3], **A. K. Tan**[1,3], and **W. Weyler**[2,3]

[1] Department of Biochemistry/Biophysics, and [2] Division of Toxicology, University of California, and [3] Molecular Biology Division, Department of Veterans Affairs Medical Center, San Francisco, California, U.S.A.

Summary. Monoamine oxidases deaminate many amines, including neurotransmitters, by oxidation followed by spontaneous breakdown of the imine product. The reduced enzyme is reoxidized slowly by oxygen, but in the presence of amines, the rate of reoxidation is markedly enhanced. The extent of enhancement depends on the amine substrate, kynuramine enhancing the rate 125-fold, but 5-hydroxytryptamine only 6-fold. Here we describe the properties of human liver monoamine oxidase A which has been cloned into and overexpressed in yeast. The purified enzyme has a higher K_m for oxygen than does the placental enzyme, but the steady-state parameters for the endogenous amines are the same. Tertiary amines are oxidized at slightly different rates by the two enzymes. The consequences of the branched pathway mechanism with substrate-dependent enhancement of reoxidation for the steady-state levels of the various enzyme species is discussed.

Introduction

The oxidation of amines by monoamine oxidases has been studied intensively for many years (for recent reviews, see Singer, 1991, and Weyler et al., 1990a). Although the role in peripheral tissues may be a general one, the monoamine oxidases in brain play a crucial role in maintaining low levels of transmitter amines and their by-products. Monoamine oxidase inhibitors result in elevated levels of brain amines and thus are used in antidepressants and as adjuncts in the treatment of Parkinson's disease. It is the search for the perfect MAO inhibitor, as well as the need to understand how the two forms of MAO (A and B) work on neurotransmitters in different parts of the brain, that makes it essential to understand the kinetic mechanism of these enzymes.

Most steady-state studies of MAO suggest a simple ping-pong mechanism but detailed analysis of the steady-state (Pearce and Roth, 1985; Husain et al., 1982) and presteady-state (Husain et al., 1982; Ramsay et al., 1987; Ramsay, 1991) kinetics revealed that both binary and ternary complex

R. R. Ramsay et al.

pathways can occur with any given substrate (see Scheme 1). The mechanism followed in any given turnover depends on the relative rate constants and the dissociation constants for the complexes.

For both MAO A and MAO B, the free reduced enzyme (E_R) is oxidized at a rate less than the observed steady-state rate. The reduced enzyme-substrate complex (E_R-S) is reoxidized dramatically faster, particularly for MAO A (Ramsay et al., 1987; Ramsay, 1991). However the extent of the enhancement of the rate was different for the two substrates examined for each enzyme (Ramsay et al., 1987; Ramsay, 1991). We have now established that the enhancement of the oxidative half-reaction is dependent on the particular substrate used (Tan and Ramsay, submitted) and here discuss some consequences of this feature of the mechanism.

The enzyme used in this work was human liver MAO A cloned and overexpressed in yeast as described by Weyler et al. (1990b).

Material and methods

Enzyme

Human Liver Monoamine Oxidase A was overexpressed in yeast as described by Weyler et al. (1990b) and purified as described by Tan et al. (1991). As a result of the overexpression, 287 mg of pure enzyme was obtained from 10 L of yeast cells. This large quantity of enzyme was necessary for the stopped-flow experiments where the oxidation or reduction of the flavin is measured at 450 nm ($\varepsilon_o = 10,800 \, M^{-1} cm^{-1}$ (Weyler and Salach, 1985)).

Reagents

Amine substrates and detergents were purchased from Sigma Chemical Company. MPTP, $MPDP^+$ and MPP^+ were purchased from Research Biochemicals, Inc.

Steady-state assays were performed spectrophotometrically (kynuramine, benzylamine, MPTP, $MPDP^+$) or polarographically (all other substrates) at 30°C in 50 mM Na^+ phosphate buffer, pH 7.2, containing 0.5% (w/v) Brij 35. The values given for the kinetic constants are averages of at least 3 separate determinations.

Stopped-flow measurements were carried out as described by Ramsay (1991) except that all measurements were made at 30°C in 50 mM Na^+ phosphate – 0.8% octyl-β-D-glucopyranoside.

The reactions studied were:

Reduction
$$E_{ox} + S \underset{k_2}{\overset{k_2}{\rightleftharpoons}} E_{ox}S \overset{k_3}{\rightarrow} E_R P$$

Reoxidation
$$E_R S + O_2 \overset{k_{ox}}{\rightarrow} E_{ox}S + H_2O_2$$

It should be noted that during turnover, the oxidative half-reaction can take either of the pathways described in Scheme 1. In the experiments reported here, reduced enzyme is premixed anaerobically with substrate to measure only the reoxidation of the reduced enzyme-substrate complexes.

Results

Steady-state kinetics

Table 1 lists the steady-state rate constants (k_{cat} or turnover number) and K_m values for the purified human liver MAO A with various substrates. The data (with a few corrections for turnover numbers calculated without taking account of the presence of catalase) are taken from Tan et al. (1991). All the natural substrates have very similar k_{cat} values but the low K_m for tryptamine makes it the best substrate. Tryptamine excepted, the MPTP derivatives ($k_{cat}/K_m > 20$) are all better substrates than the natural primary amines ($k_{cat}/K_m = 8 - 18$). MPTP itself turns over fairly slowly despite a low K_m, suggesting that all the substitutions tested make the pyridine ring more susceptible to oxidation. The last column in Table 1 compares the efficacy of the liver and placental enzymes with each substrate as a ratio of the k_{cat}/K_m values for liver/placenta. For the endogenous amines this ratio is 1, but for the neurotoxic MPTP derivatives it is >1.5, indicating that the liver enzyme catalyzes the oxidation of these substrates more efficiently than does placental MAO A.

The second substrate for MAO A is oxygen. Figure 1 shows the effect of oxygen concentration on rate of oxidation of kynuramine by the liver and placental enzymes. Despite having the same V_{max} and very similar K_m values for kynuramine, the two enzymes are affected quite differently by the oxygen concentration. The K_m values for oxygen from Fig. 1 and 6 μM for the placental enzyme (in agreement with Ramsay, 1991) are 60 μM for the liver enzyme. Thus there are significant kinetic differences between the two enzymes.

Table 1. Substrate specificities of MAO A from human liver

Substrate	k_{cat} (s^{-1})	K_m (mM)	k_{cat}/K_m (s^{-1}M^{-1})	k_{cat}/K_m (liver) / k_{cat}/K_m (placenta)
Kynuramine	2.65	0.147	18	0.92
Tryptamine	2.65	0.032	83	0.93
Serotonin (5-HT)	2.78	0.251	11	1.06
Dopamine	1.83	0.234	7.8	0.98
Tyramine	3.63	0.244	15	1.09
MPTP	0.20	0.081	2.5	1.46
2′-n-propyl MPTP	6.10	0.277	22	1.89
2,6-dimethyl MPTP	3.72	0.161	23	1.76
4-(4′-iodobenzyl) MPTP	1.23	0.033	37	1.76
4-(1-naphthylmethylene)-MTP[a]	13.1	0.182	72	1.52

Data are taken from [9] with some corrections
[a] 4-(1-naphthylmethylene)-N-methyltetrahydropyridine

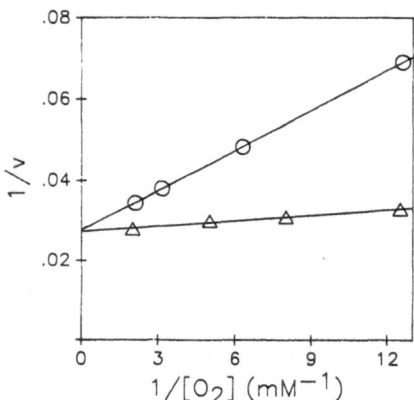

Fig. 1. Dependence on oxygen of the steady-state rate of oxidation of kynuramine by MAO A from human liver and placenta. The oxidation of kynuramine (1 mM) was followed spectrophotometrically at 316 nm. The assay cuvette contained 50 mM Na$^+$ phosphate, pH 7.2, 0.2% Triton X-100 at 30°C and MAO purified from human liver (○) or placenta (△). Closed cuvettes were equilibrated with various concentrations of oxygen before the reaction was initiated by the addition of kynuramine

Half reactions of MAO A

The rate of oxidation of the FAD cofactor of MAO A by substrate depends on the nature of the amine. Three indole substrates which differ only at the 5 position remote from the amine group were examined, namely tryptamine, 5-hydroxytryptamine (serotonin) and 5-methoxytryptamine. Figure 2 shows that although all give the same value for k_3 (the rate constant for $E_{ox}S \rightarrow E_RP$, (see Scheme 1), which is obtained from the reciprocal of the y intercept), the K_D values differ greatly (see Table 2). At low amine concentrations, tryptamine oxidation is fastest, then 5-methoxytryptamine, then 5-HT. The differences in K_D (tryptamine, 0.07 mM; 5-MT, 0.16 mM; 5-HT, 0.40 mM) indicate that substitutions at the 5 position hinder binding, and that the OH group is worse than the methoxy group.

Table 2 lists the kinetic constants for all the amines studied. Note that in all cases, the k_2 values obtained for the reductive half-reaction are very close to the k_{cat} values for the steady state reaction. For all these amines, the reductive half-reaction appears to be the rate limiting step in turnover.

In the oxidative half-reaction, all the amines strongly stimulated the observed rate of the reaction between the reduced enzyme and oxygen. Figure 3A shows the slow first order reaction of the free enzyme with oxygen in contrast to the much faster rates observed in the presence of indoleamines. MPTP (40-fold) and kynuramine (125-fold) stimulate so strongly that the rate constant scale is 0–250 s^{-1} in Fig. 3B in contrast to 0–15 s^{-1} in Fig. 3A. The second order rate constants (Table 2, last column), show that the substrates can be divided arbitrarily into two groups, namely those that enhance strongly, such as kynuramine, benzylamine, MPTP and MPDP, and those that enhance modestly such as the indoleamines. These

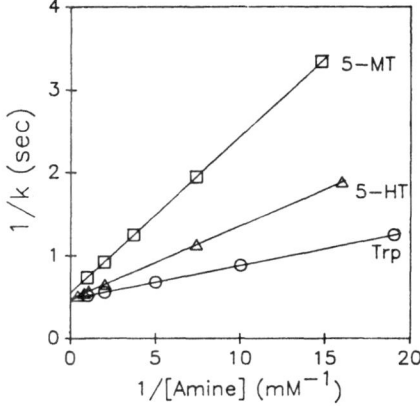

Fig. 2. Rate constants for the reduction of MAO A by indole substrates. The reduction of the flavin was followed at 450 nm as described in the Methods section after rapid mixing of the anaerobic enzyme and substrate solutions. The final enzyme concentration was 8 μM and the substrates (2 mM) were tryptamine (○), 5-hydroxytryptamine (△), and 5-methoxytryptamine (□)

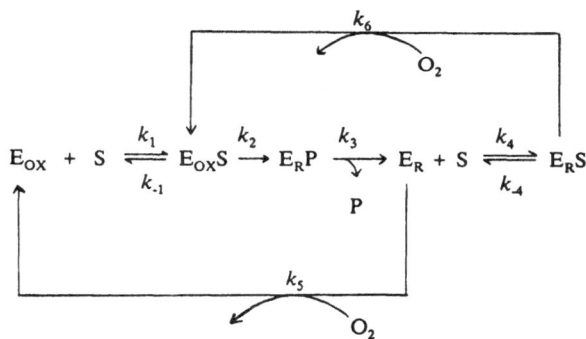

Scheme 1. Kinetic pathways for monoamine oxidase

differences in rate will effect the regeneration of oxidized enzymes ready for the next catalytic cycle. In all cases, $E_R S$ reacts more rapidly with oxygen than does free enzyme (E_R) alone. The flux through the accelerated $E_R S$ reoxidation pathway will depend not only on the relative rate constants but also on the rate of association of the substrate with E_R. Since the oxidative half-reaction for MAO A is apparently first order, K_D values for $E_R S$ cannot be extracted from the data reported here. Further experiments are in progress to see if the K_D for $E_R S$ differs from the K_D for $E_{ox}S$ determined in the reductive half-reaction. For MAO B, the K_D values are quite different. For MPTP, the K_D for the E_{ox}-MPTP complex is 0.04 mM, whereas that for the E_R-MPTP complex is greater than 2 mM (Ramsay et al., 1987).

Table 2. Kinetic parameters for human liver MAO A from steady-state and stopped-flow half-reaction experiments

Substrate	Steady-State		Reduction		Oxidation	
	k_{cat} (s^{-1})	K_m (mM)	k_2 (s^{-1})	K_D (mM)	k_{app}* (s^{-1})	k_{ox} $(mM^{-1}s^{-1})$
Kynuramine	2.65	0.15	3.1	0.58	120	508
Benzylamine	0.02	0.90	0.06	0.23	23	106
Phenylethylamine	0.75	0.50	1.10	0.90	12	48
Tryptamine	2.65	0.03	2.1	0.07	8	29
Serotonin (5-HT)	2.80	0.40	2.1	0.40	5.7	24
5-methoxytryptamine	1.81	0.184	1.72	0.161	2.8	11
MPTP	0.20	0.09	0.2	0.04	40	166
MPDP$^+$	0.018	0.024	0.017	0.10	29	130
MPP$^+$					No Reoxidation	
NONE					0.94	4.0

* At 0.238 mM O_2. All rates were measured at 30°C

Turnover experiments

The relative rates of the reductive and oxidative half-reactions (and the relevant dissociation constants) will determine the distribution of the enzyme between the oxidized and reduced states. Figure 4 shows stopped-flow monitored turnover experiments for liver (panel A) and placental (panel B) MAO A with an excess of several amine substrates. In each experiment, there is an initial rapid bleaching of the flavin absorbance at 450 nm, followed by slower changes between the oxidized and reduced pools of enzyme in the steady state. Finally, as the oxygen concentration decreases, the flavin becomes fully reduced. The extent of reduction of the flavin in the steady state is an indication of the difference between the rate of reduction and the rate of oxidation for each substrate. With MPTP, which has a low rate of turnover and reduces the flavin only slowly (Ramsay, 1991), most (98%) of the flavin remains oxidized. With kynuramine approximately 80% remains oxidized, but for 5-HT only 10% is oxidized in the steady-state (Fig. 4). Kynuramine and 5-HT are oxidized by MAO A at the same rate (2.65 and 2.80 s^{-1} respectively), and the reductive half-reaction is actually faster for kynuramine (3.1 versus 2.1 s^{-1}). The more rapid accumulation of the reduced enzyme species with 5-HT indicates that the reduction alone does not determine the distribution of the enzyme between the oxidized and reduced forms and that the rate of reoxidation must be taken into account. For kynuramine, the ratio of the rate constants for oxidation and reduction (k_{app}/k_2) is 39, whereas for 5-HT the ratio is only 2.7, so that although the substrates are turned over at the same rate, the proportion of reduced flavin accumulates faster with 5-HT.

Although the amount of enzyme used in each panel is different (28 μM in A, 17 μM in B) and the K_m value for O_2 is different for the two enzymes

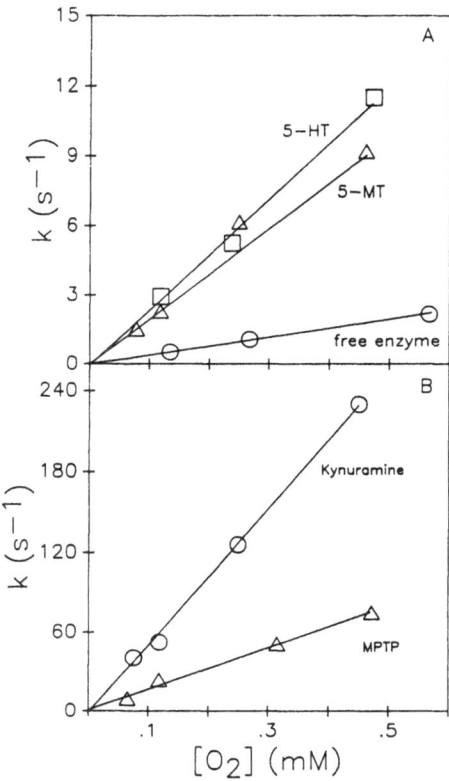

Fig. 3. The enhancement of the rate constants for the oxidative half-reaction by substrates. MAO A was reduced as described in the Methods and incubated anaerobically with substrate for five minutes before mixing with oxygen in the stopped-flow spectrophotometer. The substrates were, in **A**, no amine present (O), 2 mM serotonin (5-HT, □), 2 mM 5-methoxytryptamine (△); in **B** (note the change in scale), 2 mM kynuramine (O), 2 mM MPTP (△). The standard deviations are smaller than the size of the symbols

(60 μM in A, 6 μM in B), internal comparisons can be made relative to the MPTP curve. For MPTP, the k_2 and oxidative k_{app} are the same for the liver (0.2 and 40 s^{-1}) and placental (0.19 and 36 s^{-1}) enzymes. In relative terms, the placental enzyme remains oxidized longer than the liver enzyme with kynuramine but is reduced faster with 5-HT. Further information about the molecular mechanism must be sought to explain these differences.

Discussion

The kinetic properties of human liver MAO A are not identical to those for the placental enzyme. In particular, the K_m value of O_2 for liver enzyme (60 μM) is 10-fold higher than that for the placental enzyme (6 μM) when assayed with kynuramine. The rates for the oxidation of primary amines are virtually the same for both, but differences are found for the tertiary

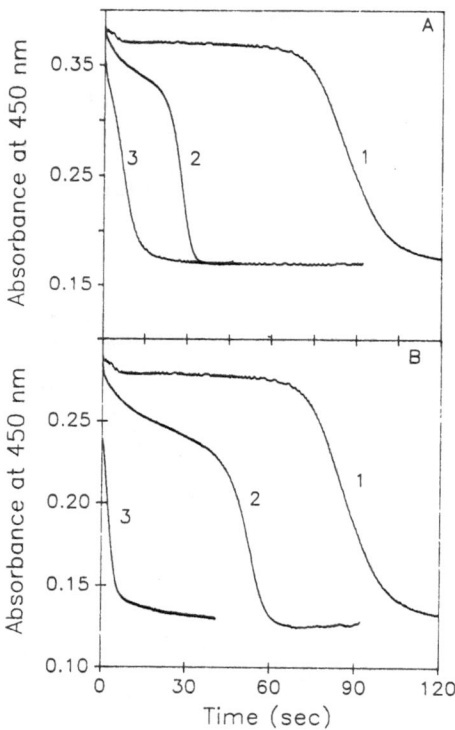

Fig. 4. Variation in the steady-state concentration of oxidized MAO A during stopped-flow monitored turnover in the presence of substrates. MAO A (final concentration 8 μM) was mixed with 10 mM MPTP (*1*), kynuramine (*2*) or serotonin (5-HT, *3*) in the presence of 0.238 mM oxygen and the decrease in the absorbance at 450 nm was monitored. When the O_2 is exhausted, the enzyme becomes fully reduced. **A** shows human liver MAO A; **B** shows human placental MAO A

amines. The k_{cat}/K_m values for the liver enzyme oxidizing MPTP and its derivatives are all at least 50% higher than those for the placental enzyme. However, the K_m values are only slightly altered and K_i values for the pyridinium products are the same for both enzymes, suggesting that binding is not affected. Rather, the differences seem to be kinetic. The high yield of pure recombinant human liver MAO A obtained from yeast facilitated extensive stopped-flow studies to determine the rate constants for the reductive and oxidative half-reactions. For all the substrates investigated, the reductive half-reaction is rate-limiting for MAO A, in contrast to MAO B where the oxidative half-reaction is at least partly rate-determining for some substrates (Ramsay et al., 1987). The rate of the oxidative half-reaction was different for all 8 substrates studied, confirming that the difference between the rates for kynuramine and MPTP reported earlier (Ramsay, 1991) was indeed substrate-induced. From the data in Table 2, we conclude that:

1. The reduced enzyme-substrate complex (E_RS) is reoxidized faster than free enzyme alone.

2. The stimulation of the rate of reoxidation is dependent on the nature of the amine substrate.

3. The product (MPP^+) inhibits the reoxidation.

The inhibition by product and differential stimulation by substrate suggest that the reactivity of the reduced flavin with oxygen is altered. However, the best substrates for the reductive half-reaction are not best for the enhancement of the oxidative half-reaction. For example, MPTP reduces MAO A at a tenth of the rate observed with 5-HT, yet the E_R-MPTP complex is reoxidized 5 times faster than the E_R-5-HT complex. Further investigation into the mechanism of the stimulation of reoxidation is necessary, but one can speculate that some conformation change alters the reactivity of the FAD with O_2. This could be a consequence, for example, of an altered redox potential, promotion of O_2 binding, or stabilization of the one electron transfer intermediate.

Overall, those complex properties and the alternate pathway mechanism shown in Scheme 1 mean that complex steady state kinetics may be observed for this enzyme. A good example of this is the biphasic kinetics observed for the oxidation of trans-stylbasole by MAO A (Bachurin et al., 1989) which was originally interpreted as due to two binding sites. Current studies in our laboratory (S. O. Sablin and T. P. Singer, unpublished) show that it is rather a consequence of the branched mechanism where the K_D of E_RS is very different from that of $E_{ox}S$ so that at low substrate concentrations reoxidation of the free enzyme predominates whereas at high substrate concentration, the faster $E_RS \rightarrow E_{ox}S$ pathway determines the steady-state rate. Although not so obvious for the more usual substrates, the differences in the steady-state redox state of the enzyme with different amines and potentially different inhibitor binding characteristics for the oxidized and reduced enzyme mean that simple inhibition studies must be analyzed with caution.

Lastly, we would like to point out general consequences of the branched mechanism with substrate-dependent stimulation of oxidation which are of relevance to the pharmacology of the enzyme and to the design of inhibitors. Our initial data suggest that the K_D for E_RS is much higher than that for $E_{ox}S$ (Ramsay et al., 1987; Ramsay, 1991; Sablin and Singer, unpublished). Thus at high substrate concentrations an apparent activation of MAO can occur. At low substrate concentration the slower E_R path will predominate whereas at high concentrations the fast E_RS path will become important. The redox state of MAO A during turnover is predominantly oxidized (Fig. 4), but that of MAO B is much more reduced (Husain et al., 1982). For example, with phenylethylamine, MAO B is completely reduced in the steady-state. The consequence of this is a potential heterogeneity of the observed response of cells with different complements of amine substrates to a given inhibitor, assuming that the binding of inhibitors, like that of substrate, is different for the reduced and oxidized forms of the enzyme. Further data are needed to support this hypothesis, but if it holds it might explain puzzling discrepancies in the potency of MAO inhibitors on different cell populations.

Acknowledgements

This work was supported by the National Institutes of Health (HL-16251), the National Science Foundation (DMB-9020015), and the Department of Veterans Affairs.

References

Bachurin SO, Sablin SO, Grishina GV, Gaydorova EL, Dubova LG, Zubor ND (1989) Kinetics of biotransformation of physiologically active 1-methyl-4-aryl-1,2,3,6-tetrahydropyridines by monoamines oxidase. Bioorganicheskaya khimia (russ) 15: 620–626

Husain M, Edmondson DE, Singer TP (1982) Kinetic studies on the catalytic mechanism of liver monoamine oxidase. Biochemistry 21: 595–600

Pearce LB, Roth JA (1985) Human brain monoamine oxidase type B: mechanism of deamination as probed by steady-state methods. Biochemistry 24: 1821–1826

Ramsay RR, Koerber SC, Singer TP (1987) Stopped-flow studies on the mechanism of oxidation of N-methyl-4-phenyltetrahydropyridine by bovine liver monoamine oxidase B. Biochemistry 26: 3045–3050

Ramsay RR (1991) Kinetic mechanism of monoamine oxidase A. Biochemistry 30:- 4624–4629

Singer TP (1991) Monoamine oxidases. In: Müller F (ed) Chemistry and biochemistry of flavoenzymes, vol 2. CRC Press, Boca Raton, pp 437–470

Tan AK, Ramsay RR (1993) Substrate-specific enhancement of the oxidative half-reaction of monoamine oxidase. Biochemistry 32: 2137–2143

Tan AK, Weyler W, Salach JI, Singer TP (1991) Differences in substrate specificities of monoamine oxidase A from human liver and placenta. Biochem Biophys Res Comm 181: 1084–1088

Weyler W, Hsu Y-PP, Breakefield XO (1990a) Biochemistry and genetics of monoamine oxidase. Pharmacol Ther 47: 391–417

Weyler W, Titlow CT, Salach JI (1990b) Catalytically active monoamine oxidase type A from human liver expressed in Saccharomyces cerevisiae contains covalent FAD. Biochem Biophys Res Commun 173: 1205–1211

Weyler W, Salach JI (1985) Purification and properties of mitochondrial monoamine oxidase Type A from human placenta. J Biol Chem 260: 13199–13207

Authors' address: Dr. R. R. Ramsay, Molecular Biology Division, 151-S, Veterans Administration Medical Center, 4150 Clement Street, San Francisco, CA 94121, U.S.A.

J Neural Transm (1994) [Suppl] 41: 27–33
© Springer-Verlag 1994

Identification of human monoamine oxidase (MAO) A and B gene promoters

J. C. Shih, Q.-S. Zhu, J. Grimsby, and **K. Chen**

Department of Molecular Pharmacology and Toxicology, School of Pharmacy,
University of Southern California, Los Angeles, California, U.S.A.

Summary. The promoter of human monoamine oxidase (MAO) A and B genes have been identified. The core promoter region of MAO A is comprised of two 90 bp repeats each of which contains two Sp1 elements and lacks a TATA box. The MAO B core promoter region contains two sets of overlapping Sp1 sites which flank a CACCC element all upstream of a TATA box. The different organization of the MAO A and B promoters may underlie their different cell and tissue specific expression.

Introduction

Monoamine oxidase A and B (MAO A and B; flavin-containing deaminating amine:oxygen oxidoreductase, EC 1.4.3.4) catalyze the oxidative deamination of a number of neurotransmitters, dietary amines and xenobiotics including the Parkinsonism-producing neurotoxin 1-methyl-4-phenyl-1,2,3,6-tetrahydropyridine (Chiba et al., 1984). Both forms are located in the outer mitochondrial membrane and are distinguished by their different substrate preference and sensitivity to inhibitors. Cloning of the cDNAs for MAO A and B demonstrates that these two forms of the enzyme are coded by different genes (Bach et al., 1988; Hsu et al., 1988; Lan et al., 1989; Shih, 1990) which were derived from the same ancestral gene (Grimsby et al., 1991). Both genes are closely linked and located on the X chromosome at Xp11.23 to Xp22.1 and are deleted in some patients with Norrie's disease (Ozelius et al., 1988; Lan et al., 1989).

MAO A and B transcripts are coexpressed in most human tissues examined (Grimsby et al., 1990). However, they do show different tissue and cell distribution, and they are regulated differently during development. In addition, abnormal MAO activity may be associated with mental disorders and MAO inhibitors have been used for the treatment of Parkinsonism and mental depression. In order to understand the mechanisms controlling the expression of these two forms of MAO, it is essential to characterize the promoter regions of their genes. This report shows that the immediate 5' flanking sequences of both MAO A and B genes contain cis-elements

J. C. Shih et al.

needed for active transcription, but the organization of these elements are different.

Materials and methods

The 5' flanking sequences of MAO A and B genes are isolated from corresponding genomic clones and sequenced after being subcloned into M13. On the basis of sequence analysis, various DNA fragments from the presumed promoter regions are isolated and measured for promoter activity by inserting them into the promoterless expression vector pOGH which contains human growth hormone as the reporter gene. These constructs are transfected into SHSY-5Y (human neuroblastoma) and NIH3T3 (mouse fibroblast) cells. The human growth hormone synthesized is secreted into the medium and is measured with a Kit from Nichols Institute Diagnostics. MAO catalytic activities are assayed as previously described (Chen et al., 1984), using serotonin and phenylethylamine as substrate for MAO A and B, respectively.

Results and discussion

Sequence analysis of the 5' flanking region of MAO A and B exon 1 revealed potential DNA sequences involved in regulating transcription (Table 1). The first 200 bp 5' of the cDNA start site for MAO A and B are

Table 1. Potential cis-elements in the promoter of MAO A and B genes. These sites are numbered from 3' to 5' as in Fig. 1–3. The location of these sites are indicated by the distance to the A of the translation initiation codon ATG, which is defined as +1. The negative sign indicates that they are upstream of the ATG

	MAO A			MAO B	
site	sequence	location (bp)	site	sequence	location (bp)
1	CTCCGCCC	−94	1	GGGCGGG	−82
2	CCCGCCC	−142	2	TAATATA	−146
3	CTCCGCCC	−184	3	GGGCGGG	−181
4	TCCGCCC	−237	4	GGGCGGG	−185
5	TAATAA	−269	5	AGGCGGG	−189
6	CACCC	−316	6	CACCC	−210
7	CACCC	−404	7	GGGCGGTG	−226
8	CCAAT	−443	8	GGGCGGG	−233
9	CACCC	−521	9	CCCGCCC	−267
10	CACCC	−531	10	CACCC	−270
11	CCAAT	−570	11	CTCCGCCC	−278
12	TGACCTCA	−618	12	GGGCGGG	−310
13	CACCC	−670	13	CCAAT	−405
14	CCCGCCC	−729	14	CACCC	−409
15	CCAAT	−777	15	AATTGG	−430
16	CCAAT	−920	16	CACCC	−442
17	CACCGCCC	−1,325	17	GGGTG	−709
18	CCCGCCC	−1,418	18	TGATGTCA	−807
			19	TGACTCA	−1,340

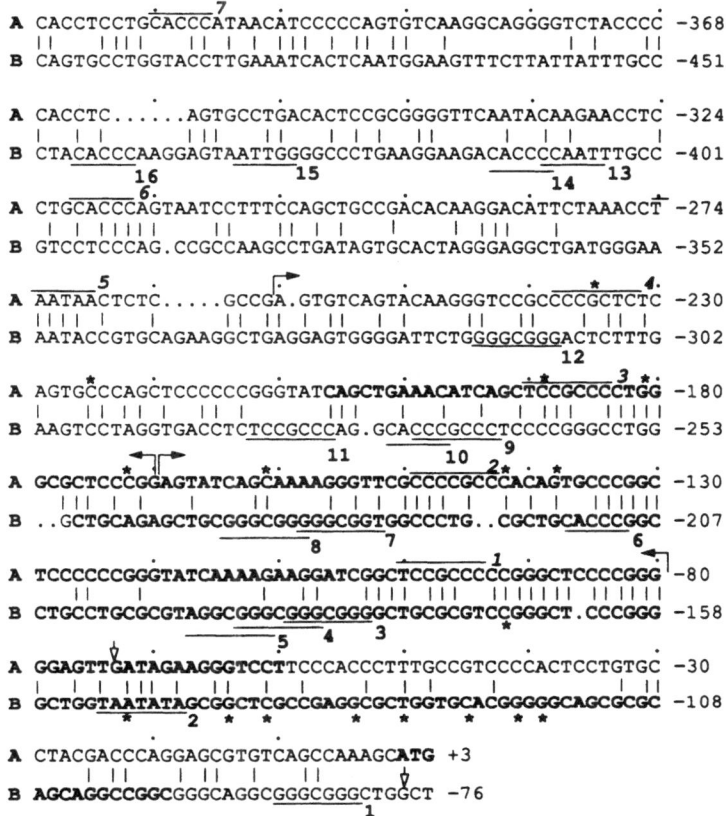

Fig. 1. Sequence comparison of the core promoter regions of MAO A (A) and B (B) genes. The position of the two promoter regions are aligned to yield the highest degree of homology as determined by computer analysis. The potential cis-elements 1–7 of MAO A and 1–16 of MAO B as shown in Table 1 are marked. The sequences of A0.14 and B0.15 fragments are in bold type. The two 90 bp repeats of MAO A are marked by solid arrows (↦ ↤). The cDNA start site is indicated by an open arrow (↓). Transcription initiation sites as determined by primer extension analysis (figure not shown) are marked with asterisks (*)

comprised of 68% and 82% GC residues respectively. These two regions, later shown to contain the core region of promoter activity, also share the highest homology (61%, Fig. 1), suggesting that these two promoters may be derived from a common ancestral gene. However, none of the transcription factor binding sequences are conserved at their corresponding positions (Fig. 1) suggesting that these two promoters have functionally diverged during evolution and thus their expression are differently regulated (Zhu et al., 1992).

Promoter activity measurements with DNA fragments obtained from both 5' and 3' deletions show that the MAO A promoter activity is located in the two 90 bp direct repeat region (Fig. 2, each repeat is represented with an arrow). These two 90 bp repeats share 83% sequence identity and each contains two Sp1 elements in reversed orientation (Fig. 2, site 1–4). Dele-

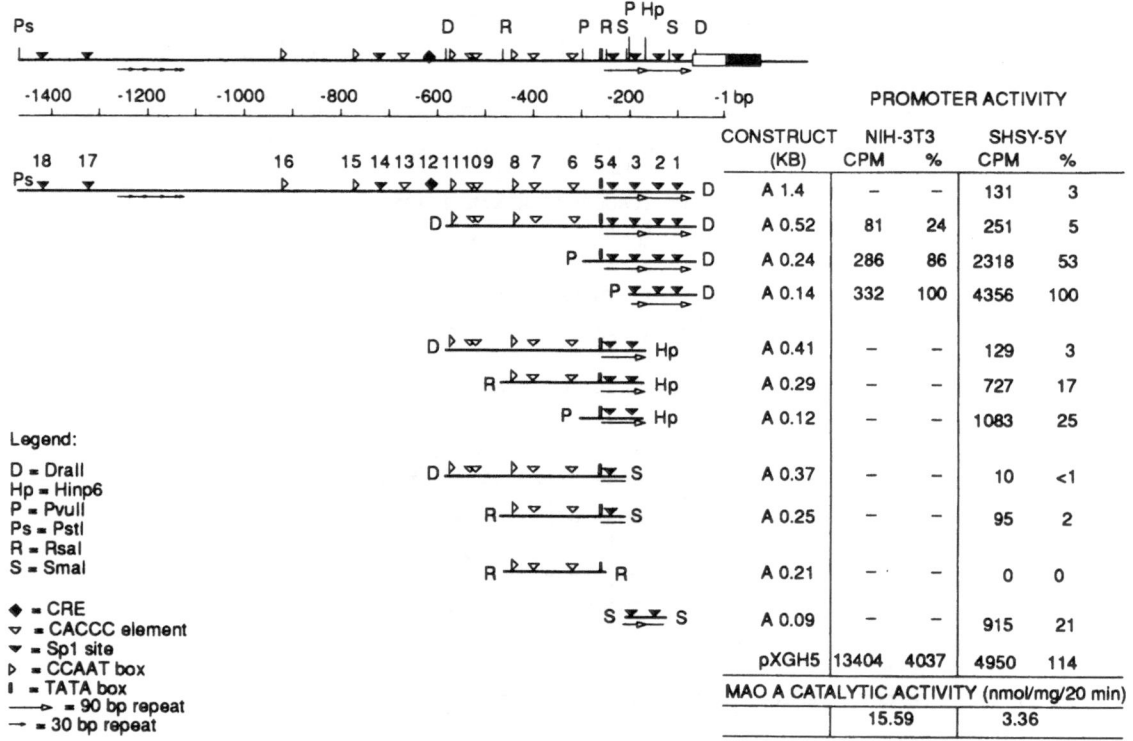

Fig. 2. Restriction map of the MAO A promoter region and promoter activity measurements. At the top is the restriction enzyme map, where only restriction enzyme sites used for subcloning for promoter activity measurement are marked. The open box on the top line represents untranslated region and the closed box for the coding sequence of MAO exon 1. The potential transcription factor binding sites are numbered as in Table 1. The A of the codon ATG is defined as +1bp. The pOGH constructs containing DNA fragments to be tested are named according to the size of the inserts (kb). *A* denotes DNA fragments of MAO A gene. The activity of the A0.14 construct is taken as 100% for each cell lines. pXGH5 represents the plasmid containing mouse metallothionein promoter instead of the MAO promoter fragments, which was used to monitor the transfection efficiency. The last line shows MAO A enzymatic activity measured in each cell line

tion of the 3' 90 bp repeat from A0.24, a 0.24 kb PvuII/DraII fragment, which contains both 90 bp repeat and an upstream TAATAA sequence, decreases promoter activity by approximately 50% (compare A0.24 (53%) and A0.12 (25%) in Fig. 2). However, deletion from the 5' end which removes the TAATAA sequence and the Sp1 site 4, results in the most active fragment (A0.14) which contains Sp1 sites 1–3 and its activity is taken as 100%. A0.09 and A0.12 containing Sp1 sites 2–3 and 3–4 are capable of transcription activation although with only 20–25% the activity of A0.14. This means that at least three Sp1 sites or yet undetermined elements located within this region work cooperatively to activate transcription. DNA fragments containing only one Sp1 site (site 4, see A0.37 and

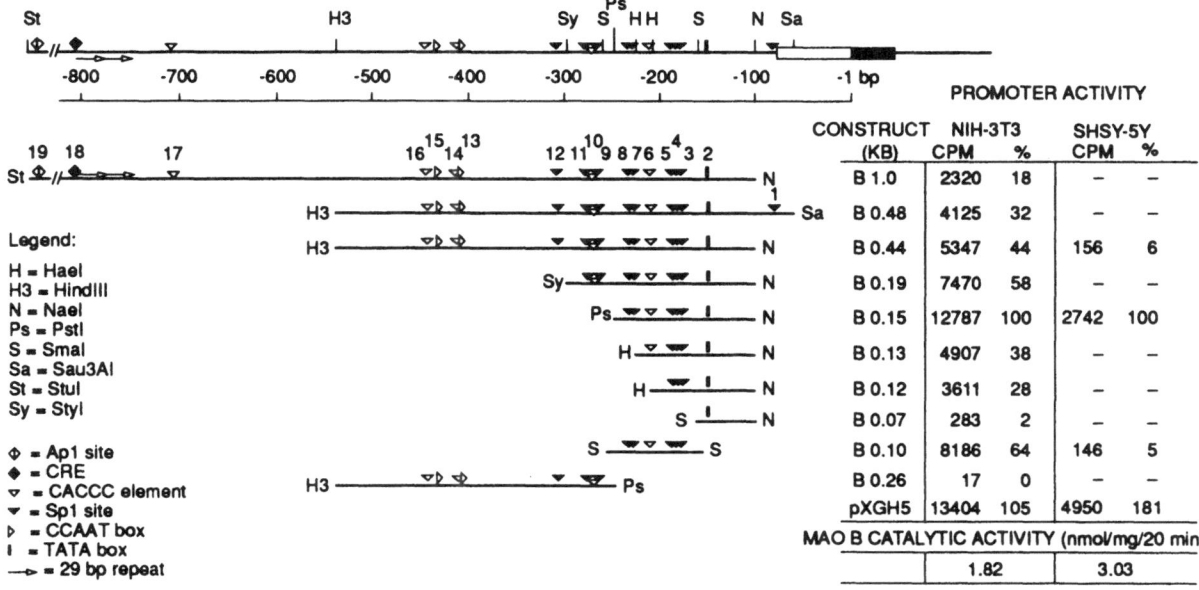

Fig. 3. Restriction map of the MAO B promoter region and promoter activity measurements. The symbols used in this figure are the same as in Fig. 2. The potential transcription factor binding sites are numbered as in Table 1 (1–19). *B* denotes DNA fragment of MAO B gene. The last line shows MAO B enzymatic activity measured in each cell line

A0.25) display very low promoter activity. Lack of a Sp1 site (A0.21) has no detectable promoter activity. The decreased promoter activity by upstream sequences that contain possible binding sites for transcription factors (compare A1.4, A0.52, A0.24 and A0.14), (A0.41, A0.29 and A0.12) and (A0.37 and A0.25) may result from down regulating transcription factors or competition of transcription factors for the activating sites on the polymerase complex.

Figure 3 shows that the 0.15 kb PstI/NaeI fragment exhibits the highest MAO B promoter activity. This fragment contains a TAATATA box (site 2), three overlapping Sp1 elements (site 3, 4 and 5), a CACCC element (site 6) and two overlapping Sp1 elements (site 7 and 8) therefore has a structure:

$$5'\text{-(Sp1)}_2\text{-CACCC-(Sp1)}_3\text{-TAATATA-}3'$$

Deletion from the 5′ end of the B0.15 kb fragment, resulting in a stepwise loss of 5′ Sp1 sites (site 7, 8: B0.13), the CACCC element (site 6: B0.12) and the 3′ Sp1 sites (site 3, 4 and 5: B0.07) resulted in 38%, 28% and 2% of the promoter activity, respectively. Deletion of the TATA box (B0.10) results in 64% activity thus demonstrating the positive role of the TATA box. Like the MAO A promoter, inclusion of further upstream sequences results in a stepwise decrease in the maximal promoter activity observed with B0.15 in NIH3T3 cells (B1.0: 18%, B0.48: 32%, B0.44: 44%

and B0.19: 58%). B0.26 has no promoter activity despite the presence of potential cis-elements.

The fact that the core region of both MAO A and B promoters are GC rich, contain potential Sp1 binding sites and can activate transcription initiation in the absence of a TATA box or a CCAAT box demonstrates these two promoters are housekeeping like. On the other hand, the differences in their cis-element organization and their behavior in human (SHSY-5Y) and mouse (NIH3T3) cells indicate that the mechanisms regulating MAO A and B expression are quite different.

In the human cell line SHSY-5Y the core MAO A and B promoter activities (4,356 cpm for A0.14 and 2,742 cpm for B0.15) correlate well with catalytic activities (3.36 nmol/mg/20 min for MAO A and 3.03 nmol/mg/20 min for MAO B). The unusually high MAO B promoter activity in NIH3T3 cells (12,787 cpm) compared with moderate mouse MAO B enzymatic activity (1.82 nmol/mg/20 min) suggests that the mouse fibroblasts contain sufficient activating factors or less inhibiting factors for human MAO B constructs. The poor correlation observed in mouse (NIH3T3) cells between measured human MAO promoter activities (332 cpm for A0.14 and 12,787 cpm for B0.15) and mouse catalytic activities (15.59 nmol/mg/29 min for MAO A and 1.82 nmol/mg/20 min for MAO B) suggests that mouse MAO promoters are different from humans.

Abnormal levels of MAO activity have been reported in a number of mental disorders and may result from promoter sequence changes. It will be interesting to investigate whether the varied MAO levels in disease states is caused by promoter modification (cis-element changes), or by changes of transcription factors required for MAO expression. The knowledge of the mechanism for MAO expression regulation can also be used to design new generation of MAO inhibitors based on altering gene transcription. This work opens up a new area of research concerning the molecular basis of MAO gene expression in disease states.

Acknowledgements

This work was supported by grant R01 MH37020, R37 MH39085 (Merit Award), and Research Scientist Award K05 MH00796, from the National Institute of Mental Health. Support from the Boyd and Elsie Welin Professorship is also appreciated.

References

Bach AWJ, Lan NC, Johnson DL, Abell CW, Bembenek ME, Kwan S-W, Seeberg PH, Shih JC (1988) cDNA cloning of human monoamine oxidase A and B: molecular basis of differences in enzymatic properties. Proc Natl Acad Sci USA 85: 4934–4938
Chen S, Shih JC, Xu QP (1984) Interaction of N-(2-nitro-4-azidophenyl)-serotonin with two types of monoamine oxidase in rat brain. J Neurochem 43: 1680–1687
Chiba K, Trevor A, Castagnoli N (1984) Metabolism of the neurotoxic tertiary amine, MPTP by brain monoamine oxidase. Biochem Biophys Res Commun 120: 574–578

Grimsby J, Chen K, Wang LJ, Lan NC, Shih JC (1991) Human monoamine oxidase A and B genes exhibit identical exon-intron organization. Proc Natl Acad Sci USA 88: 3637–3641

Grimsby J, Lan C, Neve R, Chen K, Shih JC (1990) Tissue distribution of human MAO A and B mRNA. J Neurochem 55: 1166–1169

Hsu YP, Weyler W, Chen S, Sims KB, Rinehart WB, Utterback M, Powell JF, Breakefield XO (1988) Structural features of human monoamine oxidase A elucidated from cDNA and peptide sequences J Neurochem 51: 3121–1324

Lan NC, Chen CH, Shih JC (1989) Expression of functional human monoamine oxidase (MAO) A and B cDNA in mammalian cells. J Neurochem 52: 1652–1654

Lan NC, Heinzmann C, Gal A, Klisak I, Orth U, Lai E, Grimsby J, Sparkes RS, Mohandas T, Shih JC (1989) Human monoamine oxidase A and B genes map to Xp11.23 and are deleted in a patient with Norris disease. Genomics 4: 552–55

Ozelius L, Hsu YP, Bruns G, Powell JF, Chen S, Weyler W, Utterback M, Zucker D, Haines J, Trofalter JA, Conneally PM, Gusella JF, Breakefield XO (1988) Human monoamine oxidase gene (MAOA): chromosome position (Xp21–p11) and DNA polymorphism. Genomics 3: 53–58

Shih JC (1990) Molecular basis of human MAO A and B. Neuropsychopharmacology 4: 1–7

Zhu QS, Grimsby J, Chen K, Shih JC (1992) Promoter organization and activity of human monoamine oxidase (MAO) A and B genes. J Neurosci 12(11): 4437–4446

Authors' address: Dr. J. C. Shih, Department of Molecular Pharmacology and Toxicology, School of Pharmacy, University of Southern California, 1985 Zonal Avenue, Los Angeles, CA 90033, U.S.A.

J Neural Transm (1994) [Suppl] 41: 35–39

Some problems associated with measuring monoamine oxidase activity in the presence of sodium azide

C. J. Barwell and **S. A. Ebrahimi**

School of Pharmacy and Biomedical Sciences, University of Portsmouth,
Portsmouth, United Kingdom

Summary. The colourimetric assay of monoamine oxidase activity, as hydrogen peroxide production, normally requires the use of sodium azide to inhibit breakdown of hydrogen peroxide by catalase. Sodium azide was shown to act as an uncompetitive inhibitor of benzylamine deamination with an inhibitor constant of 1.5 mM. Catalase activity of isolated rat liver mitochondria could be eliminated with the irreversible inhibitor of catalase, 3-amino-1,2,4-triazole. The treatment did not affect benzylamine deaminating activity. The catalase-free preparation could be used to assay monoamine oxidase activity colourimetrically, as hydrogen peroxide production, in the absence of sodium azide.

Introduction

Monoamine oxidase (MAO) can be assayed by a variety of methods (Tipton and Youdim, 1983). A colourimetric method described by Szutowicz et al. (1984) is both sensitive and economical. It is based upon measurement of hydrogen peroxide, produced in the MAO catalysed reaction, with peroxidase. The method has become popular as it has a sensitivity similar to the widely used but more expensive radiometric assay. Preparations of MAO, such as tissue homogenates and isolated mitochondria, normally contain high catalytic activities of catalase, which must be inhibited in order to conserve hydrogen peroxide produced by MAO. Inhibition of catalase is commonly achieved by including sodium azide in the assay at a concentration of 1 mM to 10 mM, which has been reported (Szutowicz et al., 1984) and is assumed to have little or no effect upon MAO activity.

When benzylamine is used as the MAO substrate, enzyme activity may be assayed spectrophotometrically by determining the increase in absorbance at 250 nm due to production of benzaldehyde (Tabor et al., 1954). This method does not require the inclusion of sodium azide in the assay. During studies of benzylamine deamination we have measured enzyme activity as both hydrogen peroxide production, in the presence of 5 mM azide, and benzaldehyde production, in the absence of azide. The activity measured in the presence of azide was normally approximately one half that in its absence. The reason for this discrepancy was investigated and found to be due to inhibition of MAO by azide.

Materials and methods

Male Wistar rats were killed by asphyxiation with carbon monoxide. Livers were removed, rinsed and homogenized in 0.3 M sucrose with a Thomas homogenizer. Mitochondria were isolated by centrifugation and washed twice by homogenization and centrifugation in 10 mM potassium phosphate buffer, pH 7.4.

Monoamine oxidase activity was assayed with 5 mM benzylamine at 37°C in 50 mM sodium phosphate buffer pH 7.4, both as hydrogen peroxide production (Szutowicz et al., 1984) and by determining the increase in absorbance at 250 nm due to production of benzaldehyde (Tabor et al., 1954). For kinetic investigations the concentration of benzylamine was varied and kinetic constants were determined from Hanes plots of substrate concentration divided by initial velocity against substrate concentration (S/v against S) with lines of best fit calculated by linear regression.

Catalase activity was assayed with 50 μM hydrogen peroxide at 37°C in 50 mM sodium phosphate buffer pH 7.4, using peroxidase and 2,2'-azinodi(3-ethylbenzthiazo-line-6-sulphonic acid) to measure the hydrogen peroxide remaining.

Isolated mitochondria were treated with aminotriazole by suspending the washed mitochondria from 1 g liver in 10 mL of 10 mM potassium phosphate buffer pH 7.4 containing 100 mM aminotriazole and 4 mM ascorbic acid then incubating at 37°C for 30 min (Margoliash et al., 1960). The treated mitochondria were recovered and washed twice by centrifugation.

Results

In the method which measures MAO activity as hydrogen peroxide production no activity was detectable when sodium azide was omitted from the assay. When azide was included activity became detectable and was apparently optimal at concentrations between 2 mM and 10 mM in the assay. Benzylamine deamination, measured as hydrogen peroxide production in the presence of 5 mM azide, was approximately one half that measured, as benzaldehyde production in the absence of azide. A similar relationship was found when enzyme activity was measured as benzaldehyde production, both in the absence and presence of 5 mM azide.

The apparent inhibitory effect of azide was investigated using the assay based upon benzaldehyde production. At 5 mM, sodium azide reduced both the Michaelis constant (K_m) and maximum velocity (V_{max}). The ratio (K_m/V_{max}) of these kinetic constants was essentially the same in the absence and presence of azide. From the kinetic data an inhibitor constant of 1.5 mM was calculated. A typical experiment is illustrated as a Hanes plot (S/v against S) in Fig. 1.

A mitochondrial preparation with a catalase activity of 50,000 nmol h^{-1} mL^{-1} and benzylamine deaminating activity, measured as benzaldehyde production, of 1,630 nmol h^{-1} mL^{-1} was treated with aminotriazole. The treatment inhibited more than 99% of catalase but had no statistically

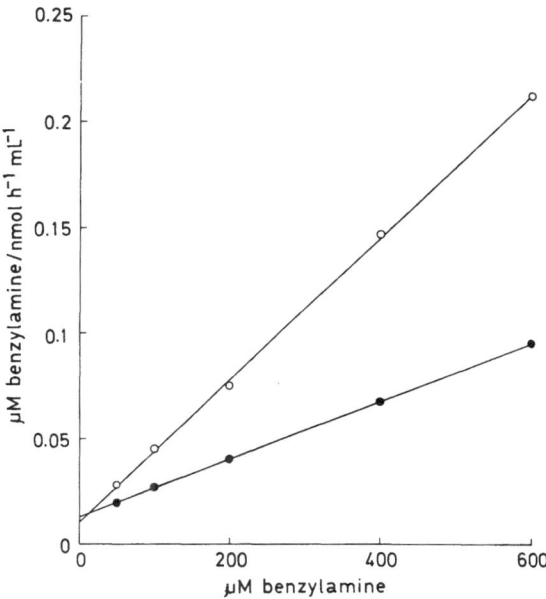

Fig. 1. Effect of sodium azide upon benzylamine deamination. Activity was measured as benzaldehyde production at pH 7.4 and 37°C. Reactions were started, after 5 min preincubation, by addition of benzylamine. With 5 mM azide (○), without azide (●)

Table 1. Effect of sodium azide upon kinetic parameters for benzylamine deamination measured as hydrogen peroxide production

Condition	Kinetic parameter		
	K_m (μM)	V_{max} (nmol h^{-1} mL^{-1})	K_m/V_{max} ($\times 10^3$)
No azide	132	1,760	75
5 mM azide	75	1,060	71

Isolated rat liver mitochondria were treated with 3-amino-1,2,4-triazole to inhibit catalase, as described in Materials and methods. Benzylamine deamination was measured as hydrogen peroxide production at pH 7.4 and 37°C

significant effect upon benzylamine deaminating activity. The aminotriazole-treated preparation could be used to measure benzylamine deaminating activity as hydrogen peroxide production, in the absence of sodium azide. The activity, measured as hydrogen peroxide (1,500 ± 13 nmol h^{-1} mL^{-1}) was essentially the same as that measured as benzaldehyde production (1,630 ± 52 nmol h^{-1} mL^{-1}), (mean ± SE, n = 10). Using the assay which measures hydrogen peroxide production, the kinetics of benzylamine deamination were measured with and without 5 mM azide in the assay. Results were similar to those obtained when activity was measured as benzaldehyde production. The values in Table 1 show that both the Michaelis constant and maximum velocity were reduced by azide and the ratio K_m/V_{max} was similar in the absence and presence of 5 mM azide.

Discussion

Colorimetric measurement of hydrogen peroxide with peroxidase and a reduced dye provides a simple and economical assay of MAO. With ABTS as the dye, measurement of 1 nmol product is possible, due to its high absorption coefficient of 24,600 M^{-1} cm^{-1} (Szutowicz et al., 1984). This sensitivity is similar to that of the widely used but more expensive radiometric assay (Tipton and Youdim, 1983).

Catalase occurs in most mammalian tissues and when MAO is assayed in tissue homogenates and isolated mitochondrial fractions, this enzyme will normally be present. Consequently it must be inhibited to prevent loss of hydrogen peroxide and underestimation of MAO activity. Sodium azide is commonly used as a catalase inhibitor and the concentration required for a particular preparation would be found by using a fixed concentration of enzyme preparation and measuring colour formation, due to hydrogen peroxide production, in the presence of various azide concentrations. In our experience, a concentration in the range 1 mM to 10 mM is required for apparently optimal recovery of hydrogen peroxide. It has been reported that such concentrations of azide have little or no effect upon MAO activity (Szutowicz et al., 1984). However, our observations show that sodium azide inhibits benzylamine deamination and therefore at least MAO-B. The form of Hanes plots and the essentially equal effect upon the Michaelis constant and Maximum velocity indicates that azide acts as an uncompetitive inhibitor. The inhibitor constant was 1.5 mM. This means that azide concentrations commonly used in the colorimetric assay would result in apparent MAO activities which are approximately half the true value.

Aminotriazole was used successfully to essentially eliminate catalase activity from isolated mitochondria preparations which could be used to measure benzylamine deamination, as hydrogen peroxide production, without azide in the assay. The conditions used to inhibit catalase did not affect benzylamine deaminating activity. Aminotriazole is an irreversible inhibitor of catalase (Margoliash et al., 1960), therefore, catalase activity is not regained during manipulation of the treated preparations and their dilution for subsequent assays. Treatment of MAO preparations, such as tissue homogenates, with aminotriazole could be used to eliminate the requirement for azide in assays which measure MAO activity as hydrogen peroxide production. The catalytic activities and substrate affinities measured would be more representative of physiological values and would be obtained using a simple and economical assay.

Acknowledgement

We are grateful to the Ministry of Medical Education of the Islamic Republic of Iran for a grant to support S. A. Ebrahimi and in support of this work.

References

Margoliash E, Novogrodsky A, Schejter A (1960) Irreversible reaction of 3-amino-1:2:4-triazole and related inhibitors with the protein of catalase. Biochem J 74: 339–348

Szutowicz A, Kobes RD, Orsulak PJ (1983) Colorimetric assay for monoamine oxidase in tissues using peroxidase and 2,2′-azinodi(3-ethylbenzthiazoline-6-sulphonic acid) as chromogen. Anal Biochem 138: 86–94

Tabor S, Tabor H, Rosenthal SM (1954) Purification of amine oxidase form beef plasma. J Biol Chem 208: 645–661

Tipton KF, Youdim MBH (1983) Methods in biogenic amine research, 1st edn. Elsevier, Amsterdam, p 441

Authors' address: Dr. C. J. Barwell, School of Pharmacy and Biomedical Sciences, University of Portsmouth, King Henry 1 Street, Portsmouth, PO1 2DZ, United Kingdom.

J Neural Transm (1994) [Suppl] 41: 41–45
© Springer-Verlag 1994

Some kinetic properties of guinea pig liver monoamine oxidase

C. J. Barwell and **S. A. Ebrahimi**

School of Pharmacy and Biomedical Sciences, University of Portsmouth,
Portsmouth, United Kingdom

Summary. Titration of monoamine oxidase activity in isolated guinea pig liver mitochondria with clorgyline and assay of remaining activity with tyramine yielded biphasic inhibition curves. The position of the plateaus obtained with mitochondria from four animals, indicated that the B form of monoamine oxidase accounted for 30% to 70% of the tyramine deaminating activity. Benzylamine deamination was selectively inhibited by (−)-deprenyl. However, benzylamine and other amines which are selective substrates for the B form of monoamine oxidase from the rat, were deaminated at only low rates by the guinea pig liver enzyme. Guinea pig liver contains a monoamine oxidase-B which is unusual in that although it exhibits apparently normal sensitivity to selective irreversible inhibitors, it has a low catalytic activity with substrates which the enzyme from rat liver deaminates rapidly.

Introduction

The monoamine oxidase (MAO) of guinea pig liver has often been used in studies upon substrate deamination. Generally it has been found that tyramine, a substrate for both MAO-A and MAO-B and 5-hydroxytryptamine (5-HT), a selective substrate for MAO-A are deaminated rapidly and at similar rates. In contrast, deamination of amines, which are now normally regarded as selective for MAO-B (Dostert et al., 1989), occurs either at low rates or is not detectable (Alles and Heegaard, 1943; Randall, 1946). Such studies upon substrate selectivity indicate that guinea pig liver contains little or no activity of the B form of monoamine oxidase. However, studies with selective irreversible inhibitors indicate the occurrence of MAO-B and that it constitutes approximately one half of the total tyramine deaminating activity (Squires, 1972; Das and Guha, 1980). We have carried out an investigation of guinea pig liver MAO in an attempt to resolve the apparent contradiction between results obtained with substrate selectivity and inhibitor sensitivity.

Materials and methods

Male guinea pigs and Wistar rats were killed by asphyxiation with carbon monoxide. Livers were removed, rinsed and homogenized, in 0.3 M sucrose, with a Thomas homogenizer. Mitochondria were isolated by centrifugation and washed twice by homogenization and centrifugation in 10 mM potassium phosphate buffer, pH 7.4.

Monoamine oxidase activity was assayed at 37°C in 50 mM sodium phosphate buffer, pH 7.4. Deamination of 5-HT (2 mM) was measured with an oxygen electrode. Deamination of the following amines was measured colorimetrically as hydrogen peroxide production (Szutowicz et al., 1984): tyramine (5 mM), N,N-dimethyltyramine (5 mM), benzylamine (5 mM), and N,N-dimethylphenethylamine (0.5 mM). Under the conditions, the method detected 1 nmol hydrogen peroxide in the assay.

Deamination of benzylamine (5 mM) was also measured using an HPLC assay in which benzaldehyde was converted to benzaldehyde semicarbazone by addition of semicarbazide. The semicarbazone was separated on a C_{18} reverse phase column and detected at 280 nm The method detected 0.01 nmol on the column (Barwell and Ebrahimi, unpublished).

Enzyme preparations were incubated with clorgyline, selective irreversible inhibitor of MAO-A, and (−)-deprenyl, selective irreversible inhibitor of MAO-B, (see Dostert et al., 1989) for 30 min at pH 7.4 and 37°C, the remaining activity was then assayed by addition of 5 mM substrate.

Results

Tyramine deaminating activity of both rat and guinea liver mitochondria was inhibited in a biphasic manner, with a plateau between 10^{-8} M and 10^{-7} M. This is illustrated in Fig. 1, for preparations containing ap-

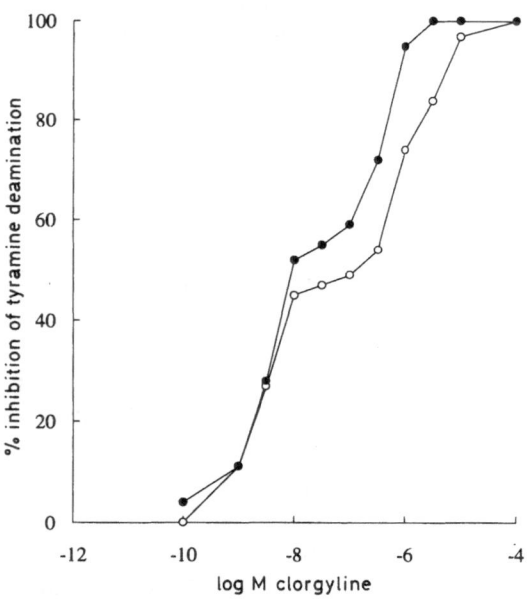

Fig. 1. In vitro effect of clorgyline upon tyramine deaminating activity of rat (○) and guinea pig (●) liver mitochondria. Enzyme preparation was incubated with clorgyline for 30 min at pH 7.4 and 37°C then remaining activity assayed with 5 mM tyramine

Table 1. Some properties of rat and guinea pig liver monoamine oxidase

| Mitochondria preparation | Deaminating activity (nmol h^{-1} g liver^{-1}) | | Benzylamine as percentage tyramine | %MAO-B* |
	Benzylamine	Tyramine		
Guinea pig				
A	4,428	44,800	10	30
B	4,476	58,856	8	70
C	4,374	53,356	8	85
D	—	41,200	—	50
Rat				
A	8,600	24,800	35	47
B	5,200	13,800	38	50
C	4,098	12,428	33	55

Deaminating activities were measured at pH 7.4 and 37°C with 5 mM benzylamine and tyramine. *The percentage of monoamine oxidase B in each preparation was determined from the plateau obtained by titration with clorgyline and assaying remaining activity with 5 mM tyramine, see Fig. 1

proximately equal activities of MAO-A and MAO-B, as defined by inhibitor sensitivity. The proportion of tyramine deaminating activity due to MAO-B in mitochondria isolated from different animals is shown in Table 1. The proportion in preparations from three rats was similar and approximately 50%. In contrast the proportion in four preparations from guinea pig varied from 30% to 85%.

Each preparation deaminated both tyramine and benzylamine. Benzylamine deaminating activity, as a percentage of tyramine deaminating activity, was four times lower in guinea pig. It was inhibited in an essentially monophasic manner and (−)-deprenyl was the most potent inhibitor, as is shown in Fig. 2.

A guinea pig preparation containing 50% MAO-B activity as defined by its sensitivity to clorgyline with tyramine as substrate, deaminated tyramine and 5-HT at similar rates of; 52,300 ± 4,100 and 47,000 ± 3,000 nmol h^{-1} g liver^{-1}, respectively. Benzylamine deamination was 4,200 ± 200 nmol h^{-1} g liver^{-1} but deamination of N,N-dimethylphenethylamine and N,N-dimethyltyramine was not detectable and therefore less than 800 nmol h^{-1} g liver^{-1}.

Discussion

Catalytic activities of the isoenzymes of monoamine oxidase may be differentiated kinetically by their substrate specificity and inhibitor sensitivity (Dostert et al., 1989). MAO-A selectively deaminates 5-HT and is selectively inhibited by clorgyline. MAO-B selectively deaminates benzylamine and is selectively inhibited by (−)-deprenyl. Both isoenzymes deaminate

44   C. J. Barwell and S. A. Ebrahimi

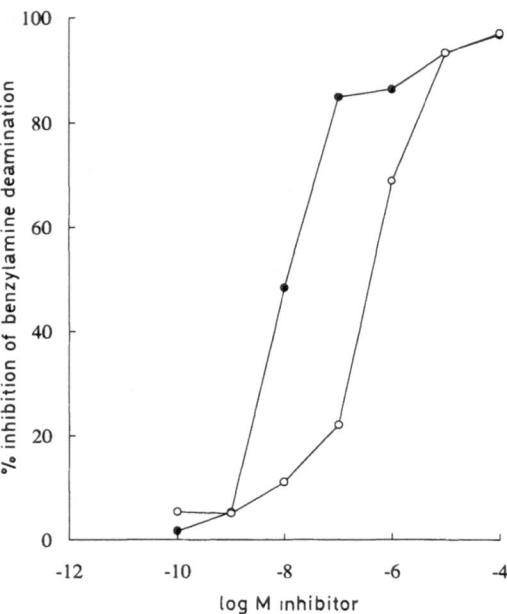

Fig. 2. In vitro effect of clorgyline and deprenyl upon benzylamine deaminating activity of guinea pig liver mitochondria. Enzyme preparation was incubated with either clorgyline (○) or deprenyl (●) for 30 min at pH 7.4 and 37°C then remaining activity assayed with 5 mM benzylamine

tyramine. The catalytic activity of MAO-A with 5-HT and tyramine is similar and the catalytic activity of MAO-B with benzylamine and tyramine is similar. It follows that a preparation containing approximately equal catalytic activities of the two isoenzymes would deaminate 5-HT and benzylamine at similar rates and at approximately half the rate of tyramine. These kinetic characters apply to "typical" monoamine oxidases which are represented by the isoenzymes found in rat liver. However the enzymes of some species and tissues can exhibit different kinetic properties (Fowler et al., 1981).

In this study, with rat liver mitochondria and tyramine as substrate, clorgyline yielded typical biphasic inhibition curves, indicating approximately equal catalytic activities of MAO-A and MAO-B, and tyramine and benzylamine were deaminated at similar rates. Therefore the isolation procedure for mitochondria and conditions used to investigate kinetic characters yielded typical results.

Titration of tyramine deaminating activity of guinea pig liver mitochondria with clorgyline also yielded biphasic inhibition curves, indicating the occurrence of two forms of MAO and with similar inhibitor sensitivities to the isoenzymes of the rat. Thus the preparations contained both MAO-A and MAO-B as defined by inhibitor sensitivity. Benzylamine deaminating activity of guinea pig was selectively inhibited by (−)-deprenyl and therefore was an apparently typical MAO-B. However, compared to the rat, benzylamine was deaminated at a low rate by preparations containing 4 times greater tyramine deaminating activity of which 70% to 85% was apparently due to an MAO-B (see data in Table 1). The tertiary amines, N,N-dimethylphenethylamine and N,N-dimethyltryamine, are highly selective substrates for the MAO-B of rat liver and are deaminated at approximately one half the rate of benzylamine (Barwell et al., 1988, 1989). Deamination of these amines by guinea pig preparations was not detectable.

The results presented here indicate that guinea pig liver contains an MAO-B which is similar to the rat isoenzyme with regard to both inhibitor sensitivity and selectivity for benzylamine. However, the guinea pig MAO-B is unusual in that it exhibits a relatively low catalytic activity with benzylamine. Results obtained by earlier workers who measured little or no activity with benzylamine and other amines now known to be highly selective substrates for MAO-B, can be explained by the low activity of guinea pig liver MAO-B with such amines, together with the low sensitivity of the assay methods used in the investigations.

Acknowledgement

We are grateful to the Ministry of Medical Education of the Islamic Republic of Iran for a grant to support S. A. Ebrahimi and in support of this work.

References

Alles GA, Heegaard EV (1943) Substrate specificity of amine oxidase. J Biol Chem 147: 487–503

Barwell CJ, Basma AN, Canham CA, Williams C (1988) Evaluation of N,N-dimethylphenethylamine, and N,N-dimethyltyramine as substrates for monoamine oxidase B. Pharmacol Res Commun 20 [Suppl] 4: 101–102

Barwell CJ, Basma AN, Lafi MAK, Leake LD (1989) Deamination of hordenine by monoamine oxidase and its action of vasa deferentia of the rat. J Pharm Pharmacol 41: 421–423

Das PK, Guha SR (1980) MAO types in guinea pig liver mitochondria. Biochem Pharmacol 29: 2049–2053

Dostert PL, Strolin Benedetti M, Tipton KF (1989) Interactions of monoamine oxidase with substrates and inhibitors. Med Res Rev 9 1: 45–89

Fowler CJ, Oreland L, Callingham BA (1981) The acetylenic monoamine oxidase inhibitors clorgyline, deprenyl, pargyline and J-508: their properties and applications. J Pharm Pharmacol 33: 341–347

Randall LO (1946) Oxidation of phenethylamine derivatives by amine oxidase. J Pharmacol Exp Ther 88: 216–220

Squires RF (1972) Multiple forms of monoamine oxidase in intact mitochondria as characterised by selective inhibitors and thermal stability: a comparison of eight mammalian species. Adv Biochem Psychopharmacol 5: 355–370

Szutowicz A, Kobes RD, Orsulak PJ (1983) Colorimetric assay for monoamine oxidase in tissues using peroxidase and 2,2′-azinodi(3-ethylbenzthiazoline-6-sulphonic acid) as chromogen. Anal Biochem 138: 86–94

Authors' address: Dr. C. J. Barwell, School of Pharmacy and Biomedical Sciences, University of Portsmouth, King Henry 1 Street, Portsmouth, PO1 2DZ, United Kingdom.

J Neural Transm (1994) [Suppl] 41: 47–53

Estimation of monoamine oxidase concentrations in soluble and membrane-bound preparations by inhibitor binding

M. C. Anderson and **K. F. Tipton**

Department of Biochemistry, Trinity College, Dublin, Ireland

Summary. A modification of the [³H]-pargyline labelling technique is presented for determining the active-site concentration of monoamine oxidase in soluble preparations. Kinetic considerations show that the rate of reaction of MAO-A with low concentrations of free pargyline will be very much slower than that of MAO-B. Failure to use adequate reaction times for the concentration of pargyline added can lead to gross underestimation of the quantity of MAO-A present.

Introduction

The determination of the irreversible binding of [³H]-labelled pargyline to monoamine oxidases A and B provides a simple, accurate and sensitive method for determining the active-site concentrations of these enzymes in tissue preparations (see Parkinson and Callingham, 1980; Gomez et al., 1986; O'Carroll et al., 1989). It can be applied even to preparations of the enzyme as crude as tissue homogenates.

In these procedures separation of enzyme-bound from free pargyline can be readily achieved by centrifugation or filtration. Such an approach cannot, however, be applied to preparations of the enzyme that have been rendered soluble. In this work we describe a simple procedure whereby separation of free from bound pargyline from soluble preparations of the enzyme can be effected by treatment with charcoal. The incubation times necessary to achieve complete labelling of both MAO-A and -B are also estimated from the known kinetic parameters for the irreversible inhibition of both enzymes by pargyline.

Materials and methods

Monoamine oxidase-B was partially purified from ox liver by the procedure of Salach (1979). Activity was determined radiochemically (Tipton, 1985) or spectrophotometrically (Tabor et al., 1954). Labelled pargyline hydrochloride (phenyl-3, benzyl-[³H]) was obtained from New England Nuclear and diluted to the required specific activity with unlabelled pargyline HCl.

Pargyline binding was determined by the following procedure. Enzyme samples were incubated at 37°C with different concentrations of [^{3}H]-pargyline (1.25 Ci.mol^{-1}) for 60 min. The reaction was then terminated by removing the unbound pargyline by treating the incubation mixtures with 1% (w/v) activated charcoal and 2% (w/v) Bovine Serum Albumin followed by rapid centrifugation.

Non-specific pargyline binding was determined by preincubating the enzyme sample at 37°C for 60 min with 2 mM unlabelled pargyline before determining the binding of labelled pargyline as described above. The alternative procedures of pretreating the enzyme samples with low concentrations of l-deprenyl, or with higher concentrations of clorgyline in order to inhibit the activity of MAO-B (see Gomez et al., 1986; O'Carroll et al., 1989) gave results that were not significantly different.

Results and discussion

The procedure for use with insoluble preparations of MAO is compared with that for soluble preparations in Schemes 1 and 2.

Scheme 1. Radioactive pargyline binding to mitochondrial MAO preparations

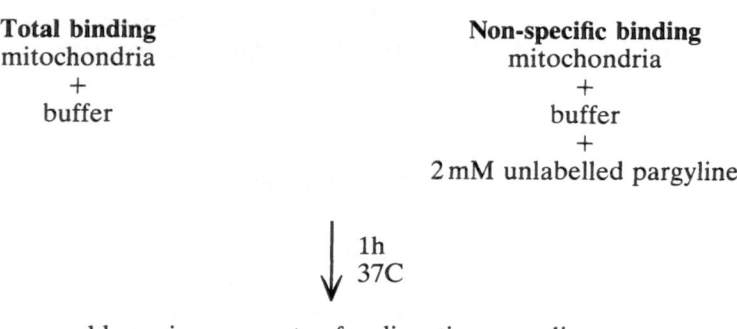

Total binding
mitochondria
+
buffer

Non-specific binding
mitochondria
+
buffer
+
2 mM unlabelled pargyline

1h
37C

add varying amounts of radioactive pargyline

1h
37C

cool on ice, add 1 ml buffer

centrifuge

resuspend pellet, wash as before

centrifuge

resuspend pellet in buffer add scintillant
and count for radioactivity

The procedures are described in detail in Gomez et al. (1986) and O'Carroll et al. (1989)

Scheme 2. Radioactive pargyline binding to soluble MAO preparations

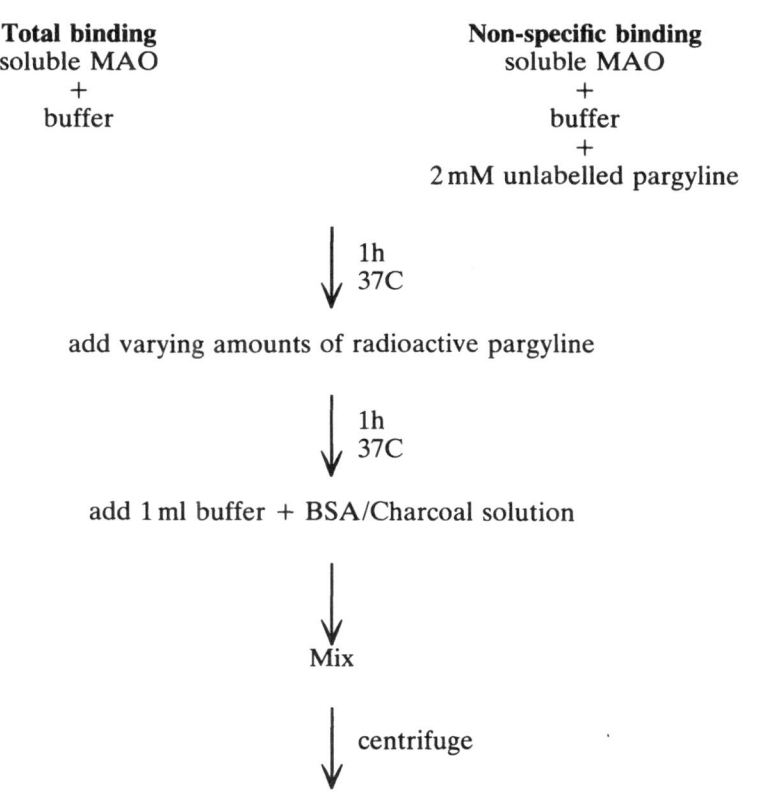

Total binding
soluble MAO
+
buffer

Non-specific binding
soluble MAO
+
buffer
+
2 mM unlabelled pargyline

↓ 1h
  37C

add varying amounts of radioactive pargyline

↓ 1h
  37C

add 1 ml buffer + BSA/Charcoal solution

↓

Mix

↓ centrifuge

remove aliqot of supernatent
add to scintillant and
count for radioactivity

Further details are given in the text

Figure 1 shows the determination of the concentration of MAO in a soluble and partly purified preparation of ox liver MAO-B by this latter procedure. As might be expected the degree of non-specific binding of labelled pargyline was considerably lower than that obtained in previous studies with crude tissue preparations.

To assess the performance of the procedure, the active-site concentrations of soluble preparations of the enzyme from ox liver (Salach, 1979) of different degrees of purity were determined by the pargyline-binding procedure. Despite the disparities in specific activities, the turnover numbers (mol.product.active site^{-1}.min^{-1}) were the same, with a value of 375 ± 42.

Since the above procedure relies on the separation of the insoluble chacoal-bound inhibitor from the soluble enzyme, it is unsuitable for use with the insoluble preparations of the enzyme in crude tissue preparations. In such cases the procedures relying on direct filtration or centrifugation to

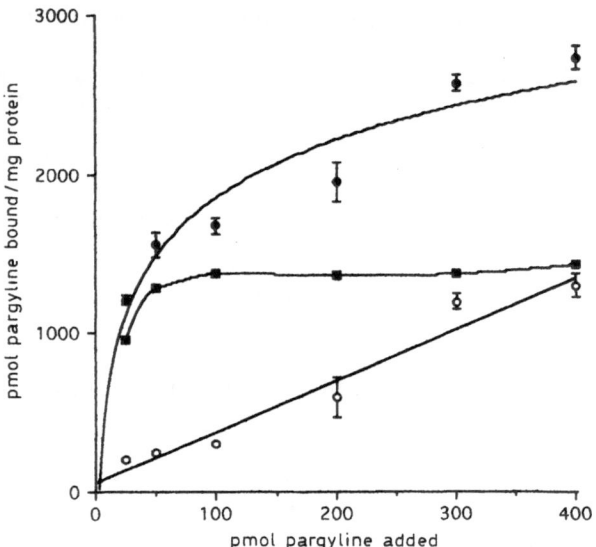

Fig. 1. The binding of [³H]-labelled pargyline to a partially purified soluble preparation of ox liver monoamine oxidase-B. The enzyme sample was prepared by the procedure of Salach (1979). Charcoal treatment was used for separating free from irreversibly enzyme-bound pargyline. Other experimental details were as described in the text. Non-specific binding (○) was determined after preincubation of the sample for 60 min at 37°C with 2 mM unlabelled pargyline before incubation for a further 60 min at 37°C with [³H]-labelled pargyline. The preincubation step was omitted for determination of total binding (●). Specific binding (■) was calculated as the difference between the total and non-specific binding values. Data are given as mean values ± s.e.m.

separate bound and free inhibitor (see Parkinson and Callingham, 1980; Gomez et al., 1986; O'Carroll et al., 1989) would be appropriate.

If the pargyline binding procedure is to give accurate measures of the active-site concentrations of MAO-A and -B, it is essential that the conditions under which the enzyme preparation and labelled-pargyline are incubated are sufficient to ensure complete reaction of both enzyme forms. Examination of the published results that have used this procedure suggest that in some instances the concentration of MAO-A may have been underestimated by failure to ensure that labelling conditions were adequate.

Pargyline is a mechanism-based inhibitor of MAO for which the inhibitory process may be described by the kinetic pathway shown below (Fowler et al., 1982):

$$E + I \underset{k_{-1}}{\overset{k_{+1}}{\rightleftharpoons}} E.I \overset{k_{+2}}{\longrightarrow} E\text{-}I$$

where E and I represent the enzyme and inhibitor, respectively, and E.I. and E-I represent the initial, noncovalent, enzyme-inhibitor complex and the irreversibly inhibited enzyme, respectively. The dissociation constant (K_i) for initial, non-covalent, enzyme.inhibitor complex (E.I) formation is given by the relationship:

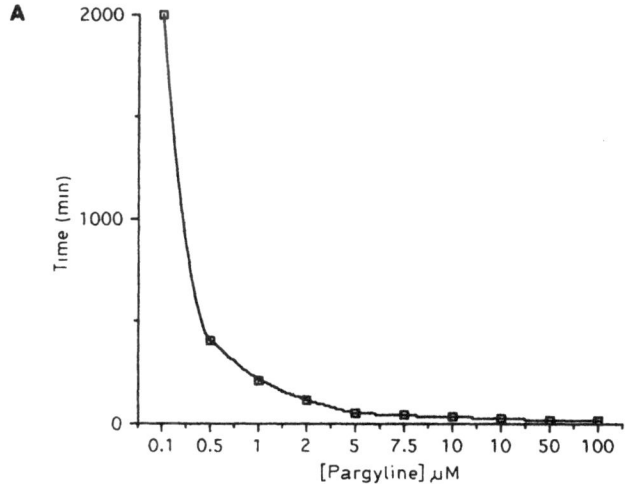

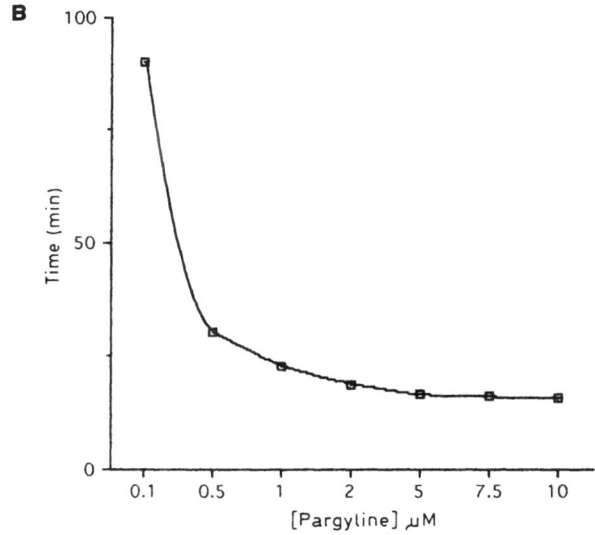

Fig. 2. Times taken for the mechanism-based irreversible inhibition of MAO by pargyline to reach 95 percent of maximum. Values were calculated with respect to the free pargyline concentration for MAO-A (**A**) and MAO-B (**B**) using the reaction mechanism and kinetic constants given in the text

$$K_i = k_{-1}/k_{+1}$$

The kinetic constants defining the inhibition of MAO-A and -B, at 30°C, according to this mechanism have been previously reported (Fowler et al., 1982). Unlike clorgyline and deprenyl, the rates of irreversible enzyme-inhibitor adduct formation depend solely on the K_i values (13 ± 5 and 0.5 ± 0.08 µM for MAO-A and -B, respectively), since the values of the first-order rate constants k_{+2} ($0.2 \pm 0.06\,\text{min}^{-1}$) are the same for both enzymes. Because of this, the rates of irreversible inhibition of MAO-A and -B will be the same when the pargyline concentrations are high enough to saturate both of them (>10 times the K_i value for MAO-A). Under such conditions the half-life for enzyme labelling will tend to the minimum value of about 3–4 min (Fowler et al., 1982).

As the concentrations of pargyline are decreased the rates of irreversible reaction with both enzymes will decline but those for MAO-A will decrease much more sharply than those for MAO-B. This is illustrated in Fig. 2 where the values were calculated from the data of Fowler et al. (1982), with the reasonable assumption (see Gomez et al., 1986; O'Carroll et al., 1989) that the enzyme concentration was much less than that of the inhibitor. It can be seen that, although MAO-B will be essentially fully labelled within 60 min at a pargyline concentration as low as 0.5 µM, a concentration of 5 µM would be required for the labelling of MAO-A to reach 95% of the maximum within this time. Even at free pargyline concentrations as high as 10 µM a 30 min incubation time would inadequate for complete labelling of MAO-A.

Such data indicate the importance of using sufficiently high pargyline concentrations and incubation times to ensure complete reaction of both MAO-A and -B. The pargyline concentrations in these calculations represents the *free* concentration. The relatively high proportions of non-specific binding of pargyline in crude tissue preparations (see Gomez et al., 1986; O'Carroll et al., 1989) would reduce the effective concentration so that even longer times would be required to achieve complete labelling of MAO. Thus the only effective way of ensuring that the incubation time used is adequate for complete labelling is to determine the time-courses of incorporation of label. Any loss of enzyme function, for example as a result of denaturation or proteolysis, during extended incubation periods would also result in an underestimation of the quantity of enzyme originally present, since the irreversible reaction with mechanism-based inhibitors relies on the retention of the oxidative catalytic function. The use of higher concentrations of labelled pargyline could minimize this problem, but this may be limited by the need to ensure that the specific radioactivity is high enough for adequate sensitivity and also by cost considerations.

References

Fowler CJ, Mantle TJ, Tipton KF (1982) The nature of the inhibition of rat liver monoamine oxidase types A and B by the acetylenic inhibitors clorgyline, l-deprenyl and pargyline. Biochem Pharmacol 31: 3555–2561

Gomez N, Unzeta M, Tipton KF, Anderson MC, O'Carroll A-M (1986) Determination of monoamine oxidase concentration in rat liver by inhibitor binding. Biochem Pharmacol 35: 4467–4472

O'Carroll A-M, Anderson MC, Tobbia I, Phillips JP, Tipton KF (1989) Determination of the absolute concentrations of monoamine oxidase A and B in human tissues. Biochem Pharmacol 38: 901–905

Parkinson D, Callingham BA (1980) The binding of [^{3}H] pargyline to rat liver mitochondrial monoamine oxidase J Pharm Pharmacol 32: 49–54

Salach J (1979) Monoamine oxidase from beef liver mitochondria: simplified isolation procedure, properties and determination of its cysteinyl flavin content. Arch Biochem Biophys 192: 128–137

Tabor CW, Tabor H, Rosenthal SM (1954) Purification of amine oxidase from beef plasma. J Biol Chem 208: 645–661

Tipton KF (1985) Determination of monoamine oxidase. Meth Find Exp Clin Pharmacol 7: 361–367

Authors' address: Prof. Dr. K. F. Tipton, Department of Biochemistry, Trinity College, Dublin 2, Ireland.

Monoamine oxidase: functions

J Neural Transm (1994) [Suppl] 41: 57–67

Monoamine oxidase and catecholamine metabolism

I. J. Kopin

National Institute of Neurological Disorders and Stroke, National Institutes of
Health, Bethesda, Maryland, U.S.A.

Summary. The enzyme which has come to be known as monoamine oxidase was discovered in liver over 60 years ago as tyramine oxidase (Hare, 1928). Almost 10 years later, Blaschko et al. (1957a,b) established that epinephrine, norepinephrine and dopamine were also substrates for this enzyme. Zeller (1938) distinguished monoamine oxidase as different from several other amine oxidases, such as diamine oxidase. Although it was generally assumed that catecholamines were metabolized by MAO, this was not established until isotopically labelled epinephrine and an MAO inhibitor became available. Schayer (1951) found that after administration of N-methyl-^{14}C-epinephrine, only about 50% of the radioactivity appeared in the urine, whereas when the ^{14}C label was incorporated into the β-position on the side chain, almost all of the radioactivity could be recovered. One year later, Zeller et al. (1952) discovered that isonicotinic acid hydrazide (iproniazid) inhibited MAO. When animals pretreated with the MAO inhibitor were administered N-methyl-^{14}C-epinephrine, almost all of the radioactivity was recovered (Schayer et al., 1955), indicating that the enzyme was responsible for the metabolism of about half of the administered catecholamine. Schayer et al. (1952, 1953) had found that five urinary metabolite products of β-labelled-^{14}C-norepinephrine could be separated by paper chromatography, but the chemical structures of these compounds were not known.

Armstrong et al. (1957) showed that 3-methoxy-4-hydroxymandelic acid (vanillyl mandelic acid, VMA) was the major metabolite of norepinephrine and Shaw et al. (1957) demonstrated that large amounts of homovanillic acid (HVA) were excreted in urine after administration of 3,4-dihydroxyphenylalanine (DOPA). These observations led Axelrod to examine the possibility that O-methylation might precede deamination and to his discovery of catechol-O-methyl transferase (Axelrod, 1957, 1959). At that time it became apparent that there were two possible routes for metabolism of norepinephrine to VMA — either deamination followed by O-methylation or O-methylation and subsequent deamination. The relative roles of these two pathways in terminating the physiological actions of catecholamines then became a focus of attention. Biochemical methods were used to access directly the relative importance of the two metabolic pathways. Physiological methods, based on the effects of drugs which alter metabolism of the catecholamine, were used to examine the role of MAO and COMT in terminating the actions of administered or endogenously released catecholamines.

Biochemical assessment of catecholamine metabolism

As indicated above, the demonstration by Schayer et al. (1955) that inhibition of MAO enhanced recovery of N-methyl-^{14}C-norepinephrine provided the first direct in vivo evidence that MAO was important in the metabolism of catecholamines and the discovery of VMA by Armstrong et al. (1957) confirmed this conclusion. When ^{3}H-epinephrine was administered to humans about 40% of the administered radioactivity was recovered in the urine as metanephrine (LaBross et al., 1958). When corrected for conversion of the O-methylated amine to VMA, it was calculated that initial O-methylation amounted to about 2/3 of the administered catecholamine in humans (Kopin, 1960) and rats (Kopin et al., 1961). This is also the case for most of intravenously administered ^{3}H-norepinephrine (Kopin and Gordon, 1962). Administered norepinephrine, however, is metabolized differently from the endogenous compound. Whereas, initially (0–3 hrs) excreted metabolites of the labelled catecholamine were mostly norepinephrine and normetanephrine, at later times (10–13 hrs) the deaminated metabolites predominated (Kopin and Gordon, 1962). Furthermore, when reserpine was administered, the increment in tritium excreted from the released catecholamine was predominantly as deaminated compounds, whereas after tyramine administration, the pattern of metabolite excretion resembled that seen during the initial interval. This was also apparent in isolated perfused rat hearts previously exposed to ^{3}H-norepinephrine (Kopin et al., 1962). Activity of drugs promoting intraneuronal metabolism by MAO versus extraneuronal release and metabolism by COMT could be distinguished by the patterns of metabolite excretion (Kopin and Gordon, 1963). This established that O-methylation is the predominant means of metabolic inactivation of released norepinephrine, whereas deamination is the metabolic route of intraneuronal norepinephrine metabolism (Kopin and Axelrod, 1963).

Physiological assessment of the role of MAO and COMT

After the introduction of iproniazid, as a potent inhibitor of MAO, it was shown that inhibition of this enzyme potentiated the actions of tyramine and phenylethylamine, but not that of norepinephrine (Greisemer et al., 1953). Also, inhibition of MAO did not slow the disappearance of administered catecholamines from the blood (Celander and Mellander, 1955) nor increase urinary catecholamine excretion (Corne and Graham, 1957). The effects of nerve stimulation on release of norepinephrine were not potentiated by MAO inhibition (Brown and Gillespie, 1957). All of these results were consistent with the notion that MAO was not the means for inactivating released norepinephrine.

When pyrogallol, which was known to potentiate the effects of administered catecholamines, was shown to be a COMT inhibitor (Axelrod and Laroche, 1959), it appeared that COMT might be responsible for physiological inactivation of released catecholamines. Although MAO inhibition appeared to prolong the effects of sympathetic nerve stimulation in some tissues (Wylie et al., 1960), inhibition of both COMT and MAO failed to alter cardiovascular responses to intravenously administered catecholamines (Crout, 1961). The observation that reuptake of norepinephrine from the synapse into the presynaptic noradrenergic terminal is the major means for terminating the action of norepinephrine physiologically released from the sympathetic nerve terminals (Hertting and Axelrod, 1961) clarified the issue and explained the dissociation between metabolic routes and physiological effects of catecholamines.

Oxidative deamination by MAO is responsible for metabolic inactivation of cytoplasmic norepinephrine, whether derived from reuptake (although most norepinephrine captured by reuptake is stored in vesicles) or from leakage for vesicular storage sites. O-methylation is the main means of metabolism of norepinephrine which escapes recapture, whether in the tissues or after delivery by the circulation to the liver and kidney (Kopin, 1985).

In a wide variety of sympathetically innervated tissues studied in vitro using tritium labelled catecholamines, it has been shown that the predominant deaminated metabolites are ^{3}H-3,4-dihydroxy phenylethyleglycol (DHPG) from ^{3}H-norepinephrine and ^{3}H-3,4-dihydroxphenyl acetic acid (DOPAC) from ^{3}H-dopamine (Graefe and Bönisch, 1988). Endogenous DHPG had been discovered many years earlier (Kopin and Axelrod, 1960) and was considered to be the precursor of ^{3}H-3-methoxy-4-hydroxy phenylethylene glycol (MHPG). When endogenous metabolites were examined in isolated tissues, DHPG was found to account for almost all the norepinephrine metabolites (Starke et al., 1981; Majewski et al., 1982) indicating O-methylation occurs in other tissues, presumably the liver. In the intact animal also, DHPG in plasma appears to be derived mostly from intraneuronal metabolism of norepinephrine (Eisenhofer et al., 1987; Goldstein, 1988). Both DHPG and its O-methylated derivative, MHPG, are derived mainly from norepinephrine deaminated intraneuronally. During free movement, without stress, about 2/3 of rat plasma DHPG (and MHPG) appear to be derived from norepinephrine that "leaks" from the vesicular storage sites into the axoplasm and only one third of the metabolites are the products of norepinephrine recaptured after exocytotic release. Administration of reserpine markedly reduces plasma DHPG and MHPG levels, presumably as a result of blockade of vesicular uptake (and subsequent β-hydroxylation) of dopamine. This is consistent with the reserpine-induced decrease in MHPG excretion attended by an increase in HVA excretion in rats (Kopin and Weise, 1968).

60   I. J. Kopin

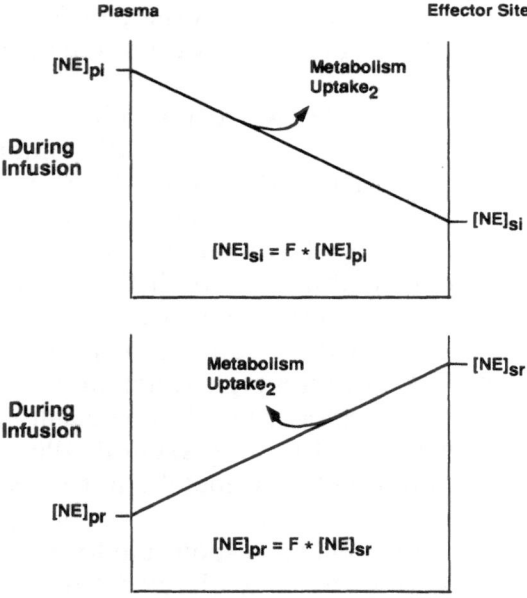

Fig. 1. Schematic representation of gradients in concentration of norepinephrine between plasma and effector sites during infusion or during release of endogenous neurotransmitter. The concentration at the receptor is reflected by a physiological response, whereas the concentration at the uptake site is similarly reflected by the rate of DHPG formation (Eisenhofer et al., 1987). It is assumed that the gradient is reciprocal. This has been supported by symmetrical effects of desipramine on the response-NE concentration relationship (see Kopin et al., 1984), but not established for plasma levels of DHPG, although the data obtained using the metabolite yield gradients which are similar to those obtained using blood pressure responses (Goldstein et al., 1988)

Since DHPG is derived almost wholly from intraneuronal metabolism of norepinephrine (Eisenhofer et al., 1987), it is possible to estimate the concentration of norepinephrine at sympathetic neuron uptake sites by relating the changes in steady state plasma DHPG levels to changes in plasma norepinephrine levels during enhanced norepinephrine release and during infusion of norepinephrine (Fig. 1). During release, changes in plasma norepinephrine levels, $\Delta[NE]_{pr}$, are a constant fraction, F, of changes in concentration of norepinephrine at the uptake site, $\Delta[NE]_{ur}$. Thus,

$$\Delta[NE]_{pr} = F . \Delta[NE]_{ur}$$

Since a constant fraction of norepinephrine taken up into the neuron is deaminated to DHPG, in a steady state, changes in DHPG plasma levels are related by a constant, K_d, to the changes in norepinephrine at the uptake site:

$$\Delta[DHPG]_p = K_d\Delta[NE]_u$$

Thus, during release of *NE*

$$\Delta[DHPG]_{pr} = K_d a\Delta[NE]_{ur} = (1/F)K_d\Delta[NE]_{pr}$$

During an infusion of *NE*, changes in plasma *NE* concentration $\Delta[NE]_{pi}$ are reduced by the same fraction, *F*, so that

$$\Delta[NE]_{ui} = F\Delta[NE]_{pi}$$

As during release, the plasma DHPG level is related to the concentration of norepinephrine at the uptake site as during release.

$$\Delta[DHPG]_{pi} = KD\Delta[NE]_{ui} = K_d . F\Delta[NE]_{pi}$$

From the above equations and, we obtain:

$$F^2 = \frac{\Delta[DHPG]_{pi}/\Delta[NE]_{pi}}{\Delta[DHPG]_{pi}/\Delta[NE]_{pr}}$$

The concentration at the uptake site is then the geometric mean of the plasma concentration during the infusion and that during release of the endogenous catecholamine. In humans, the concentration of norepinephrine at the uptake site has been estimated to be between about one fourth to one third of that in plasma (Goldstein et al., 1988). Thus, understanding the origin and disposition of the deaminated norepinephrine metabolite, DHPG, provides insights into the kinetics of norepinephrine disposition.

Considerable amounts of dopamine formed in the axoplasm of noradrenergic neurons (from the decarboxylation of DOPA) appears to escape capture into the storage vesicular and subsequent β-hydroxylation to norepinephrine. This dopamine is deaminated to form DOPAC. Thus, Anden and Gradowski-Anden (1983) could demonstrate formation of DOPAC in rat brain noradrenergic neurons. Later they reported the rapid turnover of dopamine in these neurons (Anden et al., 1985). Deamination of dopamine appears to occur also in human peripheral adrenergic neurons since there are parallel decreases in plasma levels of MHPG and of HVA in patients with peripheral autonomic failure (Kopin et al., 1988b).

The peripheral origin of HVA is a confounding variable which limits the usefulness of measurements of its levels in plasma or of its urinary excretion rate as an index of central dopaminergic functions. Attempts to improve the reliability of plasma HVA as an index of brain dopamine metabolism have been based on inhibition of peripheral HVA production using a MAO inhibitor which does not act in brain. Debrisoquin has been used for that purpose in experimental animals (Kendler et al., 1981; Sternberg et al., 1983) and in humans (Swann et al., 1980; Maas et al., 1985; Riddle et al., 1986). The affinity of debrisoquin for the norepinephrine transporter is the same as that for norepinephrine (Langeloh et al., 1987), so that high intraneuronal concentrations are attained. This results in selective inhibition of peripheral sympathetic intraneuronal MAO (Pettinger et al., 1969). After administration of debrisoquin there are parallel decrements in plasma levels of DOPAC, HVA, DHPG, MHPG as well as in DOPA. These results indicate that elevated levels of cytoplasmic intraneuronal norepinephrine, with consequent inhibition of tyrosine hydroxylation, contribute to the decline in plasma catecholamine metabolites. The decrement in plasma dopamine metabolites is relatively less than that of the norepinephrine metabolites, since brain dopamine metabolism is not affected by debrisoquin. Use of debrisoquin along with extrapolation of plasma dopamine metabolite levels to the level expected when plasma MHPG levels would be zero, has provided a more sensitive method for indirectly assessing brain dopamine metabolism (Kopin et al., 1988a).

Clinical effects of MAO inhibitors

Although MAO inhibitors do not appear to be important in the inactivation of peripheral catecholamines, they do have significant pharmacological effects and have been found to be of some therapeutic benefit. After the antidepressant effects of iproniazid were described (Crane, 1957; Klein, 1958), a host of compounds were reported to be inhibitors of this enzyme. Several of these were hydrazides, chemically related to iproniazid, but a number of highly potent irreversible non-hydrazine MAO inhibitors soon appeared. Treatment with these MAO inhibitors elevated brain levels of dopamine and norepinephrine in rabbits, rats, and mice (Spector et al., 1958, 1960) but not in cats or dogs (Spector, 1960; Vogt, 1959).

These observations provided a cornerstone of the biogenic amine hypothesis of endogenous depression (Schildkraut and Kety, 1967). The tranquilizing effects of reserpine in animals and the side effect of depression in hypertensive patients supported this hypothesis. Another important factor was the observation by Axelrod et al. (1961) that imipramine, an effective antidepressant, blocks inactivation by reuptake of norepinephrine released into the synapse.

In contrast with the central stimulatory effects of MAO inhibitors, and contrary to expectations, these drugs often produced orthostatic hypotension. MAO inhibitors were at one time considered as potential therapeutic agents for hypertension, but they have been replaced by safer and more effective anti-hypertensive agents. The mechanism of the hypotensive effect of MAO inhibitors, however, is of interest because it appears to involve displacement of norepinephrine by a false transmitter (Kopin et al., 1968). MAO inhibitors, in combination with foods which contain high levels of tyramine, will precipitate hypertensive crises (Marley and Blackwell, 1970); with an ordinary diet, however, octopamine, the β-hydroxylated derivative of tyramine, accumulates in the tissues and replaces norepinephrine in the vesicular sympathetic nerve terminal. When the nerves are stimulated, a mixture of the false transmitter and norepinephrine are released, resulting in release of less active amines (Kopin, 1968). Thus MAO is important in removing tyramine and related amines from the circulation and preventing inappropriate intraneuronal amine accumulation.

MAO subtypes

There are at least two types of MAO. Initially, these were distinguished by their substrate preference and by selective inhibitors, as well as by their physical properties (Youdim et al., 1988), MAO-A, was designated by Johnston (1968) as the form which is sensitive to clorgyline and which deaminated serotonin. The other form, MAO-B, is insensitive to clorgyline and selectively deaminates benzylamine. Deprenyl was found to be most potent in selectively and irreversibly inhibiting MAO-B (Knoll and Magyar, 1972). Tyramine is an equally good substrate for MAO-A and MAO-B, but inhibition of MAO-B with deprenyl does not potentiate the sympathomimetic actions of tyramine. This might be the result of inhibition by deprenyl of tyramine uptake by the norepinephrine transporter (Knoll, 1978) and the rapid deamination by intact MAO-A of any tyramine which does enter the nerve terminal. Predominance of MAO-A in the intestine, where it acts as an enzymatic barrier, as well as in the axoplasm of sympathetic nerves, where it protects against accumulation of amines, could account for MAO-A (or nonselective MAO) inhibition-mediated potentiation of the effects of tyramine in foods noted above.

Deprenyl treatment in Parkinson's disease

Because MAO-B was considered a "safe" MAO inhibitor, deprenyl came into use in conjunction with levodopa to potentiate the effects of the dopamine precursor in treating patients with Parkinson's disease (Birkmayer et al., 1975). An unexpected finding was that patients treated with deprenyl along with levodopa/benserazide lived longer than patients not treated with the MAO-B inhibitor (Birkmayer et al., 1985). This, combined with the discovery of MPTP toxicity and the role of MAO-B in bioactivation of MPTP to form MPP^+, gave impetus to exploring the possibility that an environmental or endogenous protoxin might be of etiological importance in Parkinson's disease and that, like MPTP, its bioactivation might be blocked by treatment with deprenyl. Oxidative stress has also been proposed as a potential pathogenetic factor in Parkinson's disease. The possibility that inhibition of MAO-B might reduce oxidative stress due to hydrogen peroxide generated during MAO-B-mediated oxidative deamination of dopamine, provided another rationale for use of deprenyl as a neuroprotective agent.

Two studies have demonstrated that deprenyl treatment delays the onset of disability of requiring DOPA therapy in early, otherwise untreated parkinsonians (Tetrud and Langston, 1989; The Parkinson Study Group, 1989). Whether this effect is due to neuroprotection or is symptomatic, is controversial and remains a subject of active research. The outcome will certainly provide new information regarding the mechanism of action of this MAO-B inhibitor.

Conclusion

MAO subtypes have an important role in the metabolism of catecholamines. Use of MAO inhibitors as pharmacological tools to further our understanding of amine metabolism and selective inhibition of subtypes of MAO provide useful diagnostic strategies as well as therapeutic opportunities.

References

Anden NE, Grabowska-Anden M (1983) Formation of deaminated metabolites of dopamine in noradrenaline neurons. Naunyn Schmiedebergs Arch Pharmacol 324: 1–6

Anden NE, Grabowska-Anden M, Lindgren S, Oweling M (1985) Very rapid turnover of dopamine in noradrenaline cell body regions. Naunyn Schmiedebergs Arch Pharmacol 329: 258–263

Armstrong MD, McMillan A, Shaw KNF (1957) 3-Methoxy-4-hydroxy-D-mandelic acid, a urinary metabolite of norepinephrine. Biochem Biophys Acta (Amst) 25: 422–423

Axelrod J (1957) O-Methylation of catecholamines in vitro and in vivo. Science 126: 400–401

Axelrod J (1959) The metabolism of catecholamines in vivo and in vitro. Pharmacol Rev 11 (Part 2): 402–408

Axelrod J, Laroche MJ (1959) Inhibitor of O-methylation of epinephrine and nore-pinephrine in vitro and in vivo. Science 130: 800–801

Axelrod, J, Whitby LG, Hertting G (1961) Effect of psychoactive drugs on the uptake of H³-norepinephrine by tissues. Science 133: 338–384

Birkmayer W, Knoll J, Riederer P, Youdim MBH, Haas V, Marton J (1985) Increased life expectancy resulting from addition of L-deprenyl to madopar treatment in Parkinson's disease: a long term study. J Neural Transm 64: 113–127

Birkmayer W, Riederer P, Youdim MBH, Linauer W (1975) Potentiation of antikinetic effect after L-dopa treatment by an inhibitor of MAO-B, L-deprenyl. J Neural Transm 36: 303–323

Blaschko H (1957a) Metabolism and storage of biogenic amines. Experimentia (Basel) 13: 9–12

Blaschko H (1957b) Formation of catecholamines in the animal body. Br Med Bull 13: 162–165

Brown GL, Gillespie JS (1957) Output of sympathetic transmitter from the spleen of the cat. J Physiol (Lond) 138: 81–102

Celander O, Mellander S (1955) Elimination of adrenaline and noradrenaline from circulating blood. Nature (Lond) 176: 973

Corne SJ, Graham JDP (1957) Effect of inhibition of monoamine oxidase in vivo on administered adrenaline, noradrenaline, tyramine and serotonin. J Physiol (Lond) 135: 339–349

Crane GE (1957) Iproniazid (Marsilid) phosphate: a therapeutic agent for mental disorders and debilitating illness. Psychiat Res Rep Am Psychiat Ass 8: 142–152

Crout JR (1961) Effect of inhibiting both catechol-O-methyl transferase and monoamine oxidase on cardiovascular responses to norepinephrine. Proc Soc Exp Biol 108: 482–484

Eisenhofer G, Goldstein DS, Stull R, Ropchak TG, Keiser HR, Kopin IJ (1987) Dihydroxyphenylglycol and dihydroxymandelic acid during intravenous infusions of noradrenaline. Clin Sci 73: 123–127

Eiesnhofer G, Ropchak TG, Stull RW, Goldstein DS, Keiser HR, Kopin IJ (1987) Dihydoxyphenylglycol and intraneuronal metabolism of endogenous and exogenous norepinephrine in the rat vas deferens. J Pharmacol Exp Ther 142: 547–553

Griesemer EC, Barsky J, Dragstedt CA, Wells JA, Zeller EA (1953) Potentiating effect of iproniazid on the pharmacological actions of sympathomimetic amines. Proc Soc Exp Biol 84: 699–701

Goldstein DS, Eisenhofer G, Stull R, Folio CJ, Kerier HR, Kopin IJ (1988) Plasma dihydroxyphenylglycol and the intraneuronal disposition of norepinephrine in humans. J Clin Invest 81: 213–220

Graefe KH, Bonisch H (1988) The transport of amines across the axonal membranes of noradrenergic and dopaminergic neurons. In: Catecholamines, vol 90. Springer, Berlin Heidelberg New York Tokyo, pp 193–245

Hare MLC (1928) Tyramine oxidase. A new enzyme system in liver. Biochem J 22: 968–979

Hertting G, Axelrod J (1961) Fate of tritiated noradrenaline at the sympathetic nerve endings. Nature 192: 172–173

Johnston JP (1968) Some observations upon a new inhibitor of monoamine oxidase in brain. Biochem Pharmacol 17: 1285–1297

Kendler KS, Heninger GR, Roth RH (1981) Brain contribution to the haloperidol-induced increase in plasma homovanillic acid. Eur J Pharmacol 71: 321–326

Klein NS (1958) Clinical experience with iproniazid (Marsilid). J Clin Exp Psychopathol 19: 72–78

Knoll J (1978) The possible mechanisms of action of (−)-deprenyl in Parkinson's disease. J Neural Transm 43: 177–198

Knoll J, Magyar K (1972) Some puzzling pharmacological effects of monoamine oxidase inhibitors. Adv Biochem Psychopharmacol 5: 393–408

Kopin IJ (1960) Technique for the study of alternative metabolic pathways: epinephrine metabolism in man. Science 131: 1372–1374

Kopin IJ (1968) False adrenergic transmitters. Ann Rev Pharmacol 8: 377–394

Kopin IJ (1985) Catecholamine metabolism: basic aspects and clinical significance. Pharmacol Rev 37: 333–364

Kopin IJ, Axelrod J (1960) 3,4-Dihydroxyphenylglycol, a metabolite of epinephrine. Arch Biochem Biophys 89: 148–149

Kopin IJ, Axelrod J (1963) The role of monoamine oxidase in the release and metabolism of norepinephrine. Ann NY Acad Sci 107: 848–855

Kopin IJ, Axelrod J, Gordon EK (1961) The metabolic fate of H^3-epinephrine and C^{14}-metanephrine in the rat. J Biol Chem 136:2109–2113

Kopin IJ, Fischer JE, Musacchio J, Horst WD (1964) Evidence for a false neurochemical transmitter as a mechanism for the hypotensive effect of monoamine oxidase inhibitors. Proc Natl Acad Sci 52: 716–721

Kopin IJ, Gordon EK (1962) Metabolism of norepinephrine-H^3 released by tyramine and reserpine. J Pharmacol 138: 351–357

Kopin IJ, Gordon EK (1963) Metabolism of administered and drug-released norepinephrine-7-H^3 in the rat. J Pharmacol 140: 207–216

Kopin IJ, Harvey-White J, Bankiewicz K (1988a) A new approach to biochemical evaluation of brain dopamine metabolism. Cell Mol Neurobiol 8: 171–179

Kopin IJ, Hertting G, Gordon EK (1962) Fate of norepinephrine-H^3 in the isolated perfused rat heart. J Pharmacol Exp Ther 138: 34–40

Kopin IJ, Oliver JA, Polinsky, RJ (1988b) Relationship between urinary excretion of homovanillic acid and norepinephrine metabolites in normal subjects and patients with orthostatic hypotension. Life Sci 43:125–131

Kopin IJ, Weise VK (1968) Effect of reserpine and metaraminol on excretion of homovanillic acid and 3-methoxy-4-hydroxyphenylglycol in the rat. Biochem Pharmacol 17: P1461–1464

Kopin IJ, Zukowska-Grojec Z, Bayorh MA, Goldstein DS (1984) Estimation of intra-synaptic noradrenaline concentrations at vascular neuroeffector junctions in vivo. Naunyn Schmiedebergs Arch Pharmacol 325: 298–305

LaBrosse EH, Axelrod J, Kety SS (1958) O-Methylation, the principal route of metabolism of epinephrine in man. Science 128: 593–594

Langeloh A, Bonisch H, Trendelenburg U (1987) The mechanism of 3H-noradrenaline releasing effect of various substrates of uptake 1: multifactorial induction of outward transport. Naunyn Schmiedebergs Arch Pharmacol 336: 603–610

Maas JW, Contreras SA, Bowden CL, Weintraub SE (1985) Effects of debrisoquin on CSF and plasma HVA concentrations in man. Life Sci 36: 165–176

Majewski H, Hedler L, Steppeler A, Starke K (1982) Metabolism of endogenous and exogenous noradrenaline in the rabbit perfused heart. Naunyn Schmiedebergs Arch Pharmacol 319: 125–129

Marley E, Blackwell B (1970) Interactions of monoamine oxidase inhibitors, amines and foodstuffs. Adv Pharmacol Chemother 8: 186–239

Pettinger WA, Korn A, Spieger H, Solomon HM, Porcelinko R, Abrams WB (1969) Debrisoquin, a selective inhibitor of intraneuronal monoamine oxidase in man. Clin Pharmacol Ther 10: 667–674

Riddle MA, Leckman JF, Cohen DJ, Anderson M, Ort SI, Caruso KA, Shaywitz BA (1986) Assessment of central dopaminergic function using plasma-free homovanillic acid after debrisoquin administration. J Neural Transm 67: 31–43

Schayer RW (1951) Metabolism of β-C^{14} DL-adrenaline. J Biol Chem 189: 301–306

Schayer RW, Smiley RL, Davis KJ, Kobayashi Y (1955) Role of monoamine oxidase in noradrenaline metabolism. Am J Physiol 182: 285–286

Schayer RW, Smiley RL, Kaplan EH (1952) Metabolism of adrenaline containing isotopic carbon (II). J Biol Chem 198: 545–551

Schayer RW, Smiley RL, Kennedy J (1953) Metabolism of epinephrine containing isotopic carbon (III). J Biol Chem 202: 425–430

Schildkraut JJ, Kety S (1967) Biogenic amines and emotion. Science 156: 21–55

Shaw KNF, McMillan A, Armstrong MD (1957) Metabolism of 3,4-dihydroxyphenylalanine. J Biol Chem 226: 255–266

Spector S, Prockop D, Shore PA, Brodie BB (1958) Effect of iproniazid on brain levels of norepinephrine and serotonin. Science 127: 704–705

Spector S, Shore PA, Brodie BB (1960) Biochemical and pharmacological effects of monoamine oxidase inhibitors, iproniazid, 1-phenyl-2-hydrazine propane (JB 516) and 1-phenyl-3-hydrazinobutane. J Pharmacol Exp Ther 128: 15–21

Starke K, Hedler L, Steppeler A (1981) Metabolism of endogenous and exogenous noradrenaline in guinea-pig atria. Naunyn Schmiedebergs Arch Pharmacol 317: 193–198

Sternberg DE, Heninger GR, Heninger RH (1983) Plasma homovanillic acid as an index of brain dopamine metabolism: enhancement with debrisoquin. Life Sci 32: 2447–2452

Swann AC, Maas JW, Hattox SE, Landis H (1980) Catecholamine metabolites in human plasma as indices of brain function: effects of debrisoquin. Life Sci 27: 1857–1862

Tetrud JW, Langston JW (1989) The effect of deprenyl (selegiline) on the natural history of Parkinson's disease. Science 41: 519–522

The Parkinson Study Group (1989) Effect of deprenyl on the progression of disability in early Parkinson's disease. N Engl J Med 321: 1364–1371

Vogt M (1959) Catecholamines in brain. Pharmacol Rev 11: 483

Wylie DW, Archer S, Arnold A (1960) Augmentation of pharmacological properties of catecholamines by O-methyl transferase inhibitors. J Pharmacol Exp Ther 130: 239–244

Youdim MBH, Finberg JPM, Tipton KF (1988) Monoamine oxidase. In: Catecholamines, vol 90. Springer, Berlin Heidelberg New York Tokyo, pp 117–192

Zeller EA (1938) Über den enzymatischem Abbau von Histamin und Diaminen. Helv Chim Acta 21: 880–890

Zeller EA, Barsky J, Berman ER, Fouts JR (1952) Action of isonicotinic acid hydrazide and related compounds on enzymes of brain and other tissues. J Lab Clin Med 40: 965–966

Author's address: Dr. I. J. Kopin, National Institute of Neurological Disorders and Stroke, National Institutes of Health, Building 10, Room 5N214 9000 Rockville Pike, Bethesda, MD 20892, U.S.A.

J Neural Transm (1994) [Suppl] 41: 69–73

Monoamine oxidase (MAO; E.C. 1.4.3.4) characteristics of platelets influenced by in vitro and in vivo ethanol on alcoholics and on control subjects

T. May and **H. Rommelspacher**

Department of Neuropsychopharmacology, Free University, Berlin,
Federal Republic of Germany

Summary. Ethanol (ETOH) in vitro displays a competitive inhibition of human platelet MAO-B with a K_i of 270 ± 30 mM. Lineweaver-Burk analyses with 6 substrate concentrations (5–160 µM kynuramine) in the presence or absence of 200 mM ETOH were performed with platelets of alcoholics before withdrawal (Alc day 1), one week (Alc day 8) and 3 months (Alc mon 3) after withdrawal as well as in control subjects without and after ETOH intake.

In all groups the K_m increases highly significantly ($p < 0.001$) but the V_{max} is unchanged by the presence of ETOH in vitro supporting the view of a competitive inhibition in each group.

The V_{max} of Alc day 1 is significantly 25% decreased, of Alc day 8 unchanged and of Alc mon 3 nonsignificantly 19% decreased in comparison with the controls. The increase of the K_m in the presence of 200 mM ETOH is significantly 15% reduced in Alc day 1 and 9% in Alc day 8 but unchanged in Alc mon 3 compared with the controls.

Introduction

The MAO in human platelets is often used as a peripheral marker of vulnerability in psychic disorders like schizophrenia and depression as well as in alcoholism (Faraj et al., 1987; Major et al., 1981; von Knorring et al., 1991; Yates et al., 1990).

Tabakoff et al. (1985) determined a competitive inhibition of human platelet MAO with [14C]phenylethylamine as substrate by ETOH in vitro with a K_i of about 170 mM. Furthermore, Tabakoff et al. (1988) described a stronger inhibition by in vitro ETOH (400 mM) of platelet MAO in alcoholics with one substrate concentration of 12 µM [14C]phenylethylamine.

The aim of the present study was to monitor the in vitro and in vivo effects of ETOH in controls as well as the in vitro ETOH effects in alcoholics before and during different abstinence periods on the characteristics (K_m and V_{max}) of platelet MAO.

Materials and methods

Subjects

The alcoholics were diagnosed by research criteria of ICD-10 and DSM-IIIR. The blood was withdrawn from the alcoholics after chronic ethanol intoxication (Alc day 1), after one week of withdrawal (Alc day 8) as well as after 3 months of abstinence (Alc mon 3). Furthermore, blood of the control subjects was used without ETOH consumption (Con − ETOH) and 4 h after drinking 1 g ETOH/kg (Con + ETOH).

Chemicals

Kynuramine and 4-hydroxyquinoline were purchased from Sigma (Munich, F.R.G.), pargyline from Serva (Heidelberg, F.R.G.), L-deprenyl and clorgyline from RBI, BIOTREND (Köln, F.R.G.). Brofaromine was a gift from Ciba Geigy (Basel, Switzerland).

Preparation of platelet membranes

30 ml venous blood were withdrawn into 2.5 ml 1.5% EDTA/0.7% NaCl and centrifuged 15 min at 4°C (200 × g). The supernatant containing the platelet rich plasma was centrifuged 10 min at 2,000 × g. The platelet pellet was frozen at −80°C until use. The thawed pellet was homogenized in 10 ml of 50 mM K_2HPO_4/KH_2PO_4, pH 7.4, 5 mM EDTA by glass/teflon Potter and centrifuged 10 min at 40,000 × g. This washing procedure was repeated once and the final pellet was homogenized in 4 ml of 0.2 M K_2HPO_4/KH_2PO_4, pH 7.4, 5 mM EDTA and used for the MAO assay.

MAO assay

The fluorometric assay was performed according to a method slightly modified from that of Kraml (1965) in Eppendorf micro tubes containing 600 µl of 115 mM K_2HPO_4/KH_2PO_4, pH 7.4, 3 mM EDTA. 50 µl homogenate (about 15 µg protein) and six concentrations in duplicate of kynuramine (5–160 µM for Lineweaver-Burk analyses) in the absence and presence of 200 mM ETOH or a single concentration (20 µM kynuramine for inhibition analyses) with several different concentrations of inhibitors (ETOH, pargyline, clorgyline, L-deprenyl, brofaromine). After 30 min preincubation of the samples at 37°C, the assay was started by the addition of 50 µl of the substrate. 30 min later the assay was stopped by the addition of 400 µl 10% trichloroacetic acid and 6 min centrifugation at 10,000 × g. 800 µl of the resultant supernatant were stirred with 2 ml of 1 M NaOH and the fluorescence of the enzyme product 4-hydroxyquinoline was measured in a fluorescence-spectrometer (excitation: 315 nm; emission: 380 nm). The results were calculated by the aid of internal 4-hydroxyquinoline (product) standards and after protein determination with bovine serum albumin as standard (Bradford, 1976) in nmol 4-hydroxyquinoline formed/mg protein/min.

Data analyses

All data were analyzed by Hewlett-Packard table computers (series 200/300). Lineweaver-Burk analyses were calculated with multipurpose computer programs. The MAO inhibition experiments were analyzed, applying a nonlinear, unweighted least squares procedure (Wiemer et al., 1982).

The results are presented as means ± S.E.M. values. Significant differences ($p <$ 0.05) were calculated by analyses of variance and two-tailed Student's t-tests.

Results and discussion

Inhibition studies ($n = 3\text{--}4$) of platelet MAO revealed with the potent and selective MAO-A inhibitors clorgyline (Johnston, 1968) and brofaromine (Waldmeier et al., 1983) IC_{50} values of about $1\,\mu M$ as well as with the potent and selective MAO-B inhibitors pargyline (Squires, 1968) and L-deprenyl (Knoll and Magyar, 1972) IC_{50} values of about $10\,nM$. Furthermore, the Hill coefficients of all four inhibitors were about unity and therefore a solely MAO subtype was found with the pharmacological profile of MAO-B.

ETOH was tested in vitro and displayed a competitive inhibition with a K_i of $270 \pm 30\,mM$ and a Hill coefficient of 1.17 ± 0.15 ($n = 4$) similar as described with [^{14}C]phenylethylamine as substrate by Tabakoff et al. (1985) but with a higher K_i value in the present study.

In the five groups of this clinical study ETOH in vitro evoked a highly significantly increase in the K_m (decrease in substrate affinity) but did not influence the V_{max} (maximum of enzyme activity), i.e., ETOH is obviously a competitive inhibitor in all groups (Table 1). The extent of the inhibition by ETOH in vitro is not influenced by the acute ETOH intake of the controls but significantly reduced, by 15%, in the intoxicated ALC day 1 and 9% in the short-term abstinent Alc day 8 but not in the long-term abstinent Alc mon 3 (Table 1). This phenomenon of a reduced inhibiting efficacy of the membrane fluidizing ETOH after chronic ETOH intoxication could be caused by a biochemical/biophysical alteration in the membrane fluidity and represents, therefore, a kind of membrane tolerance and is a possible state marker. This result of a reduced in vitro ETOH inhibiting efficacy is contradictory to results of Tabakoff et al. (1988). They found an increased inhibition from 6.1% in controls to 12.5% in alcoholics (this group combined alcoholics with 5–90 days since the last ETOH intake) with $400\,mM$ ETOH and a single substrate concentration of $12\,\mu M$ [^{14}C]phenylethylamine but not with $1.2\,\mu M$ of the substrate.

The V_{max} (in the presence and absence of ETOH in vitro) in Alc day 1 is significantly 25% reduced and nonsignificantly 19% decreased in Alc mon 3 but unchanged in Alc day 8 in comparison with the controls. The results of a decreased MAO enzyme activity in alcoholics compared with the controls (Table 1) is in accordance with many previously described clinical studies (Major et al., 1981; Faraj et al., 1987; Yates et al., 1990; von Knorring et al., 1991) but in disagreement with the study of Tabakoff et al. (1988) who found no difference between alcoholics and controls.

An increase of the MAO V_{max} during the ETOH withdrawal has also been described previously as well as a later decrease in the abstinence

T. May and H. Rommelspacher

Table 1. Effects of ETOH in vitro and in vivo on human platelet MAO affinity (K_m) and activity (V_{max})

	Alc day 1 (n = 36)	Alc day 8 (n = 36)	Alc mon 3 (n = 19)	Con − ETOH (n = 17)	Con + ETOH (n = 17)
K_m − ETOH (μM)	33.3 ± 1.5	34.9 ± 1.2	36.0 ± 1.5	36.7 ± 1.8	35.6 ± 1.4
V_{max} − ETOH (nmol/mg/min)	2.56 ± 0.18	3.48 ± 0.26	2.79 ± 0.31	3.31 ± 0.29[b]	3.51 ± 0.32[c]
K_m + ETOH (μM)	58.5 ± 1.7[a]	61.7 ± 2.1[a]	67.2 ± 2.1[a,c]	68.1 ± 2.5[a,c]	70.6 ± 2.3[a,d,e]
V_{max} + ETOH (nmol/mg/min)	2.42 ± 0.18	3.22 ± 0.24[c]	2.71 ± 0.29	3.24 ± 0.28[b]	3.40 ± 0.30[c]

The assays were performed as described in Materials and methods with 6 concentrations of the substrate kynuramine in the presence (+ ETOH) or absence (− ETOH) of ETOH in vitro (200 mM). [a] $p < 0.001$ vs samples − ETOH in vitro, [b] $p < 0.05$ vs Alc day 1, [c] $p < 0.01$ vs Alc day 1, [d] $p < 0.001$ vs Alc day 1, [e] $p < 0.05$ vs Alc day 8 significantly different in the two-tailed Student's t-test

period (Major et al., 1981). The significance of a further decrease of platelet MAO activity and, therefore, the presence of a possible trait marker will be validated in the present study with a larger sample size at Alc mon 3 and a further MAO testing after 6 mon of abstinence.

Acknowledgements

The authors thank S. Strauss and H. Damm for their excellent technical assistance. This research was supported by the Deutsche Forschungsgemeinschaft (DFG Az.: He 916/7-1).

References

Bradford MM (1976) A rapid and sensitive method for the quantitation of microgram quantities of protein utilizing the principle of protein-dye binding. Anal Biochem 72: 248–254

Faraj BA, Lenton JD, Kutner M, Camp VM, Stammers TW, Lee SR, Lolies PA, Chandora D (1987) Prevalence of low monoamine oxidase function in alcoholism. Alcoholism: Clin Exp Res 11: 464–467

Johnston JP (1968) Some observations upon a new inhibitor of monoamine oxidase in brain tissue. Biochem Pharmacol 17: 1285–1297

Knoll J, Magyar K (1972) Some puzzling pharmacological effects of monoamine oxidase inhibitors. In: Costa E, Sandler M (eds) Advances in biochemical psychopharmacology, vol 5. Raven Press, New York, pp 393–408

Kraml M (1965) A rapid microfluorimetric determination of monoamine oxidase. Biochem Pharmacol 14: 1684–1686

Major LF, Goyer PF, Murphy DL (1981) Changes in platelet monoamine oxidase activity during abstinence. J Stud Alcohol 42: 1052–1057

Squires RF (1968) Additional evidence for the existence of several forms of mitochondrial monoamine oxidase in the mouse. Biochem Pharmacol 17: 1401–1409

Tabakoff B, Lee JM, De Leon-Jones F, Hoffman PL (1985) Ethanol inhibits the activity of the B form of monoamine oxidase in human platelet and brain tissue. Psychopharmacology 87: 152–156

Tabakoff B, Hoffman PL, Lee JM, Saito T, Willard B, De Leon-Jones F (1988) Differences in platelet enzyme activity between alcoholics and nonalcoholics. N Engl J Med 318: 134–139

von Knorring A-L, Hallman J, von Knorring L, Oreland L (1991) Platelet monoamine oxidase activity in type 1 and type 2 alcoholism. Alc Alcoholism 26: 409–416

Waldmeier PC, Felner AE, Tipton KF (1983) The monoamine oxidase inhibiting properties of CGP 11305 A. Eur J Pharmacol 94: 73–83

Wiemer G, Wellstein A, Palm D, v Hattingberg HM, Brockmeier D (1982) Properties of agonist binding at the β-adrenoceptor of the rat reticulocyte. Naunyn Schmiedebergs Arch Pharmacol 321: 11–19

Yates WR, Wilcox J, Knudson R, Myers C, Kelly MW (1990) The effect of gender and subtype on platelet MAO in alcoholism. J Stud Alcohol 51: 463–467

Authors' address: Dr. T. May, Department of Neuropsychopharmacology, Free University, Ulmenallee 30, D-14050 Berlin, Federal Republic of Germany.

J Neural Transm (1994) [Suppl] 41: 75–81

The effects of ethanol on rat brain monoamine oxidase activities

L. Della Corte[1], L. Bianchi[1], A. Colivicchi[1], N. P. Kennedy[2],
and **K. F. Tipton[2]**

[1] Department of Pharmacology, University of Firenze, Italy
[2] Department of Biochemistry, Trinity College, Dublin, Ireland

Summary. In contrast to the reported behaviour of human platelet MAO-B, chronic ethanol feeding does not significantly affect the sensitivities of either MAO-A or -B from rat brain to inhibition by ethanol in vitro. The thermal stabilities of rat brain MAO-A and -B are not significantly affected by chronic ethanol feeding.

Introduction

There have been conflicting reports on the effects of chronic alcohol consumption on the activities of monoamine oxidase (EC 1.4.3.4; MAO). Relatively low activities of platelet MAO have been reported in some (see von Knorring et al., 1985; Pandey et al., 1988; Sullivan et al., 1990) but not all (Tabakoff et al., 1985) studies. It has been suggested that low platelet MAO is associated with the hereditable class of alcoholism, which is classified as Type 2 alcoholism by Cloninger (1987), rather than the more common "milieu-limited" or Type 1 alcoholism (Pandey et al., 1988; Sullivan et al., 1990) but this has not been confirmed by others (Yates et al., 1990). An alternative explanation is that individuals with low platelet monoamine oxidase represent a "sensation-seeking" personality group who may, thus, be predisposed to substance abuse (see Fowler et al., 1980a; Anokhina et al., 1988). Studies with post-mortem brain samples have also suggested the enzyme from this source to be less active in alcoholics (Oreland et al., 1983). However, chronic ethanol consumption has been reported to be without effect on the activities of brain MAO-A or -B in either rat liver or brain (Romanova, 1980; Wiberg et al., 1977).

Ethanol, in common with other lipophilic compounds (Fowler et al., 1980b), has been reported to inhibit human platelet MAO-B in vitro (Tabakoff et al., 1985) and the sensitivity of this enzyme to such inhibition was reported to be greater in In human alcoholics than in control subjects (Tabakoff et al., 1985).

Changes in the sensitivities of monoamine oxidase to inhibition by ethanol might have important implications in terms of the central actions of this drug. In the present study we have examined the effects of chronic ethanol consumption on the behaviour of rat brain MAO-A and -B with particular reference to their sensitivities to inhibition by ethanol.

Methods

Three groups of male Wistar rats (mean weight 140 g) were used:

Alcohol group (Group C)

These were fed a liquid diet (Lieber and DeCarli, 1986, 1989) containing 5%, (w/v) ethanol, representing 36% of their total calorie intake, for a period of 4 weeks after a induction period of 1 week during which the ethanol concentration in the diet was gradually increased to this level.

Pair-fed controls (Group B)

Pair-fed (isocaloric) controls received the same diet in which dextrin-maltose (1 : 1, w/w) was substituted for the ethanol. Dietary intake was monitored each day and the quantity available to this control group was adjusted to ensure that they only received the same calorie intake as the alcohol-fed group.

Ad-lib. controls (Group A)

This group had unrestricted access to the same liquid diet as the pairfed controls.

The use of two control groups was believed to be desirable, since the animals receiving the ethanol and pair-fed diets gained weight more slowly than the ad lib. group and it is thus necessary to distinguish the effects of semi-starvation from those of ethanol consumption.

Brain homogenates were prepared as previously described (Fowler and Tipton, 1982) and stored frozen until assay. Monoamine oxidase A and B activities were determined radiochemically (Tipton, 1985). MAO-A activity was determined using 100 μM 5-HT as substrate. Two concentrations of 2-phenylethylamine (PEA), 2 μM and 20 μM, were used to determine the activity of MAO-B. The former concentration was that used by Tabakoff et al. (1988) in their studies on the sensitivity of human platelet MAO-B to inhibition by ethanol. The concentration of ethanol used in the inhibition experiments was 400 mM, as used by Tabakoff et al. (1988) in their studies on the inhibition of the human platelet enzyme. Protein concentrations were determined by the method of Markwell et al. (1978). Unless otherwise indicated, data are mean values ± S.E.M. (n = 3).

Results and discussion

There was no significant difference between the activities of MAO-A from alcohol-fed rats and the ad lib. controls. However, the activity in the pair-fed controls was significantly higher than those of the other groups (Fig. 1).

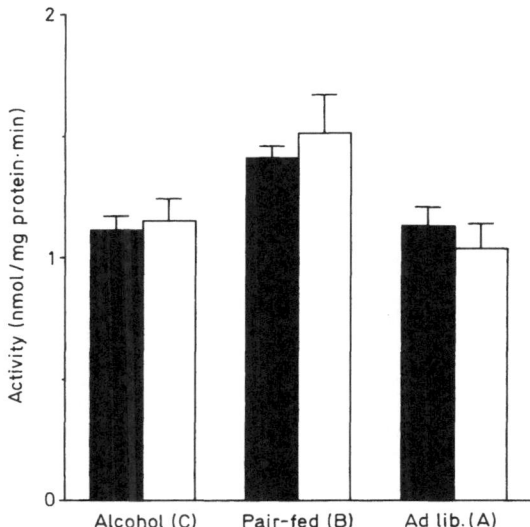

Fig. 1. The activities of monoamine oxidase-A in homogenates of brain prepared from chronic alcohol-fed (*C*), isocalorically pair-fed (*B*) and ad lib.-fed control (*A*) rats. Activities were determined with $100\,\mu M$ 5-HT as substrate. Closed columns show activities in the absence of added alcohol and open columns show the activity determined in the presence of $400\,mM$ ethanol. Experimental details are given in the text

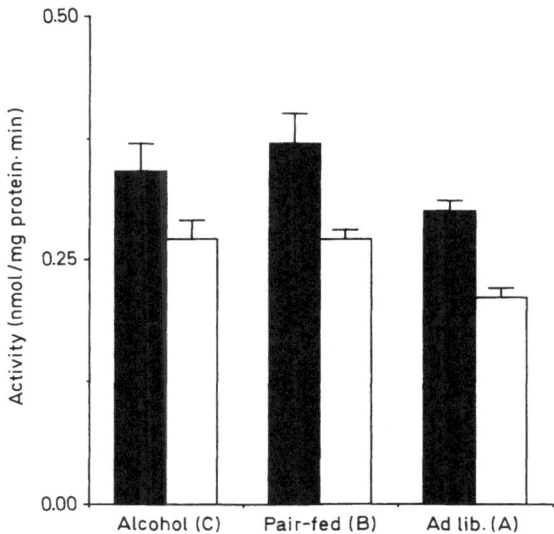

Fig. 2. The activities of monoamine oxidase-B in homogenates of brain prepared from chronic alcohol-fed (*C*), isocalorically pair-fed (*B*) and ad lib.-fed control (*A*) rats. Activities were determined with $20\,\mu M$ 2-phenylethylamine as substrate. Closed columns show activities in the absence of added alcohol and open columns show the activity determined in the presence of $400\,mM$ ethanol. Experimental details are given in the text

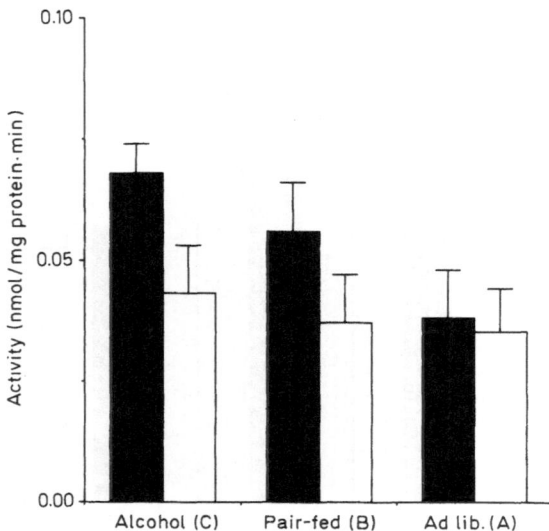

Fig. 3. The activities of monoamine oxidase-B in homogenates of brain prepared from chronic alcohol-fed (*C*), isocalorically pair-fed (*B*) and ad lib.-fed control (*A*) rats. Activities were determined with $2\,\mu M$ 2-phenylethylamine as substrate, as used by Tabakoff et al. (1988) in their studies with human platelet MAO-B. Closed columns show activities in the absence of added alcohol and open columns show the activity determined in the presence of 400 mM ethanol. Experimental details are given in the text

This may represent an effect of starvation which is countered by the metabolic effects of ethanol. Such results emphasise the importance of including both pair-fed and ad lib. controls in such studies. As shown in Fig. 1, 400 mM ethanol had no significant effects on the activities of MAO-A from any of the groups. There were no significant time-dependent changes in the activities of the samples when they were preincubated at 37°C for periods of up to 30 min with 400 mM ethanol before the addition of 5-HT to start the assay.

Figures 2 and 3 show that the basal levels of MAO-B in rat brain were not significantly different between the alcohol-fed group and either control group, whether determined at 2 or $20\,\mu M$ PEA. These results, therefore, suggest that the ethanol consumption had no significant effects on the K_m value or maximum velocity of MAO-B towards this substrate. The lack of effect of chronic ethanol consumption is consistent with the results of Wiberg et al. (1977) and Romanova (1980), although one preliminary report (Carrilho et al., 1988) has suggested that chronic ethanol consumption results in an increase in the K_m value of rat liver MAO-B.

Ethanol at a concentration of 400 mM had a significant inhibitory effect on MAO-B activity in vitro when assayed at both concentrations of PEA. However, there were no significant differences between the sensitivities of the enzyme from any of the groups. In all cases the degree of inhibition was not significantly changed by preincubation of the tissue samples at 37°C for

Table 1. Thermal denaturation of rat brain monoamine oxidase-A

Group	Preincubation 37°C activity remaining	Preincubation 60°C activity remaining	% inhibition
Isocaloric (A)	0.79 ± 0.03	0.26 ± 0.03	66.7 ± 4.0
Ad Lib. (B)	1.34 ± 0.04	0.49 ± 0.12	63.4 ± 9.0
Alcohol (C)	0.66 ± 0.07	0.23	65.2 ± 3.7

Activity determined at 37°C, after 20 min preincubation at the stated temperature, with 100 µM 5-HT. Percentage inhibition values for the samples incubated at 60°C were calculated with respect to the samples preincubated at 37°C, and are given as mean values ± standard error of ratio. Experimental details are given in the text

Table 2. Thermal denaturation of rat brain monoamine oxidase-B

Group	Preincubation 37°C activity remaining	Preincubation 60°C activity remaining	% inhibition
Isocaloric (A)	0.22 ± 0.10	0.06 ± 0.003	72.7 ± 12.5
Ad Lib. (B)	0.29 ± 0.11	0.04 ± 0.004	86.2 ± 9.0
Alcohol (C)	0.25 ± 0.09	0.07 ± 0.03	72.0 ± 3.7

Activity determined at 37°C, after 20 min preincubation at the stated temperature, with 2 µM 2-phenylethylamine. Percentage inhibition values for the samples incubated at 60°C were calculated with respect to the samples preincubated at 37°C, and are given as mean values ± standard error of ratio. Experimental details are given in the text

periods of up to 30 min with 400 mM ethanol before the addition of PEA to start the assay.

Since MAO is tightly bound to the mitochondrial outer-membrane, the membrane-perturbing effects of ethanol and subsequent adaptive changes might be expected to have some effects on its environment (see Fowler et al., 1983). To assess this, the effects of thermal denaturation were studied by incubating the enzyme preparation at 60°C for 20 min before assaying the activities at 37°C. The activities remaining were compared with samples that had been incubated for 20 min at 37°C. As shown in Table 1, there was a considerable loss of activity of MAO-A on incubation at 60°C but there was no significant difference between the thermal stabilities of the alcohol-fed group and either control group. MAO-B activity, whether assayed at 2 or 20 µM PEA was even less stable to thermal denaturation (see Table 2). Although the relatively high degree of activity loss at this temperature would make it difficult to detect small alterations in thermal stability, there were no significant differences between the sensitivities of the groups. Further studies at lower temperatures and shorter incubation times (not shown) confirmed that there were no significant differences between the thermal stabilities of the groups studied.

The results reported here suggest that changes in the levels of the biogenic amine neurotransmitters in the brain, that have been reported to follow ethanol consumption (see Tipton, 1988) do not result from any gross direct effects on the monoamine oxidase activities or on their sensitivities to inhibition by ethanol.

References

Anokhina JP, Kogan BM, Drozdov AZ (1988) Disturbances in regulation of catecholamine neuromediation in alcoholism. Alcohol Alcoholism 23: 343–350

Carrilho M, Freire A, Ponces Freire A, Amorim M, Alves P, Pinto Correia J (1988) Effects of ethanol consumption on hepatic mitochondrial enzymes. Gastroenterology Int 1 [Suppl] 1: A610

Cloninger CR (1987) Neurogenic adaptive mechanisms in alcoholism. Science 236: 410–416

Fowler CJ, Tipton KF (1982) Deamination of 5-hydroxytryptamine by both forms of monoamine oxidase in the rat brain. J Neurochem 38: 733–736

Fowler CJ, Tipton KF, MacKay AVP, Youdim MBH (1980a) Human platelet monoamine oxidase — a useful enzyme in the study of psychiatric disorders? Neuroscience 7: 1577–1594

Fowler CJ, Callingham BA, Mantle TJ, Tipton KF (1980b) The effects of lipophilic compounds upon the activity of rat liver mitochondrial monoamine oxidase-A and -B. Biochem Pharmacol 29: 1177–1183

Knorring A-L von, Bohman M, Knorring L von, Oreland L (1985) Platelet monoamine oxidase activity as a biological marker in subgroups of alcoholism. Acta Psychiatr Scand 72: 51–58

Lieber CS, DeCarli LM (1986) The feeding of alcohol in liquid diets. Alcohol Clin Exp Res 6: 523–531

Lieber CS, DeCarli LM (1989) Liquid diet technique of ethanol administration 1989 update. Alcohol Alcoholism 24: 197–211

Markwell MAK, Haas SM, Bieber LL, Tolbert NE (1978) A modification of the Lowry procedure to simplify protein determination in membrane and lipoprotein samples. Anal Biochem 87: 206–210

Oreland L, Wiberg Å, Winblad B, Fowler CJ, Gottfries G-G, Kiianmaa K (1983) The activity of monoamine oxidase-A and -B in brains from chronic alcoholics. J Neural Transm 56: 73–83

Pandey GN, Fawcett J, Gibbons R, Clark DC, Davis JM (1988) Platelet monoamine oxidase in alcoholism. Biol Psychiatry 24: 15–24

Romanova LA (1980) The effects of acute and chronic administration of ethanol on monoamine oxidase activity in rat brain and liver tissues. Vopr Med Khim Acad Med Nauk SSSR 26: 252–255

Sullivan H, Baebziger JC, Wagner DL, Rauscher FP, Nurnberger JI, Holmes JS (1990) Platelet MAO subtypes in alcoholism. Biol Psychiatry 27: 911–922

Tabakoff B, Lee JM, De Leon-Jones F, Hoffman PL (1985) Ethanol inhibits the activity of the B form of monoamine oxidase in human platelet and brain tissue. Psychopharmacology 87: 152–156

Tabakoff B, Hoffman PL, Lee JM, Saito T, Willard B, De Leon-Jones JM (1988) N Engl J Med 318: 134–139

Tipton KF (1985) Determination of monoamine oxidase. Meth Find Exp Clin Pharmacol 7: 361–367

Tipton KF (1988) Central neurotransmission and alcohol: critique. Neurochem Int 13: 301–305

Wiberg Å, Wahlström G, Oreland L (1977) Brain monoamine oxidase activity after chronic ethanol treatment. Psychopharmacology 52: 111–113

Yates WR, Wilcox J, Knudson R, Myers C, Kelly MW (1990) The effects of gender and subtype on platelet MAO in alcoholism. J Stud Alc 51: 463–467

Authors' address: Prof. Dr. L. Della Corte, Department of Pharmacology, University of Florence, Viale G. B. Morgagni 65, I-50134 Florence, Italy.

J Neural Transm (1994) [Suppl] 41: 83–87
© Springer-Verlag 1994

Species differences in changes of heart monoamine oxidase activities with age

M. Strolin Benedetti[1], **J. Thomassin**[1], **P. Tocchetti**[1], **P. Dostert**[1], **R. Kettler**[2], and **M. Da Prada**[2]

[1] Farmitalia Carlo Erba, Research and Development — Erbamont Group, Milan, Italy
[2] Pharmaceutical Research Department, F. Hoffmann-La Roche Ltd., Basel, Switzerland

Summary. It has previously been established that MAO-A activity markedly increases (about 600%) with age in the rat heart. The aim of the present study was to examine the effect of age on the activity of MAO-A and -B in the mouse heart.

In contrast to the rat heart, β-phenylethylamine is deaminated by MAO-B in the mouse heart. Heart MAO-B activity was found to significantly increase (about 70%) with age in both male and female mice, whereas MAO-A activity remained unchanged. Compared to age-matched rats, serotonin-deaminating activity was about 35 times and 204 times lower in young and old mice, respectively.

Introduction

Monoamine oxidase (MAO) has been shown to increase with age in the heart of Osborn & Mendle and Sprague Dawley rats (Novick, 1961; Horita and Lowe, 1972; Fuentes et al., 1977). Cao Danh et al. (1984) have reported that in the heart of 3-month-old male Wistar rats serotonin (5-HT) deaminating activity was about 14 times higher than β-phenylethylamine (PEA) deaminating activity. In this strain both 5-HT (+623%) and PEA (+575%) deaminating activities markedly increased in heart with ageing, in contrast with modest, although statistically significant, changes in other tissues.

The purpose of the present study was to investigate whether this phenomenon is peculiar to the rat heart. Another rodent, the C57 B1/6J mouse, was chosen and the 5-HT and PEA deaminating activities were measured in the heart of 2- and 23-month-old male and female animals as well as of 23-month-old male and female animals as well as of 23-month-old male and female animals having received acetyl-L-carnitine (AC), as AC has been shown to improve energy production and to display protection against lipid peroxidation in mitochondria.

Material and methods

Sixteen 2-month-old mice (8 males and 8 females) (group A) and 17 23-month-old mice (7 males and 10 females) (group B) were used. An additional group of 18 23-month-old mice (7 males and 11 females) (group C) was given AC (150 mg/kg/day, added in the drinking water) during the last 11 months.

Mice were killed by decapitation and whole heart rapidly removed, washed in saline, dried on filter paper, frozen in a mixture of petroleum ether and dry ice, and weighed.

MAO activity was assayed basically as described by Strolin Benedetti et al. (1983), using 5-HT (400 µM) and PEA (50 µM) as substrates. Contribution of MAO-A and -B to the metabolism of 5-HT and PEA in the mouse heart was assessed by experiments where the decrease in oxidative deamination of 400 µM 5-HT or 50 µM PEA was measured as a function of increasing concentrations of the selective MAO-A inhibitor, clorgyline (10^{-10}–10^{-3} M). Preincubation time with clorgyline was 1 h at 37°C. Hearts were homogenized in phosphate buffer 0.1 M, pH 7.4, using an Ultra Turrax (1 g tissue/16 ml buffer). Aliquots (0.1 ml) of heart homogenates were taken for the determination of MAO activity in a final volume of 0.5 ml. The reaction was started by addition of ^{14}C-PEA or ^{14}C-5-HT. After incubation at 37°C in normal air for 5 and 3 min for 5-HT and PEA, respectively, the reaction was stopped by cooling the tubes on ice and addition of 0.2 ml of 4N HCl. Blank samples were prepared with 0.1 ml of heart homogenate, 0.3 ml of buffer and acidification with 0.2 ml HCl 4N before the addition of ^{14}C-5-HT or ^{14}C-PEA. The deaminated products were extracted with 7 ml of toluene-ethyl acetate (1/1, v/v). The tubes were kept at −20°C for 1 h to allow the aqueous layer to freeze.

The organic layer was poured into a scintillation vial, and 10 ml of Instafluor was then added. The samples were counted in a Packard scintillation counter and the values obtained were corrected for the quenching, but not for the efficiency of extraction. The value of blank samples was subtracted from the value of the reaction samples. Enzyme activity was expressed as nmol.mg protein^{-1}.min^{-1}. Protein concentrations of the homogenates were determined by the method of Bradford (1976).

Results and discussion

Average whole body and heart weights for the males and females of the three groups are presented in Table 1. Both whole body and heart weights significantly increased in groups B and C as compared to group A. However, when the ratios heart weight/body weight of the old and young mice were compared, a statistically significant difference was found in females but not in males. The protein content of heart from young and old mice was not different (Table 2). The oxidative deamination of 5-HT and PEA in the heart of young mice as a function of increasing concentrations of clorgyline is presented in Fig. 1. In the mouse heart, PEA was a substrate of the B form of MAO, whereas 5-HT was deaminated essentially by the A form.

In the heart of 2-month-old male and female mice PEA deaminating activity (MAO-B) was about 7 times higher than 5-HT deaminating activity (MAO-A) (Table 3). A relatively modest, although statistically significant, increase in PEA deaminating activity (71%) was found to occur with ageing in the male mice, whereas no significant changes occurred in the 5-HT deaminating activity.

Table 1. Average weights of whole body (WB) and heart (H), and H to WB ratio for male and female mice of the three groups

		WB (g)	H (g)	H/WB × 100
Group A	Male (n = 8)	24.5 ± 1.4	0.118 ± 0.00	0.484 ± 0.039
	Female (n = 8)	18.9 ± 1.1	0.106 ± 0.014	0.560 ± 0.062
Group B	Male (n = 7)	44.2 ± 7.7**	0.189 ± 0.023**	0.433 ± 0.066
	Female (n = 10)	34.2 ± 5.8**	0.165 ± 0.025**	0.490 ± 0.092*
Group C	Male (n = 7)	45.1 ± 4.2**	0.194 ± 0.015**	0.433 ± 0.044
	Female (n = 11)	34.1 ± 5.3**	0.150 ± 0.018**	0.441 ± 0.030**

Mean ± SD
Significantly different from group A of the same sex, * $p < 0.05$, ** $p < 0.01$
ANOVA one-way followed by simple contrast (Siphar package)

Table 2. Protein content of heart from young and old mice

Group A	Male (n = 8)	134 ± 18
	Female (n = 8)	131 ± 43
Group B	Male (n = 7)	136 ± 10
	Female (n = 10)	128 ± 35

Mean ± SD
Values are expressed as mg prot/g of fresh tissue

Table 3. Average MAO activities for male and female mice of the three groups. For each heart the mean value of MAO activity was determined from triplicate measurements

		MAO activity (nmol.mg protein^{-1}.min^{-1})	
		PEA (50 µM)	5-HT (400 µM)
Group A	Male (n = 8)	0.783 ± 0.323	0.110 ± 0.046
	Female (n = 8)	0.687 ± 0.231	0.103 ± 0.083
Group B	Male (n = 7)	1.341 ± 0.459*	0.138 ± 0.045
	Female (n = 10)	1.170 ± 0.474*	0.119 ± 0.057
Group C	Male (n = 7)	1.618 ± 0.427**	0.173 ± 0.042*
	Female (n = 11)	1.156 ± 0.455*	0.117 ± 0.024

Mean ± SD
Significantly different from group A of the same sex, * $p < 0.05$, ** $p < 0.01$
ANOVA one-way followed by simple contrast (Siphar package)

Results in female mice were very similar, with a significant increase in PEA deaminating activity of 70%. AC treatment did not affect mouse heart 5-HT and PEA deaminating activities, as no significant differences were found between groups B and C (Table 3). No sex differences in 5-HT and

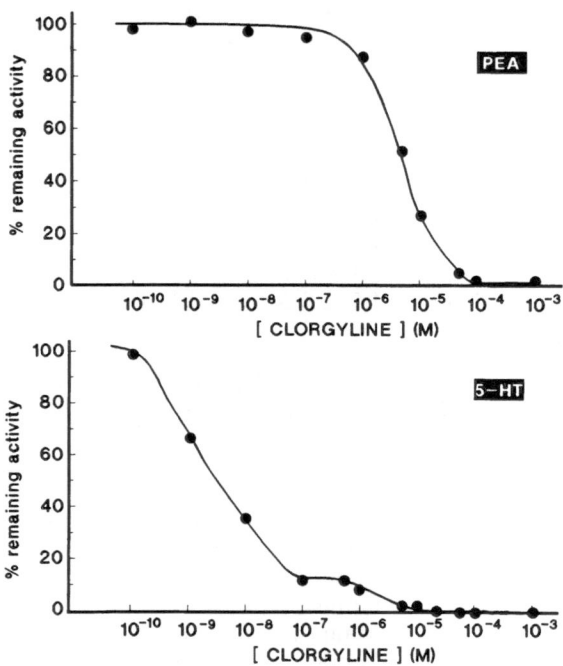

Fig. 1. Oxidative deamination of 5-HT (400 μM) and PEA (50 μM) in the heart of young mice as a function of increasing concentrations of clorgyline. Each point represents the mean of duplicate measurements from a typical experiment and is expressed as a percentage of the activity in the absence of clorgyline.

PEA deaminating activities were found in groups A and B, whereas both 5-HT and PEA deaminating activities were found to be higher in male than in female mice in group C (Student's t-test, $p < 0.05$).

Conclusion

In the mouse heart PEA is deaminated by MAO-B, in contrast with the rat heart, where it is deaminated by MAO-A (Cao Danh et al., 1984). A significant increase in PEA deaminating activity occurs in the mouse heart with age (about 70%). This effect is modest compared to the large increase in PEA deaminating activity occurring in the male rat heart with age, where it is actually due to the increase in MAO-A activity (Cao Danh et al., 1984). In the mouse heart 5-HT is deaminated by MAO-A and this activity is not affected by age. The 5-HT deaminating activity in the heart of young male rats (Cao Danh et al., 1984) is at least 35 times higher than in the heart of young male mice. The species difference is much more important in old animals, where the 5-HT deaminating activity of old male rats (Cao Danh et al., 1984) is about 204 times higher than that of old male mice. According to Parkinson and Callingham (1971) PEA is deaminated by MAO-B and 5-HT by MAO-A in the human heart, as in the mouse heart. However, the effect of age was not examined in that study.

Acknowledgement

The authors thank Dr. I. Poggesi for the statistical analyses.

References

Bradford MM (1976) A rapid and sensitive method for the quantitation of microgram quantities of protein utilizing the principle of protein-dye binding. Anal Biochem 72: 248–254

Cao Danh H, Strolin Benedetti M, Dostert P (1984) Differential changes in monoamine oxidase A and B activity in aging rat tissues. In: Tipton KF, Dostert P, Strolin Benedetti M (eds) Monoamine oxidase and disease. Prospects for therapy with reversible inhibitors. Academic Press, London, pp 301–317

Fuentes JA, Trepel JB, Neff NH (1977) Monoamine oxidase activity in the cardiovascular system of young and aged rats. Exp Gerontol 12: 113–115

Horita A, Lowe MC (1972) On the extraneuronal nature of cardiac monoamine oxidase in the rat. Adv Biochem Psychopharmacol 5: 227–242

Novick WJ (1961) The effect of age and thyroid hormones on the monoamine oxidase of rat heart. Endocrinology 69: 55–59

Parkinson D, Callingham BA (1979) Substrate and inhibitor selectivity of human heart monoamine oxidase. Biochem Pharmacol 28: 1639–1643

Strolin Benedetti M, Dostert P, Guffroy C, Tipton KF (1983) Partial or total protection from long-lasting monoamine oxidase inhibitors (MAOIs) by new short-acting MAOIs of type A MD 780515 and type B MD 780236. Mod Probl Pharmacopsychiatry 19: 82–104

Authors' address: Dr. M. Strolin Benedetti, Farmitalia Carlo Erba Research and Development, Via C. Imbonati 24, I-20159 Milan, Italy.

J Neural Transm (1994) [Suppl] 41: 89–94

Age-related changes on MAO in Bl/C57 mouse tissues: a quantitative radioautographic study

J. Saura[1], J. G. Richards[2], and N. Mahy[1]

[1] Biochemistry Unit, Faculty of Medicine, University of Barcelona, Barcelona, Spain
[2] Pharma Division, Preclinical Research, Hoffmann-La Roche, Basel, Switzerland

Summary. The distribution of MAO-A and -B in brain and peripheral tissues of Bl/C57 mice and their changes during ageing were studied by quantitative enzyme radioautography with [^{3}H]Ro41-1049 and [^{3}H]Ro19-6327. In the brain, MAO-A decreased between weeks 4 and 8 and then remained unchanged until 25 months, whereas MAO-B increased for the whole period studied. Heart also showed a continuous increase in MAO-B, but not MAO-A, with ageing, and liver showed a decrease in MAO-B in the older animals. These results show marked species differences in the distribution and age-related changes of MAO and might help to elucidate the high sensitivity of Bl/C57 mouse of MPTP, which increases with age.

Introduction

Monoamine oxidases (EC 1.4.3.4.; MAO) are the main enzymes responsible for the catabolism of monoamine neurotransmitters and trace amines. The enzyme exists in two forms, MAO-A and MAO-B, identified by their inhibitor sensitivity and substrate selectivity. In rodent brain, serotonin, dopamine and noradrenaline are preferentially deaminated by MAO-A; the trace amines phenethylamine and methylhistamine, by MAO-B; and tyramine and octopamine by both enzymes. The CNS cellular localization of the enzymes does not always correspond with the presence of their preferred substrates: MAO-A is found in adrenergic and noradrenergic neurons and, probably in low levels, in glia, whereas MAO-B is found in astrocytes, serotoninergic and histaminergic neurons. Dopaminergic neurons seem not to express MAO-A or MAO-B.

Most previous studies have shown age-related changes in both enzymes corresponding to their main cellular localization. MAO-A, mainly neuronal, shows a decrease with ageing whereas MAO-B, mainly glial, increases (Strolin Benedetti and Keane, 1980; Jossan et al., 1989). The possibility of applying new tools, [^{3}H]Ro41-1049 and [^{3}H]Ro19-6327 (Saura et al., 1992), sensitive and selective MAO radioligands, to quantify and spatially resolve MAO, led us to the study of age-related changes on MAO in a rodent strain, Bl/C57 mouse, especially interesting for its high sensitivity to the Parkinson-inducing neurotoxin MPTP (1-methyl-4-phenyl-1,2,5,6-tetra-hydropyridine), which is also age-related (Walsh and Wagner, 1989).

Materials and methods

Bl/C57 male mice aged 4 weeks (young), 9 weeks (adult), 19 months (old$_1$) and 25 months (old$_2$) were used in the experiments.

After decapitation under ether anaesthesia, brain and peripheral organs-liver, kidney, pancreas, spleen, heart and lung- were quickly removed and frozen. Tissues were kept at $-80°C$ until sectioning. 12 μm cryostat sections were obtained and mounted on gelatinized slides.

Sections were kept at $-30°C$ until incubation with either [^{3}H]-Ro41-1049 (MAO-A marker) or [^{3}H]-Ro19-6327 (MAO-B marker). Both radioligands were used at 2 concentratins: 15 nM (for measuring most structures) and 3.75 nM (for measuring structures with very high binding). Incubations were performed for 60 min at 37°C (for [^{3}H]-Ro41-1049) and for 90 min at 22°C (for [^{3}H]-Ro19-6327) followed by a 30″ + 30″ + 60″ wash in cold buffer and a quick dip in cold distilled water. A buffer solution (pH = 7.4) containing 50 mM Tris, 120 mM NaCl, 1 mM MgCl$_2$, 5 mM KCl and 0.5 mM EGTA was used in all experiments. Non-specific binding was determined by incubation in the presence of 1 μM of Clorgyline or 1-deprenyl.

After drying under a stream of cold air, sections were exposed to ^{3}H-sensitive film (Hyperfilm^{TM}-^{3}H, Amersham) for 2 weeks (total binding) or 4 weeks (non-specific binding). Films were then developed and quantified by means of plastic standards (^{3}H-micro scales, Amersham) with a computer-assisted Image Analysis System (Interdens, Microm).

The statistical significance of changes amongst age groups was analyzed by variance analysis. Homogeneity of variance was checked by Cochran's test.

Results

MAO-A, brain

In all brain structures studied, with the only exception of the molecular layer of the cerebellum, a significant decrease in MAO-A was found between young and adult animals (Fig. 1). This decrease was of 50% (corpus callosum), 40–42% (frontal cortex, caudate putamen), 29–32% (hippocampus) and 13–22% (cerebellum).

No significant differences were found between adult and old$_1$ animals.

Some structures (frontal and parietal cortex, stratum radiatum of hippocampus) showed a moderate increase (23–26%) between old$_1$ and old$_2$ animals. In other structures i.e. inferior colliculus, substantia nigra, the levels of old$_1$ and old$_2$ animals did not differ.

MAO-B, brain

A continuous age-related increase in MAO-B was found in all structures. Comparison between young and old$_2$ animals always gave a statistically

MAO-A

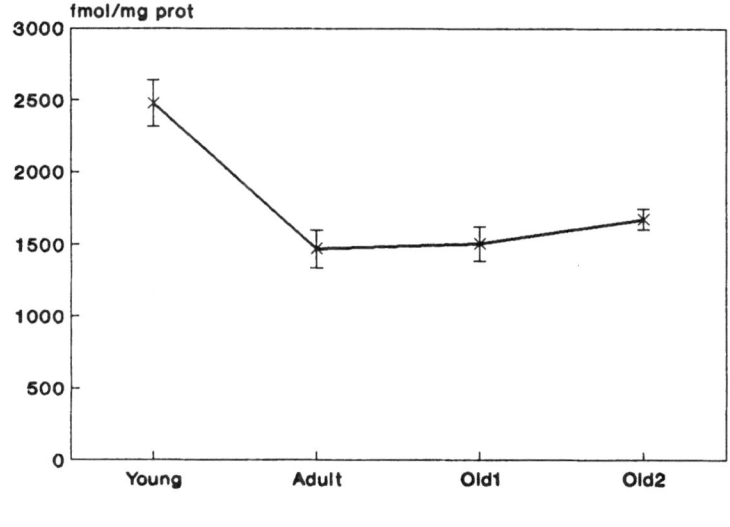

MAO-B

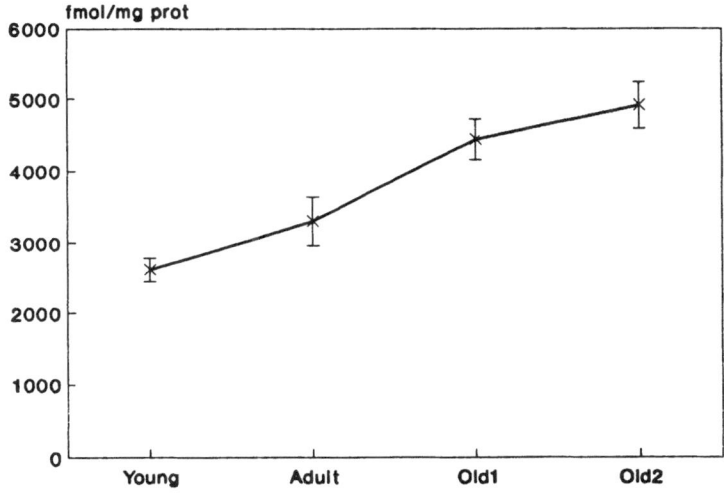

Fig. 1. Age-related changes in specific binding of [³H]Ro41-1049 (MAO-A) and [³H] Ro19-6327 (MAO-B) to Bl/C57 mouse nucleus accumbens. Note the decrease in MAO-A between young (4 week) and adult (8 week) animals and the continuous increase in MAO-B. This pattern of changes was observed in most brain structures studied. Error bars shown SEM (n = 4–8)

significant increase, varying from 20% (corpus callosum), 30% (hippocampus, substantia nigra, inferior colliculus), 60–70% (parietal cortex, caudate putamen) to 90% (nucleus accumbens) (Fig. 1) and was highest in the molecular layer of cerebellum (144%).

92 J. Saura et al.

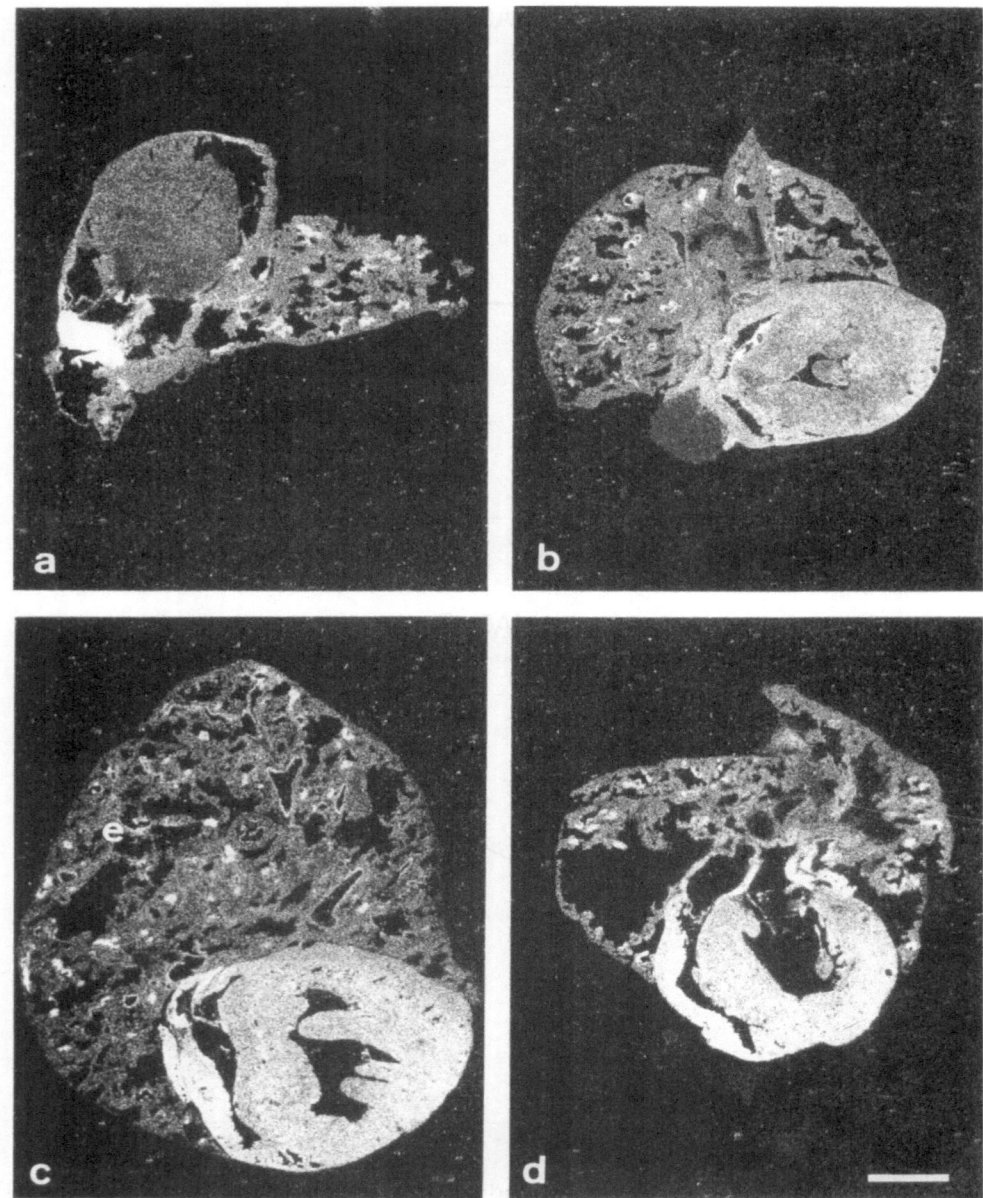

Fig. 2. [³H]Ro19-6327 binding (MAO-B) to lung and heart of Bl/C57 mice aged 4 weeks (**a**), 8 weeks (**b**), 19 months (**c**) and 25 months (**d**). Note that heart MAO-B content increases with age, whereas lung content remains unchanged. Bar: 2 mm

MAO-A, periphery

In most peripheral tissues studied (spleen, pancreas, heart and lung) no age-related changes in MAO-A were observed.

In the cortex of kidney, but not in the medulla, a significant decrease (−46%) was seen. A unique situation was observed in the liver, with 2 peaks of MAO-A in young and old_1 animals and 2 troughs in adults and old_2.

None of the peripheral organs studied followed the pattern of age-related changes in MAO-A observed in the brain.

MAO-B, periphery

The most interesting finding was the age-related increase in MAO-B (190%, young vs old_2) observed in heart (Fig. 2).

On the other hand, an age-related decrease in MAO-B was observed in liver, both in periportal and in central spaces. This decrease was especially marked in the last period of life, between old_1 and old_2 animals.

Discussion

The decrease in brain MAO-A found in the early weeks of life (young-adult) agrees with previous studies, including some dealing with human brain (Kornhuber et al., 1989). Histogenetic cell death might underlie this phenomenon. In aged animals, MAO-A appears not to change, or even to increase slightly. Most previous reports dealing with rat and human brain have found MAO-A to decrease in senescence, this being interpreted as an effect of natural neuronal death. The absence of such MAO-A decrease in Bl/C57 mouse brain could mean that in this strain, glial cells are enriched in MAO-A as compared with glial cells in other species such as man or rat.

The increase in MAO-B in the brain of old animals agrees with previous reports and is most probably due to the gliosis secondary to the neuronal death associated with ageing (Tatton et al., 1991). The increased MAO-B levels in the brain of old animals might explain the higher sensitivity to MPTP observed in these animals. At the same time, the reduced levels of MAO-B in the liver of old_2 animals could lead to an increased availability of MPTP in the brain and therefore contribute to a greater sensitivity to the neurotoxin.

Finally, the marked age-related increase of MAO-B in Bl/C57 heart parallels the age-related increase in MAO-A in rat heart (Lyles and Callingham, 1979). Moreover, whereas MAO-A is the predominant form in the rat heart, MAO-B is the most abundant in Bl/C57 mouse heart. This suggests that the physiological role of MAO-A in rat heart is played by MAO-B in Bl/C57 mouse heart. This differential content and age-related pattern of changes observed for MAO-A and -B in rat and Bl/C57 mouse heart has recently been confirmed by measuring in vitro MAO-A and -B activities (Strolin-Benedetti et al., this meeting).

Acknowledgement

This work was supported by CICYT PM88-0117.

References

Jossan SS, Sakurai E, Oreland L (1989) MPTP toxicity in relation to age, dopamine uptake and MAO-B activity in two rodent species. Pharmacol Toxicol 64: 314–318

Kornhuber J, Konradi C, Mack-Burkhardt W, Riederer P, Heinsen H, Beckmann H (1989) Ontogenesis of monoamine oxidase-A and -B in the human brain frontal cortex. Brain Res 499: 81–86

Lyles GA, Callingham BA (1979) Selective influences of age and thyroid hormones on type A monoamine oxidase of the rat heart. J Pharm Pharmacol 31: 755–760

Saura J, Kettler R, Da Prada M, Richards JG (1992) Quantitative enzyme radioautography with [^{3}H]Ro41-1049 and [^{3}H]Ro19-6327 in vitro: localization and abundance of MAO-A and MAO-B in rat CNS, peripheral organs and human brain. J Neurosci 12: 1977–1999

Strolin Benedetti M, Keane PE (1980) Differential changes in MAO-A and -B in the aging rat brain. J Neurochem 35: 1026–1032

Tatton WG, Verrier MC, Holland DP, Kwan MH, Biddle FE (1991) Different rates of age-related loss for four murine monoaminergic neuronal populations. Neurobiol Aging 12: 543–556

Walsh SL, Wagner GC (1989) Age-dependent effects of MPTP: correlation with MAO-B. Synapse 3: 308–314

Authors' address: D. J. Saura, Unitat de Bioquímica, Facultat de Medicina, Universitat de Barcelona, Diagonal 643, E-08028 Barcelona, Spain.

J Neural Transm (1994) [Suppl] 41: 95–99

The FAD dependent amine oxidases in relation to developmental state of enterocyte

W. A. Fogel and **C. Maslinski**

Department of Biogenic Amines, Polish Academy of Sciences, Lodz, Poland

Summary. Epithelial cells from bovine and guinea pig small intestines contain monoamine and polyamine oxidases with MAO-A preponderance at any maturational stage. For either species, K_m values for 5HT and N^1acetylspermine remain throughout cellular maturation on the same levels, whereas the V_{max} values do not. For serotonin, the dividing crypt cells showed in cow lower and in guinea pig higher V_{max} than the mature cells; for N^1acetylspermine, mature cells, independently of species, showed lower V_{max}.

Introduction

The mature intestine with its permanent enterocyte proliferation and replacement (Lipkin, 1981) offers a good model for studying the relation between cellular growth processes and aminergic system. Amongst amines putrescine and polyamines, spermidine and spermine, are known to regulate the activity of various proliferating and differentiating cells, enterocytes including (Heby, 1981; Luk and Baylin, 1982). There are some indications that also histamine (Tutton, 1976) and serotonin (Tutton, 1974) may influence the proliferation of gastrointestinal cells. It has been reported that diamine oxidase (DAO), the enzyme involved in regulation of cellular concentration of putrescine and histamine, is more abundant in the functional than in proliferative compartment of the crypt-villus axis (Shakir et al., 1977) and it was even postulated to use DAO as a marker of enterocyte maturation (Luk et al., 1980). However, in regard to FAD dependent amine oxidases that play a role in maintenance of appropriate levels of serotonin and polyamines such information is missing. This prompted us to examine whether monoamine and polyamine oxidase activities are dependent upon the developmental state of the enterocyte.

Materials and methods

Segments of guinea pig and cow small intestine were used to isolate epithelial cells at various developmental state. The cells were obtained from different levels of the

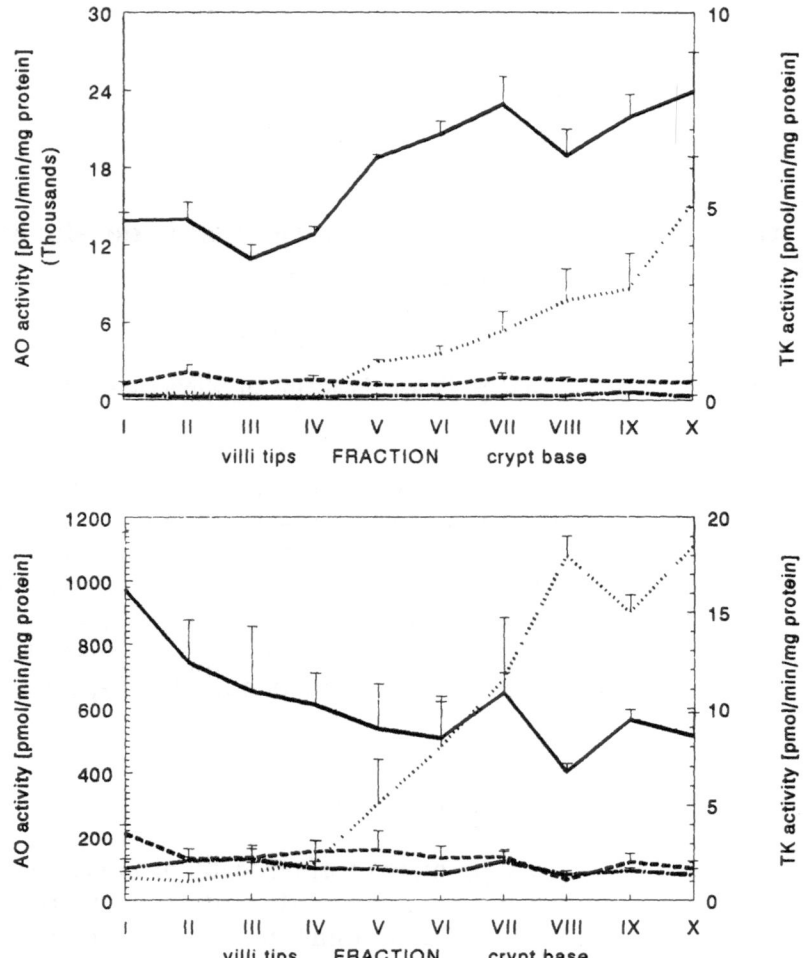

Fig. 1. Distribution of activity of FAD dependent amine oxidases: monoamine oxidase A (——*M-A*), monoamine oxidase B (– – –*M-B*) and polyamine oxidase (–.–.–*PAO*) along villi-tip crypt axis. Enterocytes were isolated in 10 sequential fractions using thymidine kinase (....*TK*) as a marker of proliferative compartment. **A** Enterocytes isolated from guinea pig small intestine (n = 4); **B** enterocytes isolated from bovine small intestine (n = 4)

villous and crypt zones by successive elutions with calcium-free EDTA buffer solutions according to Weiser (1973). Thymidine kinase activity (Ives et al., 1969) served as a marker of proliferating zone. The activity of monoamine oxidase A and B and polyamine oxidase were measured in individual cell fractions using the fixed substrate concentrations, e.g. 500 μM serotonin (5HT), 20 μM β-phenylethylamine (PEA) and 100 μM N[1]-acetylspermine (AcSp). Kinetic parameters for MAO-A, MAO-B and PAO were determined with 5HT (40–1,000 μM) in the presence of deprenyl (0.3 μM), PEA (1–40 μM) in the presence of clorgyline (0.3 μM) and AcSp (50–1,000 μM) in the presence of pargyline and semicarbazide (1 mM each), respectively. Pooled cell fractions 1–6 represented villi and fractions 7–10, crypts.

MAO activity was assayed in cell lysates by radiometric (Wurtman and Axelrod, 1963) and PAO by fluorometric (Matsumoto et al., 1982) method.

Results

All individually measured isolated cell fractions irrespectively of their origin showed activity against serotonin, phenylethylamine and acetylspermine (Fig. 1A,B). Monoamine oxidase A was the most active enzyme and at the same time the prevailing form of MAO in the material studied. In guinea pig enterocytes MAO-A activity decreases whereas in cow enterocytes, on the contrary, it tends to increase with cell maturation. Polyamine oxidase activity shows consistently lower values in mature cells the opposite being true for MAO-B. To find out whether the changes in the enzymes are related to their different kinetic behaviour as the cell gets mature, pools of fractions 1–6 and 7–10 representing villi and crypts, respectively, were used to obtain apparent kinetic constants. As seen in Table 1, K_m values of the enzymes towards their specific substrates do not change upon maturation but V_{max} do so, indicating that the observed changes in the enzyme activities are due to varying concentration of the enzymes.

Discussion

The data presented here show that the enterocyte, irrespectively of the developmental stage, is able to degrade monoamines and acetylspermine. During cellular maturation quantitative changes in the activity of FAD dependent amine oxidases occur, albeit less dramatic than those described for Cu-dependent diamine oxidase from rat intestinal mucosa (Shakir et al., 1977). It appears that maturation affects a particular enzyme in a different way in a given species. Thus, MAO-A activity was lower in the guinea pig but higher in the cow mature enterocytes as compared to proliferating crypts.

Table 1. Kinetic parameters[1] for two forms of monoamine oxidase (MAO-A, MAO-B) and polyamine oxidase (PAO) in isolated crypt and villi cells

Species/cell fraction		K_m, µM			V_{max}, nmole/min/mg protein		
		MAO-A	MAO-B	PAO	MAO-A	MAO-B	PAO
Guinea pig	crypt	270	9.8	68	15.42	0.196	0.182
	villi	275	8.4	76	3.06	0.253	0.133
Cow	crypt	144	11.2	73	1.79	0.058	0.154
	villi	139	9.0	67	2.68	0.097	0.092

[1] Apparent K_m and V_{max} values were obtained with serotonin (40–1,000 µM) as substrate and cell lysates preincubated with 0.3 µM deprenyl for MAO A; β-phenylethylamine (1–40 µM) and 0.3 µM clorgyline for MAO B; N[1]acetylspermine (50–1,000 µM) and 1 mM of each, pargyline and semicarbazide for PAO. SEM is within ±10% of the given values which are the mean of 3–4 determinations. Crypt denotes pool of fractions 7–10; villi fractions 1–6

It is worth noting that although the activity of monoamine oxidase-A is extremely high in cells isolated from guinea pig small intestine, the enzyme affinity towards serotonin is significantly lower than that exhibited by MAO-A from other sources (Gomez et al., 1986). Similarly, for partially purified polyamine oxidase from rat liver K_m value of $0.6\,\mu M$ for acetylspermine was reported (Seiler et al., 1981); yet, in enterocytes of either guinea pig or cow the corresponding value is two orders of magnitude higher. In conclusion, it appears from our data that intestinal amine oxidases, though very potent, are of low specificity and their activities may be regulated in different species in a totally different way.

Acknowledgements

We wish to thank the technical staff: K. Adach, B. Szczutkowska, W. Majcherek, E. Bechtold and E. Sarri — Ph.D. student from Dept. Biochem. Mol. Biol., Autonomous University of Barcelona, whose stay was supported by Tempus YEP 0789/91, for their excellent assistance.

References

Gomez N, Unzeta M, Tipton KF, Anderson MC, O'Carroll A-M (1986) Determination of monoamine concentrations in rat liver by inhibitor binding. Biochem Pharmacol 35: 4467–4472

Heby O (1981) Role of polyamines in the control of cell proliferation and differentiation. Differentiation 19: 1–20

Ives DH, Durham JP, Tuck VS (1969) Rapid determination of nucleoside kinase and nucleotidase activities with tritium-labeled substrates. Anal Biochem 28: 192–205

Lipkin M (1981) Proliferation and differentiation of gastrointestinal cells in normal and disease states. In: Johnson LR (ed) Physiology of the gastrointestinal tract. Raven Press, New York, pp 145–168

Luk GD, Baylin SB (1982) Ornithine decarboxylase in intestinal maturation and adaptation. In: Robinson JWL, Dowling RH, Riecken EO (eds) Mechanisms of intestinal adaptation. MTP, Lancaster, pp 65–80

Luk GD, Bayless TM, Baylin SB (1980) Diamine oxidase (histaminase) — a circulating marker for rat intestinal mucosal maturity and integrity. J Clin Invest 66: 66–70

Matsumoto T, Furuta T, Nimura Y, Suzuki O (1982) Increased sensitivity of fluorometric method of Snyder and Hendley for oxidase assays. Biochem Pharmacol 31: 2207–2209

Seiler N, Bolkenius FN, Rennert OM (1981) Interconversion, catabolism and elimination of the polyamines. Med Biol 59: 334–346

Shakir KMM, Margolis S, Baylin SB (1977) Localization of histaminase (diamine oxidase) in rat small intestinal mucosa: site of release by heparin. Biochem Pharmacol 26: 2343–2347

Tutton PJM (1974) The influence of serotonin on crypt cell proliferation in the jejunum of rat. Virchovs Arch (Cell Pathol) 16: 79–87

Tutton PJM (1976) Influence of histamine on epithelial crypt cell proliferation in the jejunum of the rat. Clin Exp Pharmacol Physiol 3: 369–373

Weiser (1973) Intestinal epithelial cell surface membrane glycoprotein synthesis. J Biol Chem 248: 2536–2541

Wurtman RJ, Axelrod J (1963) A sensitive and specific assay for the estimation of monoamine oxidase activity. Biochem Pharmacol 12: 1439–1441

Authors' address: Dr. W. A. Fogel, Department of Biogenic Amines, Polish Academy of Sciences, Tylna 3, POB 225, Lodz, Poland.

J Neural Transm (1994) [Suppl] 41: 101–105

Role of monoamine oxidase and cathecol-O-methyltransferase in the metabolism of renal dopamine*

M. H. Fernandes** and **P. Soares-da-Silva**

Department of Pharmacology and Therapeutics, Faculty of Medicine, Porto, Portugal

Summary. Incubation of slices of rat renal cortex with 50 μM L-DOPA during 15 min resulted in the formation of dopamine and of its deaminated (3,4-dihydroxyphenylacetic acid; DOPAC), methylated (3-methoxytyramine; 3-MT) and deaminated plus methylated (homovanillic acid; HVA) metabolites. The presence of pargyline (100 μM) resulted in a 90% reduction in the formation of DOPAC and HVA; levels of dopamine and 3-MT were found to be significantly increased. A concentration dependent decrease in the formation of methylated metabolites was obtained in the presence of (10, 50 and 100 μM) tropolone (10–50% reduction) and (0.1, 0.5, 1.0 and 5.0 μM) Ro 40-7592 (50–95% reduction). Ro 40-7592 was also found to significantly increase DOPAC (20–40%) and dopamine (10–30%) levels, whereas tropolone slightly increased DOPAC (10%) levels. These results show that deamination represents the major pathway in the metabolism of newly formed dopamine under in vitro experimental conditions in the rat kidney. In addition, only when MAO is inhibited does methylation appear to represent an alternative metabolic pathway.

Introduction

In kidney, a major source of dopamine results from the decarboxylation of filtered 3,4-dihydroxyphenylalanine (L-DOPA) in tubular epithelial cells (Baines and Chan, 1981). The formation of dopamine and its metabolic degradation is a matter of particular importance because of the natriuretic and diuretic effects of the amine (Siragy et al., 1989). Renal tissues are endowed with one of the largest monoamine oxidase (MAO) activity in the body (Caramona and Soares-da-Silva, 1990; Fernandes and Soares-da-Silva, 1992) and previous work has shown a relevant role of MAO in the deamination of newly-formed dopamine in kidney slices loaded with L-DOPA (Fernandes and Soares-da-Silva, 1990). Catechol-O-methyltransferase

---

* Supported by grant number CEN 1139/92 from the JNICT. ** Permanent address: Faculty of Medical Dentistry, 4200 Porto, Portugal

(COMT) is another key enzyme in the metabolic degradation of catechol-amines and its intervention has been shown in several circunstances to follow or to preeced that of MAO (Kopin, 1985). The most common situations are the methylation of the deaminated metabolite of dopamine, 3,4-dihydroxyphenylacetic acid (DOPAC), into homovanillic acid (HVA) or the methylation of dopamine itself into 3-methoxytyramine (3-MT), which can, thereafter, be also deaminated into HVA. Inhibition of MAO frequently results in an amplification of the effects of dopamine and a similar picture might be expected to occur during inhibition of COMT, namely when MAO and COMT inhibitors are employed together or when methylation of dopamine assumes particular importance.

The aim of the present work was to study the role of MAO and COMT in the metabolism of dopamine formed from L-DOPA in the rat kidney under in vitro experimental conditions and to assess the eficacy of two COMT inhibitors, tropolone and Ro 40-7592 (Zürcher et al., 1990), in the methylation of both dopamine and DOPAC.

Materials and methods

Slices of the rat (male Wistar, Biotério do Instituto Gulbenkian de Ciência, Oeiras, Portugal) renal cortex approximately 0.5 mm thick and weighing about 30 mg wet weight were prepared with a scalpel, as previously described (Soares-da-Silva and Fernandes, 1992). Thereafter, renal slices were preincubated for 30 min in 2 ml warmed (37°C) and gassed (95% O_2 and 5% CO_2) Krebs' solution. The Krebs' solution had the following composition (mM): NaCl 120, KCl 4.7, $CaCl_2$ 2.4, $MgSO_4$ 1.2, $NaHCO_3$ 25, KH_2PO_4 1.2, EDTA 0.4, asorbic acid 0.57, glucose 10 and sodium butyrate 1; 1-alpha-methyl-p-tyrosine (50 μM) and copper sulphate (10 μM) were also added to the Krebs' solution in order to inhibit the enzyme tyrosine hydroxylase and inhibit the endogenous inhibitors of dopamine β-hydroxylase, respectively. After preincubation, renal slices were incubated for 15 min in Krebs' solution with 50 μM L-DOPA. The preincubation and incubation were carried out in glass test tubes, continuously shaken throughout the experiment. In experiments in which the effects of pargyline (100 μM), tropolone (10, 50 and 100 μM) or Ro 40-7592 (0.1, 0.5, 1.0 and 5.0 μM) on the renal production of dopamine were tested, the compounds were present during the preincubation and incubation periods. At the end of incubation, the reaction was stopped by the addition of 250 μl of 2 M perchloric acid and samples were stored at 4°C until the quantification of catecholamines, within the next 24 hours.

The assay of L-DOPA, dopamine, DOPAC, 3-MT and HVA in renal tissues was performed by means of high pressure liquid chromatography with electrochemical detection, as previously descibed (Soares-da-Silva and Garrett, 1990).

Mean values ± SEM of n experiments are given. Significance of differences between one control and several experimental groups was evaluated by Tuckey-Kramer method. A P value less than 0.05 was assumed to denote a significant difference.

Dopamine hydrochloride, L-β-3,4-dihydroxyphenylalanine (L-DOPA), 3,4-dihydroxyphenylacetic acid (DOPAC), homovanillic acid (HVA), 3-methoxytyraime (3-MT) and tropolone hydrochloride were purchased at Sigma Chemical Company (St Louis, M), USA). Ro 40-7592 (3,4-dihydroxy-4'-methyl-5-nitrobenzophenone) was kindly donated by Prof. M. Da Prada, Hoffmann-La Roche (Basle, Switzerland).

Results

Incubation of slices of rat renal cortex with exogenous L-DOPA (50 μM) resulted in the formation of considerable amounts of dopamine and of its deaminated, methylated and deaminated plus methylated metabolites (Table 1). The levels of dopamine in kidney slices incubated in the absence of exogenous L-DOPA were 3,000-fold lower than in those loaded with L-DOPA (50 μM) and DOPAC, 3-MT and HVA were found not detectable. The formation of dopamine in rat kidney slices loaded with exogenous L-DOPA has been found to be dependent on the concentration of L-DOPA added to the incubation medium (Fernandes et al., 1991). In these experimental conditions, DOPAC was found to be the major metabolite and significantly lower amounts of the methylated (3-MT and HVA) compounds were found to occur. The presence of pargyline (100 μM) resulted in a 90% reduction in the formation of DOPAC and HVA; levels of dopamine and 3-MT were found to be increased by 37.3% and 100%, respectively (Table 1).

Figure 1 shows the effect of tropolone and Ro40-7592, two inhibitors of the enzyme COMT, on the formation of dopamine and its metabolites, DOPAC, 3-MT and HVA, in kidney slices loaded with 50 μM L-DOPA. A concentration dependent decrease in the formation of the methylated metabolites (3-MT and HVA) was obtained in the presence of tropolone (10 to 50% reduction) and Ro 40-7592 (50 to 95% reduction). Ro 40-7592 was also found to increase the formation of both DOPAC (20 to 40% increase) and dopamine (14 to 31% increase) levels; the increased formation of both dopamine and DOPAC by tropolone was found to be lower than that observed with Ro 40-7592 (Fig. 1).

Discussion

The results presented here show that deamination represents a major pathway in the metabolism of newly-formed dopamine under in vitro experimental conditions in the rat kidney. This suggestion is evidenced by

Table 1. Effect of pargyline (100 μM) on the accumulation of dopamine, 3,4-dihydroxyphenylacetic acid (DOPAC), 3-methoxytyramine (3-MT) and homovanillic acid (HVA) in slices of rat renal cortex loaded with 50 μM L-DOPA

	Control	Pargyline
Dopamine	70.9 ± 6.8	103.3 ± 3.9*
DOPAC	14.6 ± 1.3	1.5 ± 0.3*
3-MT	5.7 ± 0.7	11.0 ± 0.8*
HVA	4.1 ± 0.5	0.5 ± 0.2*

Values are mean ± SEM of five experiments per group
* Significantly different from corresponding control values
($P < 0.02$)

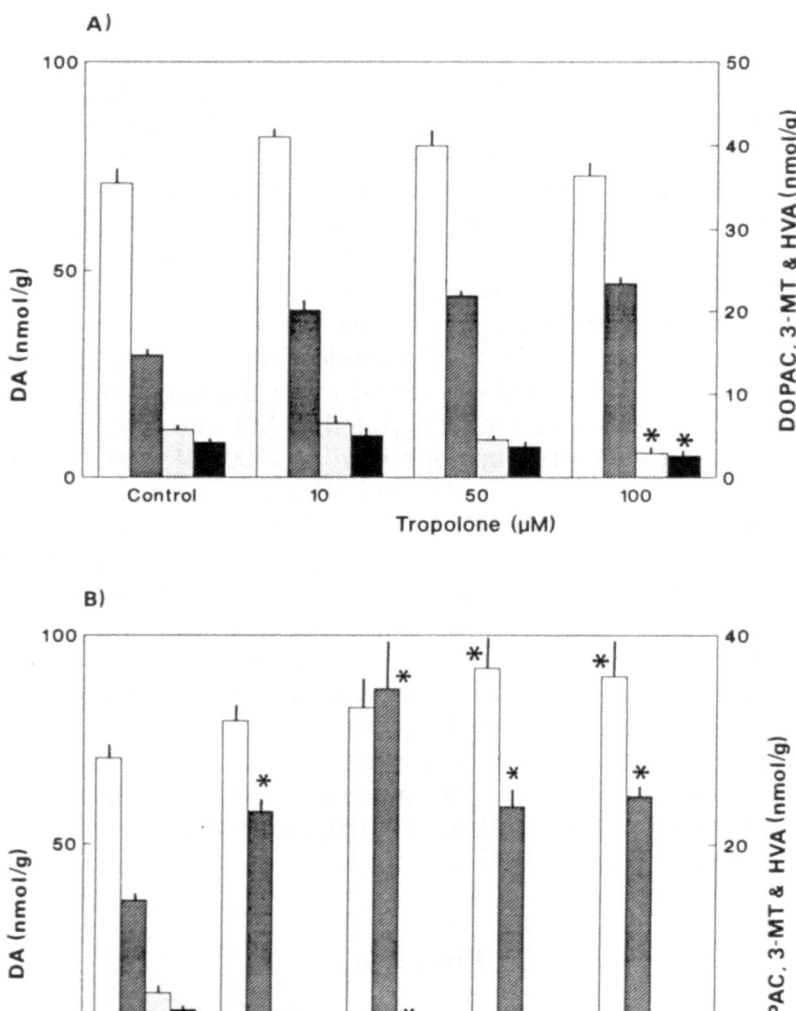

Fig. 1. Effect of **A** Tropolone (10, 50 and 100 μM) and **B** Ro40-7592 (0.1, 0.5, 1.0 and 5.0 μM) on the accumulation of dopamine (*DA*), 3,4-dihydroxyphenylacetic acid (*DOPAC*), 3-methoxytyramine (*3-MT*) and homovanillic acid (*HVA*) in slices of rat renal cortex loaded with 50 μM L-DOPA. Each column represent the mean of five experiments per group; vertical lines show SEM. *DA* open columns; *DOPAC* hatched columns; *3-MT* doted columns; *HVA* closed columns. *Significantly different from corresponding control values (P < 0.02)

two different sets of findings. Firstly, levels of DOPAC were found to be higher than the algebraic sum of the levels of the two methylated metabolites, HVA and 3-MT; some of the HVA is, however, expected to have its origin in the methylation of DOPAC or in the deamination of 3-MT. Secondly, the almost complete inhibition of MAO by pargyline (100 μM) results in a greater increase in the levels of dopamine (28.7 nmol/g, corresponding to 37% increase) than those of 3-MT (5.3 nmol/g, corresponding to 107% increase). This further suggests that only a minor amount (10%) of the total dopamine which has not been deaminated undergoes methylation into 3-MT. In addition, it might be suggested that only when MAO is inhibited, does methylation appear to represent an important alternative metabolic pathway.

The inhibitory effects of tropolone and Ro 40-7592 on the methylation of dopamine and DOPAC evidenced by the increase in their respective levels and the decrease in the production of, respectively, 3-MT and HVA, agrees well with the effects of these two compounds upon COMT. It is interesting to note the far greater effect of Ro 40-7592 in reducing the formation of 3-MT and HVA in comparison with that of tropolone. In fact, the maximal inhibitory effect of tropolone on the formation of 3-MT and HVA was achieved at 100 μM and was only of about 50% reduction. By contrast, Ro 40-7592 was found to reduce by half the formation of 3-MT and HVA at the concentration of 0.1 μM; the decrease in the formation of 3-MT and HVA was maximal (95% reduction) at 5.0 μM Ro 40-7592.

In conclusion, the results presented here favour the view that deamination in the rat kidney represents a major pathway in the metabolism of newly-formed dopamine and the relative importance of COMT is only evident during inhibiton of MAO.

References

Baines AD, Chan W (1980) Production of urine free dopamine from dopa: a micropuncture study. Life Sci 26: 253–259

Caramona MM, Soares-da-Silva P (1990) Evidence for an extraneuronal location of monoamine oxidase in renal tissues. Naunyn Schmiedebergs Arch Pharmacol 341: 411–413

Fernandes MH, Soares-da-Silva P (1990) Effects of MAO-A and MAO-B selective inhibitors Ro 41-1049 and Ro 19-6327 on the deamination of newly-formed dopamine in the rat kidney. J Pharmacol Exp Ther 255: 1309–1313

Fernandes MH, Pestana M, Soares-da-Silva P (1991) Deamination of newly-formed dopamine in rat renal tissues. Br J Pharmacol 102: 778–782

Fernandes MH, Soares-da-Silva P (1992) Type A and B monoamine oxidase activities in the human and rat kidney. Acta Physiol Scand 145: 363–367

Kopin IJ (1985) Catecholamine metabolism: basic aspects and clinical significance. Pharmacol Rev 37: 333–364

Siragy HM, Felder RA, Howell NL, Chevalier RL, Peach MJ, Carey RM (1989) Evidence that intrarenal dopamine acts as a paracrine substance at the renal tubule. Am J Physiol 257: F469–F477

Soares-da-Silva P, Garrett MC (1990) A kinetic study of the rate of formation of dopamine, 3,4-dihydroxyphenylacetic acid (DOPAC) and homovanillic acid (HVA) in the rat brain: implications for the origin of DOPAC. Neuropharmacology 29: 869–874

Zurcher G, Keller HH, Getler R, Borgulya J, Bonetti EP, Eigenmann R, Da Prada M (1990) Ro 40-7592, a novel, very potent, and orally active inhibitor of caytechol-O-methyltransferase: a pharmacological study in rats. Adv Neurol 53: 497–503

Authors' address: Dr. P. Soares-da-Silva, Department of Pharmacology and Therapeutics, Faculty of Medicine, 4200 Porto, Portugal.

J Neural Transm (1994) [Suppl] 41: 107–113

Comparison of the effect of reversible and irreversible MAO inhibitors on renal nerve activity in the anesthetized rat

G. Lavian, J. P. M. Finberg, and M. B. H. Youdim

Bruce Rappaport Faculty of Medicine, Technion, Haifa, Israel

Summary. Cardiovascular effects of the irreversible MAO-A inhibitor clorgyline and reversible MAO-inhibitors, moclobemide and brofaromine, were compared in the anesthetized rat. Electrical activity of the sympathetic renal nerve was monitored as an index of central sympathetic output. A long lasting decrease in the recorded parameters: blood pressure (BP), renal nerve activity (RNA) and heart rate (HR) was produced by acute administration of clorgyline (2 mg/Kg, IP). Acute treatment with moclobemide (10 mg/Kg, IP) or brofaromine (10 mg/Kg, IP) caused only a transient decrease in RNA. Pretreatment with the α_2 antagonist yohimbine, decreased significantly the inhibitory effect of clorgyline on all three parameters. The selective α_2 antagonist CH-38083 blocked the sympathoinhibitory effect of brofaromine. These results indicate an α_2 adrenoceptor involvement in the central sympathoinhibitory effect of MAO inhibitors, which may be manifested as a hypotensive effect, including orthostatic hypotension, in patients treated with irreversible selective MAO inhibitors.

Introduction

It is generally agreed that the therapeutic effect of antidepressant drugs is connected with their action on noradrenergic and serotonergic neurons. Noradrenaline (NA) and serotonin (5-HT) are both substrates for MAO-A (Johnston, 1968), which is found in neuronal mitochondria (Neff and Goridis, 1972; Denney and Denney, 1985). Inhibition of MAO-A increases the axoplasmatic concentration of the amine neurotransmitters and may increase their synaptic concentration in the CNS (Neff et al., 1974; Finberg and Youdim, 1983; Haefely et al., 1992).

Noradrenergic as well as serotonergic neurons are also involved in the CNS control or modulation of cardiovascular (CV) parameters (Guyenet et al., 1989; Kuhn et al., 1980; Coote et al., 1987). This may explain why cardiovascular side effects are components of treatment with antidepressant drugs. Apart from dangerous hypertensive crises following tyramine ingestion ("cheese effect") a hypotensive effect, including orthostatic hypotension may accompany MAO inhibitor therapy.

The selective reversible MAO-A inhibitors (RIMA) are new, highly effective antidepressant drugs with few side effects, including only a negligible hypertensive reaction to tyramine ingestion (Da Prada et al., 1984; Korn et al., 1988).

We assessed the hypotensive action of the irreversible selective MAO-A inhibitor clorgyline in rats, as compared with the effect of the reversible inhibitors, moclobemide and brofaromine. The hypotensive effect of classical MAO inhibitors has been attributed to both peripheral (Neff and Yang, 1974) and central (Fuentes et al., 1979) actions of the drugs. We hypothesized that the hypotensive action results from an inhibitory effect of the MAOI on the central sympathetic output, probably mediated by α_2 receptors located on noradrenergic neurons which belong to CV centers in the CNS. Monitoring the electrophysiological activity of the postganglionic sympathetic renal nerve enabled us to evaluate the central sympathetic output, and to assess a central action of the antidepressant drug, assuming a lack of ganglionic effect for the MAO inhibitor.

Materials and methods

Experiments were performed on normotensive male, Sprague Dawley rats (250–330 g), anesthetized with a mixture of Na pentobarbital (15 mg/Kg) and chloral hydrate (60 mg/Kg) with an intact baroreceptor reflex. Left carotid artery and right jugular vein catheterization was performed on every animal in order to enable systemic blood pressure recording and drug injection, respectively. The left renal nerve was exposed and electrical nerve activity was recorded via a pair of Pt-Ir electrodes. The pulse wave, heart rate and integrated renal nerve activity were displayed on a chart recorder. The electrocardiogram and rectal temperature were monitored over the time of the experiment. Body temperature was maintained between 36–38°C by placing the animal on a metal stage warmed with paraffin oil.

The different parameters were recorded after drug administration at predetermined time points over the length of the experiment. All the parameters were expressed as mean values ± S.E.M. Significance levels for the difference between the experimental and control curves were obtained by performing repeated measure analysis of variance (ANOVA) according to "SAS General Linear Models Procedure" (SAS/STAT TM 1987). For further details on methods and data analysis see Lavian et al. (1991).

Drugs used were: clorgyline (Sigma), yohimbine hydrochloride (Sigma), moclobemide (Hoffman La Roche), brofaromine (Ciba-Geigy) and CH38083 (Chinoin-Budapest).

Results and discussion

Acute treatment with clorgyline (2 mg/Kg, IP) resulted, within 30 min, in a significant reduction in BP, RNA and HR by 28%, 34% and 16%, respectively (Figs. 1a–c). As a selective MAO-A inhibitor, clorgyline prevents the deamination of both NA and 5-HT. The dose used was chosen to cause more than 85% inhibition of the enzyme (Goridis and Neff, 1971; Neff et al., 1974). The persistence of reduced values in all three parameters

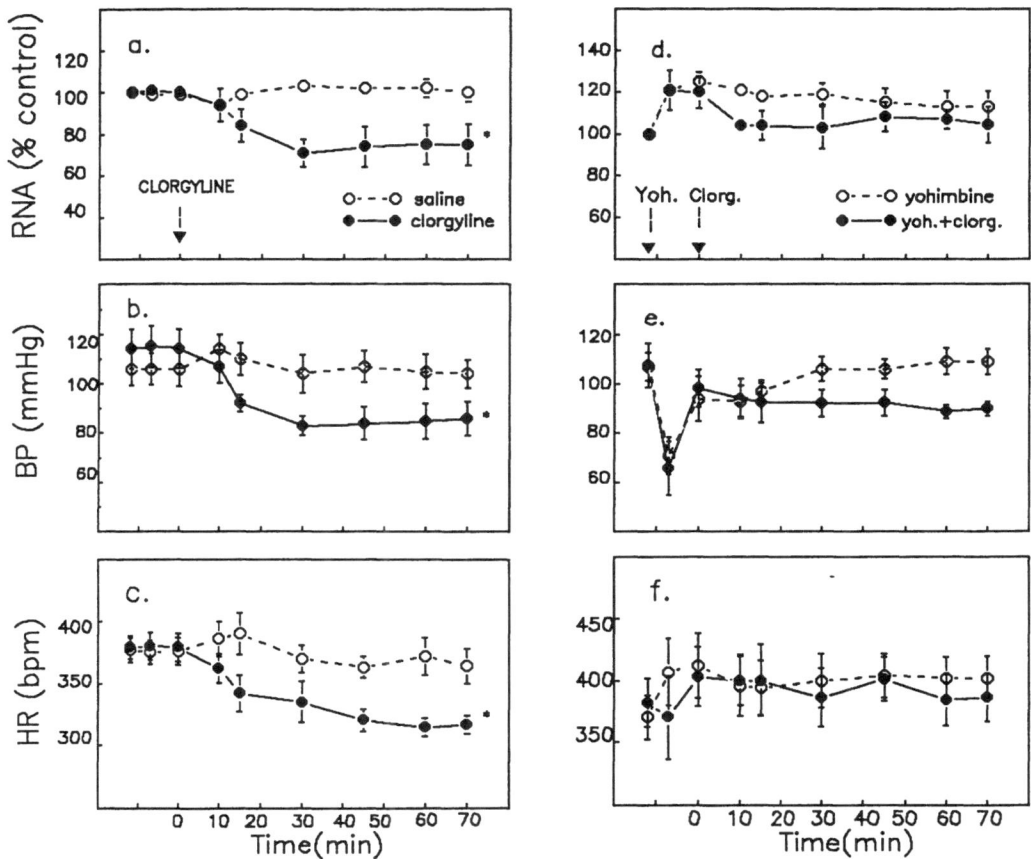

Fig. 1. Effect of acute treatment with clorgyline (2 mg/Kg, IP; n = 5) in anesthetized rats on: **a** RNA, **b** BP, **c** HR and of the same treatment after pretreatment with yohimbine (0.3 mg/Kg, IV; n = 5) on **d** RNA, **e** BP, **f** HR. In this and the following figure, asterisk indicate the significance level for the difference between drug, and control curves performed with ANOVA test (*P < 0.05)

over the time of the experiment is consistent with the irreversible nature of the drug's action (Neff et al., 1974).

Pretreatment with the α_2 adrenoceptor antagonist yohimbine (0.3 mg/Kg, IV), reduced the inhibitory effect of clorgyline on BP and RNA and abolished its bradycardic action (Figs. 1d–f).

Acute administration of moclobemide (10 mg/Kg, IP) did not reduce BP or HR in the anesthetized rats (Figs. 2b and c), but induced a transient significant (P < 0.05, t-test) reduction of 20% in the RNA, which attained a minimum within 15 min and returned to control value 30 min after drug injection (Fig. 2a). The dose of moclobemide used in our experiments inhibits MAO-A activity in rat brain by about 80% (Finberg and Youdim, 1988; Da Prada et al., 1989).

A transient inhibitory effect on RNA was also recorded after acute treatment with brofaromine (10 mg/Kg, IP). A significant (P < 0.05, t-test)

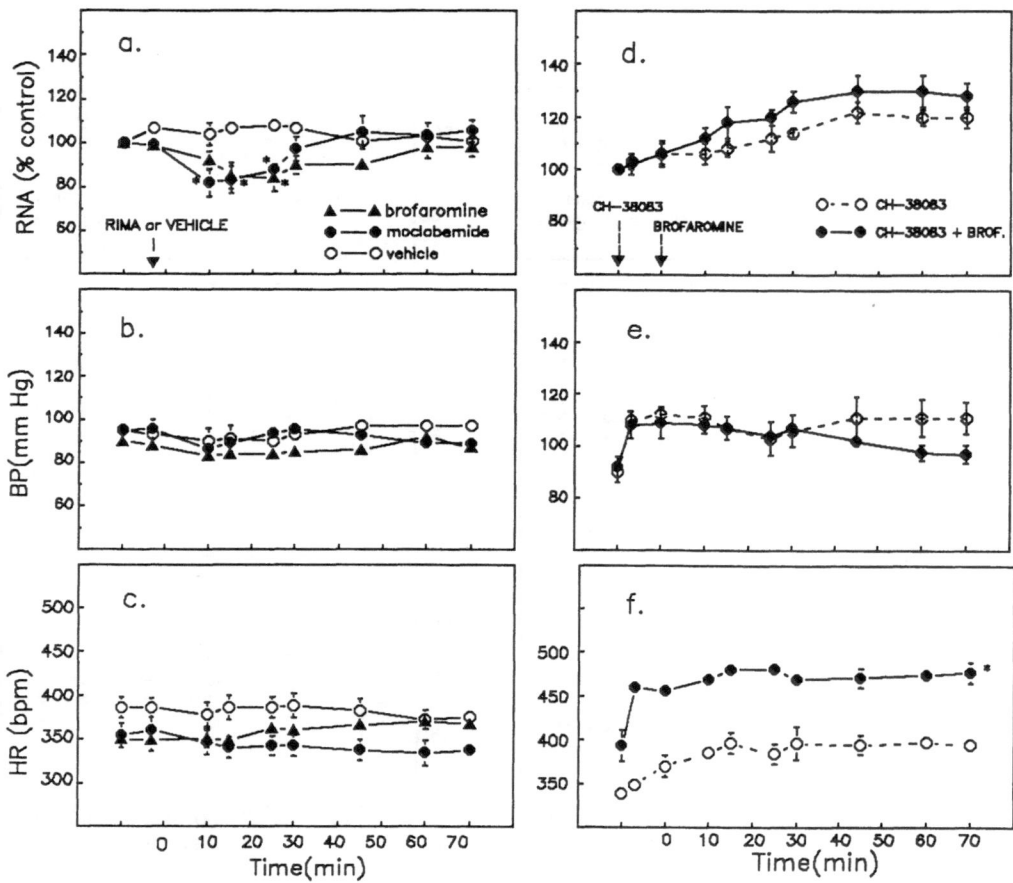

Fig. 2. Effect of acute treatment with moclobemide (10 mg/Kg, IP; n = 6) or brofaromine (10 mg/Kg, IP; n = 6) on **a** RNA, **b** BP, **c** HR and of acute treatment with the
same dose of brofaromine after pretreatment with CH-38083 (0.3 mg/Kg, IV; n = 5) on
d RNA, **e** BP and **f** HR

minimal value of 80% was recorded 15 min after drug administration. Nerve
activity returned to control value within 60 min (Fig. 2a). Acute treatment
with brofaromine did not affect BP or HR (Figs. 2b and c).

The effect of brofaromine was studied in rats pretreated with the selective α_2-adrenoceptor antagonist CH-38083, a derivative of yohimbine,
devoid of 5-HT receptor mediated sympatholytic effect (McCall et al., 1991;
Vizi et al., 1986). Pretreatment of rats with CH-38083 induced a significant
increase in RNA, BP and HR and abolished the inhibitory effect of brofaromine on RNA (Figs. 2d–f). These results indicate an involvement of α_2
adrenergic receptors in the central sympathoinhibitory effect of brofaromine.

The inhibitory effect of clorgyline and RIMA on the recorded parameters may be explained as follows: the increased axoplasmatic level of NA
and 5-HT in brain neurons, even if not entirely available for synaptic
release by exocytosis, may activate synaptic receptors following passive
diffusion or active transport over the synaptic membrane (Finberg and

Youdim, 1983; Paton, 1973). Moreover, the increased axoplasmatic NA concentration causes a reduction in its inward transport by the "uptake I" process (Graefe et al., 1971). The central sympathoinhibitory effect of the drugs may, therefore, result from the activation of α_2 adrenoceptors at noradrenergic neurons in centers such as C1, A1, A2 located in the medulla and involved in CV regulation (Brown and Guyenet, 1984; Guyenet et al., 1989). The sympathoinhibitory effect, following MAO inhibition, may also be a result of serotonergic neuronal activation. Conflicting results on the CV effects of centrally administered 5-HT were reported in early works. These may be due to use of various animal species (conscious or anaesthetized) and the dose and locus of drug administration (Kuhn et al., 1980; Ashkenazi et al., 1983). Later works, including our unpublished data on CV effects of fluvoxamin, show that in normotensive rats, increased central 5-HT concentration has a predominant sympathoinhibitory effect, leading to hypotension and bradycardia (Takahashi, 1985; Coote et al., 1987). Moreover, central and systemic administration of 8-OH-DPAT, a 5-HT_{1A} selective agonist, results in hypotension and bradycardia which can be abolished by pretreatment with an α_2 adrenergic receptor blocking agent (Fozard et al., 1987).

A ganglion blocking effect of the MAOI cannot be ruled out. Such an effect could contribute to the hypotensive action of the drugs, although no reports of a ganglionic action of selective MAO-A inhibitors, in clinically relevant doses, exist.

The reversible nature of RIMA's action is probably responsible for the only transient central sympathoinhibitory effect, recorded in our experiments, after treatment with moclobemide and brofaromine (Da Prada et al., 1989; Moeller et al., 1991). The lack of change in BP and HR may be due to peripheral pressor and tachycardic tendencies mediated by activation of α_1, α_2 and β_1 postsynaptic adrenoceptors which balance the transient reduced central sympathetic output.

The 5-HT uptake inhibitory properties of brofaromine may also contribute to its central sympathoinhibitory effect (Waldmeier and Staecklin, 1989).

In conclusion, our results are in accord with a central sympatholytic action of MAO-A inhibitors caused by an elevation of the synaptic level of NA and probably 5-HT at central neurons in areas involved in CV regulation. Alpha-2 adrenergic receptors are involved in the central sympathoinhibitory effect, caused in normotensive rats by clorgyline, moclobemide and brofaromine. The different time course displayed by the sympathoinhibitory action of different drugs may be explained by the reversible nature of the binding of RIMA to the enzyme vs. irreversible binding of clorgyline. The present data are in accord with clinical reports of a lack of hypotensive complications in patients treated with RIMA (Gasic et al., 1983; Korn et al., 1988) in contrast to severe hypotension, including orthostatic hypotension, reported in patients treated with irreversible MAO-A inhibitors. These results confirm again the advantages of RIMA as safer and better tolerated antidepressant drugs.

References

Ashkenazi R, Finberg JPM, Youdim MBH (1983) Effect of LM5008, a selective inhibitor of 5-hydroxy-tryptamine uptake, in blood pressure and responses to sympathomimetic amines. Br J Pharmacol 79: 915–922

Brown DL, Guyenet PG (1984) Cardiovascular neurons of brain stem with projections to the spinal cord. Am J Physiol 247: R1009–1016

Coote JH, Dalton DW, Feniuk FW, Humphrey PP (1987) The central site of sympatho-inhibitory action of 5-hydroxy tryptamine in the cat. Neuropharmacology 26: 147–154

Da Prada M, Kettler R, Burkard WP, Haefely WE (1984) Moclobemide, an anti-depressant with short-acting MAO-A inhibition: brain catecholamines and tyramine pressor effects in rats. In: Tipton KF, Dostert P, Strolin-Benedetti M (eds) Monoamine oxidase and disease. Prospect for therapy with reversible inhibitors. Academic Press, London, pp 137–154

Da Prada M, Kettler R, Keller HH, Burkard WP, Muggli-Maniglio D, Haefely WE (1989) Neurochemical profile of moclobemide, a short-acting and reversible inhibitor of monoamine oxidase type A. J Pharmacol Exp Ther 248: 400–414

Denney RM, Denney CB (1985) An update on the identity crisis of monoamine oxidase: new and old evidence for the independence of MAO A and MAO B. Pharmacol Ther 30: 227–259

Finberg JPM, Youdim MBH (1983) Selective MAO A and B inhibitors: their mechanism of action and pharmacology. Neuropharmacology 22: 441–446

Finberg JPM, Youdim MBH (1988) Potentiation of tyramine pressor responses in conscious rats by reversible inhibitors of monoamine oxidase. J Neural Transm [Suppl] 26: 11–16

Fozard JR, Mir AK, Middlemiss DN (1987) Cardiovascular response to 8-hydroxy-2-(di-n-propylamino) tetraline (8-OH-DPAT) in the rat: site of action and pharmacological analysis. J Cardiovasc Pharmacol 9: 328–347

Fuentes JA, Ordaz A, Neff NH (1979) Central mediation of the antihypertensive effect of pargyline in spontaneous hypertensive rats. Eur J Pharmacol 57: 21–27

Gasic S, Korn A, Eichler HC, Oberhummer I, Zapotoczky HG (1983) Cardiocirculatory effects of moclobemide (RO 11-1163), a new reversible, short acting MAO-inhibitor with preferential type A inhibition, in healthy volunteers and depressive patients. Eur J Clin Pharmacol 25: 173–177

Goridis C, Neff NH (1971) Monoamine oxidase in sympathetic nerves: a transmitter specific enzyme type. Br J Pharmacol 43: 814–818

Graefe K-H, Boenisch H, Trendelenburg U (1971) Time-dependent changes in neuronal net uptake of noradrenaline after pretreatment with pargyline and/or reserpine. Naunyn Schmiedebergs Arch Pharmacol 271: 1–28

Guyenet PG, Haselton JR, Sun MK (1989) Sympathoexcitatory neurons of the rostroventral medulla and the origin of the sympathetic vasomotor tone. In: Cirielo J, Caverson MM, Polosa C (eds) Progress in brain research, vol 91. Elsevier, p 105

Haefely W, Burkard WP, Cesura AM, Kettler R, Lorez HP, Martin JR, Richards JG, Scherschlicht R, Da Prada M (1992) Biochemistry and pharmacology of moclobemide, a prototype RIMA. Psychopharmacology 106: S6–S14

Johnston JP (1968) Some observations upon a new inhibitor of monoamine oxidase in brain tissue. Biochem Pharmacol 17: 614–627

Korn A, Da Prada M, Raffesberg W, Gasic S, Eichler HG (1988) The effect of moclobemide, a new reversible monoamine oxidase inhibitor, on absorption and pressor effect of tyramine. J Cardiovasc Pharmacol 11: 17–23

Kuhn DM, Wolf WA, Lovenberg W (1980) Review of the role of central serotonergic neural system in blood pressure regulation. Hypertension 2: 243–255

Lavian G, Dibona G, Finberg JPM (1991) Inhibition of sympathetic nerve activity by administration of the tricyclic antidepressant desipramine. Eur J Pharmacol 194: 153–159

Mc Call RB, Harris LT, King KA (1991) Sympatholytic action of yohimbine mediated by 5-HT$_{1A}$ receptors. Eur J Pharmacol 199: 263–265

Moeller HJ, Wendt G, Waldmeier P (1991) Brofaromine — a selective, reversible and short-acting MAO-A inhibitor: Review of the pharmacological and clinical findings. Pharmacopsychiatry 24: 50–54

Neff NH, Goridis C (1972) Neuronal monoamine oxidase: specific enzyme types and their rate of formation. In: Costa E, Sandler M (eds) Monoamine oxidase and the monoamine oxidase inhibitor drugs—new vistas. Raven Press, New York, p 307

Neff NH, Yang NYT (1974) Another look at the monoamine oxidase inhibitor drugs. Life Sci 14: 2061–2074

Neff NH, Yang HYT, Fuentes JA (1974) The use of selective monoamine metabolism in brain. In: Usdin E (eds) Neuropharmacology of monoamines and their regulatory enzymes. Raven Press, New York, p 49 (Adv Biochem Psychopharmacol, vol 12)

Paton DM (1973) Mechanisms of efflux of noradrenaline from adrenergic nerves in rabbit atria. Br J Pharmacol 49: 614–627

Takahashi H (1985) Cardiovascular and sympathetic response to intracarotid and intravenous injection of serotonin in rats. Arch Pharmacol 329: 222–226

Vizi ES, Harsing LG JR, Gaal J, Kapocsi J, Bernath S, Somogyi GT (1986) CH-38083, a selective, potent antagonist of alpha-2 adrenoceptors. J Pharmacol Exp Ther 238: 701–706

Waldmeier PC, Staecklin K (1989) The reversible MAO inhibitor, brofaromine, inhibits serotonin uptake in vivo. Eur J Pharmacol 169: 197–204

Authors' address: Dr. G. Lavian, Department of Pharmacology, Bruce Rappaport Faculty of Medicine, Technion, Haifa, Israel.

J Neural Transm (1994) [Suppl] 41: 115–122

Chronic administration of the antidepressant phenelzine and its N-acetyl analogue: effects on GABAergic function

K. F. McKenna[1], **D. J. McManus**[1], **G. B. Baker**[1,2], and **R. T. Coutts**[1,2]

Neurochemical Research Unit, [1] Department of Psychiatry and
[2] Faculty of Pharmacy and Pharmaceutical Sciences, University of Alberta,
Edmonton, Alberta, Canada

Summary. The MAO inhibitor phenelzine (2-phenylethylhydrazine; PLZ) is used widely in psychiatry for the treatment of depression and panic disorder. Its N-acetyl metabolite, N^2-acetylphenelzine (N^2AcPLZ) is a reasonably potent nonselective inhibitor of monoamine oxidase (MAO) that causes elevation in brain levels of the biogenic amines. In the studies reported here, PLZ (0.05 mmol/kg/day), N^2AcPLZ (0.10 mmol/kg/day) or vehicle were administered to male rats for 28 days s.c. with Alzet minipumps, and their effects on GABAergic function were examined. Whole brain concentrations of γ-aminobutyric acid (GABA) were significantly elevated in the PLZ but not in the N^2AcPLZ-treated group. PLZ was found to inhibit the anabolic enzyme glutamic acid decarboxylase (GAD) and, to a greater extent, the catabolic enzyme GABA transaminase (GABA-T). The results of these investigations suggest that the free hydrazine moiety in PLZ is crucial to producing the elevated levels of GABA, probably through inhibition of GABA-T. Despite the considerable increase in whole brain GABA levels in the PLZ-treated rats, there were no significant differences in $GABA_A$ or benzodiazepine receptor binding parameters (K_D or B_{max}) between the groups as measured using ^{3}H-muscimol and ^{3}H-flunitrazepam in radioligand binding assays.

Introduction

Phenelzine (PLZ), a monoamine oxidase (MAO) inhibitor, is efficacious in the treatment of a wide spectrum of affective and anxiety disorders. Previous attempts to explain the mechanisms of the therapeutic action of PLZ have emphasized inhibition of MAO-A and -B, resulting in increased brain levels of the amine neurotransmitters noradrenaline (NA), dopamine (DA), and 5-hydroxytryptamine [serotonin; 5-HT] (Murphy et al., 1987) and the trace amines β-phenylethylamine, tyramine and tryptamine (Philips and Boulton, 1979). Recent research has suggested actions of both antipanic and antidepressant medications on the inhibitory neurotransmitter γ-aminobutyric acid (GABA) may be contributing to their therapeutic effectiveness (Breslow et al., 1989). In clinical studies of depressed patients, reduced cerebrospinal and plasma levels of GABA have been detected (Petty et al., 1990) and the GABA agonists progabide, baclofen, muscimol and fengabine have been reported to possess antidepressant activity (Lloyd et al., 1989; Nielsen et al., 1991). Several of the benzodiazepines, including clonazepam and alprazolam, and the GABA agonist baclofen are effective in blocking panic attacks (Breslow et al., 1989). Although there is considerable controversy in this area, several antidepressants have been reported to cause an up-regulation of $GABA_B$ receptors (Lloyd et al., 1989) and a down-regulation of the $GABA_A$ receptor-benzodiazepine receptor-chloride ionophore complex (Suranyi-Cadotte et al., 1990).

PLZ has been reported to have a potent elevating effect on brain GABA (Popov and Matthies, 1969; Perry and Hansen, 1973; Baker et al., 1991) and alanine (ALA) levels (Wong et al., 1990a). In experiments examining the acute effects of PLZ in rodents, an intact hydrazine group was found necessary to affect GABA and ALA levels but not to inhibit MAO; N^2-acetylphenelzine (N^2AcPLZ) was found to be a relatively potent MAO inhibitor that elevated brain levels of NA, DA and 5-HT but had no effect on brain GABA or ALA levels (McKenna et al., 1991). The studies described in the present paper examined the effects of chronic (28 day) administration of PLZ and N^2AcPLZ on whole brain GABA levels in rats, and the activity of the enzymes glutamic acid decarboxylase (GAD) and GABA-transaminase (GABA-T). The effects of drug treatment on the affinity (K_D) and density (B_{max}) of $GABA_A$ and benzodiazepine (BZD) receptors were also studied using ^{3}H-muscimol and ^{3}H-flunitrazepam respectively as radioligands for in vitro binding studies.

Materials and methods

Subjects and drugs

Phenelzine sulfate (PLZ) was obtained from Sigma Chemical Company (St. Louis, MO, USA), and N^2-acetylphenelzine (N^2AcPLZ) was synthesized from phenylacetaldehyde and acetylhydrazine by a previously reported method (Danielson et al., 1984). Male Sprague-Dawley rats were randomly assigned to treatment with PLZ (0.05 mmol/kg/day, sulfate salt), N^2AcPLZ (0.10 mmol/kg/day, free base), or distilled water vehicle. The doses of the drugs were chosen on the basis of findings in our laboratories using acute administration of the drugs (McKenna et al., 1991). Drugs and vehicle were administered with osmotic mini-pumps (Alzet Model 2ML2, Alza Corp., Palo Alto, CA). Following 28 days of drug treatment animals were sacrificed by guillotine decapitation and the whole brain was rapidly removed and frozen solid in isopentane on solid carbon dioxide. The tissues were stored at −80°C until the time of analysis.

Analysis

An electron-capture gas chromatographic (GC) assay developed by Wong et al. (1990b) was used for the analysis of GABA concentrations in tissue samples. GABA-T activity was measured with a modification of the procedure of Sterri and Fonnum (1978) in which ^{3}H-GABA is incubated with tissue homogenates and the products extracted with a liquid anion exchanger. The activity of GAD was measured using a radiochemical assay in which $^{14}CO_2$ is formed from [1-^{14}C]-glutamate (Albers and Brady, 1958). ^{3}H-Muscimol was used as radiolabel to determine the number (B_{max}) and affinity (K_D) of GABA$_A$ binding sites in rat brain homogenates; the procedure used was a modification of that of Schwartz et al. (1986), as described by Bristow and Martin (1990). Aliquots of the final membrane suspension were incubated for 60 min at 0°C in tubes containing ^{3}H-muscimol (20 Ci/mmol) [5 nM] and concentrations of unlabelled muscimol ranging from 2 to 64 nM. Non-specific binding was defined using unlabelled GABA (1 mM) and represented approximately 10–20% of total binding. ^{3}H-Flunitrazepam was used to define the K_D and B_{max} of benzodiazepine binding sites in rat cortex and hippocampus (Kimber et al., 1987). ^{3}H-Flunitrazepam at concentrations ranging from 0.25 to 7.5 nM was used as the radioligand and nonspecific binding was defined using clonazepam (3 μM). To examine GABA-facilitation of ^{3}H-flunitrazepam binding, GABA (100 μM) and NaCl (150 mM) were added to a series of tubes.

Statistics

Data were analyzed by ANOVA followed by the Newman Keuls multiple comparison test with a probability value of $p < 0.05$ used to establish statistical significance.

Results

The effects of chronic (28 d) administration of PLZ and N^{2}AcPLZ on brain levels of GABA are shown in Fig. 1. PLZ induced sustained elevations in the levels of GABA while N^{2}AcPLZ had no effect.

The effects of PLZ treatment on the activity of GABA-T and GAD are shown in Table 1. Chronic PLZ treatment induced a significant decrease in the activity of GAD and GABA-T compared to N^{2}AcPLZ- and vehicle-treated animals. The effects of PLZ on GABA-T (33% inhibition) were more pronounced than on GAD (18% inhibition). The net effect appears to be an elevation in levels of GABA.

The effects of chronic (28 d) administration of PLZ, N^{2}AcPLZ or vehicle (saline) on GABA$_A$ high-affinity and benzodiazepine receptor binding parameters (K_D and B_{max}) in rat brain cortex and hippocampus were studied using the GABA$_A$ agonist ^{3}H-muscimol and the benzodiazepine agonist ^{3}H-flunitrazepam as radioligands respectively. There were no significant differences in GABA$_A$ binding parameters (K_D or B_{max}) between the treatment groups (Table 2). The binding parameters are similar to those reported by other investigators (Ito et al., 1988; Kimber et al., 1987).

No significant differences in the benzodiazepine binding parameters between the treatment groups were observed in either cortex or hippocampus (Tables 3 and 4). The addition of GABA to the homogenates increased the affinity of the benzodiazepine receptor (lower K_D) for ^{3}H-flunitrazepam and increased the apparent density of binding sites (increase in B_{max}). This is in keeping with the hypothesis that GABA allosterically modulates the benzodiazepine binding site (Martin and Candy, 1978).

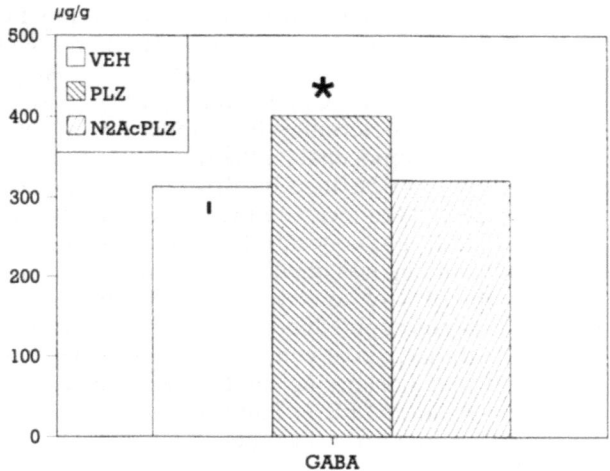

Fig. 1. Effects of 28 d administration of PLZ (0.05 mmol/kg/d) and N^{2}AcPLZ (0.10 mmol/kg/d) on whole brain levels of GABA. Results are expressed as mean concentrations (μg/g), n = 6. *p < 0.05, a significant difference from control. The line in the clear box represents the standard error derived from the MSerror term

Table 1. Effects of chronic (28 day) administration of PLZ (0.05 mmol/kg/day) and N^{2}AcPLZ (0.10 mmol/kg/day) on whole brain activity of glutamic acid decarboxylase (GAD) and GABA-transaminase [GABA-T] (μmol/g of tissue/h ± SEM)

	VEH	PLZ	N^{2}AcPLZ
GAD	10.6 ± 0.8	8.7 ± 0.2*	11.7 ± 0.6
GABA-T	28.3 ± 1.5	19.0 ± 1.0**	25.4 ± 0.9

n = 6
**p < 0.01 compared to VEH and N^{2}AcPLZ, *p < 0.05 compared to VEH and N^{2}AcPLZ
There were no significant differences between VEH and N^{2}AcPLZ

Table 2. Effects of chronic (28 d) administration of PLZ (0.05 mmol/kg/day) and N^{2}AcPLZ (0.10 mmol/kg/day) on ^{3}H-muscimol binding to membrane fractions from rat brain cortex

	K_D (nM)	B_{max} (fmol/mg protein)
VEH	12.9 ± 0.9	834 ± 77
PLZ	12.1 ± 0.8	792 ± 51
N^{2}AcPLZ	12.2 ± 0.7	831 ± 35

Results expressed as mean ± SEM (n = 6)
There are no significant effects of treatment on the K_D or B_{max}

Table 3. Effects of chronic (28 d) administration of PLZ (0.05 mmol/kg/day) and N^{2}AcPLZ (0.10 mmol/kg/day) on ^{3}H-flunitrazepam binding to membrane fractions from rat brain cortex

	K_D (nM)	B_{max} (fmol/mg protein)	+ GABA (10^{-4} M)	
			K_D (nM)	B_{max} (fmol/mg protein)
VEH	1.43 ± 0.08	971 ± 48	0.83 ± 0.04	1,108 ± 32
PLZ	1.45 ± 0.05	1,015 ± 29	0.85 ± 0.04	1,238 ± 36
N^{2}AcPLZ	1.44 ± 0.10	986 ± 36	0.80 ± 0.04	1,165 ± 35

Results expressed as mean ± SEM (n = 6)
The two columns on the right demonstrate the effects of adding GABA to the incubation mixture

Table 4. Effects of chronic (28 d) administration of PLZ (0.05 mmol/kg/day) and N^{2}AcPLZ (0.10 mmol/kg/day) on ^{3}H-flunitrazepam binding to membrane fractions from rat brain hippocampus

	K_D (nM)	B_{max} (fmol/mg protein)	+ GABA (10^{-4} M)	
			K_D (nM)	B_{max} (fmol/mg protein)
VEH	1.32 ± 0.05	795 ± 47	0.48 ± 0.02	969 ± 50
PLZ	1.39 ± 0.08	826 ± 63	0.54 ± 0.02	1,039 ± 81
N^{2}AcPLZ	1.35 ± 0.11	812 ± 65	0.49 ± 0.07	985 ± 50

Results expressed as mean ± SEM (n = 6)
The two columns on the right demonstrate the effects of adding GABA to the incubation mixture

Discussion

Chronic administration of PLZ, but not N^{2}AcPLZ, elevated whole brain levels of GABA. These findings are consistent with acute studies in which only treatment with PLZ elevated GABA levels (McKenna et al., 1991). In keeping with the effects on GABA, only treatment with PLZ had an effect on enzyme activity, decreasing GABA-T to 67% of vehicle activity. Presumably the greater effect of PLZ on the transaminase enzyme than on the anabolic enzyme (GAD) is responsible for the elevation in GABA levels observed. Investigations in our laboratory have examined the effects of several other antidepressants on GABA levels and GABA-T activity and found that chronic administration of tranylcypromine, imipramine or desmethylimipramine had no effect (McManus et al., 1992).

The presence of an unsubstituted hydrazine group apppears to be crucial for PLZ to affect GABA levels and enzyme activity. Substitution of an acetyl group at the N2 position (N^{2}AcPLZ) blocks the elevation of GABA but has minimal effects on the ability of PLZ to inhibit MAO. Likewise, the substituted hydrazines iproniazid and nialamide are MAO inhibitors that do not change GABA levels or GABA-T activity after acute administration (Todd et al., 1991). As GABA-T requires pyridoxal 5'-phosphate as a cofactor, the ability of PLZ to bind to this cofactor, forming an inactive hydrazone complex, may be the mechanism through which PLZ elevates GABA levels (Yu and Boulton, 1992).

A series of radioligand binding experiments were conducted to determine if the sustained elevation in GABA levels was associated with changes in $GABA_A$ or benzodiazepine binding parameters (K_D and B_{max}). Previous studies from our laboratories had demonstrated no change in $GABA_B$ receptor function after chronic administration of PLZ (McManus and Greenshaw, 1991a,b). No significant differences among treatment groups were noticed with regard to effects on $GABA_A$ receptors or benzodiazepine receptors in the present study. The $GABA_A$ high affinity binding site in cortical tissue was studied initially. The protocol did not allow characterization of the low affinity binding site, probably due to rapid dissociation of the ligand from the receptor during the washing steps. There were no significant differences among the treatment groups with regard to the high affinity site. The benzodiazepine receptor was studied in cortical and hippocampal tissue and the modulating effects of GABA were also examined. We were interested in the modulating effects of GABA as this would reflect the function of the low affinity $GABA_A$ binding site. It is generally thought that the lower affinity $GABA_A$ site is functionally linked to the benzodiazepine receptor (Unnerstahl et al., 1981). If there were no differences in the benzodiazepine binding parameters produced by the drugs, but a difference appeared with the addition of GABA it would be suggestive that the $GABA_A$ low affinity site was being altered by the drug treatment. In both cortical and hippocampal tissue, there were no significant differences in benzodiazepine receptor binding parameters between the drug-treated and vehicle-treated groups even after the addition of GABA. As reported previously, the addition of GABA and NaCl to the binding suspension resulted in a significant increase in affinity (decrease in K_D) of the benzodiazepine receptor for flunitrazepam and a significant increase in the density of binding sites [B_{max}] (Martin and Candy, 1978), but this was uniform among the treatments in the present study.

In summary, the antidepressant/antipanic drug PLZ elevated brain GABA levels and inhibited GABA-T (and to a lesser extent, GAD) activity following chronic administration but did not affect $GABA_A$ or benzodiazepine receptor binding parameters. Further studies, e.g. on GABA-stimulated chloride uptake in synaptoneurosomes or using molecular biological techniques to investigate isoforms of $GABA_A$ receptor subunits, are warranted to understand more clearly the implications of the elevated brain GABA concentrations produced by this drug.

Acknowledgements

Funds were provided by the Medical Research Council of Canada and the Alberta Heritage Foundation for Medical Research. The authors wish to thank Drs. I. L. Martin and A. J. Greenshaw for their valuable advice and discussions on this work. They are grateful to Ms. S. Omura and Ms. R. Sherry-McKenna for help in preparing this manuscript.

References

Albers RW, Brady RO (1958) The distribution of glutamic decarboxylase in the nervous system of the rhesus monkey. J Biol Chem 234: 296–298

Baker GB, Wong JTF, Yeung JM, Coutts RT (1991) Effects of the antidepressant phenelzine on brain levels of γ-aminobutyric acid (GABA). J Affect Dis 21: 207–211

Breslow MF, Faukhauser MP, Potter RL, Meredith KE, Misiaszek J, Hope DG (1989) Role of γ-aminobutyric acid in antipanic drug efficacy. Am J Psychiatry 146: 353–356

Bristow DR, Martin IL (1990) Biochemical characterization of an isolated and functionally reconstituted γ-aminobutyric acid/benzodiazepine receptor. J Neurochem 54: 751–761

Danielson TJ, Coutts RT, Baker GB, Rubens M (1984) Studies in vivo and in vitro on N-acetylphenelzine. Proc West Pharmacol Soc 27: 507–510

Ito Y, Lim DK, Hoskins B, Ho IK (1988) Bicuculline up-regulation of $GABA_A$ receptors in rat brain. J Neurochem 51: 145–152

Kimber JR, Cross JA, Horton RW (1987) Benzodiazepine and $GABA_A$ receptors in rat brain following chronic antidepressant drug administration. Biochem Pharmacol 36: 4175–4176

Lloyd KG, Zivkovic B, Scatton B, Morselli PL, Bartholini G (1989) The GABAergic hypothesis of depression. Prog Neuropsychopharmacol Biol Psychiatry 13: 341–351

Martin IL, Candy JM (1978) Facilitation of benzodiazepine binding by sodium chloride and GABA. Neuropharmacology 17: 993–998

McKenna KF, Baker GB, Coutts RT (1991) N^2-Acetylphenelzine: effects on rat brain GABA, alanine and biogenic amines. Naunyn Schmiedebergs Arch Pharmacol 343: 478–482

McManus DJ, Baker GB, Martin IL, Greenshaw AJ, McKenna KF (1992) Effects of the antidepressant/antipanic drug phenelzine on GABA concentrations and GABA-transaminase activity in rat brain. Biochem Pharmacol 43: 2486–2489

McManus DJ, Greenshaw AJ (1991a) Differential effects of chronic antidepressants in behavioural tests of β-adrenergic and $GABA_B$ receptor function. Psychopharmacology (Berl) 103: 204–208

McManus DJ, Greenshaw AJ (1991b) Differential effects of antidepressants on $GABA_B$ and β-adrenergic receptors in rat cerebral cortex. Biochem Pharmacol 42: 1525–1528

Murphy DL, Aulakh CS, Garrick NA, Sunderland T (1987) Monoamine oxidase inhibitors as antidepressants: implications for the mechanism of action of antidepressants and the psychobiology of the affective disorders and some related disorders. In: Meltzer HY (ed) Psychopharmacology: the third generation of progress. Raven Press, New York, pp 545–552

Nielsen EB, Suzdak PD, Andersen KE, Knutsen LJS, Sonnewald U, Braestrup C (1991) Characterization of tiagabine (NO-328), a new potent and selective GABA uptake inhibitor. Eur J Pharmacol 196: 257–266

Perry TL, Hansen S (1973) Sustained drug-induced elevation of brain GABA in the rat. J Neurochem 21: 1167–1175

Petty F, Kramer GL, Dunnam D, Rush AJ (1990) Plasma GABA in mood disorders. Psychopharmacol Bull 26: 157–161

Philips SR, Boulton AA (1979) The effect of monoamine oxidase inhibitors on some arylalkylamines in rat striatum. J Neurochem 33: 159–167

Popov N, Matthies H (1969) Some effects of monoamine oxidase inhibitors on the metabolism of γ-aminobutyric acid in rat brain. J Neurochem 16: 899–907

Schwartz RD, Skolnick P, Seale TW, Paul SM (1986) Demonstration of GABA-barbiturate-receptor-mediated chloride transport in rat brain synaptoneurosomes: functional assay of GABA receptor coupling. Adv Biochem Psychopharmacol 41: 33–49

Sterri SH, Fonnum F (1978) Isolation of organic anions by extraction with liquid anion exchangers and its application to micromethods for acetylcholinesterase and 4-aminobutyrate aminotransferase. Eur J Biochem 91: 215–222

Suranyi-Cadotte BE, Bodnoff SR, Welner SA (1990) Antidepressant-anxiolytic interaction: involvement of the benzodiazepine-GABA and serotonin systems. Prog Neuropsychopharmacol Biol Psychiatry 14: 633–654

Todd KG, Yamada N, Takahashi S, McKenna KF, Baker GB, Coutts RT (1991) Comparison of the effects of 2-phenylethylhydrazine (phenelzine) and some substituted hydrazine MAO inhibitors on monoaminergic and GABAergic mechanisms in rat brain. Proc. 15th Ann Meet Can Coll Neuropsychopharmacol, Saskatoon, Canada, p W-13

Unnerstahl JR, Kuhar MJ, Niehoff DL, Palacios JM (1981) Benzodiazepine receptors are coupled to a subpopulation of γ-aminobutyric acid (GABA) receptors: evidence from a quantitative autoradiographic study. J Pharmacol Exp Ther 218: 797–804

Wong JTF, Baker GB, Coutts RT, Dewhurst WG (1990a) Long-lasting elevation of alanine in brain produced by the antidepressant phenelzine. Brain Res Bull 25: 179–181

Wong JTF, Baker GB, Coutts RT (1990b) A rapid, sensitive assay for γ-aminobutyric acid in brain using electron-capture gas chromatography. Res Commun Chem Path Pharmacol 70: 115–124

Yu PH, Boulton AA (1992) A comparison of the effect of brofaromine, phenelzine and tranylcypromine on the activities of some enzymes involved in the metabolism of different neurotransmitters. Res Commun Chem Path Pathol 16: 141–153

Authors' address: Dr. K. F. McKenna, Neurochemical Research Unit, Department of Psychiatry, University of Alberta, Edmonton, Alberta, Canada, T6G 2B7.

J Neural Transm (1994) [Suppl] 41: 123–125

Modification of cerebral cortical noradrenaline release by chronic inhibition of MAO-A

J. P. M. Finberg[1], **K. Pacak**[2], **D. S. Goldstein**[2], and **I. J. Kopin**[2]

[1] Bruce Rappaport Faculty of Medicine, Technion, Haifa, Israel
[2] N.I.N.D.S., N.I.H., Bethesda, Maryland, U.S.A.

Summary. Chronic treatment of rats with clorgyline (1 mg/kg i.p. daily for 21 days) caused a highly significant increase in the concentration of noradrenaline in microdialysate from the frontal cortex of the awake animal. Acute (one injection, 2 mg/kg) or subacute (1 mg/kg daily for 3 days) treatment did not lead to a significant increase in microdialysate noradrenaline. Concentrations of deaminated metabolites (DPHG, MHPG, DOPAC) in the microdialysate decreased with time of treatment, reaching a minimum after 21 days.

Introduction

One of the basic tenets of the amine hypothesis of depression is that administration of antidepressant drugs, including the MAO inhibitors, results in an increase in synaptic levels of noradrenaline (NA) and 5-HT. Although brain and peripheral tissue levels of noradrenaline are elevated following MAO inhibition, most biochemical studies have been unable to demonstrate an increase in neuronal NA release (see Youdim and Finberg, 1985). Whole body NA release is also reduced in patients under clorgyline treatment (Linnoila et al., 1982). We have previously studied NA release from peripheral sympathetic nerves as a model of release from CNS noradrenergic nerves (Finberg and Kopin, 1986; Ari and Finberg, 1992). These studies have shown that endogenous NA release is enhanced following chronic (3 weeks) but not acute (one dose) selective inhibition of MAO-A with clorgyline. The purpose of the present study was to see whether cerebral cortical NA release, as measured by microdialysis technique, was also enhanced by chronic, selective inhibition of MAO-A. We have studied the effect of a small, selective, dose of clorgyline, given over 3 weeks, since this is the period of time required for down-regulation of β-adrenoceptors in the cerebral cortex of rats, and for appearance of antidepressant effect in patients. The full results of this study are published elsewhere (Finberg et al., 1993).

Methods

Rats (250–350 g) were treated acutely (1 × 2 mg/kg), sub-acutely (3 × 1 mg/kg) or chronically (21 × 1 mg/kg) with clorgyline, or physiological saline (all injections i.p.). On the penultimate day, the animals were anesthetised with pentobarbitòne, and a 4 mm concentric cannula was implanted in the frontal cortex under stereotactic control. The cannulae were perfused with artificial c.s.f. (1 μl/min), and the rats were kept in a cylindrical plexiglass cage overnight. On the final treatment day, collections of microdialysate were commenced from the awake animals 2 hours after the final injection of clorgyline (or the single injection in the acute treatment group). The content of NA and metabolites in the microdialysate was determined by HPLC, following alumina adsorption.

Results and discussion

The basal NA concentration in cortical microdialysate was slightly, but not significantly, increased following the acute and subacute treatments, but was markedly and significantly increased ($P < 0.001$) after the chronic treatment (79 ± 7 fmol/30 min in the saline control as compared to 297 ± 31 fmol/30 min in the chronic treatment group). Together with the increase in microdialysate NA, the concentrations of the deaminated metabolites, DHPG, MHPG and DOPAC, were reduced to a greater extent following the chronic as compared with the acute and subacute treatments.

The findings in this study are consistent with an increased NA release from cortical neurons following chronic, but not acute or subacute, clorgyline treatment. The degree of inhibition of MAO-A and MAO-B, as measured by the ex vivo assay using ^{14}C-5-HT and ^{14}C-phenylethylamine was similar in all animals treated with clorgyline, i.e. 95% or more inhibition of MAO-A with little or no inhibition of MAO-B. On the other hand, the concentration of deaminated NA metabolites was significantly less following the chronic treatment as compared to acute and subacute. One possible factor in explaining the increased release could therefore be a more complete inhibition of MAO-A with time. Cassis et al. (1986) demonstrated that the difference between 95% and 100% inhibition of MAO on metabolite formation may be profound. Such a change would not have been detected by the ex vivo assay.

How does the more complete inhibition of MAO with time affect neuronal NA release? NA level in the extracellular space is the product of neuronal firing rate, release per pulse and clearance from the extracellular space. One possibility is that more complete MAO inhibition is accompanied by reduction in net reuptake (Graefe et al., 1971). Release per pulse could also be enhanced. Another possiblity is that neuronal discharge rate is reduced by acute treatment but is normal or elevated in the chronic treatment group, reflecting an adaptation of the cells to modulatory input. Once such adaptive change could be the down-regulation of inhibitory somato-dendritic α_2-adrenoceptors. The firing rate of locus coeruleus cells, however, was reduced to the same extent by acute and chronic clorgyline treatment in chloral hydrate-anesthetised rats (Blier and De Montigny, 1985).

It is interesting that acute clorgyline administration in anesthetised rats, at the same dose and time used in the present study, effectively reduces renal nerve activity and blood pressure (Lavian et al., this symposium). Thus in the central, as opposed to the peripheral, nervous system, acute MAO inhibition produces immediate effects on noradrenergic neuronal transmission. One explanation for this difference could be the limited diffusability of NA from extracellular space into blood vessels in the CNS as compared with the periphery.

Further study of the detailed mechanisms behind the changes detected in this study will have to include some measurement of neuronal discharge rate, but the results of our current study show that chronic MAO inhibition is accompanied by a large increase in cortical NA in the extracellular space, which presumably indicates a similar increase in synaptic NA. These results are, therefore, compatible with the hypothesis that adrenoceptor down-regulation in the cortex is the result of a compensatory response to elevated NA concentrations.

References

Ari G, Finberg JPM (1992) Effects of clorgyline on the release of noradrenaline from rat vas deferens. Eur J Pharmacol 219: 89–96

Blier P, De Montigny C (1985) Serotoninergic but not noradrenergic neurons in rat central nervous system adapt to long-term treatment with monoamine oxidase inhibitors. Neuroscience 16: 949–955

Cassis L, Ludwig J, Grohmann M, Trendelenburg U (1986) The effect of partial inhibiton of monoamine oxidase on the steady-state rate of deamination of ^{3}H-catecholamines in two metabolizing systems. Naunyn Schmiedebergs Arch Pharmacol 333: 253–261

Finberg JPM, Kopin IJ (1986) Chronic clorgyline treatment enhances release of nor-epinephrine following sympathetic stimulation in the rat. Naunyn Schmiedebergs Arch Pharmacol 332: 236–242

Finberg JPM, Ari G, Lavian G, Hovevey-Sion D (1990) Modification of alpha-2 presynaptic receptor activity and catecholamine release following chronic MAO inhibition. J Neural Transm [Suppl] 32: 405–412

Finberg JPM, Pacak K, Kopin IJ, Goldstein DS (1993) Chronic inhibition of mono-amine oxidase type A increases noradrenaline release in rat frontal cortex. Naunyn Schmiedebergs Arch Pharmacol 347: 500–505

Graefe K-H, Bonisch H, Trendelenburg U (1971) Time-dependent changes in neuronal net uptake of noradrenaline after pretreatment with pargyline and/or reserpine. Naynyn Schmiedebergs Arch Pharmacol 271: 1–28

Linnoila M, Karoun F, Potter WZ (1982) Effect of low dose clorgyline on 24-hour binary monoamine excretion in patients with rapidly cycling bipolar affective disorder. Arch Gen Psychiatry 39: 513–516

Youdim MBH, Finberg JPM (1985) Monoamine oxidase inhibitor antidepressants. In: Grahame-Smith DG, Cowen PJ (eds) Psychopharmacology, vol 1. Excerpta Medica, Amsterdam, pp 38–70

Authors' address: Dr. J. P. M. Finberg, Bruce Rappaport Faculty of Medicine, Technion, Haifa, Israel.

J Neural Transm (1994) [Suppl] 41: 127–134

Comparisons of the actions of high and low doses of the MAO inhibitor tranylcypromine on 5-HT$_2$ binding sites in rat cortex

D. B. Goodnough and **G. B. Baker**

Neurochemical Research Unit, Department of Psychiatry, University of Alberta, Edmonton, Alberta, Canada

Summary. Tranylcypromine (TCP) is a commercially available antidepressant drug, and recent literature reports suggest that high doses of this drug may be particularly effective in treating refractory depression. Down-regulation of 5-HT$_2$ receptors in rat cortex is an effect produced after chronic administration of several antidepressants, and we have conducted a chronic study comparing low- and high-dose TCP in this regard. Male Sprague-Dawley rats were administered TCP (0.5 or 2.5 mg/kg/day) or vehicle (distilled water) via Alzet minipumps implanted subcutaneously in the dorsal thoracic area. Groups of rats were killed 4, 10 or 28 days after pump implantation and whole cortex was dissected out and utilized for preparation of a membrane fraction. Binding studies were performed with this fraction using ^{3}H-ketanserin as the radioligand. Down-regulation (decrease in B$_{max}$) of the 5-HT$_2$ binding site was observed in high-dose animals after 10 and 28 days but not after 4 days. Low-dose TCP had no effect on 5-HT$_2$ densities at any time interval. The affinity of ^{3}H-ketanserin for the 5-HT$_2$ site was not affected by either dose at any time interval. These results suggest that down-regulation of the 5-HT$_2$ site may contribute to the efficacy of high-dose TCP in the treatment of refractory depression.

Introduction

Tranylcypromine (TCP) is a nonselective monoamine oxidase (MAO) inhibitor that has been used for many years as an antidepressant. A dose of approximately 0.5 mg/kg/day of TCP is sufficient to produce a level of MAO inhibition normally required for antidepressant efficacy (Ferris et al., 1975; Robinson et al., 1978; Giller and Lieb, 1980; Giller et al., 1982). Amsterdam and Berwish (1989) reported that a much higher dose of TCP (90–170 mg, or approximately 1.3–2.4 mg/kg/day) was effective in treating refractory depressives. This treatment had been reported to be effective in earlier case reports (Robinson, 1983; Guze and Baxter, 1987; Pearlman, 1987).

5-HT$_2$ receptors have been implicated in the etiology of depression. They are down-regulated following chronic treatment with several antidepressants (Baker and Greenshaw, 1989; Eison et al., 1991; Lafaille et al., 1991) and they have been recently reported to be elevated in postmortem brain tissue from drug-free depressives (Yates et al., 1990). TCP, at doses much higher (on a mg/kg basis) than those used in the clinical setting, has been reported to cause 5-HT$_2$ receptor down-regulation (Kellar et al., 1981; Goodwin et al., 1984). The present study compares the effects of low (0.5 mg/kg/day)- and high (2.5 mg/kg/day)-dose TCP on 5-HT$_2$ receptor density and affinity in rat cortex.

Methods

Animals

Male Spraque-Dawley rats were implanted in the dorsal thoracic area with Alzet 2ML2 osmotic minipumps loaded to administer the following doses of drugs: TCP (0.5 mg/kg/day), TCP (2.5 mg/kg/day), or vehicle (distilled water). At time intervals of 4, 10, or 28 days groups of animals (8–10 per group) were sacrificed by decapitation and the brains removed. Whole cortex was dissected out and a membrane fraction prepared and employed for studying 5-HT$_2$ receptor number and affinity using ^{3}H-ketanserin as the radioligand (Leysen et al., 1982). The rest of brain (whole brain minus cortex, hippocampus and striatum, which were removed for other neurochemical studies) was retained for analysis of brain amine concentrations by HPLC (Baker et al., 1987), monoamine oxidase (A and B) activity using a radiochemical procedure (Wurtman and Axelrod, 1963), and TCP levels by electron capture gas chromatography (Nazarali et al., 1987).

Drugs

TCP, mianserin, and β-phenethylamine (hydrochlorides) and 5-HT creatinine sulphate were obtained from Sigma Chemicals. β-[Ethyl-1-^{14}C]-phenethylamine hydrochloride (50.8 mCi/mmol), 5-[2-^{14}C]-hydroxytryptamine binoxalate (54.7 mCi/mmol) and ^{3}H-ketanserin hydrochloride (60.0 Ci/mmol) were purchased from NEN chemicals. Tris (hydroxymethyl)aminomethane and poly(ethylenimine) were obtained from Fischer Scientific and Aldrich Chemicals, respectively.

^{3}H-ketanserin binding

Eight point saturation curves were obtained using a modification of the method described by Eison et al. (1990). Displaceable ^{3}H-ketanserin binding was determined from the difference between total binding and binding in the presence of 10 μM mianserin (Goodnough and Baker, 1993). Bmax and Kd values were determined from Scatchard analysis of the data. Protein content was determined by the method of Lowry et al. (1951).

Statistical analysis

Data were subjected to ANOVA followed where necessary by the Neuman-Keuls multiple comparisons test. Treatment groups were considered to be significantly different when $p < 0.05$.

Results

5-HT$_2$ receptor density and affinity

The low (0.5 mg/kg/day) dose of TCP did not result in a significant down-regulation of the 5-HT$_2$ receptor site at any time interval. The high (2.5 mg/kg/day) dose of TCP produced a down-regulation after 10 and 28 days of chronic treatment, but not after only 4 days (Table 1). A significant change in the affinity (Kd) of ^{3}H-ketanserin for the 5-HT$_2$ site did not occur after drug treatment.

Neurotransmitter amine levels

Both doses of TCP produced significantly greater levels of the neurotransmitter amines noradrenaline (NA) and 5-hydroxytryptamine (5-HT) in rat brain than did vehicle at the 4, 10 and 28 day intervals (Figs. 1 and 2). Animals receiving the high dose of TCP had increases in these amines which were significantly higher than those observed in animals receiving the low-dose over the time intervals studied.

MAO activity

The high-dose of TCP inhibited MAO to a greater extent than the low-dose at all time intervals (Figs. 3 and 4).

Table 1. 5-HT$_2$ receptor density and affinity

Drug treatment	Duration	B_{max} (fmol/mg protein)	Kd (nM)
Vehicle (dist. water)	4 days	198 ± 7	0.50 ± 0.01
TCP (0.5 mg/kg day)	4 days	185 ± 6	0.49 ± 0.02
TCP (2.5 mg/kg/day)	4 days	186 ± 5	0.52 ± 0.02
Vehicle (dist. water)	10 days	252 ± 15	0.58 ± 0.03
TCP (0.5 mg/kg/day)	10 days	235 ± 9	0.58 ± 0.04
TCP (2.5 mg/kg/day)	10 days	$207 \pm 15^*$	0.64 ± 0.04
Vehicle (dist. water)	28 days	203 ± 11	0.54 ± 0.03
TCP (0.5 mg/kg/day)	28 days	195 ± 13	0.58 ± 0.02
TCP (2.5 mg/kg/day)	28 days	$147 \pm 11^*$	0.65 ± 0.05

Values represent mean $\pm$ SEM (n = 8–10)
* denotes significant difference ($p < 0.05$) from control

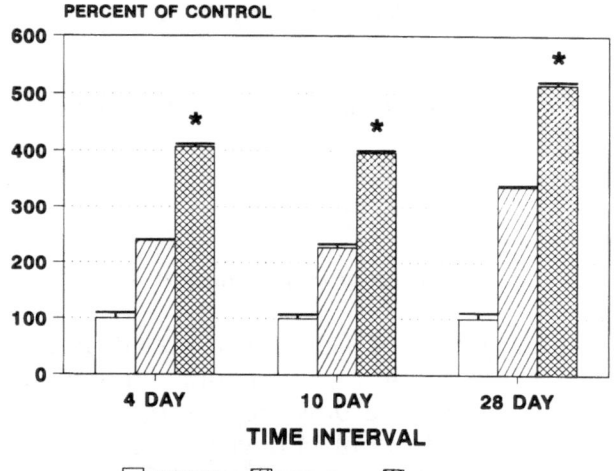

Fig. 1. Levels of 5-HT in rest of brain. Values represent means ± SEM (n = 8–10) and are expressed as % of values in rats treated with vehicle for the same number of days. Doses are expressed as mg/kg/day. All drug treatment groups were significantly different from control. *denotes significant difference (p < 0.05) from TCP (0.5 mg/kg/day). Control values (in ng/g tissue): 210 ± 18, n = 28

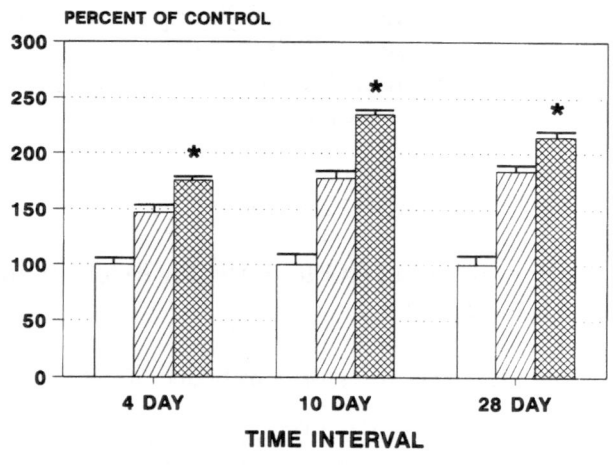

Fig. 2. Levels of NA in rat rest of brain. Values represent means ± SEM (n = 8–10) and are expressed as % of values in rats treated with vehicle for the same number of days. Doses are expressed as mg/kg/day. All drug treatment groups were significantly different from control. *denotes significant difference (p < 0.05) from TCP (0.5 mg/kg/day). Control values (in ng/g tissue): 335 ± 28, n = 28

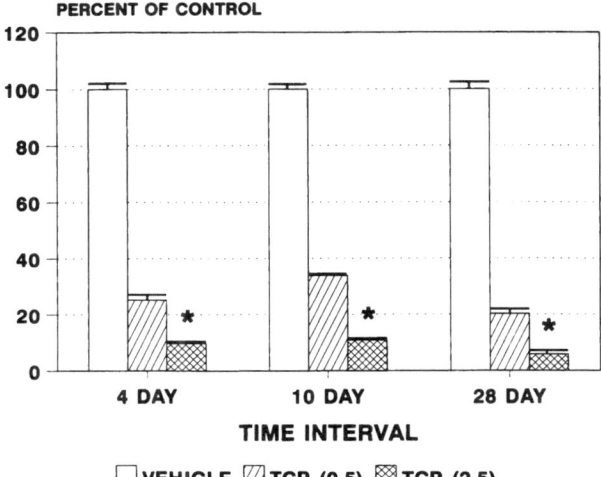

Fig. 3. Activity of MAO-A in rat rest of brain. Values represent means ± SEM (n = 8–10) and are expressed as % of values in rats treated with vehicle for the same number of days. Doses are expressed as mg/kg/day. All drug treatment groups were significantly different from vehicle. *denotes significant difference (p < 0.05) from TCP (0.5 mg/kg/day). Data are adapted from Goodnough et al. (1994)

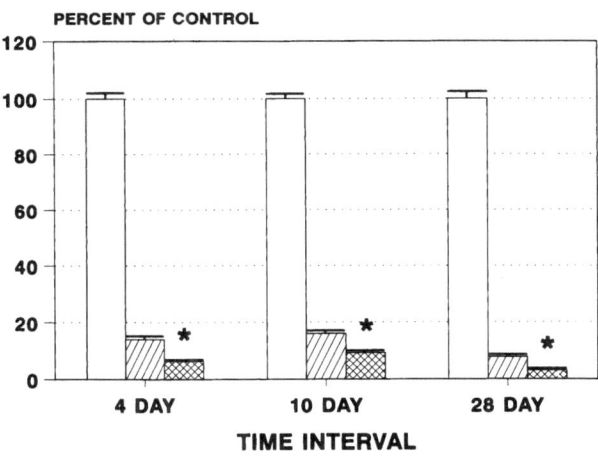

Fig. 4. Activity of MAO-B in rat rest of brain. Values represent means ± SEM (n = 8–10) and are expressed as % of values in rats treated with vehicle for the same number of days. Doses are expressed as mg/kg/day. All drug treatment groups were significantly different from vehicle. *denotes significant difference (p < 0.05) from TCP (0.5 mg/kg/day). Data are adapted from Goodnough et al. (1994)

Discussion

Down-regulation of the $5\text{-}HT_2$ receptor occurs after chronic treatment with several antidepressants (review: Baker and Greenshaw, 1989). The 0.5 mg/kg/day dose of TCP is similar, on a mg/kg basis, to that used in the clinical setting and has been demonstrated to cause down-regulation of β-noradrenergic (Sherry-McKenna et al., 1992), α_2-noradrenergic (Greenshaw et al., 1988), and tryptamine receptors (Mousseau et al., 1992) after chronic administration. TCP, at the 0.5 mg/kg/day dose, produced a level of MAO inhibition thought to be required for antidepressant efficacy in the clinical situation (Robinson et al., 1978; Giller and Lieb, 1980; Giller et al., 1982) but did not result in a significant down-regulation of the $5\text{-}HT_2$ site at 4, 10, or 28 days. The high dose of TCP caused a significantly greater inhibition of MAO-A and -B than did the low dose, resulting in levels of 5-HT that are almost twice those seen at the low dose. The high dose of TCP had less of an effect on NA levels (approximately 20% more NA is seen at the high dose than at the low dose). The dramatic increase in 5-HT seen with the high dose over that seen at the low dose probably accounts for the $5\text{-}HT_2$ down-regulation seen at 10 and 28 days.

TCP has been shown to possess neurochemical properties other than MAO inhibition. In addition to being a potent MAO inhibitor, TCP has moderately strong effects on the uptake and release of catecholamines in nerve terminals (Hendley and Snyder, 1968; Schildkraut, 1970; Baker et al., 1978, 1980; Reigle et al., 1980). With administration of the 2.5 mg/kg/day dose, levels of TCP in the present study were found to be approximately $1\,\mu M$ after 4 days and approximately $1.25\,\mu M$ after 10 and 28 days in rest of brain. At this concentration TCP will have significant effects on NA and DA uptake. It is unclear at this time if these effects on catecholamines are contributing to the change in $5\text{-}HT_2$ receptors observed in the present study.

Acknowledgements

Funding was provided by the Alberta Mental Health Research Fund, the Medical Research Council and the University of Alberta Faculty of Medicine 75th Anniversary Scholarship. DBG is a recipient of an Alberta Mental Health Research Fund Scholarship. The authors would like to thank Dr. A. J. Greenshaw for his useful discussions on topics related to the data presented here.

References

Amsterdam JD, Berwish NJ (1989) High dose tranylcypromine treatment for refractory depression. Pharmacopsychiatry 22: 21–25
Baker GB, Hiob LE, Dewhurst WG (1980) Effects of monoamine oxidase inhibitors on release of dopamine and 5-hydroxytrypamine from rat striatum in vitro. Cell Mol Biol 26: 182–186

Baker GB, McKim HR, Calverley DG, Dewhurst WG (1978) Effects of the monoamine oxidase inhibitors tranylcypromine, phenelzine, and pheniprazine of the uptake of catecholamines in slices from rat brain regions. Proc Eur Soc Neurochem 1: 536

Baker GB, Coutts RT, Rao TS (1987) Neuropharmacological and neurochemical properties of N-(cyanoethyl)-2-phenethylamine, a prodrug of 2-phenethylamine. Br J Pharmacol 92: 243–255

Baker GB, Greenshaw AJ (1989) Effects of long-term administration of antidepressants on receptors in the central nervous system. Cell Mol Biol 9: 1–44

Eison AS, Yocca FD, Gianutsos G (1991) Effect of chronic administration of antidepressant drugs on 5-HT$_2$-mediated behavior in the rat following noradrenergic or serotonergic devervation. J Neural Transm 84: 19–32

Ferris RM, Howard JL, White HL (1975) A relationship between clinical efficacy and various biochemical parameters of monoamine oxidase inhibitors (MAOIs). Pharmacologist 17: 451

Giller EG, Lieb J (1980) Monoamine oxidase inhibitors and platelet monoamine oxidase inhibition. Comm Psychopharmacol 4: 79–82

Giller EG, Bialos D, Riddle M, Sholomskas A, Harkness L (1982) Monoamine oxidase inhibitor-responsive depression. Psychiatry Res 6: 41–48

Goodnough DB, Baker GB (1993) Tranylcypromine does not enhance the effects of amitriptyline on 5-HT$_2$ receptors in rat brain. J Pharm Sci (in press)

Goodnough DB, Baker GB, Mousseau DD, Greenshaw AJ, Dewhurst WG (1994) Effects of low- and high-dose tranylcypromine on [^{3}H]-tryptamine binding sites in the rat hippocampus and striatum. Neurochem Res (in press)

Goodwin GM, Green AR, Johnson P (1984) 5-HT$_2$ receptor changes in frontal cortex and 5-HT$_2$-mediated head-twitch behavior following antidepressant treatment to mice. Br J Pharmacol 83: 235–242

Greenshaw AJ, Nazarali AJ, Rao TS, Baker GB, Coutts RT (1988) Chronic tranylcypromine treatment induces functional α_2-adrenoreceptor down-regulation in rats. Eur J Pharmacol 154: 67–72

Guze BH, Baxter LR, Jr (1987) Refractory depression treated with high doses of a monoamine oxidase inhibitor. J Clin Psychiatry 48: 31–32

Hendley ED, Snyder SH (1968) Relationship between the action of monoamine oxidase inhibitors on the noradrenaline uptake system and their antidepressant efficacy. Nature 220: 1330–1331

Kellar KJ, Cascio CS, Butler JA, Kurtze RN (1981) Differential effects of electroconvulsive shock and antidepressant drugs on serotonin-2 receptors in rat brain. Eur J Pharmacol 69: 515–518

Lafaille F, Welner SA, Suranyi-Casotte BE (1991) Regulation of serotonin type 2 (5-HT$_2$) and β-adrenergic receptors in rat cerebral cortex following novel and classical antidepressant treatment. J Psychiatr Neurosci 16: 209–214

Leysen JE, Niemegeers CJE, Van Neuten JM, Laduron P (1982) ^{3}H-Ketanserin (R41468), a selective ^{3}H-ligand for serotonin-2 receptor binding sites. Mol Pharmacol 21: 301–314

Lowry OH, Rosenbrough NJ, Farr AL, Randall RJ (1951) Protein measurements with the Folin phenol reagent. J Biol Chem 193: 265–275

Mousseau DD, McManus DJ, Baker GB, Juorio AV, Dewhurst WG, Greenshaw AJ (1993) Effects of age and of chronic antidepressant treatment on ^{3}H-tryptamine binding to rat cortical membranes. Cell Mol Neurobiol 13: 3–13

Nazarali AJ, Baker GB, Coutts RT, Yeung JM, Rao TS (1987) Rapid analysis of β-phenethylamine in tissues and body fluids utilizing pentaflourobenzoylation followed by electron-capture gas chromatography. Prog Neuropsychopharmacol Biol Psychiatry 11: 251–258

Pearlman C (1987) High dose tranylcypromine in refractory depression. J Clin Psychiatry 48: 424

Reigle TG, Orsulak PJ, Avni J, Platz PA, Schildkraut JJ (1980) The effects of tranylcypromine isomers on ^{3}H-norepinephrine metabolism in rat brain. Psychopharmacology 69: 193–199

Robinson DS, Nies A, Ravaris CL, Ives JO, Barlett D (1978) Clinical pharmacology of phenelzine: MAO activity and clinical response. In: Lipton M, Dimascio A, Killam KF (eds) Psychopharmacology: a generation of progress. Raven Press, New York

Robinson DS (1983) High-dose monoamine oxidase-inhibitor therapy. J Am Med Assoc 250: 2212

Schildkraut JJ (1970) Tranylcypromine: effects on norepinephrine metabolism in rat brain. Am J Psychiatry 126: 925–931

Sherry-McKenna RL, Baker GB, Mousseau DD, Coutts RT, Dewhurst WG (1992) 4-Methoxytranylcypromine, a monoamine oxidase inhibitor: effects on biogenic amines in rat brain following chronic administration. Biol Psychiatry 31: 881–888

Wurtman RJ, Axelrod J (1963) A sensitive and specific assay for the estimation of monoamine oxidase. Biochem Pharmacol 12: 1439–1441

Yates M, Leake A, Candy JM, Fairbairn AF, McKeith IG, Ferrier IN (1990) 5-HT$_2$ receptor changes in major depression. Biol Psychiatry 27: 489–496

Authors' address: Dr. D. B. Goodnough, Neurochemical Research Unit, Department of Psychiatry, 1E7.44 W Mackenzie Health Sciences Centre, Edmonton, Alberta, Canada, T6G 2B7.

J Neural Transm (1994) [Suppl] 41: 135–139

Clorgyline effect on pineal melatonin biosynthesis in adrenalectomized rats pretreated with 6-hydroxydopamine

S. Reuss[1], **P. J. Requintina**[2], **R. Riemann**[1], and **G. F. Oxenkrug**[2]

[1] Department of Anatomy, Johannes Gutenberg-University, Mainz, Federal Republic of Germany
[2] Pineal Research Laboratory, Department of Psychiatry and Human Behavior, Brown University, and Psychiatry Service, VAMC, Providence, Rhode Island, U.S.A.

Summary. The response to administration of the specific monoamine oxidase A (MAO-A) blocker clorgyline was investigated in adult male Sprague-Dawley rats which were adrenalectomized four days prior to treatment or were additionally sympathectomized as newborns by injection of 6-hydroxydopamine. In both groups, the contents of pineal indoles melatonin and N-acetylserotonin were augmented, and the contents of 5-hydroxyindoleacetic acid and 5-hydroxyindoletryptophol decreased 90 min following clorgyline injections when compared to rats receiving saline. The observed responses were less pronounced in rats both adrenalectomized and sympathectomized. The results are in line with the hypothesis that preservation from oxidation of both MAO-A substrates, noradrenaline and serotonin, upon clorgyline administration contributes to the observed increase in melatonin biosynthesis thought to be associated with the anti-depressant effects of MAO inhibition.

Introduction

Selective inhibition of monoamine oxidase-A type (MAO-A), but not of MAO-B, stimulates pineal melatonin (MEL) biosynthesis in rodents, primates and humans (Oxenkrug et al., 1984; for review see Oxenkrug, 1991). These findings might be of clinical importance since selective MAO-A (but not MAO-B) inhibitors exert clinical antidepressive effects (Lipper et al., 1979).

MEL synthesis is driven by postganglionic sympathetic fibres stemming from the superior cervical ganglia (SCG; Reuss and Moore, 1989). Removal of the SCG resulted in a loss of pineal MAO-A activity (Snyder et al., 1965).

Experimental data suggest that the effect of MAO-A inhibitors might be mediated via stimulation of the physiological pathway of melatonin synthesis since ganglionectomy (McIntyre et al., 1985) but not adrenal demedullation (Oxenkrug and McCauley, 1987) attenuated clorgyline-induced stimulation of rat pineal melatonin synthesis.

The biochemical mechanism of melatonin synthesis stimulation induced by selective inhibition of MAO-A, however, is not clear. Both MAO-A substrates (serotonin, 5-HT; and noradrenaline, NA) are intimately involved in melatonin synthesis: 5-HT is a substrate of melatonin biosynthesis while NA (released from postganglionic sympathetic fibers) is an important activator of the first (and, probably, rate-limiting) reaction of melatonin synthesis: N-acetylation of 5-HT (see Lewy, 1983). The in vitro study suggested that stimulation of melatonin synthesis induced by the selective MAO-A inhibitor clorgyline depended upon preservation of NA rather than 5-HT (see Oxenkrug, 1991) while the attenuation of clorgyline-induced stimulation of melatonin synthesis in ganglionectomized animals (McIntyre et al., 1985) might be related to the 50% reduction of pineal 5-HT content caused by this procedure (Pellegrino de Iraldi et al., 1963; McIntyre et al., 1985).

In a previous study (Reuss and Oxenkrug, 1989), in which we used chemical sympathectomy by 6-hydroxydopamine (6-OHDA) of newborn rats to achieve sympathetic denervation prior to the study of selective MAO-A inhibition effects on pineal melatonin synthesis in adult animals, sympathectomized animals exhibited similar — although less pronounced — responses compared to intact animals (i.e., augmentation of pineal indoles MEL, serotonin and 5-hydroxytryptophan 90 min following clorgyline injections). Since sympathectomy does not eliminate the exposure of pinealocytes to noradrenaline produced by the adrenal medulla, in the present study we sought to investigate the effects of combined chemical sympathectomy and surgical adrenalectomy on clorgyline-induced stimulation of pineal melatonin synthesis.

Materials and methods

Twenty-eight male Sprague-Dawley rats used for the present study were maintained with food and water ad libitum in a light-controlled (LD 12:12), temperature-regulated room. For chemical sympathectomy (Angeletti, 1971), 14 of them were given for the first 5 days of life s.c. injections of the false neurotransmitter 6-hydroxydopamine (6-OHDA; 50 µg/g b.w.) dissolved in 0.2% ascorbic acid (volume 0.05 ml) with a 32-gauge needle. Fourteen control animals received vehicle injections only.

At the age of ten weeks, all animals were bilaterally adrenalectomized (AdX) under deep tribromethanol anesthesia, then returned to their original photoperiods and provided with drinking water containing 1% saline. Four days following AdX, half of the animals of both, sympathectomized and vehicle-treated groups received injections of clorgyline (Sigma, 2.5 mg/kg b.w. in physiological saline), the other halfs received saline injections. Ninety min following treatment (middle of light period), rats were killed by decapitation, the pineals quickly removed and transferred to liquid nitrogen, then stored at −70°C until assayed.

Pineal indole concentrations were determined by a high-pressure-liquid-chromatographic-fluorometric system (see Reuss and Oxenkrug, 1989). Group differences were analyzed by one-way analysis of variance and Student's t-test. A $p < 0.05$ was regarded as statistically significant.

Results

As shown in Table 1 (comparison of groups I and II), the application of clorgyline in adrenalectomized rats resulted in increased pineal contents of melatonin (MEL), N-acetyl-serotonin (NAS), and serotonin (5-HT), accompanied by reductions in pineal 5-hydroxyindoleacetic acid (5-HIAA) and 5-hydroxyindoletryptophol (5-HIOL).

In animals which were sympathectomized by 6-OHDA injections as newborns and adrenalectomized as adults (group III vs. IV), pineal MEL content was increased, although to a minor degree, upon injection of clorgyline. NAS was drastically augmented, while 5-HT and 5-HTP were not influenced. 5-HIAA and 5-HIOL were reduced to less than 20% by clorgyline administration, similarly to the effects seen in groups I/II.

Discussion

Various studies have shown that inhibition of the MAO-A by application of the specific blocker clorgyline stimulates melatonin production in vivo (for review see Oxenkrug, 1991).

Since a previous study (Reuss and Oxenkrug, 1989) suggested that the preservation of both MAO-A substrates, 5-HT and NA, by MAO-A blocking contributes to the augmentation of pineal melatonin synthesis, the objective of the present study was to compare clorgyline effects in rats in which adrenal medullary noradrenaline is depleted and in animals that were, in addition to adrenalectomy, sympathectomized chemically as newborns. Animals of the latter group do not develop a sympathetic system so that the pineal gland is deprived of NA-containing nerve terminals.

The present results demonstrate that the above mentioned effects are also present in adrenalectomized rats. It is seen that, in adrenalectomized but not sympathectomized animals, pineal MEL, NAS and 5-HT contents increase to various degrees while 5-HIAA and 5-HIOL contents are reduced upon clorgyline administration. In animals that were sympathectomized as newborns and adrenalectomized as adults, a similar situation was found, although the effects were observed to be smaller. Considering the higher baseline contents of MEL (and NAS) in rats that were sympathectomized and adrenalectomized, however, clorgyline-induced stimulation of melatonin biosynthesis was less pronounced in these animals.

In conclusion, the present results indicate that the selective MAO-A blocker clorgyline may increase pineal melatonin content, albeit to a minor degree, also in total absence of noradrenaline and are thus in line with the suggestion that the preservation from degradation of both MAO-A substrates, serotonin and noradrenaline, contribute to the antidepressant effects of MAO inhibition (Oxenkrug et al., 1986b).

S. Reuss et al.

Table 1. Effect of MAO inhibitor clorgyline (clorg) or saline (sal) injections on pineal indoles melatonin (MEL), N-acetyl-serotonin (NAS), serotonin (5-HT), 5-hydroxytryptamine), 5-hydroxytryptophan (5-HTP), 5-hydroxyindoleacetic acid (5-HIAA) and 5-hydroxyindoletryptophol (5-HIOL) in adrenalectomized (AdX) rats that were chemically sympathectomized (OHDA) or had received vehicle injections

Grp.	Treatment	MEL	NAS	5-HT	5-HTP	5-HIAA	5-HIOL
I	AdX/veh/sal	0.22 ± 0.12[1]	nd	124.4 ± 16.5[3]	3.8 ± 1.6	26.6 ± 11.5[3]	0.7 ± 0.5[3]
II	AdX/veh/clorg	0.63 ± 0.26	1.29 ± 1.49[2]	163.3 ± 48.5[2]	3.8 ± 2.3	4.7 ± 2.9[2]	0.09 ± 0.06
III	AdX/OHDA/sal	0.43 ± 0.18	0.59 ± 0.45[2]	92.4 ± 17.6	2.2 ± 0.8	14.1 ± 2.8	0.3 ± 0.1[2]
IV	AdX/OHDA/clorg	0.72 ± 0.17	3.93 ± 1.92	88.9 ± 31.6	3.3 ± 2.4	2.25 ± 1.4	0.05 ± 0.04

Means ± S.E.M., ng/pineal
Statistically significant differences were observed in comparison to the corresponding values in [1] groups II–IV, [2] group IV, [3] groups II, III

Acknowledgement

The authors wish to thank A. Thomas-Semm for her technical assistance.

References

Angeletti PU (1971) Chemical sympathectomy in newborn animals. Neuropharmacology 10: 55–59

Bieck PR, Antonin KH, Balon R, Oxenkrug GF (1988) Brofaromine and pargyline effect on human plasma melatonin. Prog Neuropsychopharmacol Biol Psychiatry 12: 93–101

Lewy A (1983) Biochemistry and regulation of mammalian melatonin production. In: Relkin R (ed) The pineal gland. Elsevier, New York, pp 77–128

Lipper S, Murphy D, Slater S, Buchbaum M (1979) Comparative behavioral effects of clorgyline and pargyline in man: a preliminary evaluation. Psychopharmacology 62: 123–128

McIntyre IM, McCauly R, Murphy S, Goldman H, Oxenkrug GF (1985) The effect of superior cervical ganglionectomy on melatonin stimulation by MAO-A inhibitor. Biochem Pharmacol 34: 3394–3395

Murphy DL, Tamarkin L, Sunderland T, Garrick NA, Cohen R (1986) Human plasma melatonin is elevated during treatment with the monoamine oxidase inhibitors clorgyline and tranylcypromine but not deprenyl. Psychiatry Res 17: 119–127

Oxenkrug GF (1991) The acute effect of monoamine oxidase inhibitors on serotonin conversion to melatonin. In: Sandler M, Coppen A, Harnett S (eds) 5-Hydroxytryptamine in psychiatry. A spectrum of ideas. Oxford University Press, pp 98–109

Oxenkrug GF, McCauley RB (1987) The effect of MAO inhibitors on melatonin synthesis: mechanism and clinical implication. Pharmacol Toxicol 60 [Suppl] 1: 36

Oxenkrug GF, McCauley RB, McIntyre IM, Filipowicz C (1984) Effect of clorgyline and deprenyl on rat pineal melatonin. J Pharm Pharmacol 36: 55W

Oxenkrug GF, McCauley RB, Fontana DJ, McIntyre IM, Commissaris RL (1986a) Possible melatonin involvement in the hypotensive effect of MAO inhibitors. J Neural Transm 66: 271–280

Oxenkrug GF, Balon R, Jain AK, McIntyre IM, Appel D (1986b) Single dose of tranylcypromine increases human plasma melatonin. Biol Psychiatry 21: 1085–1089

Pellegrino de Iraldi A, Zieher LM, De Robertis E (1963) The 5-hydroxytryptamine content and synthesis of normal and denervated pineal gland. Life Sci 2: 691–696

Reuss S, Moore RY (1989) Neuropeptide Y-containing neurons in rat superior cervical ganglion: projections to the pineal gland. J Pineal Res 6: 307–316

Reuss S, Oxenkrug GF (1989) Chemical sympathectomy and clorgyline-induced stimulation of rat pineal melatonin synthesis. J Neural Transm [Gen Sect] 78: 167–172

Snyder SH, Fisher J, Axelrod J (1965) Evidence for the presence of MAO in sympathetic nerve endings. Biochem Pharmacol 14: 363–365

Authors' address: Doz. Dr. S. Reuss, Department of Anatomy, University of Mainz, Saarstrasse 19–21, D-55099 Mainz, Federal Republic of Germany.

J Neural Transm (1994) [Suppl] 41: 141–144

Synergistic sedative effect of selective MAO-A, but not MAO-B, inhibitors and melatonin in frogs

P. J. Requintina[1], G. F. Oxenkrug[1], A. Yuwiler[2], and A. G. Oxenkrug[1]

[1] Pineal Research Laboratory, Psychiatry Service, VAMC, and Department of Psychiatry and Human Behavior, Brown University School of Medicine, Providence, Rhode Island, and
[2] Neurochemistry Laboratory, West Los Angeles, Brentwood, VAMC, and Department of Psychiatry and Biobehavioral Sciences, UCLA, Los Angeles, California, U.S.A.

Summary. Total suppression of righting reflex in frogs (Rana pipiens, 25–35 mg b.w.) was observed after combined administration of melatonin (12.5 mg/kg) and selective inhibitors of MAO-A: clorgyline (2.5 mg/kg) and moclobemide (50 mg/kg) but not MAO-B: selegiline (25 mg/kg) and Ro-19-6327 (50 mg/kg). None of these drugs alone affected the righting reflex. Clorgyline and selegiline selectively inhibited brain MAO-A and MAO-B activity (by more than 90%), resp. Frogs might represent a convenient model to study the selective MAO-A and B type inhibitors since they provide the opportunity to correlate behaviour and biochemical changes induced by MAO inhibitors.

Introduction

The first screening tests for antidepressants (both tricyclic and MAO inhibitors) were based on their ability to counteract the sedative effect of reserpine in rodents. Antagonism between antidepressants and reserpine was suggested to be mediated via potentiation of catecholaminergic neurotransmission (Schildkraut, 1985). Contrary to their effects in rodents, antidepressants potentiate reserpine-induced sedation in frogs (Lapin et al., 1968). The synergism between antidepressants and reserpine in the frog was explained by the positive serotoninergic effects of antidepressants (Lapin and Oxenkrug, 1969). The frog model was suggested for the screening of antidepressants with serotonin-positive (thymoleptic) action (Lapin and Oxenkrug, 1970). Using this model Skene and Potgieter (1981) found that melatonin potentiated the sedative effect of reserpine in frogs pretreated with the non-selective MAO inhibitor, pargyline.

Since selective MAO-A (but not MAO-B) inhibitors stimulate pineal melatonin biosynthesis in rodents, primates and humans (Oxenkrug et al., 1984; for rev. see Oxenkrug, 1991), we were interested to see if melatonin potentiation of the reserpine-induced sedation in frogs (suppression of righting reflex: turning from supine to prone position) requires MAO-A or MAO-B inhibition. During preliminary experiments we have found that the non-selective MAO inhibitor, phenelzine, in combination with melatonin induced sedation in frogs even without reserpine administration (see below). Therefore, our final goal was to determine the type of MAO inhibition (A or B) required for melatonin suppression of the righting reflex in frogs.

Materials and methods

Frogs (Rana pipiens, male and female, 20–40 g b.w.) were kept at 4°C for at least two weeks before the experiments so as to control for seasonal differences in the frog's sensitivity to antidepressants (Harri, 1974). Drugs were injected into the submandibular and suprafemoral lymph sacs. The non-selective MAO inhibitor, phenelzine, the selective MAO-A inhibitors, clorgyline and moclobemide, and the selective MAO-B inhibitors, selegiline (l-deprenyl) and Ro 19-6327 were used in the doses indicated below.

We initially intended to follow previous protocols, injecting MAO inhibitors 90 min before melatonin and 120 min before reserpine administration (Skene and Potgieter, 1981). However, in preliminary experiments phenelzine-pretreated frogs showed inhibition of the righting reflex 30 min after melatonin injection, even before reserpine administration. Therefore, reserpine administration was eliminated in the final experiments. Frogs were injected with a MAO inhibitor and 30 min later with melatonin. Righting reflexes (ten trials) were evaluated before the injections and 30 min after melatonin administration. Two sets of the experiments were performed: one in December and the other in June with identical results. Brain MAO-A and B activities were evaluated in frogs treated with phenelzine, clorgyline and selegiline by the method of Wurtman and Axelrod (1963). Each experimental group consisted of six frogs. Data were statistically treated according to one way analysis of variance and student's t-test.

Results

Melatonin alone (12.5 mg/kg) did not sedate the animals. Total suppression of the righting reflex was observed after melatonin was injected into frogs pretreated with phenelzine (5 mg/kg), clorgyline (2.5 mg/kg) or moclobemide (50 mg/kg). Selegiline (25 mg/kg) and Ro 19-6327 (50 mg/kg) did not affect the righting reflex. None of the MAO inhibitors alone produced sedation.

Clorgyline specifically inhibited brain MAO-A (by more than 90%). Selegiline markedly inhibited MAO-B but also produced a 16% inhibition of MAO-A. Phenelzine (5 mg/kg), inhibited only MAO-A at the dose used. Moclobemide inhibited only MAO-A activity, while Ro 19-6327 had no effect on either MAO-A or B activity. The lack of effect of Ro 19-6327 on MAO-B activity differs from data obtained in rodents and might be explained by species differences. This issue needs further investigation.

Table 1. Brain MAO activity and righting reflex in frogs treated with melatonin and MAO inhibitors

	Number of rightings	MAO-A*	MAO-B*
Saline	10	10.44 ± 0.29	3.95 ± 0.11
Melatonin	10		
Saline + melatonin	10	10.34 ± 0.67	2.95 ± 0.53
Phenelzine	10		
Phenelzine + melatonin	0	2.81 ± 0.92**	2.94 ± 1.03
Clorgyline	10		
Clorgyline + melatonin	0	1.27 ± 0.11**	4.02 ± 0.10
Deprenyl	10		
Deprenyl + melatonin	10	8.63 ± 0.40***	0.20 ± 0.01**
Moclobemide	10		
Moclobemide + melatonin	0	7.18 ± 0.72***	3.71 ± 0.32
Ro 19-6327	10		
Ro 19-6327 + melatonin	10	9.48 ± 2.70	3.66 ± 0.70

* mean + S.E. (μmoles/hr/mg); ** $p < 0.0001$; *** $p < 0.01$ vs saline

Discussion

The present study adds Rana pipiens to the list of frog species (Rana temporaria: Harri, 1974; Lapin et al., 1968, and Xenopus laevis: Skene and Potgieter, 1981) in which drug-induced sedation was demonstrated. Our experiments revealed that sedation could be induced in frogs by the combination of MAO inhibitors and melatonin without reserpine administration. The synergism between antidepressants and reserpine in frogs was explained by the positive serotoninergic effects of antidepressants (Lapin and Oxenkrug, 1969). Similarly, melatonin potentiation of the sedative effect of reserpine in pargyline-pretreated frogs was attributed to the serotonin-positive effect of melatonin (Skene and Potgieter, 1981). There is ample evidence supporting 5HT involvement in the control of muscle tone in amphibia (Popova et al., 1984). However, the possibility of an influence of melatonin on the righting reflex cannot be excluded since pharmacological interventions stimulating serotonin biosynthesis might increase production of melatonin as well.

The other new finding of the present study is that selective MAO-A, but not MAO-B, inhibitors synergise with melatonin to produce sedation in the frog. The doses of MAO inhibitors were selected on the bases the literature data obtained in rodents. Our results show that clorgyline and selegiline exert the same selectivity in frogs. Further experiments are needed to evaluated the selectivity of moclobemide, Ro 19-6327 and other recently produced selective MAO inhibitors on MAO activity and behaviour in the frog. Observation of synergism between selective MAO-A inhibitors and melatonin in frogs might be both of theoretical and practical interest. The frog might represent a convenient model to study selective MAO-A and B type inhibitors since it allows the simultaneous observation of behavioural and biochemical changes.

Acknowledgements

The authors wish to thank Profs. Da Prada and Haefely for the generous gift of moclobemide and Ro 19-6327.

References

Harri MNE (1974) The dependence of imipramine-induced sedation upon central 5-hydroxytryptamine-like activity in the frog. J Pharm Pharmacol 26: 73–74

Lapin IP, Osipova SV, Uskova NV, Stabrovski EM (1968) Synergism of imipramine and desmethylimipramine with reserpine in the frog. Interaction with 5-hydroxy-tryptophan and 2-bromolysergid diethylamide (BOL-148). Arch Int Pharmacodyn 174: 37–49

Lapin IP, Oxenkrug GF (1969) Intensification of the central serotoninergic processes as a possible determinant of the thymoleptic effect. Lancet i: 32–39

Lapin IP, Oxenkrug GF (1970) The frog as a subject for screening thymoleptic drugs. J Pharm Pharmacol 22: 781–782

Oxenkrug G, McCauley R, McIntyre I, Filipowicz C (1984) Effect of clorgyline and deprenyl on rat pineal melatonin. J Pharm Pharmacol 36: 55W

Oxenkrug GF (1991) The acute effect of monoamine oxidase inhibitors on serotonin conversion to melatonin. In: Sandler M, Coppen A, Harnett S (eds) 5-Hydroxytry-ptamine in psychiatry. A spectrum of ideas. Oxford University Press, pp 98–109

Popova NK, Lobacheva II, Karmanova IG, Shilling NV (1984) Serotonin in the control of the sleep-like states in frogs. Pharmacol Biochem Behav 20: 653–657

Schil dkraut JJ (1965) The catecholamine hypothesis of affective disorders: a review of supporting evidence. Am J Psychiatry 122: 509–522

Skene DJ, Potgieter B (1981) Investigation of two animal models of depression. S African J Sci 77: 180–182

Wurtman RJ, Axelrod J (1963) A sensitive and specific assay for the estimation of monoamine oxidase. Biochem Pharmacol 62: 1439–1440

Authors' address: P. J. Requintina, T-20, VAMC, 830 Chalkstone Ave., Providence, RI 02908, U.S.A.

J Neural Transm (1994) [Suppl] 41: 145–148

Clorgyline effect on pineal melatonin biosynthesis in roman high- and low-avoidance rats

P. J. Requintina[1], **P. Driscoll**[2], and **G. F. Oxenkrug**[1]

[1] Pineal Research Laboratory, Psychiatry Service, VAMC and Department of
Psychiatry and Human Behavior, Brown University School of Medicine, Providence,
Rhode Island, U.S.A.
[2] Behavior Biology Laboratory, ETHZ, Zürich, Switzerland

Summary. Pineal melatonin and related indoles levels were higher in
Roman high- than in Roman low-avoidance rats, while 5-HIAA/5-HT ratio,
as an index of MAO activity was higher in low- than in high-avoidance rats.
Clorgyline stimulated pineal melatonin biosynthesis in both lines of rats.
However, melatonin and N-acetylserotonin levels remained higher and 5-
HIAA levels remained lower in the high avoidance rats treated with low
dose (0.5 mg/kg) while treatment with 1.0 mg/kg of clorgyline eliminated the
differences in melatonin production between high- and low-avoidance rats.

Introduction

The Swiss line of Roman high-avoidance (RHA/Verh) rats was selected and
bred for the rapid acquisition of active, two-way avoidance in the shuttle
box, whereas the corresponding Roman low-avoidance (RLA/Verh) line
was selected and bred for the non-acquisition of that response (Driscoll and
Baettig, 1982). Besides the differences in the behavioural response the two
lines also differ in neurochemical and morphological parameters with a
higher basal turnover rate of serotonin (5-HT) in the hypothalamus in
RHA/Verh than in RLA/Verh rats (see Driscoll, 1988). 5-HT is a substrate
of the melatonin biosynthesis in the pineal gland which was found in
RHA/Verh to be twice as large as in RLA/Verh (Rivest et al., 1988). Pineal
levels of 5-HT, N-acetylserotonin (NAS), intermediate product of melatonin
biosynthesis, and melatonin as well as activity of the both enzymes of
melatonin biosynthesis were higher in RHA/Verh than in RLA/Verh rats
(Seidel et al., 1990) and a recent cross-breeding study has indicated that this
finding may correlate with the avoidance behaviour in these lines (Lipp and
Heinzeller, unpublished).

Besides N-acetylation (which initiates melatonin biosynthesis from 5-HT)
the other major way of 5-HT metabolism in the pineal gland is oxidative
deamination (catalyzed by monoamine oxidase-A; MAO-A) resulting in

the formation of 5-hydroxyindoleacetic acid (5-HIAA) and 5-hydroxy-tryptophol (5-HTOL) (Mefford et al., 1983). The lower rate of melatonin formation from 5-HT in RLA/Verh than in RHA/Verh rats suggests the higher availability of 5-HT as a substrate for MAO-A in RLA/Verh than in RHA/Verh animals. Selective MAO-A inhibitors were shown to stimulate pineal melatonin biosynthesis in rodents, primates and humans (Oxenkrug et al., 1984; for rev. see Oxenkrug, 1991). Therefore, one might expect the activation of pineal melatonin biosynthesis in RLA/Verh rats after administration of the selective MAO-A inhibitors.

In the present study we have compared the response of melatonin biosynthesis to the selective MAO-A inhibitor, clorgyline, in RHA/Verh and RLA/Verh rats.

Methods

2.5 Month old males of each line were maintained at 12 hrs light: 12 hrs dark schedule for at least two weeks before the experiments. In the first experiment animals were decapitated in the middle of the day (HD and LD) or night (HN and LN) phase. In the second experiment clorgyline (0.5 or 1.0 mg/kg, s.c.) was injected during the light phase and rats were decapitated 90 min. after the injections. Pineals were removed, frozen and stored at $-70°C$ until melatonin and related indoles levels were determined by the HPLC-fluorimetric procedure (see Oxenkrug, 1991). Each experimental group consisted of seven rats. Results are expressed as mean ± S.D. (ng/pineal). The differences between the groups were evaluate according to one way analysis of variance and Student's t-test.

Results

Both lines revealed strong diurnal rhythms with N-acetylserotonin (NAS) and melatonin higher during the dark while 5-HT and 5-HIAA higher during the light phase.

HD and HN rats had higher melatonin, NAS, and serotonin (5-HT) levels than did LD and LN rats, resp. In addition HD, but not HN, rats also had higher levels of 5-HT's metabolite, 5-HIAA. The 5-HIAA/5-HT ratio (index of MAO activity) was higher in LD and LN than in HD and HN rats, resp (Table 1).

Clorgyline (both doses) increased pineal NAS and melatonin levels and decreased 5-HIAA levels in both lines. After 0.5 mg/kg of clorgyline, NAS and melatonin levels remained higher and 5-HIAA levels remained lower in RHA/Verh than in RLA/Verh rats (as in the untreated rats). Administration of 1.0 mg/kg of clorgyline eliminated the differences between RHA/-Verh and RLA/Verh rats in pineal MEL, NAS and 5-HIAA levels seen at the baseline and after administration of 0.5 mg/kg of clorgyline (Table 1).

Table 1. Clorgyline effect on the pineal melatonin and related indoles in high and low-avoidance rats

Drug (mg/kg)	Melatonin**	NAS**	5-HT**	5-HIAA**	Ratio
Night (saline):					
High-avoidance	2.87 ± 0.50	12.5 ± 2.09	46.43 ± 14.2	3.89 ± 0.95	8.5
Low-avoidance	1.79 ± 0.33*	4.68 ± 0.80*	25.29 ± 7.28*	3.2 ± 0.36	12.8
Day (saline):					
High-avoidance	0.35 ± 0.03	0.10 ± 0.04	202.47 ± 31.4	10.9 ± 1.06	5.4
Low-avoidance	0.17 ± 0.01*	0.05 ± 0.01*	78.71 ± 12.8*	6.67 ± 1.07*	8.5
Day (Clorg-0.5):					
High-avoidance	0.47 ± 0.12	0.23 ± 0.11	231.9 ± 37.8	4.82 ± 1.33	2.0
Low-avoidance	0.28 ± 0.03*	0.06 ± 0.02*	77.87 ± 14.6*	3.1 ± 0.82*	3.9
Day (Clorg-1.0):					
High-avoidance	0.60 ± 0.19	0.28 ± 0.13	215.2 ± 34.7	4.34 ± 1.4	2.0
Low-avoidance	0.40 ± 0.20	0.26 ± 0.08	85.55 ± 5.9*	3.39 ± 1.8	3.9

*p < 0.01 vs high-avoidance; ** mean + S.E. (ng/pineal)

Discussion

Our results confirmed the higher rate of melatonin production in RHA/-Verh than in RLA/Verh rats and the existence of the diurnal rhythm of melatonin biosyntheis in both lines reported by Seidel et al. (1990). It is noteworthy that the animals used in our study were much younger (2.5 vs. 7.5 months old) and that we have used a different method of melatonin evaluation (HPLC vs. RIA). However, we have measured the 5-HT metabolite, 5-HIAA by the same HPLC-fluorimetric procedure.

Evaluation of 5-HIAA simultaneously with 5-HT allowed us to assess the 5-HIAA/5-HT ratio as an indirect index of MAO activity. The higher 5-HIAA/5-HT ratio in RHA/Verh than in RLA/Verh rats suggests the more active monoamine oxidation processes in low avoidance rats. Since these rats are considered to be an experimental model of depression it is noteworthy that elevated MAO activity was detected post-mortem in the brains of depressed patients and that enhanced brain MAO activity was suggested as a cause of psychotic depression (Mandell, 1973). The results of the clorgyline administration might promote further the role of MAO-A in regulating melatonin biosynthesis in these rats. A high (1.0 mg/kg) dose of clorgyline eliminated the differences in the rate of melatonin biosynthesis between RHA/Verh and RLA/Verh rats. It is not known whether the elimination of the biochemical differences would be accompanied by a corresponding change in behaviour. The investigation of the effect of clorgyline on the behavior of RHA/Verh and RLA/Verh rats is in progress.

References

Driscoll P (1988) Hypothalamic serotonin turnover in rat lines selectively bred for diffrences in two-way avoidance behaviour. Adv Biosci 70: 55–58

Driscoll P, Baettig K (1982) Behavioral, emotional and neurochemical profiles of rats selected for extreme differences in active two way avoidance performance. In: Liebich E (ed) Genetics of the brain. Elsevier, Amsterdam, pp 95–123

Mandell AJ (1973) Neurobiological barriers to euphoria. Am Sci 61: 565–573

Mefford IN, Chang P, Klein DC, Namboodiri MAA, Sugden D, Barchas J (1983) Reciprocal day/night relationship between serotonin oxidation and N-acetylation products in the rat pineal gland. Endocrinology 113: 1582–1586

Overstreet DH (1992) Genetic animal models of endogenous depression. In: Driscoll P (ed) Genetically defined animal models of neurobehavioral dysfunctions. Birkhauser, Boston, pp 253–275

Oxenkrug GF (1991) The acute effect of monoamine oxidase inhibitors on serotonin conversion to melatonin. In: Sandler M, Coppen A, Harnett S (eds) 5-Hydroxytryptamine in psychiatry. A spectrum of ideas. Oxford University Press, pp 98–109

Oxenkrug G, McCauley R, McIntyre I, Filipowics C (1984) Effect of clorgyline and deprenyl on rat pineal melatonin. J Pharm Pharmacol 36: 55W

Rivest RW, Scherrer A, Rivest-Navoratil MF, Aubert ML, Sizonenko PC, Driscoll P (1988) Roman low (RLA) and roman high (RHA) avoidance rats: inverse correlation between pituitary and pineal weights, and effect on sexual maturation. Experientia 44: A32

Seidel A, Neto JAS, Huesgen A, Vollrath L, Manz B, Gentsch C, Lichtsteiner M (1990) The pineal complex in roman high avoidance and roman low avoidance rats. J Neural Transm [GenSect] 81: 73–82

Authors' address: Dr. P. J. Requintina, T-20, VAMC, 830 Chalkstone Ave., Providence, RI 02908, U.S.A.

J Neural Transm (1994) [Suppl] 41: 149–153
© Springer-Verlag 1994

Reboxetine prevents the tranylcypromine-induced increase in tyramine levels in rat heart

P. Dostert, M. G. Castelli, P. Cicioni, and **M. Strolin Benedetti**

Farmitalia Carlo Erba, Research and Development — Erbamont Group, Milan, Italy

Summary. This study aimed to examine whether the increase in heart radioactivity levels after intravenous injection of ^{14}C-tyramine to rats pretreated with the irreversible MAO inhibitor tranylcypromine could be antagonized by reboxetine, a potent and selective noradrenaline uptake blocker.

Reboxetine was found totally to abolish the effect of tranylcypromine. Heart radioactivity levels after reboxetine and tranylcypromine were very similar to those found when tyramine was injected after reboxetine only.

These results suggest that reboxetine might be advantageously combined with tranylcypromine, or any MAO inhibitor, in depressed patients unresponsive of either treatment given alone.

Introduction

Among the more or less well-documented adverse events associated with the use of monoamine oxidase (MAO) inhibitors in therapy, much attention is paid to the so-called "cheese effect", i.e. the hypertensive reaction which sometimes occurs in patients treated with MAO inhibitors in response to the ingestion of indirectly acting pressor amines, such as tyramine (TYR), contained in foodstuffs. The incidence of hypertensive crises and their severity are reputed to be particularly high with the irreversible MAO inhibitor tranylcypromine (TRA) (for review see Dostert, 1984). In contrast, the risk of severe cheese reaction appears to be considerably reduced when the MAO inhibitors of the new generation are used (Dollery et al., 1984; Korn et al., 1988; Bieck and Antonin, 1989).

Administration of amitriptyline, a noradrenaline (NA) uptake blocker of moderate potency (Richelson and Pfenning, 1984), has been shown to diminish the pressor response to intravenous TYR in patients treated with TRA (Pare et al., 1982). Assuming that the increase in cardiac radioactivity levels after intravenous injection of ^{14}C-TYR to rats pretreated with cimoxatone, a potent and reversible MAO-A inhibitor devoid of inhibiting properties on the monoamine uptake systems (Kan and Strolin Benedetti,

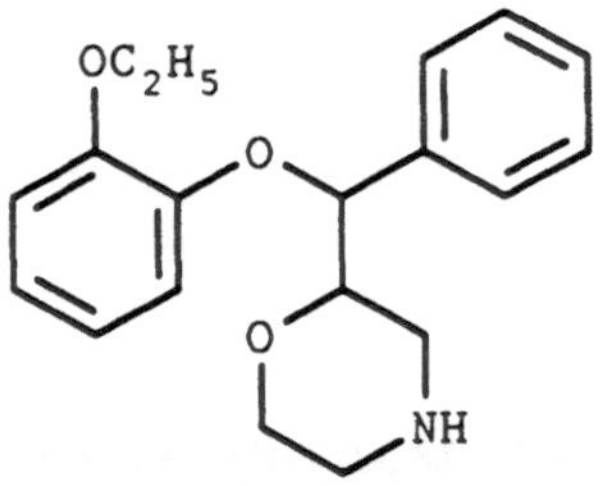

P. Dostert et al.

Fig. 1. Structure of reboxetine [RS, RS-2-(α-(2-ethoxyphenoxy)benzyl)morpholine]

1981), can be taken as a model of tyramine potentiation, Dostert et al. (1981) showed that imipramine, amitriptyline and desipramine antagonize the increase in cardiac radioactivity levels caused by cimoxatone, in keeping with Pare's results (1982).

Reboxetine (Fig. 1) is a potent and selective NA uptake blocker (Melloni et al., 1984) currently under clinical evaluation as antidepressant. The effect of reboxetine on heart radioactivity levels after intravenous injection of ^{14}C-TYR in rats treated with TRA was determined as a first step to examine whether reboxetine and TRA could be usefully combined in depressed patients unresponsive to either treatment given alone.

Material and methods

Male Sprague-Dawley rats (Charles River, Italy) weighing 200–220 g were used. Before experiment animals fasted overnight with free access to water. Five to six rats per group were used.

— 1st experiment: four groups of rats received: TYR alone, or reboxetine (5, 10, 30 mg/kg) and TYR 1 h later.
— 2nd experiment: five groups of rats received: TYR alone, or TRA (1 mg/kg) and TYR 30 min later, or reboxetine (5, 10, 30 mg/kg), 30 min later TRA (1 mg/kg) and TYR 30 min after TRA.

A 1 mg/kg oral dose of TRA was shown to inhibit rat heart MAO-A by 94% (Strolin Benedetti et al., 1983). Reboxetine doses were selected taking into account the ED_{50} value (9.9 mg/kg p.o.) of reboxetine antagonizing reserpine-induced blepharospasm in the rat (Riva et al., 1989). TRA and reboxetine were administered orally in aqueous solution as hydrochloride and methanesulfonate, respectively. Doses refer to the bases. [Side chain-1-^{14}C]tyramine hydrochloride (specific activity 53.3 mCi/mmol) was injected i.v. at a dose of 10 μg/kg in saline (0.9% NaCl) (0.6 μCi for a rat of 200 g).

Rats were killed by decapitation 15 min after TYR injection. Hearts were rapidly removed, rinsed in saline, blotted on filter paper, weighed and dissolved in 250 ml Soluene 350 (55–60°C, 48 h). Then 0.2 ml of 35% H_2O_2 and 0.5 ml of isopropylalcohol were added and vials were heated at 55°C for 2 h. Radioactivity was measured by liquid scintillation counting using 15 ml of Hionic Fluor (Packard) as scintillation cocktail.

In both experiments, the group of rats given only TYR was taken as control group and was compared to the other groups of animals using Dunnett's test.

Results

When reboxetine was given 1 h before TYR, a dose-dependent decrease in heart radioactivity levels was found (Table 1).

Administration of TRA 30 min before injection of TYR resulted in a 56% increase in heart radioactivity (Table 2). When the same doses of reboxetine as those used in the first experiment were given 30 min prior to TRA, heart radioactivity levels were found to decrease significantly, compared to controls, by 36%, 47% and 65% for the 5, 10 and 30 mg/kg dose of reboxetine, respectively. The dose-dependent decrease in heart radioactivity levels caused by reboxetine in this experiment was very similar to that observed in the absence of TRA (1st experiment, Table 1).

Marked tachycardia was noted after injection of TYR to TRA-treated rats. This effect developed to a lesser degree when the lowest dose of reboxetine was given prior to TRA dosing, and was not observed when the two other doses of reboxetine were used.

Table 1. Effect of reboxetine on rat heart radioactivity levels after i.v. injection of ^{14}C-tyramine (TYR; 10 μg/kg)

Treatment (dose of reboxetine mg/kg)	% of injected radioactivity/g of cardiac tissue (mean ± S.D.)
TYR	1.00 ± 0.21 (controls)
Reboxetine (5) + TYR	0.62 ± 0.17*
Reboxetine (10) + TYR	0.48 ± 0.13*
Reboxetine (30) + TYR	0.29 ± 0.09*

Reboxetine was given orally 1 h before tyramine
*$P \leq 0.01$ (Dunnett's test)

Table 2. Effect of reboxetine on heart radioactivity levels after i.v. injection of ^{14}C-tyramine (TYR; 10 μg/kg) in tranylcypromine (TRA)-treated rats

Treatment (dose of reboxetine in mg/kg)	% of injected radioactivity/g of cardiac tissue (mean ± S.D.)
TYR	1.15 ± 0.18 (controls)
TRA + TYR	1.80 ± 0.21*
Reboxetine (5) + TRA + TYR	0.74 ± 0.22*
Reboxetine (10) + TRA + TYR	0.61 ± 0.13*
Reboxetine (30) + TRA + TYR	0.40 ± 0.13*

Reboxetine and TRA (1 mg/kg) were given orally 1 h and 30 min before TYR, respectively
*$P \leq 0.01$ (Dunnett's test)

Discussion

In addition to its MAO inhibiting properties, TRA has been shown also to be a NA uptake blocker (Hendley and Snyder, 1968). Its potency as NA uptake blocker at a dose of 1 mg/kg, sufficient to block almost totally MAO-A in the rat heart (Strolin Benedetti et al., 1983), was clearly not enough to inhibit substantially the uptake of TYR in that tissue, as shown by the increase of heart radioactivity in rats pretreated with TRA (Table 2). However, the increase in heart radioactivity levels caused by TRA (56%) was less than that observed (134%) when cimoxatone was used as an inhibitor of MAO (Dostert et al., 1981), suggesting that the effect of a 1 mg/kg dose of TRA on the inhibition of TYR uptake in the rat heart was not negligible.

Reboxetine plasma concentrations of 320 nM were determined (Strolin Benedetti, unpublished results) in rats 1 h after administration of a 23 mg/kg dose of reboxetine, while IC_{50} values in the nM range were found for reboxetine on NA uptake in rat hypothalamic synaptosomes (Melloni et al., 1984; Dostert, unpublished results). Both parameters, NA uptake inhibition potency and plasma levels, have to be considered to account for the prevention of TYR uptake by reboxetine. In this respect, it is worth noting that, already for the lowest dose of reboxetine, the decrease in heart radioactivity in the presence of TRA (36%, Table 2) was very similar to that found (38%, Table 1) in the absence of the MAO inhibitor.

The results of the present study suggest that the high potency of reboxetine as a NA uptake blocker might be used to advantage for the prevention of cheese effect in depressed patients on MAOI therapy, although human studies are clearly needed to confirm the rat data as important species differences have been observed in reboxetine metabolism (Cocchiara et al., 1991).

References

Bieck PR, Antonin KH (1989) Tyramine potentiation during treatment with MAO inhibitors: brofaromine and moclobemide vs irreversible inhibitors. J Neural Transm [Suppl] 28: 21–31

Cocchiara G, Battaglia R, Pevarello P, Strolin Benedetti M (1991) Comparison of the disposition and of the metabolic pattern of reboxetine, a new antidepressant, in the rat, dog, monkey and man. Eur J Drug Metab Pharmacokinet 16: 231–239

Dollery CT, Brown MJ, Davies DS, Strolin Benedetti M (1984) Pressor amines and monoamine oxidase inhibitors. In: Tipton KF, Dostert P, Strolin Benedetti M (eds) Monoamine oxidase and disease. Prospects for therapy with reversible inhibitors. Academic Press, London, pp 429–441

Dostert P (1984) Myth and reality of the classical MAO inhibitors. Reasons for seeking a new generation. In: Tipton KF, Dostert P, Strolin Benedetti M (eds) Monoamine oxidase and disease. Prospects for therapy with reversible inhibitors. Academic Press, London, pp 9–24

Dostert P, Strolin Benedetti M, Sontag N (1981) Some biochemical aspects of the potential benefit of associating MD 780515 with tricyclic antidepressants. J Pharm Pharmacol 33: 639–643

Hendley ED, Snyder SH (1968) Relationship between the action of monoamine oxidase inhibitors on the noradrenaline uptake system and their antidepressant efficacy. Nature 220: 1330–1331

Kan JP, Strolin Benedetti M (1981) Characteristics of the inhibition of rat brain monoamine oxidase in vitro by MD 780515. J Neurochem 36: 1561–1571

Korn A, Da Prada M, Raffesberg W, Allen S, Gasic S (1988) Tyramine pressor effect in man: studies with moclobemide, a novel, reversible monoamine oxidase inhibitor. J Neural Transm [Suppl] 26: 57–71

Melloni P, Carniel G, Della Torre A, Bonsignori A, Buonamici M, Pozzi O, Ricciardi S, Rossi AC (1984) Potential antidepressant agents. α-Aryloxy-benzyl derivatives of ethanolamine and morpholine. Eur J Med Chem 19: 235–242

Pare CMB, Hallstrom C, Kline N, Cooper TB (1982) Will amitriptyline prevent the "cheese" reaction of monoamine oxidase inhibitors? Lancet ii: 183–186

Richelson E, Pfenning M (1984) Blockade by antidepressants and related compounds of biogenic amine uptake into rat brain synaptosomes: most antidepressants selectively block norepinephrine uptake. Eur J Pharmacol 104: 277–286

Riva M, Brunello N, Rovescalli AC, Galimberti R, Carfagna N, Carminati P, Pozzi O, Ricciardi S, Roncucci R, Rossi A, Racagni G (1989) Effect of reboxetine, a new antidepressant drug, on the central noradrenergic system: behavioural and biochemical studies. J Drug Dev 1: 243–253

Strolin Benedetti M, Dostert P, Guffroy C, Tipton KF (1983) Partial or total protection from long-lasting monoamine oxidase inhibitors (MAOIs) by new short-acting MAOIs of type A MD 780515 and type B MD 780236. Mod Probl Pharmacopsychiatry 19: 82–104

Authors' address: Dr. P. Dostert, Farmitalia Carlo Erba, Research and Development, Via C. Imbonati 24, I-20159 Milan, Italy.

J Neural Transm (1994) [Suppl] 41: 155–163

Monoamine oxidase inhibitors: effects on tryptophan concentrations in rat brain

R. L. Sherry-McKenna, J. T. F. Wong, P. R. Paetsch, G. B. Baker, D. D. Mousseau, K. F. McKenna, R. T. Coutts, and A. J. Greenshaw

Neurochemical Research Unit, Department of Psychiatry, University of Alberta, Edmonton, Alberta, Canada

Summary. It has been suggested that inhibition of tryptophan (Trp) pyrrolase and a subsequent elevation of brain Trp may contribute to the actions of antidepressant drugs. In our laboratories, we have conducted a series of experiments measuring brain Trp levels in the rat after both acute and chronic administration of several monoamine oxidase (MAO) inhibitors. The drugs studied during the course of the long-term (28 day) experiments were phenelzine, N^2-acetylphenelzine, tranylcypromine, 4-fluorotranylcypromine, 4-methoxytranylcypromine and (−)-deprenyl. High-pressure liquid chromatography with electrochemical detection was employed to measure Trp levels in brains of both MAO inhibitor- and vehicle-treated animals. No significant increases in brain Trp levels were observed as a consequence of MAO inhibitor treatment. Acute time-response (up to 24 h) and dose-response studies were conducted following the administration of phenelzine and tranylcypromine. Only after administration of high doses of these drugs was an elevation in brain Trp observed and the increase was relatively short-lived. These results suggest that elevation of brain Trp may be an important factor in the actions of MAO inhibitors only at high doses of these drugs.

Introduction

The amino acid tryptophan (Trp) is a precursor for 5-hydroxytryptamine (5-HT) and tryptamine, two amines which have been implicated in the etiology and pharmacotherapy of depressive disorders (Baker and Dewhurst, 1985). There have been numerous studies conducted in which Trp levels have been compared in body fluid samples from depressed patients and normal subjects, but conflicting results have been reported (Grahame-Smith, 1989). However, some reports (Møller et al., 1980, 1986; DeMyer et al., 1981; Maes et al., 1987) suggest that the plasma ratio of Trp to the amino acids which compete with it for transport through the blood-brain barrier may be a useful marker in depressed patients.

Trp has been also tested as an antidepressant drug but there have been varying results reported (Chouinard et al., 1979; Lundberg et al., 1979; Thomson et al., 1982; van Praag, 1984a,b; Baldessarini, 1984; Young, 1984). van Praag (1984a) has discussed some of the studies that have been carried out with Trp and 5-hydroxytryptophan (5-HTP) and suggested that further research should be conducted using larger doses and longer periods of administration, and that a therapeutic "window" effect should be considered. Dietary investigations have reported that depletion of Trp causes a rapid lowering of mood in normal males (Young et al., 1985) and that such depletion can reverse antidepressant-induced remission (Delgado et al., 1990). The latter workers reported a gradual (24–48 h) return to the remitted state on return to regular food intake and that free plasma Trp levels were negatively correlated with depression scores during acute Trp depletion. Although comprehensive studies on the effects of chronic administration of Trp or 5-HTP on 5-HT receptors have not been conducted, Blier et al. (1990) have speculated that these two amino acids would produce modifications of the 5-HT autoreceptor similar to those produced by monoamine oxidase (MAO) inhibitors. There are several reports indicating that Trp potentiates the antidepressant effects of MAO inhibitors (review: Young, 1991), tricyclic antidepressants (Shaw et al., 1975; Roos, 1976; Chouinard et al., 1979; Walinder et al., 1980; Thomson et al., 1982) and lithium (Worall et al., 1979; Brewerton and Reus, 1983). There have also been reports (Barker and Eccleston, 1984; Hale et al., 1987) indicating that lithium and Trp in combination with an MAO inhibitor or a tricyclic antidepressant are useful in treating otherwise resistant depression. However, the area of effectiveness of Trp in combination with other drugs is somewhat controversial, as indicated in reviews by Baldessarini (1984) and Young (1991).

Acute administration of different types of antidepressants (tricyclics, MAO inhibitors, novel antidepressants) has been reported to produce elevated levels in brain Trp in laboratory animals (Grahame-Smith, 1971; Tabakoff and Moses, 1976; Valzelli et al., 1980; Badawy and Evans, 1982; Edwards and Sorisio, 1988). Badawy and Evans (1981, 1982) investigated the acute effects of 19 antidepressants of different types at two doses (10 mg/kg and 0.5 mg/kg) on liver Trp pyrrolase activity and on brain Trp concentration and found significant inhibition of the pyrrolase activity and, presumably secondary to the enzyme inhibition, elevation of brain levels of Trp. Badawy and Morgan (1991) reported that acute oral administration of paroxetine resulted in increased brain Trp, and Badawy et al. (1991) observed that acute and chronic administration of lofepramine and its metabolite desipramine caused an elevation in brain levels of Trp. Tabakoff and Moses (1976) reported that the MAO-inhibiting antidepressant tranylcypromine (TCP) caused increases in brain Trp levels in mice in a dose-dependent manner 2 h after injection. The doses chosen were from 5 to 50 mg/kg; there was, however, no significant difference in brain levels of 5-HT between animals (mice) treated with 5 and 50 mg/kg doses of TCP. Grahame-Smith (1971) reported that 90 min after injecting 20 mg/kg i.p. of

TCP to rats, brain Trp was significantly increased while plasma Trp was significantly decreased.

Since a time period of 2–3 weeks or longer is often required before clinical improvement is observed with antidepressants, we have measured rat brain levels of Trp after chronic (28-day) administration of several MAO-inhibiting drugs. A short-term time and dose study was also conducted on TCP and phenelzine (PLZ), the most frequently prescribed MAO-inhibiting antidepressants.

Material and methods

TCP and PLZ were purchased from Sigma Chemicals and (−)-deprenyl from Research Biochemicals Inc. In the long-term (28 day) studies, the drugs were administered subcutaneously via osmotic minipumps (Alzet) implanted in the dorsal thoracic region. In the short-term time studies, PLZ (15 mg/kg) or TCP (10 mg/kg) was administered intraperitoneally (i.p.) to male Sprague-Dawley rats, and groups of rats (5 per group) were killed at 1, 2, 4, 8, 16 or 24 h and the whole brain removed. In the short-term dose studies, varying doses of the drugs or the vehicle were administered i.p., and the rats were killed 4 h later. 4-Methoxy-TCP, 4-fluoro-TCP and N^2-acetyl-PLZ were synthesized in our laboratories (Coutts et al., 1987, 1990; Sherry-McKenna et al., 1992) and represent analogues of PLZ and TCP under investigation as potential MAO-inhibiting antidepressants (Coutts et al., 1987; Sherry et al., 1990; McKenna et al., 1991, 1992). (−)-Deprenyl, an MAO inhibitor used in pharmacotherapy of Parkinson's disease, was also included in these investigations for comparative purposes.

At the appropriate time intervals, the rats were killed by guillotine decapitation and the brain tissue was immediately removed and frozen in isopentane over solid carbon dioxide. The procedures used in these studies had been approved by the Health Sciences Animal Welfare Committee, University of Alberta.

For analysis of Trp concentrations, HPLC with electrochemical detection was utilized. The procedure was a modification of the method described by Baker et al. (1987); an applied potential of 0.85 volt was chosen for these experiments.

Results and discussion

An elevation of brain Trp above concentrations in vehicle-treated controls was not evident in any of the drug-treated animals after 28 days of administration (Table 1). This finding is in agreement with the chronic study of Paetsch and Greenshaw (1991) in which no changes in plasma or brain Trp were found after 28-day administration of PLZ sulfate (10 mg/kg/day); a similar effect was reported in that study with desipramine HCl at a dose of 10 mg/kg/day. The doses of TCP and PLZ used in the present investigation were sufficient to cause down-regulation of β-adrenergic and/or α_2-adrenergic receptors (Greenshaw et al., 1988; McManus and Greenshaw, 1991; McKenna et al., 1992; Mousseau et al., 1993; Sherry-McKenna et al., 1992); such down-regulation is a characteristic proposed to be shared by many antidepressants (see Baker and Greenshaw, 1989 for review). At a dose of TCP considerably higher (2.5 mg/kg/day) than those presented here,

Table 1. Effects of chronic administration (28 days) of several MAO inhibitors on rat brain concentrations of Trp

Drug	Daily dose (mg/kg)	Brain region (N)	Brain tryptophan (% of control)
TCP	0.5	Whole (6)	97 ± 6[a]
TCP	0.5	Pons-medulla (8)	100 ± 4[b]
TCP	0.5	Hypothalamus (8)	90 ± 15[c]
TCP	0.5	Hippocampus (8)	109 ± 9[d]
F-TCP	0.57	Pons-medulla (9)	102 ± 4[b]
F-TCP	0.57	Hypothalamus (9)	90 ± 7[c]
F-TCP	0.57	Hippocampus (9)	101 ± 6[d]
MeO-TCP	0.61	Hypothalamus (8)	92 ± 4[e]
PLZ	5.8	Whole (8)	108 ± 6[f]
PLZ	13.6	Whole (8)	94 ± 5[f]
N[2]-Acetyl-PLZ	18.0	Whole (8)	94 ± 9[f]
(−)-Deprenyl	0.83	Whole less cortex (7)	94 ± 3[g]
(−)-Deprenyl	1.7	Whole less cortex (8)	88 ± 2[g]
(−)-Deprenyl	3.3	Whole less cortex (8)	92 ± 3[g]
(−)-Deprenyl	6.6	Whole less cortex (8)	93 ± 3[g]

Doses of all drugs are expressed as those of the free base
Control values (μg/g, means ± SEM) for Trp for each of the studies: [a] 3.5 ± 0.6; [b] 3.2 ± 0.5; [c] 4.9 ± 0.3; [d] 4.0 ± 0.3; [e] 5.1 ± 0.7; [f] 3.2 ± 0.2; [g] 4.7 ± 0.1
Abbreviations: *TCP* tranylcypromine; *F-TCP* 4-fluoro-TCP; *MeO-TCP* 4-methoxy-TCP; *PLZ* phenelzine

we also did not observe any increase in Trp concentrations in brain at 4, 10 or 28 days of drug administration (Goodnough and Baker, unpublished).

Comprehensive short-term studies on the MAO inhibitor TCP and PLZ indicated that both drugs elevated brain Trp, but the effects were produced only at high doses of the drugs (Figs. 1 and 2) and were short-lived. At a dose of PLZ of 15 mg/kg, a significant elevation (29 ± 4% above control) of Trp was observed only at 4 h, while at the 10 mg/kg dose of TCP, there was a significant elevation (40 ± 10% above control) at 1 h only.

The elevations observed may be due to inhibition of Trp pyrrolase, as shown by Badawy and Evans (1981, 1982). However, the structurally similar drug (+)-amphetamine also causes an elevation of brain Trp (Schubert and Sedvall, 1972) but is reported to have no effect on Trp pyrrolase (Schubert and Sedvall, 1972) or Trp hydroxylase (Schubert and Sedvall, 1972; Trulson and Jacobs, 1980) after acute administration. It has been suggested that (+)-amphetamine elevates brain Trp by stimulation of the uptake/transport system of Trp (Schubert and Sedvall, 1972; Valzelli et al., 1980). Given the structural similarity among amphetamine, TCP and PLZ, it is possible that these two MAO inhibitors may have a direct effect on Trp transport into the brain at high doses. Campbell et al. (1978) noted an increase in uptake of Trp into cortical synaptosomes after a single dose (15 mg/kg) of PLZ, but this effect was not evident at 1, 2 or 3 weeks after administration of this dose every 48 h. Recent reports have suggested that

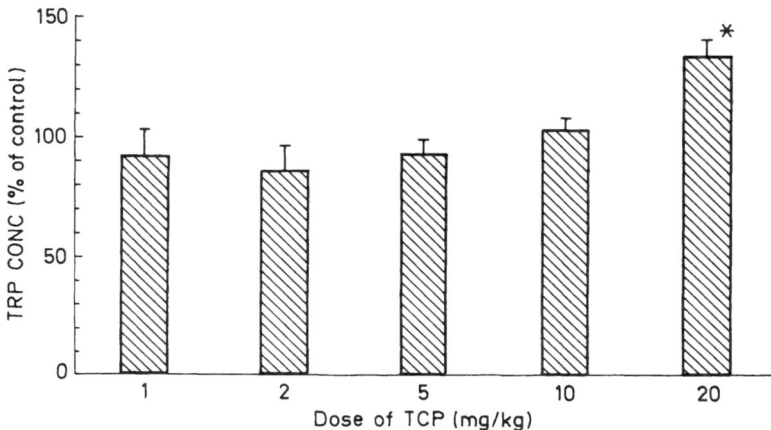

Fig. 1. Trp levels in rat whole brain 4 h after administration of various doses of TCP (doses expressed as free base). Values for Trp are expressed as % of values obtained in vehicle-treated control rats and represent means ± SEM (n = 6). *p < 0.05, compared to control values

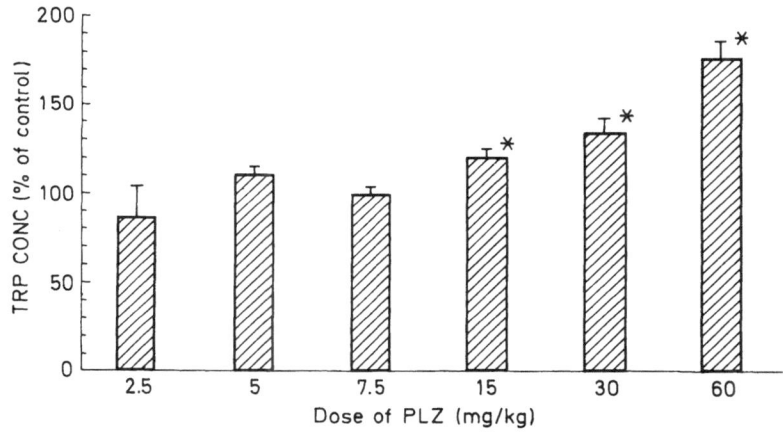

Fig. 2. Trp levels in rat whole brain 4 h after administration of various doses of PLZ (doses expressed as free base). Values for Trp are expressed as % of values obtained in vehicle-treated control rats and represent means ± SEM (n = 6). *p < 0.05, compared to control values

concentrations of Trp and other large amino acids in brain and plasma may be regulated by a β-adrenoceptor-mediated mechanism (Edwards and Sorisio, 1988; Eriksson and Carlsson, 1988; Paetsch and Greenshaw, 1991). Based on their acute studies with propranolol and imipramine, Edwards and Sorisio (1988) suggested that Trp increases in brain produced by imipramine are mediated by β-adrenoceptor stimulation, but Paetsch and Greenshaw (1991), in an investigation of 28-day administration of PLZ and desipramine, were unable to find an effect of these antidepressants on the increase in brain Trp concentrations produced by β-agonists. PLZ has also been reported to inhibit several enzymes in addition to MAO (Yu and Boulton, 1992) although, to our knowledge, it has not been reported to inhibit Trp hydroxylase. In fact, Campbell et al. (1978) found an increase in Trp hydroxylase activity in brains of rats treated chronically with PLZ at a dose of 15 mg/kg/48 h and Waldmeier (personal communication) has found that a high dose of PLZ (200 mg/kg s.c./2 h) also produced an increase in Trp hydroxylase activity in rat brain.

There is discrepancy between our findings on the effects of TCP and PLZ on whole brain Trp levels at 4 h after injection and those of Badawy and Evans (1982) at 3.5 h following injection. The latter authors found that brain Trp was elevated significantly at doses of these drugs as low as 0.5 mg/kg. We found that doses at least 20 times greater than these were required before Trp elevations were noted with these drugs. The reason for this discrepancy is not clear at this time.

Acknowledgements

Funding was provided by the Medical Research Council of Canada, the Alberta Provincial Mental Health Advisory Council (PMHAC) and the Alberta Heritage Foundation for Medical Research (AHFMR). RLS-M is the recipient of an Alberta Mental Health Scholarship, PRP has a MRC studentship, and KFM was the recipient of an AHFMR Clinical Fellowship. The authors are grateful to Dr. P. Waldmeier of Ciba-Geigy, Basel, Switzerland, for helpful discussions and to Ms. S. Omura for typing this manuscript.

References

Badawy AA, Evans M (1981) Inhibition of rat liver tryptophan pyrrolase activity and elevation of brain tryptophan concentration by administration of antidepressants. Biochem Pharmacol 30: 1211–1216

Badawy AA, Evans M (1982) Inhibition of rat liver tryptophan pyrrolase activity and elevation of brain tryptophan concentration by acute administration of small doses of antidepressants. Br J Pharmacol 77: 59–67

Badawy AA, Morgan CJ, Dacey A, Stoppard T (1991) The effects of lofepramine and desmethylimipramine on tryptophan metabolism and disposition in the rat. Biochem Pharmacol 42: 921–929

Badawy AA, Morgan CJ (1991) Effects of acute paroxetine administration on tryptophan metabolism and disposition in the rat. Br J Pharmacol 102: 429–433

Baker GB, Coutts RT, Rao TS (1987) Neuropharmacological and neurochemical properties of N-(2-cyanoethyl)-2-phenylethylamine, a prodrug of 2-phenylethylamine. Br J Pharmacol 92: 243–255

Baker GB, Dewhurst WG (1985) Biochemical theories of affective disorders. In: Dewhurst WG, Baker GB (eds) The pharmacotherapy of affective disorders: theory and practice. Croom Helm, London, pp 1–59

Baker GB, Greenshaw AJ (1989) Effects of long-term administration of antidepressants and neuroleptics on receptors in the central nervous system. Cell Mol Neurobiol 9: 1–44

Baldessarini RJ (1984) Treatment of depression by altering monoamine metabolism: precursors and metabolic inhibitors. Psychopharmacol Bull 20: 224–239

Barker WA, Eccleston D (1984) The treatment of chronic depression — an illustrative case. Br J Psychiatry 144: 317–319

Blier P, de Montigny C, Chaput Y (1990) A role for the serotonin system in the mechanism of action of antidepressant treatments: preclinical evidence. J Clin Psychiatry 51: 14–20

Brewerton TD, Reus VI (1983) Lithium carbonate and L-tryptophan in the treatment of bipolar and schizoaffective disorders. Am J Psychiatry 140: 757–760

Campbell IC, Colburn R, Walker MN, Lovenberg W, Murphy DL (1978) Norepinephrine and serotonin metabolism in the rat brain: effects of chronic phenelzine administration. In: Deniker P, Radouco-Thomas C, Villeneuve A, Baronet-Lacroix D, Garcin F (eds) Neuropsychopharmacology, vol 1. Pergamon Press, Oxford

Chouinard G, Young SN, Annable L, Sourkes TL (1979) Tryptophan, nicotinamide, imipramine and their combination in depression. Acta Psychiatr Scand 59: 395–414

Coutts RT, Mozayani A, Pasutto FM, Baker GB, Danielson TJ (1990) Synthesis and pharmacological evaluation of acyl derivatives of phenelzine. Res Commun Chem Pathol Pharmacol 67: 3–15

Coutts RT, Rao TS, Baker GB, Micetich RG, Hall TWE (1987) Neurochemical and neuropharmacological properties of 4-fluorotranylcypromine. Cell Mol Neurobiol 7: 271–290

Delgado PL, Charney DS, Price LH, Aghajanian GK, Landis H, Heninger GR (1990) Serotonin function and the mechanism of antidepressant action: reversal of antidepressant-induced remission by rapid depletion of plasma tryptophan. Arch Gen Psychiatry 47: 411–418

DeMeyer MK, Shea PA, Hendrie HC, Yoshimura NN (1981) Plasma tryptophan and five other amino acids in depressed and normal subjects. Arch Gen Psychiatry 38: 642–646

Edwards DJ, Sorisio DA (1988) Effects of imipramine on tyrosine and tryptophan are mediated by β-adrenoceptor stimulation. Life Sci 42: 853–862

Eriksson T, Carlsson A (1988) β-Adrenergic control of brain uptake of large neutral amino acids. Life Sci 42: 1583–1589

Grahame-Smith DG (1971) Studies in vivo on the relationship between brain tryptophan, brain 5-HT synthesis and hyperactivity in rats treated with a monoamine oxidase inhibitor and L-tryptophan. J Neurochem 18: 1053–1066

Grahame-Smith DG (1989) Serotonin function in affective disorders. Acta Psychiatr Scand 80 [Suppl] 350: 7–12

Greenshaw AJ, Nazarali AJ, Rao TS, Baker GB, Coutts RT (1988) Chronic tranylcypromine treatment induces functional α_2-adrenoceptor down-regulation in rats. Eur J Pharmacol 154: 67–72

Hale AS, Procter AW, Bridges PK (1987) Clomipramine, tryptophan and lithium in combination for resistant endogenous depression: seven case studies. Br J Psychiatry 151: 213–217

Lundberg D, Ahlfors HG, Dencker SJ, Fruensgaard K, Hansten S, Jensen K, Ose E, Pihkanen TA (1979) Symptom reduction in depression after treatment with L-tryptophan or imipramine: item analysis of Hamilton rating scale for depression. Acta Psychiatr Scand 60: 287–294

Maes M, De Ruyter M, Suy E (1987) Prediction of subtype and severity of depression by means of dexamethasone suppression test, L-tryptophan: competing amino acids ratio, and MHPG flow. Biol Psychiatry 22: 177–188

McKenna KF, Baker GB, Coutts RT (1991) N^2-Acetylphenelzine: effects on rat brain GABA, alanine and biogenic amines. Naunyn Schmiedebergs Arch Pharmacol 343: 478–482

McKenna KF, Baker GB, Coutts RT, Greenshaw AJ (1992) Chronic administration of the antidepressant/antipanic drug phenelzine and its N-acetyl analogue: effects on monamine oxidase, biogenic amines and α_2-adrenoceptor function. J Pharm Sci 81: 832–835

162 R. L. Sherry-McKenna et al.

McManus DJ, Greenshaw AJ (1991) Differential effects of antidepressants on GABA$_B$ and β-adrenergic receptors in rat cerebral cortex. Biochem Pharmacol 42: 1525–1528

Møller SE, Kirk L, Honoré P (1980) Relationship between plasma ratio of tryptophan to competing amino acids and the response to L-tryptophan treatment in endogenously depressed patients. J Affect Dis 2: 47–59

Møller SE, de Beurs P, Timmerman L, Tan BK, Leijnse-Ybema HJ, Cohen Stewart MH, Hopfner Petersen HE (1986) Plasma tryptophan and tyrosine ratios to competing amino acids in relation to antidepressant response to citalopram and maprotiline. A preliminary study. Psychopharmacology 88: 96–100

Morgan CJ, Badawy AA (1989) Effects of a suppression test dose of dexamethasone on tryptophan metabolism and disposition in the rat. Biol Psychiatry 25: 359–362

Mousseau DD, McManus DJ, Baker GB, Juorio AV, Dewhurst WG, Greenshaw AJ (1993) Effects of age and of chronic antidepressant treatment on ^{3}H-tryptamine binding to rat cortical membranes. Cell Mol Neurobiol 13: 3–13

Paetsch PR, Greenshaw AJ (1991) β-Adrenergic effects on plasma and brain large neutral amino acids are unaltered by chronic administration of antidepressants. J Neurochem 56: 2027–2032

Roos GE (1976) Tryptophan, 5-hydroxytryptophan and tricyclic antidepressants in the treatment of depression. Monogr Neural Sci 3: 23–25

Schubert J, Sedvall G (1972) Effect of amphetamines on tryptophan concentrations in mice and rats. J Pharm Pharmacol 24: 53–62

Shaw DM, McSweeney DA, Hewland R, Johnson AL (1975) Tricyclic antidepressants and tryptophan in unipolar depression. Psychol Med 5: 276–278

Sherry RL, Baker GB, Coutts RT (1990) Effects of low-dose 4-fluorotranylcypromine on rat brain monamine oxidase and neurotransmitter amines. Biol Psychiatry 28: 539–543

Sherry-McKenna RL, Baker GB, Mousseau DD, Coutts RT, Dewhurst WG (1992) 4-Methoxytranylcypromine, a monoamine oxidase inhibitor: effects on biogenic amines in rat brain following chronic administration. Biol Psychiatry 31: 881–888

Tabakoff B, Moses F (1976) Differential effects of tranylcypromine and pargyline on indoleamines in brain. Biochem Pharmacol 25: 2555–2560

Thomson J, Rankin H, Ashcroft GW, Yates GM, McQueen JK, Cummings SW (1982) The treatment of depression in general practice: a comparison of L-tryptophan amitriptyline and a combination of L-tryptophan and amitriptyline with placebo. Psychol Med 12: 741–751

Trulson ME, Jacobs BL (1980) Chronic amphetamine administration decreases brain tryptophan hydroxylase activity in cats. Life Sci 26: 329–335

Valzelli I, Bernasconi S, Coen E, Petkov VV (1980) Effects of different psychoactive drugs on serum and brain tryptophan levels. Neuropsychobiology 6: 224–229

van Praag HM (1984a) Depression, suicide and serotonin metabolism in the brain. In: Post RM, Ballenger JC (eds) Neurobiology of mood disorders. Williams & Wilkins, Baltimore, pp 601–608

van Praag HM (1984b) Studies on the mechanism of action of serotonin precursors in depression. Psychopharmacol Bull 20: 599–602

Walinder J, Carlsson A, Persson R, Wallin L (1980) Potentiation of the effect of antidepressant drugs by tryptophan. Acta Psychiatr Scand 28: 243–249

Worrall EF, Moody JP, Peet M, Dick P, Smith A, Chambers C, Adams M, Naylor GJ (1979) Controlled studies on the acute antidepressant effect of lithium. Br J Psychiatry 135: 255–262

Young SN (1984) Monoamine precursors in the affective disorders. In: Burrows GD, Werry JW (eds) Advances in human psychopharmacology, vol 3. JAI Press, Greenwich, CT, pp 251–285

Young SN (1991) Use of tryptophan in combination with other antidepressant treatments: a review. J Psychiatr Neurosci 16: 241–246

Young SN, Smith SE, Pihl RO, Ervin FR (1985) Tryptophan depletion causes a rapid lowering of mood in normal males. Psychopharmacology 87: 173–177

Yu PH, Boulton AA (1992) A comparison of the effect of brofaromine, phenelzine and tranylcypromine on the activities of some enzymes involved in the metabolism of different neurotransmitters. Res Commun Pyschol Psychiat Behav 16: 141–153

Authors' address: R. L. Sherry-McKenna, Neurochemical Research Unit, Department of Psychiatry, University of Alberta, Edmonton, Alberta, Canada, T6G 2B7.

Deprenyl, MAO-B and neurodegenerative conditions

J Neural Transm (1994) [Suppl] 41: 167–175

Behaviour of (−)-deprenyl and its analogues

K. Magyar

Department of Pharmacodynamics, Semmelweis University of Medicine,
Budapest, Hungary

Summary. A number of new deprenyl analogues were synthesized during the last decades and structure-activity relationship studies were carried out with the compounds. Among these derivatives U-1424 [N-methyl-N-pro-pargyl-(2-furyl-1-methyl)-ethyl ammonium] and J-508 [N-methyl-N-pro-pargyl-(1-indanyl) ammonium] preserved the selectivity to MAO-B, but the former is slightly less potent inhibitor of the enzyme, while J-508 is more effective than the parent compound. The studies led us to the conclusion that, in the case of a selective and irreversible inhibitor, it is not a proper aim to search for a more potent inhibitor than deprenyl. Nevertheless, the effects of the new derivatives independent of the enzyme inhibitory potency can be beneficial. In this respect p-fluoro-deprenyl (PFD) seems to be promising.

In addition to the enzyme inhibitory action, the compounds possess reversible effects e.g. inhibition of uptake and release of the synaptic processes. The fate of the drugs in the body including metabolism is also an important aspect of drug action. PFD is slightly less potent inhibitor of MAO-B in vitro than deprenyl but it maintains a more prolonged concentration in tissues. Its metabolites (p-fluoro-amphetamine and p-fluoro-methamphetamine) are more effective inhibitors of uptake than the parent compound. In respect of the release of transmitter amines, the (+)-isomers of the metabolites are more potent but we did not find significant differences between the uptake inhibitory potencies of the stereoisomers. PFD is more effective to protect the neurodegenerative effects of the noradrenergic neurotoxin DSP-4, compared to deprenyl.

Introduction

Deprenyl was firstly described in 1965 (Knoll et al., 1965) and, almost since that time, a continuous effort has been made in our laboratories to develop chemical derivatives of deprenyl having better qualities than the original compound (Knoll et al., 1978; Magyar et al., 1980). More than 200 congeners of deprenyl were synthesised by the chemists of the Chinoin Pharmaceutical Works (Budapest) and structure-activity relationship studies were carried out in our laboratories.

Results and discussion

Structure activity relationship studies with deprenyl derivatives revealed:

1. The substitution of deprenyl in the side chain at the alpha-position with ethyl-, isopropyl- or benzyl-group decreased the monoamine oxidase (MAO) inhibitory potency of the compounds. Beta-substitution (methyl-dimethyl-) of the side chain resulted also in similar changes.
2. Elongation or shortening of the side chain of deprenyl was also not preferable in respect to increase the inhibition of MAO.
3. Omitting the methyl group of deprenyl in the side chain at alpha-position [N-methyl-N-propargyl-(2-phenyl)-ethyl-ammonium.HCl; TZ-650] did not influence significantly the inhibitory potency and the B-type selectivity of the new derivative.
4. Alteration of the phenyl-ring (halogenation, saturation, methoxy-substitution) diminished the inhibitory potency of the new derivatives. The only exception in this group was (−)-p-fluoro-deprenyl (PFD), which compound proved to be promising, having the chemical structure, potency and B-type selectivity similar to deprenyl.
5. Replacement of the phenyl-ring with different-mainly with heterocyclic-ring were the most frequent chemical alteration on deprenyl structure. The phenyl-ring of deprenyl was replaced with furanyl-, thienyl-, cyclopentyl-, benzothienyl-, benzofuranyl-, indanyl-, naphtalene-, isoquinoline-rings. Furan [N-methyl-N-propargyl-(2-furyl-1-methyl-ethyl-ammonium.HCl; U-1424] or indane substitution [N-methyl-N-progargyl-(1-indanyl)-ammonium.HCl; J-508] resulted in potent inhibitors to MAO-B. U-1424 proved to be slightly less potent, but J-508 exceeded the inhibitory potency of deprenyl with one order of magnitude.
6. Halogenation of the propargyl-group with any halogen and at any position totally abolished the inhibitory action of the new structures.

The structure-activity relationship studies lead us at least to two important conclusions: 1. It is useless to develop a more effective MAO-B inhibitor than deprenyl. An inhibitor of an irreversible type, even it is weaker than deprenyl, in a proper dose or in a repeated treatment in vivo leads to the same degree of inhibition. 2. When we want to retain the complexity of the effect of deprenyl, we have to preserve the amphetaminergic structure of the new congener molecule. That is the reason why PFD seems to be a promising new structure.

Deprenyl cannot be considered as a simple MAO-B inhibitor in spite of the fact that in the lowest effective concentration ($IC_{50} = 10^{-8}$ M) it inhibits most potently the oxidative deamination of phenylethylamine (PEA).

On the basis of the knowledge which has been accummulating during the last decades on the complexity of deprenyl effect the following could be concluded. In addition to its B-type selective irreversible MAO inhibitory potency deprenyl is stereoselective. Its (−)-enantiomer is more potent inhibitor of the enzyme than the (+)-one (Magyar et al., 1967). It has also been proved that deprenyl has some inhibitory effect on the synaptosomal uptake of ^{3}H-noradrenaline (^{3}H-NA) (Knoll and Magyar, 1972). During long-term treatment (3 weeks) when deprenyl was administered in a fairly low dose (0.05 mg/kg) the B-type selective inhibitory spectrum can be preserved (Ekstedt et al., 1978). Deprenyl has a special distribution in the brain. Due to its high lipid solubility after intravenous treatment it penetrates rapidly to the central nervous system (30 sec). The whole body autoradiography in mice revealed that the rapid rise in ^{14}C-deprenyl concentration is followed by a sudden decrease in radioactivity of the brain. The 5 min autoradiogram shows almost background radioactivity in the brain, while body lipids and other organs contained fairly high level at this time (Magyar et al., 1968; Magyar and Szüts, 1982).

It has been confirmed in many laboratories, like in ours, that deprenyl is metabolized to methylamphetamine (MA) and amphetamine (Reynolds et al., 1978; Magyar and Szüts, 1982; Magyar and Tóthfalusi, 1984; Heinonen et al., 1989). In spite of the formation of amphetamines, deprenyl did not prove to be a potent releaser of the biogenic amines. When rats were injected with ^{3}H-NA (100 μCi) intravenously, deprenyl pretreatment (30-min i.p., prior to ^{3}H-NA administration) dose-dependently increased the concentration of ^{3}H-NA in the rat heart synaptosomes prepared in sucrose density gradient. In contrast to deprenyl, tranylcypromine, pargyline and clorgyline markedly decreased the ^{3}H-NA content of the same synaptosomal preparation due to their strong releasing potency (Knoll and Magyar, 1972).

The lack of the releasing effect of (−)-deprenyl can be due firstly to the fact that from (−)-deprenyl only the (−) isomers of amphetamines can be formed by the metabolism which have lower releasing potency than the (+)-forms. It was proved earlier in parkinsonian patients that racemase did not convert the (−)-amphetamines to their (+)-forms (Schachter et al., 1980). This might secondly be due to the fact that not directly the amphetamines, but rather their further metabolites [(+)-p-hydroxynorephedrine] are responsible for the release of NA from the depot granules (Brodie et al., 1970). Only the (+)-p-hydroxyamphetamine is converted to (+)-p-hydro-xynorephedrine in vivo [the (−)-p-hydroxyamphetamine is not a substrate for beta-hydroxylase]. This can explain the difference between the releasing potencies of the different enantiomers (Goldstein and Anagnoste, 1965).

It was published recently that in mice (−)-deprenyl (selegiline) but not MDL-72974 protected against the neurodegenerative effect of the nora-drenergic neurotoxin, DSP-4, when the inhibitors were administered i.p., one h before DSP-4 treatment (Finnegan et al., 1990). Since both selegiline and MDL-72974 produce comparable degrees of MAO-B inhibition it seems doubtful that MAO-B activity plays any significant role in the protection of DSP-4 induced neurotoxicity.

Similar results were obtained in our laboratory in rats treated with selegiline and especially with its para-fluoro-derivative (Magyar, 1991). It is noteworthy that, when deprenyl was administered 24 h prior to DSP-4 treatment it was almost ineffective, while PFD pretreatment preserved a remarkable potency to protect DSP-4 induced neurotoxicity in this experimental schedule. Mainly the amphetamine-like potential metabolites of the two MAO-B inhibitors were found to be potent in the prevention of DSP-4 induced noradrenaline depletion in the rat hippocampus (Table 1).

Since the positively charged aziridium ion — spontaneously formed from DSP-4 in an aqueous medium — is responsible for the toxic effect of DSP-4, which is actively taken up by noradrenargic neurons, potent blockers of NA uptake can effectively prevent the DSP-4 induced toxicity (Ross, 1976). MDL-72974 lacks reuptake blocking action and fails to inhibit the DSP-4 effect (Finnegan et al., 1990).

The MPTP induced neurodegeneration can be prevented by any MAO-B blockers, but a line of evidence proves that potent inhibitors of dopamine (DA) uptake (mazindol) can also prevent MPP^+ caused neurodegeneration (Langston, 1980). Because two ways are existing to prevent MPTP induced neurodegeneration, we selected the DSP-4 model where only the compounds with uptake inhibitory properties are proved to be effective.

We have been studying the uptake inhibitory effect of deprenyl and PFD and their potential metabolites on the synaptosomal uptake of ^3H-NA, ^3H-dopamine (^3H-DA) and ^3H-serotonin (^{3}-5-HT) in the rat hypothalamus, striatum and hippocampus, respectively, using the method of Snyder and Coyle in vitro (1969). As it is shown by the IC_{50} values on Table 2, none of the compounds administered, inhibited the presynaptic uptake of ^3H-5-HT in an acceptable concentration. Deprenyl and PFD poorly inhibited the uptake both of ^3H-NA and ^3H-DA, but their metabolites are more effective in this respect with one order of magnitude. The (+)-MA is more potent to inhibit the uptake of both NA and DA, but both enentiomers of p-fluoro-methylamphetamine (PFMA) are equally effective on the inhibition of the uptake of ^3H-NA. Concerning ^3H-DA uptake the (+)-form of PFMA is also more effective.

The question arises whether the quantity of the amphetamine-like metabolites potentially formed from deprenyl and PFD in the rat can or cannot be responsible for the protection against DSP-4 toxicity as a consequence of their uptake inhibitory potency.

To elucidate this problem we studied the fate of deprenyl and PFD in rats by using their alternatively and positionally labelled radioisomers. The rings were tritiated and the propargyl side-chain was ^{14}C-labelled in both of the inhibitors. By detecting the ^3H-label we were able to follow the fate of the moieties, such as the parent compound, p- or m-hydroxylated deprenyl, methylamphetamine, amphetamine, desmethyl-deprenyl and metabolites without even an amino group. By using the ^{14}C-label (propargyl-^{14}C) we could follow the fate of the parent compound, desmethyl-deprenyl, p- or m-hydroxylated deprenyl, the propargyl group as well as of the metabolite(s) of this last residue.

The rats were treated orally with the mixture of the two radiolabels (1.5 mg/kg) of deprenyl or PFD and the radioactivity of both labels were simultaneously measured in the plasma and 15 brain regions (Fig. 1). The molar concentrations of the inhibitors in the tissues calculated on the basis of the specific activities represented by the ^3H and ^{14}C separately, should be equal provided that the substances are unchanged. A considerable

Table 1. DSP-4 induced depletion of NA in rat hippocampus at various time after deprenyl, p-fluoro-deprenyl (*PFD*), methylamphetamine (*MA*) and p-fluoro-methylamphetamine (*PFMA*) pretreatment

Pretreatment	Treatment	N	NA content	MAO-B inhibition
			in % of control ± S.D.	
Saline	saline	16	100	0
Deprenyl	saline	8	108 ± 6.3	75 ± 8.5
Saline	DSP-4	8	10 ± 6.7	8 ± 1.7
Deprenyl	DSP-4	7	90 ± 13.7	58 ± 1.8
Deprenyl	DSP-4 (24 h later)	7	26 ± 7.4	63 ± 6.3
PFD	saline	7	104 ± 12.8	39 ± 2.9
PFD	DSP-4	7	95 ± 9.8	63 ± 2.4
PFD	DSP-4 (24 h later)	7	45 ± 9.1	72 ± 3.9
Saline	DSP-4*	7	42 ± 5.8	16 ± 4.4
(+)-MA	DSP-4*	7	105 ± 1.3	16 ± 2.3
(−)-MA	DSP-4*	7	116 ± 2.1	16 ± 1.3
(+)-PFMA	DSP-4*	7	147 ± 8.4	18 ± 4.9
(−)-PFMA	DSP-4*	7	130 ± 9.8	16 ± 2.1

Rats were pretreated i.p. with deprenyl, PFD (10 mg/kg), MA and PFMA (5 mg/kg). The pretreatment was followed by DSP-4 administration (50 mg/kg, i.p.) 1 h after the injection (if not otherwise stated). Rats were decapitated 7 days after DSP-4 treatment and the content of NA and the MAO-B inhibition was determined in the hippocampus NA content of the hippocampus of control rats (100%) = 0.32 ± 0.06 ng/mg tissue ± S.D. * Dose: 30 mg/kg

Table 2. The effect of deprenyl, p-fluoro-deprenyl (*PFD*) and methylamphetamine (*MA*) derivatives on the synaptosomal uptake in vitro in rats

Compounds	IC_{50} in M		
	NA Hypothalamus	DA Striatum	5-HT Hippocampus
(−)-deprenyl	5.1×10^{-5}	1.0×10^{-4}	5.0×10^{-3}
(−)-deprenyl	1.7×10^{-5}	2.4×10^{-5}	3.6×10^{-2}
(−)-PFD	1.3×10^{-5}	2.9×10^{-5}	1.4×10^{-3}
(+)-PFD	6.1×10^{-6}	1.6×10^{-5}	6.0×10^{-4}
(−)-MA	3.5×10^{-6}	4.2×10^{-5}	—
(+)-MA	3.5×10^{-7}	6.0×10^{-7}	1.9×10^{-2}
(−)-PFMA*	7.5×10^{-6}	3.0×10^{-4}	—
(+)-PFMA*	7.7×10^{-6}	2.3×10^{-5}	1.7×10^{-3}

* *PFMA* p-fluoro-methylamphetamine

difference found soon in the early phase of the experiments suggests that the molecule undergoes a rapid metabolic disintegration.

Figure 1 shows the plasma and brain levels in rats treated orally with 1.5 mg/kg of deprenyl, 45 min after the ingestion on the double labelled compound. In the plasma the ^{14}C, whilst in all of the cerebral tissues the

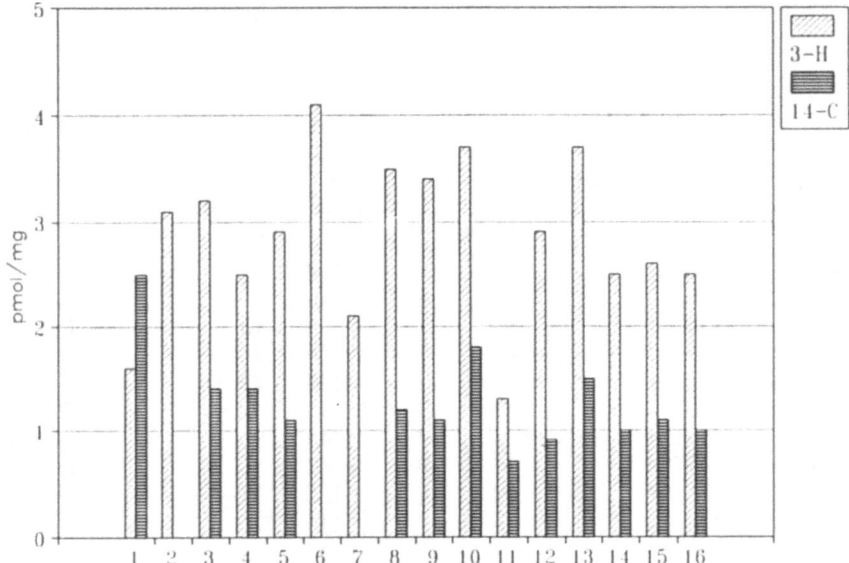

Fig. 1. The plasma and brain levels of orally administered alternatively and positionally labelled (−)-deprenyl (1.5 mg/kg) to rats, 45 min after treatment. The radioactive doses from ^{3}H and ^{14}C-labels were 0.86 and 0.19 MBq/100 g, respectively. The tissue levels were calculated on the basis of the specific activities of the ^{3}H and ^{14}C labelled radioisomers. *1* Plasma, *2* corpus pineale, *3* bulbus olfactorius, *4* hypothalamus, *5* tuberculum olfactorium, *6* substantia nigra, *7* corpus mamillare, *8* frontal cortex, *9* parietal cortex, *10* hypophysis, *11* corpus striatum, *12* hippocampus, *13* colliculus superior, *14* medulla oblongata, *15* cerebellum, *16* pons + colliculus inferior

^{3}H-tracer dominates. Only the simultaneous presence of the equal amounts of the two radiolabels indicates the integrity of the parent compound. Consequently the results suggest the formation of a considerable quantities of metabolites (MA, amphetamine) and their presence in the brain. Similar situation, the surplus of the ^{3}H-label persists during 48 h in the brain areas. The metabolites should be taken into account while evaluating the pharmacological effects of deprenyl.

Similar results were obtained with PFD (Fig. 2), but at least threefold amount of surplus ^{3}H-label was measured in the brain regions compared to deprenyl. The highest concentration of both deprenyl and PFD was reached in the substantia nigra.

The time related changes of ^{3}H and ^{14}C radioactivity in the frontal cortex is documented after deprenyl treatment (Fig. 3). The extra amount of ^{3}H-lablel compared to the ^{14}C-one is remarkable during the whole period of the experiment (96 h). Similar pattern of distribution can be demonstrated in the case of PFD as an example in the corpus striatum (Fig. 4).

These experiments indicate that though pmole quantities per mg tissue of amphetamines are present in the brain regions because of the uneven distribution their levels in certain brain areas can play a role in the pharmacological effect of deprenyl and mainly in that of PFD. The concentra-

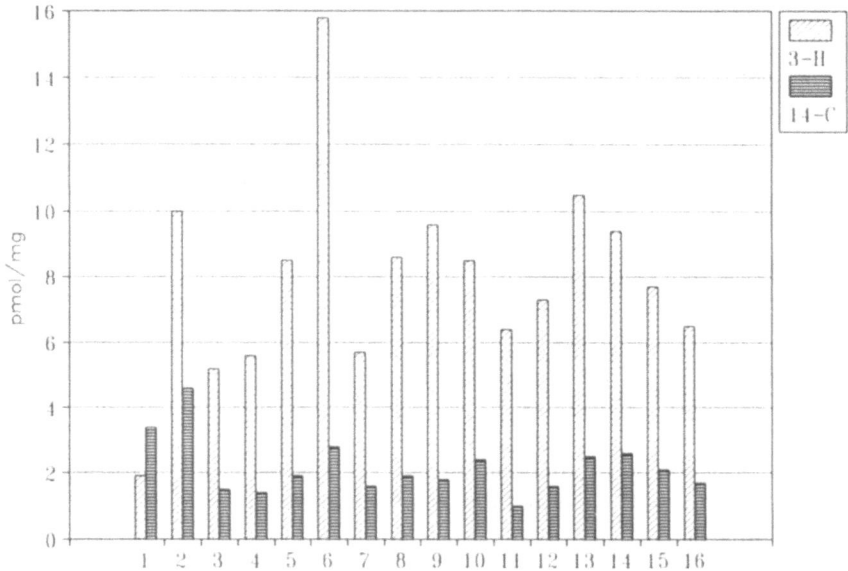

Fig. 2. The plasma and brain levels of orally administered alternatively and positionally labelled PFD (1.5 mg/kg) to rats, 45 min after treatment. The radioactive doses from ^{3}H and ^{14}C-labels were 1.47 and 0.32 MBq/100 g, respectively. The tissue levels were calculated on the basis of the specific activities of the ^{3}H and ^{14}C labelled radioisomers. Codification of the plasma and brain regions as in Fig. 1

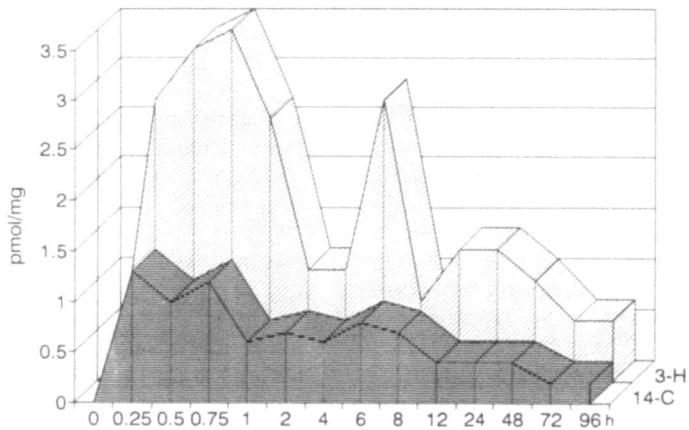

Fig. 3. Time related changes in the concentrations of orally given (−)-deprenyl, detected in the frontal cortex. Treatment as in Fig. 1. Time scale is not linear

tion of the amphetamine-like metabolites after PFD treatment are threefold of that wich was found following the administration of the same dose of deprenyl. This could be due to the altered metabolism of PFD in the body, different from deprenyl. The substitution of PFD in p-position prevents the p-hydroxylation of the molecule. Because of that the concentration of PFD

174 K. Magyar

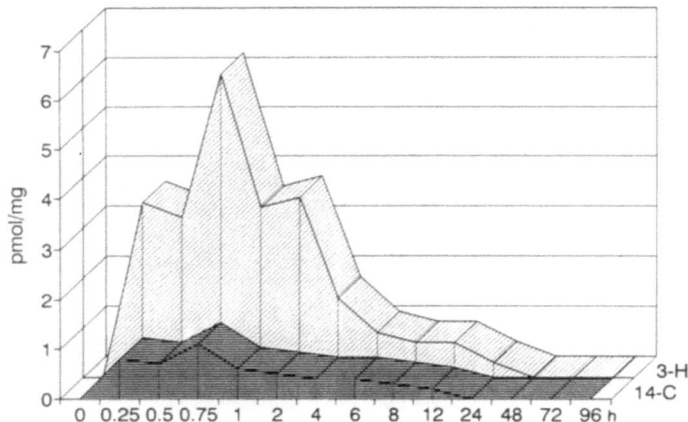

Fig. 4. Time related changes in the concentrations of orally given PFD, detected in the corpus striatum. Treatment as in Fig. 2. Time scale is not linear

can be higher and its presence is more prolonged in the body. This enhances the reversible effects of PFD not related to MAO-B inhibition.

References

Brodie BB, Cho AK, Gessa GL (1970) Possible role of p-hydroxynorephedrine in the depletion of norepinephrine induced by d-amphetamine and in tolerance to this drug. In: Costa E, Garatini S (ed) Amphetamines and related compounds. Raven Press, New York, pp 217–230

Ekstedt B, Magyar K, Knoll J (1978) Does the B form selective monoamine oxidase inhibitor lose selectivity by long term treatment? Biochem Pharmacol 28: 919–923

Finnegan KT, Skratt JJ, Irwin I, DeLanney LE, Langston JW (1990) Protection against DSP-4-induced neurotoxicity by deprenyl is not related to its inhibition of MAO-B. Eur J Pharmacol 184: 119–126

Goldstein M, Anagnoste B (1965) The conversion in vivo of d-amphetamine to (+)-p-hydroxy-norephedrine. Biochim Biophys Acta 107: 166–168

Heinonen EH, Myllyla, V, Sotaniemi K, Lammintausta R, Salonen JS, Anttila M, Savijärvi M, Kotila M, Rinne UK (1989) Pharmacokinetics and metabolism of selegiline. Acta Neurol Scand 126: 93–99

Knoll J (1983) (−)-Deprenyl (selegiline): the history of its development and pharmacological action. Acta Neurol Scand [Suppl] 95: 57–80

Knoll J, Magyar K (1972) Some puzzling pharmacological effects of monoamine oxidase inhibitors. In: Costa E, Sandler M (eds) Monoamine oxidases. New vistas. Raven Press, New York, pp 393–408 (Adv Biochem Psychopharmacol, vol 5)

Knoll J, Ecseri Z, Kelemen K, Nievel J, Knoll B (1965) Phenylisopropylmethylpropinylamine (E-250), and new spectrum psychic energizer. Arch Int Pharmacodyn Ther 155: 154–164

Knoll J, Ecseri Z, Magyar K, Sátory É (1978) Novel (−)-deprenyl — derived selective inhibitors of B-type monoamine oxidase. The relation of structure to their action. Biochem Pharmacol 27: 1739–1747

Langston JW (1980) Selegiline as neuroprotectiv therapy in Parkinson's disease: concepts and controversies. Neurology 40 [Suppl 3]: 61–66

Magyar K (1991) Neuroprotective effect of deprenyl and p-fluor-deprenyl. In: Fazekas F, Schmutzharol E, Zeiler K (eds) Paneuropean Society of Neurology, Second Congress, Vienna, p 26 (Abstract)

Magyar K, Szüts T (1982) The fate of (−)-deprenyl in the body. Preclinical studies. In: Proceedings of the International Symposium on (−)-deprenyl, Jumex, Szombathely, Hungary, 1982. Chinoin, Budapest, pp 25–31

Magyar K, Ecseri Z, Bernáth G, Sátory É, Knoll J (1980) Structure-activity relationship of selective inhibitors of MAO-B. In: Magyar K (ed) Monoamine oxidases and their selective inhibition. Pergamon Press, Budapest, pp 11–21

Magyar K, Tóthfalusi L (1984) Pharmacokinetic aspects of deprenyl effects. Pol J Pharmacol Pharm 36: 373–384

Magyar K, Vizi ES, Ecseri Z, Knoll J (1967) Comparative pharmacological analysis of the optical isomers of phenyl-isopropyl-methyl-propinylamine (E-250). Acta Physiol Hung 32: 377–387

Magyar K, Skolnik J, Knoll J (1968) Radiopharmacological analytic studies with deprenyl-^{14}C. In: Leszkowszky GP (ed) V. Conferencia Hungarica pro Therapia et Investigatione in Pharmacologia. Budapest Publ House. Acad Sci, Budapest, pp 103–109

Reynolds GP, Elsworth JD, Blau K, Sandler M, Lees AJ, Stern GM (1978) Deprenyl is metabolized to methamphetamine and amphetamine in man. Br J Clin Pharmacol 6: 542–554

Ross SB (1976) Long-term effects of N-(2-chloroethyl)-N-ethyl-2-bromobenzylamine hydrochloride on noradrenergic neurons in the rat brain and heart. Br J Pharmacol 58: 521–527

Snyder SH, Coyle JT (1969) Regional differences in ^{3}H-norepinephrine and ^{3}H-dopamine uptake into rat brain homogenates. J Pharmacol Exp Ther 165: 78–86

Author's address: Dr. K. Magyar, Department of Pharmacodynamics, Semmelweis University of Medicine, Budapest, Hungary.

J Neural Transm (1994) [Suppl] 41: 177–188

Is selegiline neuroprotective in Parkinson's disease?

M. Gerlach[1,2], M. B. H. Youdim[3], and P. Riederer[1]

[1] Clinical Neurochemistry, Department of Psychiatry, University of Würzburg,
Würzburg, and
[2] Clinical Neurochemistry, Department of Neurology, University of Bochum,
St. Joseph Hospital, Bochum, Federal Republic of Germany
[3] Department of Pharmacology, Technion, Haifa, Israel

Summary. Recent findings emphasize the significance of oxidative mechanisms, involving the activity of monoamine oxidase (MAO) and the formation of free radicals, in the pathogenesis of Parkinson's disease. The possible role of such mechanisms in the degeneration of neurones in the substantia nigra has led to clinical trials aimed at preventing or slowing the progressively disabling course of the disease. However, conclusive clinical evidence of a neuroprotective effect in PD is still lacking. In this paper, we discuss possible mechanisms by which selegiline manifests neuroprotective effects in experimental and clinical situations. Besides MAO-B inhibition, which above all explains the prevention of protoxin activation and substrate oxidation by MAO-B, selegiline appears to exhibit other mechanisms of action (induction of superoxide dismutase, stimulation of neurotrophic factor synthesis, antagonistic modulation of the polyamine binding site of the NMDA-receptor) which are independent of its action on MAO-B.

Introduction

Parkinson's disease (PD), characterized by a marked loss of nigro-striatal dopaminergic neurones (Jellinger, 1989), is generally a disorder of progressive disability. It can be accompanied by nonspecific or age-related brain pathology and a variety of other coincidental lesions elsewhere in the central nervous system (Jellinger, 1989, 1991). There is now a lot of evidence that degeneration of the dopaminergic nigro-striatal neurones and the resulting striatal dopamine(DA)-deficiency syndrome (Hornykiewicz, 1973) are responsible for its classic motor symptoms (for a review see Gerlach and Riederer, 1993). The cause of chronic nigral cell death and the underlying mechanisms remain elusive. Many hypotheses have been advanced including viral infection, aberrant DA metabolism and the involvement of neuromelanin (for a review see Duvoisin, 1982; Langston, 1988). However, none of these concepts has been substantiated and recently interest has centered on the possibility that inherited predisposition to environmental or endogenous toxic agents might be one cause of PD (for a review see Poirier et al., 1991; Tanner, 1989). Interest in this area stems from the discovery of the ability of 1-methyl-4-phenyl-1,2,3,6-tetrahydropyridine (MPTP) to distroy nigral DA-containing cells in man and other primate species, causing motor —, biochemical — and pathological deficits which closely resemble those seen in patients with PD (for a review see Gerlach et al., 1991). Neurotoxins that selectively destroy the dopaminergic neurones in the substantia nigra, such as 6-hydroxydopamine and MPTP, appear to act via oxidant stress (imbalance between the formation of cellular oxidants and the antioxidative processes). Oxidant stress, due to the formation of hydrogen peroxide and oxygen-derived free radicals, can cause cell damage due to chain reactions of membrane lipid peroxidation (Halliwell, 1992). Although proof that oxidant stress actually causes the loss of dopaminergic neurones in patients with PD is lacking, there is a considerable body of evidence from studies in both animals and humans that support the concept (for a review see Götz et al., 1990; Fahn and Cohen, 1992; Lange et al., 1992). Because of the suspicion that free radicals or similar mechanisms may be involved in the pathogenesis of PD, there has been a strong interest for investigating various antioxidant strategies to determine whether the disease can be halted.

The introduction of selegiline (l-deprenyl) has influenced research to the aetiopathogenesis of PD and other neurodegenerative disorders more than any other drug available for its treatment. Based on its blockade of monoamine oxidase, subtype B (MAO-B), and the assumed increase of DA, selegiline improves the efficacy of the DA-substitution therapy with L-DOPA (L-3,4-dihydroxyphenylalanine) (for a review see Wessel, 1993). In addition to this combination therapy, clinical trials have been performed with selegiline to elucidate the influence of selegiline treatment on the progression of the disease (for a review see LeWitt, 1993). The rationale for these studies is based on the neuroprotectice properties of this substance in experimental models and on retrospective studies showing that patients who received both L-DOPA and selegiline lived longer than patients who were treated with L-DOPA alone (Birkmayer et al., 1985). This paper will briefly review some of the clinical and experimental evidence for a neuroprotective effect of selegiline.

Rationale for a neuroprotective therapy of PD with selegiline

A protective treatment should either prevent or at least slow down the rate of nigral dopaminergic cell degeneration. Hence, a meaningful protective therapy for PD depends on the understanding of the mechanisms underlying the neurodegeneration and the various factors influencing the development and progression of the disease (Table 1). Although environmental toxins may contribute to the aetiology of PD, chronic intoxication by selective exposure to a specific agent is unlikely because PD is widespread throughout the world (see Tanner, 1989). As none of the above mentioned theories

Table 1. Factors influencing the pathogenesis of Parkinson's disease

— Generation of oxygen-derived free radicals: endogenous (related to the dopamine metabolism) or from the environment (MPTP-like substances)
— Involvement of endogenous (tetrahydroisoquinolines, β-carbolines) or exogenous neurotoxins (pesticides, manganese)
— Cumulative effect of normal ageing
— Genetic factors (possibly causing enhanced susceptibility to neurotoxins)
— Problems with long-term L-DOPA treatment

Table 2. Evidence supporting a state of oxidative stress in the substantia nigra in Parkinson's disease (modified according to Lange et al., 1992)

— Disturbed mitochondrial respiratory function with reduction in Complex I and III activities
— Altered cellular calcium homeostasis with decrease in calcium-binding protein
— Decreased glutathione and glutathione peroxidase activity leading to a reduced ability to scavenge hydrogen peroxide derived from oxidative deamination and auto-oxidation of dopamine
— Increased iron content resulting in a potential excess of radical-generating free iron
— Increased iron (III) content indicating an enhanced hydroxyl radical generation (Fenton reaction)
— Increased mitochondrial superoxide dismutase activity, perhaps reflecting an attempt to compensate for oxidative stress
— Increased peroxidation of membrane lipids inducing membrane damage and cell death

alone explains the aetiology of PD, evidence suggests that several components may be involved, including a genetic predisposition. It is quite possible that PD is the result of an interplay of environmental toxins and the ageing factor combined with a specific genetic predisposition favouring the development of the disease process. Thus, PD could be caused by an MPTP-like endogenous or exogenous neurotoxin(s) which accumulate(s) over the years and destroy(s) the dopaminergic neurones due to a state of local oxidative stress. Indeed, recent research on the biochemical pathology of PD indicates that free radicals generated from oxidation reactions are important for the neuronal loss in the substantia nigra in PD. Table 2 summarizes the evidence for supporting a state of oxidative stress in the substantia nigra in PD.

Neuroprotective effects of selegiline in experimental models

MAO-B inhibition confers protection against the damaging effects of neurotoxins

The most obvious and, indeed, well-known way in which selegiline exert its neuroprotective effects is by preventing the conversion of protoxins such as MPTP to their corresponding active neurotoxin. The experimental and clinical backgrounds of MPTP have been extensively reviewed (see for example Gerlach et al., 1991), so that detailed consideration would be supererogatory.

Major steps in the expression of neurotoxicity involve the conversion of MPTP to the toxic agent 1-methyl-4-phenylpyridinium ion (MPP$^+$) by MAO-B in the glia, specific uptake of MPP$^+$ into the nigro-striatal dopaminergic neurones, the intraneuronal accumulation of MPP$^+$, and the neurotoxic action of MPP$^+$. This is exerted mainly through the inhibition of the enzymes of the respiratory chain (Complex I), the disturbance of calcium homeostasis, and possibly by the formation of free radicals. The neurotoxic activity of MPTP leads, in experimental animals, to a relatively selective loss of dopaminergic cells in the pars compacta of the substantia nigra and an impairment of dopaminergic neurotransmission in the striatum, characterised by a reduction in the concentrations of DA and its metabolites, in the diminished activity of tyrosine hydroxylase, and in the decrease in the density of DA-receptors. These neurotoxic effects of MPTP are prevented by pretreatment with selegiline in vivo, which blocks the metabolism of MPTP to its toxic metabolite MPP$^+$ (Chiba et al., 1984; Cohen et al., 1984; Heikkila et al., 1984; Langston et al., 1984; Markey et al., 1984). Beside this MAO-B-inhibiting component there appear, however, to be other mechanisms which contribute to its neuroprotective action. Tatton and Greenwood (1991) briefly reported that selegiline protects neurones showing immunoreactivity towards tyrosine hydroxylase against the neurotoxic action of MPTP even when it is given three days after MPTP-treatment (i.e. at a time when all the MPTP had been metabolised or excreted).

A wide variety of prodrug neurotoxic analogues of MPTP is now known (Langston and Irwin, 1989). However, many of the analogues of MPTP that have been investigated are also substrates for MAO-A (Gibb et al., 1987; Basma et al., 1990). In an attempt to identify a possible endogenously-generated neurotoxin, Nagatsu and Hirata (1987) have extensively screened compounds with an analogous structure to MPTP and found that tetrahydroisoquinoline is a candidate to produce parkinsonism. β-Carbolines (2- methyl-norharman) have also been suggested as potential endo- or exoneurotoxins (Neafsey et al., 1989). However, in no case has it yet been proved that the potential toxin is present in sufficient quantity in vivo, in particular individuals, in order to have a parkinsonian effect. Nor is there yet any convincing epidemiological evidence to implicate any particular dietary factors. Furthermore, it is questionable whether a selective dose (as used therapeutically) of selegiline would be able to prevent neurotoxicity of the above mentioned examples.

Inhibition of MAO-B: effect on hydrogen peroxide generation

Selegiline may decrease the generation of hydrogen peroxide associated with DA and protoxin catabolism through its action as an MAO-B inhibitor

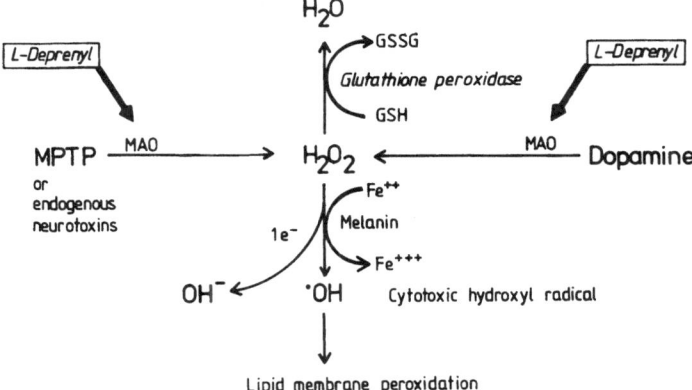

Fig. 1. Mechanism for the formation, removal and neurotoxic effect of hydrogen peroxide (H_2O_2) (Taken, with permission, from Gerlach et al., 1993). *GSH* reduced glutathione; *GSSG* oxidized glutathione; *MAO* monoamine oxidase; *MPTP* 1-methyl-4-phenyl-1,2,3,6-tetrahydropyridine

and, hence, slow the progression of PD by reducing the death of substantia nigra neurones induced by endogenous neurotoxic free radicals (Fig. 1). Cohen and Spina (1989) showed that the concentration of oxidized glutathione (GSSG) in the mouse striatum is tripled by the injection of haloperidol, which increases DA turnover. Treatment with selegiline (2.5 mg/kg) 18 hours before the haloperidol injection suppressed this rise in GSSG by 71.9%. The authors concluded that selegiline suppresses oxidant stress associated with increased monoamine turnover. However, in the rodent brain, DA is predominantly metabolized by MAO-A (Waldmeier et al., 1976), so that selective inhibition of MAO-B (as it was assumed to be with the dose used) is unlikely to have a significant effect on hydrogen peroxide formation, although a recent in vitro study (Werner and Cohen, 1991) has shown that above all MAO-B activity evokes a rise in GSSG in isolated mitochondria.

Inhibition of MAO-B: a possible indirect anti-glutamatergic effect

Excitatory amino acids such as L-glutamate and L-aspartate are neuro-transmitters in the central nervous system of mammals (Fonnum, 1984). Excessive activity of excitatory amino acids has been postulated to play a role in a variety of neurodegenerative diseases including PD (for a review see Olney, 1989; Meldrum and Garthwaite, 1990). This hypothesis is based on findings showing neurotoxic properties of both L-glutamate and gluta-mate receptor agonists which can be prevented by selective antagonists. According to current thinking, the N-methyl-D-aspartate (NMDA)-receptor, a subtype of the glutamate receptor, is above all involved in the neurotoxic effects of excitatory amino acids and excitatory toxins (Olney, 1989; Meldrum and Garthwaite, 1990). This receptor is linked to a Na^+/Ca^{2+} ion channel

M. Gerlach et al.

NMDA - RECEPTOR

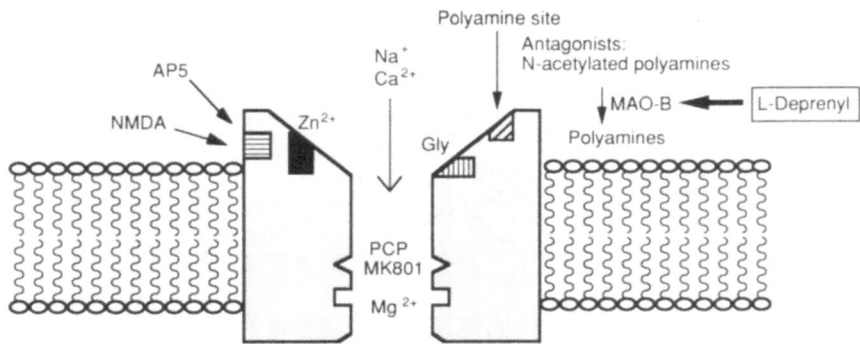

Fig. 2. Diagram illustrating the indirect effect of selegiline (l-deprenyl) on the function of the NMDA-receptor (Taken, with permission, from Gerlach et al., 1993). **AP5** 2-amino-5-phosphonopentanoate *Gly* glycine; **MAO-B** monoamine oxidase, subtype B *MK801* (+)-5-methyl-10,11-dihydro-5H-dibenzo[a,d]cyclohepten-5, 10-imine maleate; *NMDA* N-methyl-D-aspartate; *PCP* phencyclidine. For details see text

(Fig. 2) that has a much higher Ca^{2+} conductance than the ion channels associated with other excitatory amino acid receptor subtypes (MacDermott et al., 1986) and is gated by Mg^{2+} in a voltage-dependent fashion. The NMDA-receptor channel is blocked by phencyclidine and MK801 and the complex is regulated at three modulatory sites via glycine, Zn^{2+} and polyamines. Thus, as Fig. 2 illustrates, the NMDA-receptor system is a remarkably complex entity, the normal function of which depends on a dynamic equilibrium among multiple facilitative and inhibitory factors. It follows that a pathological process affecting any given factor might create an imbalance, rendering the entire system malfunctional. For example, increased presynaptic activity of L-glutamate or glycine would render this receptor system hyperfunctional and prone to an expression of excitotoxicity.

The cationic polyamines spermidine and spermine enhance the binding of agonists to the NMDA-receptor, above that induced by glutamate and glycine, via an increase in binding affinity (for a review, see Lodge and Johnson, 1990): structure-affinity relationship studies suggest that this positive modulation is receptor mediated, and studies with competitive antagonists suggest that occupation of this receptor is obligatory for NMDA-receptor function. Polyamines are mainly catabolized via mono-acetylation and oxidation. Since N-acetylated polyamines such as N-acetylputrescine, N-acetylspermidine and N-acetylspermine are good substrates of MAO-B (Youdim, unpublished observation), there is every reason to believe that under treatment with MAO-B inhibitors these compounds will be present in the brain in increased concentrations. It therefore seems very possible that they will exert a neuroprotective effect via an antagonistic modulation of the polyamine binding site of the NMDA-receptor (Fig. 2).

Effect of selegiline on superoxide dismutase induction

The role of superoxide dismutase (SOD) is to protect cells from the deleterious effects of the superoxide free radical. However, the radical-scavenging effect of SOD, to be effective, has to be followed by the actions of catalase or of glutathione peroxidase, because SOD generates hydrogen peroxide, which in the presence of transition metals is also capable of forming highly reactive radicals (Fig. 1).

Knoll (1988) originally reported that a 3-week course of daily selegiline injections (2 mg/kg subcutaneously) leads to an increased activity of SOD in the striatum of rats, to a 10-fold level compared to controls. Clorgyline, the selective MAO-A inhibitor, actually resulted in a significant decrease in SOD activity. These findings argue strongly against the inhibition of MAO being the mechanism via which SOD is induced. Carrillo et al. (1991) have reported that selegiline causes a significant 3-fold increase of both cytosolic and particulate forms of SOD in rat striata, but not elsewhere in the brain. There was also an increase in the activity of catalase 1.7 times of the control values, but not of glutathione peroxidase (Carrillo et al., 1991). Clow et al. (1991) confirmed that selegiline at 2 but not at 0.25 mg/kg causes a significant increase in SOD activity. However, the effect was confined to the soluble, cytosolic form of the enzyme (Cu/Zn-dependent), with the level of the particulate, mitochondrial form (Mn-dependent) being quite unchanged.

These findings with SOD may well be relevant to the apparent MAO-B independent neuroprotective properties of selegiline. There is a considerable literature linking SOD activity with longevity in different species (Tolmasoff et al., 1980), and in different strains of the same species (Kellog and Fridovich, 1976; Sohal et al., 1987). SOD has also been shown to protect mice against MPTP toxicity (Przedborski et al., 1990), as well as preventing sympathetic denervation caused by 6-hydroxydopamine (Albino-Teixeira et al., 1991).

Clinical trials

While there is a long legacy of clinical trials for symptomatic treatments against parkinsonism, there was little prior experience of studying treatments for protecting against the development of the disease. Several studies have focused upon the mortality of PD after the introduction of L-DOPA or other therapies, but the relationship between premature death and PD is obviously a complex matter. The first indication that selegiline might be neuroprotective came from Birkmayer et al. (1985). In a retrospective study, they evaluated the effect of 9 years of treatment in patients receiving L-DOPA plus selegiline, compared with those taking L-DOPA alone. The former lived significantly longer. However, similar results have also been obtained with bromocriptine (a DA agonist) in a controlled clinical trial with parkinsonian patients in earlier stages (Przuntek et al., 1992). Those

M. Gerlach et al.

authors reported that the mortality risk associated with L-DOPA therapy was reduced by more than 50% by its combination with bromocriptine.

Recently, two randomized, prospective, double-blind studies have compared selegiline with placebo in otherwise untreated subjects with early PD (Parkinson Study Group, 1989; Tetrud and Langston, 1989). Both studies demonstrated that selegiline produced a prolongation of the period before systematic therapy was required. A more recently report from the Parkinson Study Group (1993) shows that there was no beneficial effect of tocopherol (vitamin E) or any interaction between tocopherol and selegiline. Selegiline appears to delay the onset of disability by slowing the rate of progression of PD in newly diagnosed patients. However, conclusive evidence of a neuroprotective effect in PD is still lacking, because

1. the small but definite ameliorating influence of selegiline that was observed on the motor ratings of PD hampers a clear-cut detection of potentially protective actions of this MAO-B inhibitor and
2. comparable results as obtained with selegiline in the DATATOP study could also be found in a controlled prospective study with the DA agonist lisuride (Runge and Horowski, 1991).

Conclusions

The cause of degeneration of DA-containing neurones in PD is still unknown. Recent findings, however, emphasize the significance of oxidative

Table 3. Possible mechanisms for the neuroprotective effect of selegiline

Inhibition of MAO-B	diminished dopamine catabolism[1,2]
	prevents β-phenylethylamine oxidation[3,4] (endogenous "amphetamine-like" tonic effect)
	prevents neurotoxic actions of endo- (β-carbolines, tetrahydroisoquinolines)[5] and exotoxins (DSP-4, MPTP)[6,7]
	prevents generation of hydrogen peroxide[8,9]
Other direct actions	inhibits in vivo uptake of monamines into catecholaminergic nerve endings at higher doses[10]
	increases turnover rate of dopamine and rate of efflux of dopamine[10]
Indirect action	increases superoxide dismutase activity, especially at higher doses[11-13]
	substitutes for or stimulates production of neurotrophic factor(s)[14,15]
	may antagonize the NMDA-receptor subtype of the glutamate receptor[10]

[1] Glover et al., 1980; [2] Riederer and Youdim, 1986; [3] Reynolds et al., 1978; [4] Riederer et al., 1984; [5] Yoshida et al., 1990; [6] Cohen et al., 1984; [7] Gibson, 1987; [8] Cohen and Spina, 1989; [9] Werner and Cohen, 1991; [10] for a review see Gerlach et al., 1992; [11] Knoll, 1988; [12] Carrillo et al., 1991; [13] Clow et al., 1991; [14] Tatton and Greenwood, 1991; [15] Salo and Tatton, 1992

stress and free radical formation in the pathogenesis of the disease. Selegiline appears to have some remarkable properties, particularly in experimental models (Table 3). Furthermore, there is evidence that it prolongs life expectancy. Altogether, these various actions of selegiline contribute to a further understanding of neurodegenerative processes and their treatment strategies. Beside MAO-B inhibition, which above all explains the clinical effectiveness of selegiline in the treatment of PD, selegiline in particular appears to exhibit other mechanisms of action that are independent of its action on MAO-B (Table 3). Although a conclusive clinical evidence of a neuroprotective effect in PD is still lacking, there exists a lot of experimental data to support the neuroprotective potency in the long-term application of selegiline. The question of whether selegiline truly prevents or slow the rate of progression of PD might well be settled by employing longitudinal Gompertzian analysis, a relatively "simple method of detecting and distinguishing between symptomatic (competitive) and protective (intrinsic and environmental) influences on disease mortality at the population level after the introduction of new therapies" (Riggs, 1991).

References

Albino-Teixeira A, Azevedo I, Martel F, Osswald W (1991) Superoxide dismutase partially prevents sympathetic denervation by 6-hydroxydopamine. Arch Pharmacol 344: 36–40

Basma AN, Heikkila E, Nicklas WJ, Giovanni A, Geller HM (1990) 1-Methyl-4-phenyl-1,2,3,6-tetrahydropyridine- and 1-methyl-4-(2'-ethylphenyl-1,2,3,6-tetrahydropyridine-induced toxicity in PC12 cells: role of monoamine oxidase A. J Neurochem 55: 870–877

Birkmayer W, Knoll J, Riederer P, Youdim MBH, Hars V, Marton J (1985) Increased life expectancy resulting from addition of L-deprenyl to Madopar^R treatment in Parkinson's disease: a long-term study. J Neural Transm 64: 113–127

Carrillo M-C, Kanai S, Nokubo M, Kitani K (1991) (−)-Deprenyl induces activities of both superoxide dismutase and catalase but not of glutathione peroxidase in the striatum of young male rats. Life Sci 48: 517–521

Chiba K, Trevor A, Castagnoli Jr N (1984) Metabolism of the neurotoxic tertiary amine, MPTP, by brain monoamine oxidase. Biochem Biophys Res Commun 120: 574–578

Clow A, Hussain T, Glover V, Sandler M, Dexter DT, Walker M (1991) (−)-Deprenyl can induce soluble superoxide dismutase in rat striata. J Neural Transm [Gen Sect] 86: 77–80

Cohen G, Spina MB (1989) Deprenyl suppresses the oxidant stress associated with increased dopamine turnover. Ann Neurol 26: 689–690

Cohen G, Pasik P, Cohen B, Leist A, Mytilineou C, Yahr MD (1984) Pargyline and deprenyl prevent the neurotoxicity of 1-methyl-4-phenyl-1,2,3,6-tetrahydropyridine (MPTP) in monkeys. Eur J Pharmacol 106: 209–210

Duvoisin RC (1982) The cause of Parkinson's disease. In: Marsden CD, Fahn S (eds) Neurology, vol 2. Butterworth Scientific, London, pp 8–24

Fahn S, Cohen G (1992) The oxidant stress hypothesis in Parkinson's disease — evidence supporting it. Ann Neurol 32: 804–812

Fonnum F (1984) Glutamate: a neurotransmitter in mammalian brain. J Neurochem 42: 1–11

Gerlach M, Riederer P, Przuntek H, Youdim MBH (1991) MPTP mechanisms of neurotoxicity and their implications for Parkinson's disease. Eur J Pharmacol 208: 273–286

Gerlach M, Riederer P, Youdim MBH (1992) The molecular pharmacology of L-deprenyl. Eur J Pharmacol 226: 97–108

Gerlach M, Riederer P (1993) The pathophysiological basis of Parkinson's disease. In: Szelenyi I (ed) Series of new drugs, vol 1. Inhibitors of monoamine oxidase B. Birkhäuser, Basel, pp 25–50

Gerlach M, Riederer P, Youdim MBH (1993) The mode of action of MAO-B inhibitors. In: Szelenyi I (ed) Series of new drugs, vol 1. Inhibitors of monoamine oxidase B. Birkhäuser, Basel, pp 183–200

Gibb C, Willoughby J, Glover V, Sandler M, Testa B, Jenner P, Marsden CD (1987) Analogues of 1-methyl-4-phenyl-1,2,3,6-tetrahydropyridine as monoamine oxidase substrates: a second ring is not necessary. Neurosci Lett 76: 316–322

Gibson C (1987) Inhibition of MAO B, but not MAO A, blocks DSP-4 toxicity on central NE neurons. Eur J Pharmacol 141: 135–138

Glover V, Elsworth JD, Sandler M (1980) Dopamine oxidation and its inhibition by (−)-deprenyl in man. J Neural Transm [Suppl] 16: 163–172

Götz ME, Freyberger A, Riederer P (1990) Oxidative stress: a role in the pathogenesis of Parkinson's disease. J Neural Transm [Suppl] 29: 241–249

Halliwell B (1992) Reactive oxygen species and the central nervous system. J Neurochem 59: 1609–1623

Heikkila RE, Manzino L, Cabbat FS, Duvoisin RS (1984) Protection against the dopaminergic neurotoxicity of 1-methyl-4-phenyl-1,2,3,6-tetrahydropyridine by monoamine oxidase inhibitors. Nature 311: 467–469

Hornykiewicz O (1973) Parkinson's disease: from brain homogenate to treatment. Fed Proc 32: 183–190

Jellinger K (1989) Pathology of Parkinson's syndrome. In: Calne DB (ed) Handbook of experimental pharmacology, vol 88. Springer, Berlin Heidelberg New York Tokyo, pp 47–112

Jellinger K (1991) Pathology of Parkinson's disease: changes other than the nigro-striatal pathway. Mol Chem Neuropathol 14: 153–197

Kellog EW, Fridovich I (1976) Superoxide dismutase in the rat and mouse as a function of age and longevity. J Gerontol 4: 405–408

Knoll J (1988) The striatal dopamine dependency of life span in male rats, longevity study with (−)deprenyl. Mech Ageing Dev 46: 237–262

Lange KW, Youdim MBH, Riederer P (1992) Neurotoxicity and neuroprotection in Parkinson's disease. J Neural Transm [Suppl] 38: 27–44

Langston JW (1988) The etiology of Parkinson's disease: new directions for research. In: Jankovic J, Tolosa E (eds) Parkinson's disease and movement disorders. Urban & Schwarzenberg, Baltimore Munich, pp 75–85

Langston JW, Irwin I (1989) Pyridine toxins. In: Calne (ed) Drugs for the treatment of Parkinson's disease. Springer, Wien New York, pp 205–226

Langston JW, Irwin I, Langston EB, Forno LS (1984) Pargyline prevents MPTP-induced parkinsonism in primates. Science 225: 1480–1482

LeWitt PA (1993) Neuroprotective effects of MAO-B inhibition: clinical studies in Parkinson's disease. In: Szelenyi I (ed) Series of new drugs, vol 1. Inhibitors of monoamine oxidase B. Birkhäuser, Basel, pp 289–299

Lodge D, Johnson KM (1990) Noncompetitive excitatory amino acid receptor antagonists. TIPS [The Pharmacology of Excitatory Amino Acids: A Special Report 1991]: 13–18

MacDermott AB, Mayer ML, Westbrook GL, Smith SJ, Barker JL (1986) NMDA-receptor activation increases cytoplasmic calcium concentration in cultured spinal cord neurones. Nature 321: 519–522

Markey JP, Johannessen JN, Chiueh CC, Burns RS, Herkenham MA (1984) Intra-neuronal generation of a pyridinium metabolite may cause drug-induced parkinson-ism. Nature 311: 464–467

Meldrum B, Garthwaite J (1990) Excitatory amino acid neurotoxicity and neurode-generative disease. TIPS [The Pharmacology of Excitatory Amino Acids: A Special Report 1991]: 54–62

Nagatsu T, Hirata Y (1987) Inhibition of the tyrosine hydroxylase system by MPTP, 1-methyl-4-phenylpyridinium ion (MPP$^+$) and the structurally related compounds in vitro and in vivo. Eur Neurol 26 [Suppl 1]: 11–15

Neafsey EJ, Drucker G, Raikoff K, Collins MA (1989) Striatal dopaminergic toxicity following intranigral injection in rats of 2-methyl-norharman, a analog of N-methyl-4-phenylpyridinium ion (MPP$^+$). Neurosci Lett 105: 344–349

Olney JW (1989) Excitatory amino acids and neuropsychiatric disorders. Biol Psychiatry 26: 505–526

Parkinson Study Group (1989) Effect of deprenyl on the progression of disability in early Parkinson's disease. N Engl J Med 321: 1364–1371

Parkinson Study Group (1993) Effects of tocopherol and deprenyl on the progression of disability in early Parkinson's disease. N Engl J Med 328: 176–184

Poirier J, Kogan S, Gauthier S (1991) Environment, genetics and idiopathic Parkinson's disease. Can J Neurol Sci 18: 70–76

Przedborski S, Kostic V, Jackson-Lewis V, Carlson E, Epstein CJ, Cadet JL (1990) Transgenic mice expressing the human SOD gene are resistant to MPTP-induced toxicity. Soc Neurosci Abstr 16: 1260

Przuntek H, Welzel D, Blümner E, Danielcyzk W, Letzel H, Kaiser H-J, Kraus PH, Riederer P, Schwarzmann D, Wolf H, Überla K (1992) Bromocriptine lessens the incidence of mortality in L-Dopa-treated parkinsonian patients: prado-study discontinued. Eur J Clin Pharmacol 43: 357–363

Reynolds GP, Riederer P, Sandler M, Jellinger K, Seemann D (1978) Amphetamine and 2-phenylethylamine in post-mortem parkinsonian brain after (−)-deprenyl administration. J Neural Transm 43: 271–277

Riederer P, Jellinger K, Seemann D (1984) Monoamine oxidase and parkinsonism. In: Tipton K, Dostert P, Strolin-Benedetti M (eds) Monoamine oxidase and disease. Academic Press, London, pp 403–415

Riederer P, Youdim MBH (1986) Monoamine oxidase activity and monoamine metabolism in brains of parkinsonian patients treated with L-deprenyl. J Neurochem 46: 1359–1365

Riggs JE (1991) Parkinson's disease: an epidemiologic method for distinguishing between symptomatic and neuroprotective treatments. Clin Pharmacol 14: 489–497

Runge I, Horowski R (1991) Can we differentiate symptomatic and neuroprotective effects in parkinsonism? J Neural Transm [PD Sect] 4: 273–283

Salo PT, Tatton WG (1992) Deprenyl reduces the death of motorneurons caused by axotomy. J Neurosci Res 31: 394–400

Sohal RS, Farmer KJ, Allen RG (1987) Correlates of longevity in two strains of the housefly, Musca domestica. Mech Ageing Dev 40: 171–179

Tanner CM (1989) The role of environmental toxins in the etiology of Parkinson's disease. Trends Neurosci 12: 49–54

Tatton WG, Greenwood CE (1991) Rescue of dying neurons: a new action for deprenyl in MPTP parkinsonism. J Neurosci Res 30: 666–672

Tetrud J, Langston W (1989) The effect of deprenyl (selegiline) on the natural history of Parkinson's disease. Science 245: 519–522

Tolmasoff JM, Ono T, Cutler RG (1980) Superoxide dismutase: correlation with life-span and specific metabolic rate in primate species. Proc Natl Acad Sci USA 77: 2777–2781

Waldmeier PC, Delini-Stula A, Maitre L (1976) Preferential deamination of dopamine by an A type monamine oxidase B substrate in rat brain. Naunyn Schmiedebergs Arch Pharmacol 292: 9–14

Werner P, Cohen G (1991) Intramitochondrial formation of oxidized glutathione during the oxidation of benzylamine by monoamine oxidase. FEBS Lett 280: 44–46

Wessel K (1993) MAO-B inhibitors in neurological disorders with special reference to selegiline. In: Szelenyi I (ed) Series of new drugs, vol 1. Inhibitors of monoamine oxidase B. Birkhäuser, Basel, pp 253–275

Yoshida M, Niwa T, Nagatsu T (1990) Parkinsonism in monkeys produced by chronic administration of an endogenous substance of the brain, tetrahydroisoquinoline: the behavioral and biochemical changes. Neurosci Lett 109–113

Authors' address: Prof. Dr. P. Riederer, Clinical Neurochemistry, Department of Psychiatry, Füchsleinstrasse 15, D-97080 Würzburg, Federal Republic of Germany.

J Neural Transm (1994) [Suppl] 41: 189–196

Suppression of hydroxyl radical formation by MAO inhibitors: a novel possible neuroprotective mechanism in dopaminergic neurotoxicity

C. C. Chiueh, S.-J. Huang, and **D. L. Murphy**

Laboratory of Clinical Science, National Institute of Mental Health, NIH,
Bethesda, Maryland, U.S.A.

Summary. Prior studies concluded that 1-methyl-4-phenyl-1,2,3,6-tetrahydropyridine (MPTP, a toxin causing parkinsonism) and its analogues are bioactivated by monoamine oxidase (MAO) to toxic pyridinium metabolites. Recently, a dissociation between the neuroprotective effects of deprenyl and its MAO inhibiting effects has been proposed. Furthermore, we have demonstrated that pyridinium metabolites of MPTP stimulate dopamine efflux and the formation of cytotoxic hydroxyl free radicals ($\cdot$OH) in the striatum. Therefore, we investigated possible neuroprotective mechanisms of propargyl MAO inhibitors by studying their effects on the formation of oxygen free radicals produced by dopamine autoxidation. Our recent in vivo results indicate that deprenyl and clorgyline given systemically suppressed the generation of $\cdot$OH that followed administration of 2'-methyl-MPTP. Combined deprenyl and clorgyline pretreatment are needed to block dopamine neurotoxicity elicited by 2'-methyl-MPTP. The present in vitro studies reveal that propargyl MAO inhibitors suppress non-enzymatic dopamine autoxidation and associated free radical production. Thus, $\cdot$OH generation evoked by MPTP analogues may be due mainly to a burst increase in iron-catalyzed autoxidation of released dopamine in the basal ganglia where high levels of iron and oxygen are present. Our present in vitro and prior in vivo results suggest that a novel antioxidant property of propargyl MAO inhibitors may contribute to protection against nigral lesions elicited by dopamine autoxidation following the administration of MPTP analogues.

Introduction

During the past decade, the MPTP model of parkinsonism (Burns et al., 1983; Markey et al., 1984; Chiueh, 1988) has been employed to develop neuroimaging ligands, to evaluate neurotransplantation, to test antiparkinsonian agents, and to study the pathogenesis of nigral degeneration. Prevention of MPTP toxicity by deprenyl (Cohen et al., 1984; Mytilineou and Cohen, 1985) also led to clinical testing of potential neuroprotective therapy (Parkinson Study Group, 1989; review of Murphy and Sunderland, 1993) to slow the progression of extrapyramidal dysfunction in Parkinson's disease (Birkmayer et al., 1985) and possibly senile or Alzheimer dementia (Tariot et al., 1987; Knoll, 1989).

Regarding molecular mechanism underlying the "selective" dopaminergic neurotoxicity of MPTP, we recently proposed (Chiueh et al., 1993a) and demonstrated that ·OH free radicals are generated in vivo during the sustained "biphasic" dopamine efflux elcited by MPTP analogues in the basal ganglia (Chiueh et al., 1992a,b; Obata and Chiueh, 1992). Moreover, MAO-A and MAO-B type propargyl inhibitors (clorgyline and deprenyl) suppress the enhanced ·OH production (Chiueh et al., 1992a) and the dopaminergic neurotoxicity (Heikkila et al., 1988) caused by the 2'-methyl analog of MPTP. Additionally, MPP$^+$-stimulated free radical production (Chiueh et al., 1992b; Obata and Chiueh, 1992) and nigral injury are also blocked by propargyl MAO inhibitors in vivo (Wu et al., 1993).

In the present study, therefore, we investigated the effects of propargyl MAO inhibitors on the non-enzymatic autoxidation of dopamine in vitro which is known to produce cytotoxic ·OH and semiquinone free radicals (Donaldson et al., 1981; Graham, 1984; Poirier et al., 1985; Fornstedt et al., 1989, 1990; Ben-Shachar et al., 1991; Chiueh et al., 1992b; Chiueh et al., 1993b). The present data demonstrate that propargyl MAO inhibitors, similar to ·OH scavengers (Chiueh et al., 1993b), can suppress ·OH free radical production by inhibiting dopamine autoxidation and melanin formation.

Methods and materials

Spectrometric assay of dopamine autoxidation or melanin formation

Prior studies have shown that dopamine autoxidation led to generation of ·OH and semiquinone free radicals, and formation of melanin pigment (Graham, 1978; Donaldson et al., 1981; Poirier et al., 1985; Chiueh et al., 1992b; Chiueh et al., 1993b). The rate of dopamine (5 mM) autoxidation in the pH 7.4 Earle's balanced salt solution without glucose was measured spectrometrically by assaying melanin formation in borosilicate glass vials (Kimble disposable vial, Toledo, Ohio). Owing to the sensitivity limits of the spectrometer for this in vitro assay of dopamine melanin, millimolar concentrations of dopamine were used in this study, although the extracellular concentrations of dopamine in the striatum measured in vivo by intracerebral microdialysis procedure are less than millimolar. The relative optical density (450 nm wavelength) in 300 μl triplicate samples was intermittently measured for up to 120 hours by a spectrophotometer equipped with a 96 well microplate reader (Bio-Rad model 3550). The scanner was calibrated by using different concentrations of melanin pigments derived from dopamine or serotonin. Experiments were repeated at least three times. The results are depicted as means ± 1 S.E.M.

Results

Effects of ascorbate on autoxidation of dopamine

The addition of dopamine (5 mM) to a physiological buffer solution caused a progressive autoxidation of dopamine and the formation of melanin pigments in the test tube (Graham, 1978; Donaldson et al., 1981; Graham, 1984; Poirier et al., 1985; Chiueh et al., 1993b). The spontaneous dopamine melanin formation in the borosilicate glass vial is concentration-dependently suppressed by ascorbate (0 to 2.5 mM; Fig. 1). Thus, in the physiological buffer solution, ascorbate may reduce dopamine quinones back to dopamine and thus suppress the formation of dopamine melanin.

Effects of deprenyl, clogyline and pargyline on dopamine melanin formation

Deprenyl and clorgyline have been shown to prevent hydrogen peroxide formation during dopamine oxidation by MAO-B and MAO-A, respectively (Spina and Cohen, 1989). Similar to that of ascorbate, both deprenyl and clorgyline (2.5 mM) completely blocked dopamine melanin formation in the absence of MAO (Fig. 2). Other propargyl MAO inhibitors, such as pargyline, are also active in inhibiting melanin formation. The relative in vitro antioxidant potency of propargyl MAO inhibitors against dopamine (5 mM) autoxidation is summarized in Table 1.

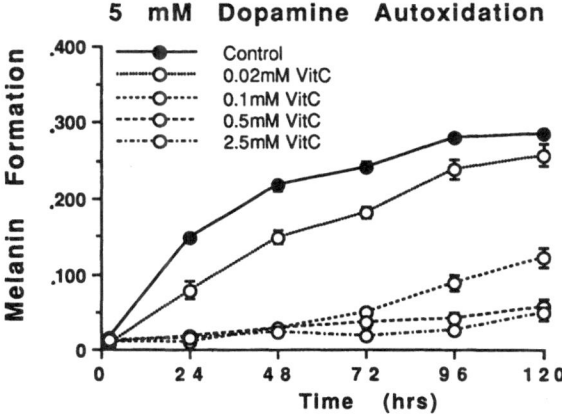

Fig. 1. Effects of vitamin C on formation of dopamine melanin (5 mM). Different concentrations of vitamin C (Vit-C) were added to Kimble borosilicate glass vials containing 5 mM dopamine in Earle's balanced salt solution (pH 7.4) at room temperature. Nonenzymatic dopamine autoxidation rate reflected by melanin formation in triplicate samples were measured by a spectrophotomenter equipped a microplate reader (450 nm wavelength). Experiments were repeated at least three times. Results shown are means ± 1 S.E.M. (N = 3–6). Results were used to determine the IC_{50} of this water-soluble antioxidant to inhibit dopamine autoxidation as depicted in Table 1

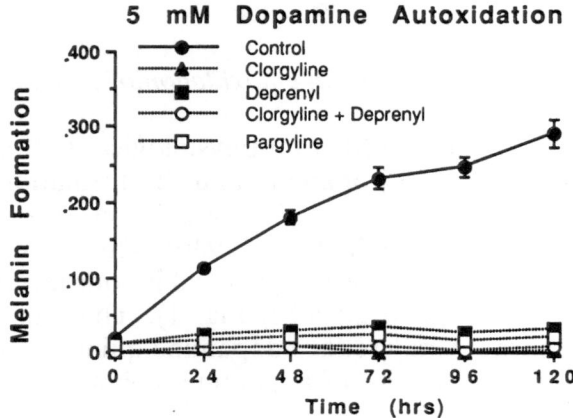

Fig. 2. Effects of propargyl MAO inhibitors on dopamine autoxidation reflected by melanin formation. Propargyl MAO inhibitors (2.5 mM clorgyline, deprenyl, clorgyline plus deprenyl and pargyline) were added to the dopamine solution (5 mM) as described in Methods and in Fig. 1 (N = 4). Different concentrations of propargyl MAO inhibitors were also used (data not shown) to calculate the antioxidant effects of MAO inhibitors (IC$_{50}$) on dopamine autoxidation presented in Table 1

Table 1. Relative antioxidant potency of propargyl MAO inhibitors versus vitamin C in suppression of dopamine autoxidation and free radical formation reflected by melanin formation in vitro

	IC$_{50}$* (mM)	Relative potency
Vitamin C	0.05	1 ×
Clorgyline	0.11	1/2 ×
Pargyline	0.26	1/5 ×
Deprenyl	0.65	1/13 ×

*Concentrations needed to suppress 50% dopamine autoxidation (5 mM) in borosilicate glass vial containing Earle's balanced salt solution (pH 7.4). Dopamine autoxidation leads to not only formation of melanin but also production of ·OH (Graham, 1978; Chiueh et al., 1992b; Chiueh et al., 1993b) and semiquinone free radicals

Discussion

The non-enzymatic oxidation of dopamine is known to be catalyzed by oxygen and transition metals, such as Fe^{2+}, Cu^{2+}, and Mn^{2+} (Donaldson et al., 1989; Graham, 1984; Poirier et al., 1985). In addition to the formation of semiquinone radicals and melanin (Graham, 1978), in vivo dopamine autoxidation leads to the production of cytotoxic ·OH free radicals in the iron-rich basal ganglia (Chiueh et al., 1992b; Chiueh et al., 1993b) which may cause nigral degeneration (Cohen, 1988; Hirsch et al., 1988; Spina and

Cohen, 1989; Halliwell, 1989; Fornstedt et al., 1989, 1990; Dexter et al., 1989; Youdim et al., 1989; Riederer et al., 1989; Jenner et al., 1990; Ben-Shachar et al., 1991; Chiueh et al., 1993a). Notably, the formation of dopamine melanin can be blocked by ·OH scavengers (Chiueh et al., 1993b) which indicates that dopamine melanin pigments in the substantia nigra neurons are a reliable in vivo marker for oxidative stress.

In the present in vitro study, we demonstrated that dopamine melanin formation is also suppressed by ascorbate and by propargyl MAO inhibitors including deprenyl, pargyline, and clorgyline. The antioxidant effects of propargyl MAO inhibitors are comparable to that of salicylate, an ·OH scavenger (Chiueh et al., 1993b). Therefore, propargyl MAO inhibitors inhibit not only enzymatic oxidation but also non-enzymatic oxidation of dopamine.

This novel antioxidant property of progargyl MAO inhibitors may contribute to their neuroprotective effects against nigral loss elicited by MPTP analogues which may be mediated by ·OH free radicals (Obata and Chiueh, 1992; Chiueh et al., 1992b; Chiueh et al., 1993a) leading to oxidative damage of mitochondrial complex I (Cleeter et al., 1992) and membrane ion channels (Chiueh et al., 1993a; Chiueh et al., in preparation). We recently demonstrated that MAO inhibitors suppress the production of ·OH free radicals elicited by 2'-methyl MPTP in the iron-rich striatum in vivo (Chiueh et al., 1992a). Moreover, dopaminergic toxicity elicited by 2'-methyl MPTP can be blocked by deprenyl given 30 to 60 min after (Sziraki et al., 1992) but not 24 hours prior to 2'-metyl MPTP (Heikilla et al., 1988). Our recent observations indicate that deprenyl (μM) suppress ·OH production elicited by MPP^+ (mM) in vivo (Wu et al., 1993) which support an early report that deprenyl protected against MPP^+ toxicity in a primary culture of midbrain dopamine neurons (Mytilineou and Cohen, 1985). It may provide a clue to explain why deprenyl decreases Lewy bodies and increases the number of surviving neurons in the medial substantia nigra of levo-dopa treated parkinsonian patients (Rinne et al., 1991).

The dissociation of neuro-protective or neuro-rescue effects of deprenyl and pargyline from that of MAO-B inhibition was recently proposed for MPTP and 2'-methyl-MPTP neurotoxicity (Tatton and Greenwood, 1991; Chiueh et al., 1992a; Sziraki et al., 1992) and also for Alzheimer's disease (Murphy and Sunderland, 1993). After binding to uniquely distributed sites of MAO-A and MAO-B, clorgyline and deprenyl may provide a site-specific protection against oxidative damage elicited by free radicals in different brain regions. By suppressing oxidative stress in widely distributed astroglial cells (MAO-B sites), deprenyl and pargyline may have broad spectrum neuroprotective actions which are certainly not limited to the protection or rescue of more vulnerable nigral neurons. In fact, deprenyl is being used to protect surgically lesioned young motor neurons (Salo and Tatton, 1991), immuno-suppressed athymic animals (Freisleben et al., 1992) and DSP-4 noradrenergic neurotoxocity (Magyar et al., 1992).

In summary, increasing evidence suggests that propargyl MAO inhibitors may protect nigral neurons from degeneration via mechanisms of action other than MAO inhibition. Moreover, the present data indicate that propargyl MAO inhibitors possess an antioxidant property which is similar to that of ascorbate (reducing agent) and/or salicylate ($\cdot$OH scavenger) in inhibiting dopamine oxidation and free radical production. Propargyl MAO inhibitors may prevent oxidative damage to nigral neurons caused by free radicals through the suppression of not only the intraneuronal hydrogen peroxide formation produced by MAO oxidation of cytosomal dopamine but also the extraneuronal free radical generation caused by the autoxidation of "released" dopamine as demonstrated in vivo (Wu et al., 1993).

References

Burns RS, Chiueh CC, Markey SP, Ebert MH, Jacobowitz DM, Kopin IJ (1983) A primate model of parkinsonism: selective destruction of dopaminergic neurons in pars compacta of the substantia nigra by N-methyl-4-phenyl-1,2,3,6-tetrahydropyridine. Proc Natl Acad Sci USA 80: 4546–4550

Ben-Shachar D, Riederer P, Youdim MBH (1991) Iron-melanin interaction and lipid peroxidation: implications for Parkinson's disease. J Neurochem 57: 1609–1614

Birkmayer W, Knoll J, Riederer P, Youdim MBH, Hars V, Martin J (1985) Improvement of life expectancy due to l-deprenyl addition to madopar treatment in Parkinson's disease: a long term study. J Neural Transm 64: 113–127

Chiueh CC (1988) Dopamine in the extrapyramidal motor function: a study based upon the MPTP-induced primate model of parkinsonism. Ann NY Acad Sci 515: 226–238

Chiueh CC, Huang S-J, Murphy DL (1992a) Enhanced hydroxyl radical generation by 2'-methyl analog of MPTP: suppression by clorgyline and deprenyl. Synapse 11: 346–348

Chiueh CC, Krishna G, Tulsi P, Obata T, Lang K, Huang S-J, Murphy DL (1992b) Intracranial microdialysis of salicylic acid to detect hydroxyl radical generation through dopamine autoxidation in the caudate nucleus: effects of MPP^+. Free Radic Biol Med 13: 581–583

Chiueh CC, Miyake H, Peng MT (1993a) Role of dopamine autoxidation, hydroxyl radical generation, and calcium overload in underlying mechanisms involved in MPTP-induced parkinsonism. In: Narabayashi H, Nagatsu T, Yanagisawa N, Mizuno Y (eds) Parkinson's disease: from basic research to treatment. Raven Press, New York, pp 251–258 (Adv Neurol 60)

Chiueh CC, Murphy L, Miyake H, Lang K, Tulsi PK, Huang S-J (1993b) Hydroxyl free radicals ($\cdot$OH) formation reflected by salicylate hydroxylation and neuromelanin: in vivo markers for oxidant injury of nigral neurons. Ann NY Acad Sci 629: 370–375

Cleeter MWJ, Cooper JM, Schapira AHV (1992) Irreversible inhibition of mitochondrial complex I by 1-methyl-4-phenylpyridinium: evidence for free radical involvement. J Neurochem 58: 786–789

Cohen G, Pasik P, Cohen B, Leist A, Mytilineou C, Yahr MD (1984) Pargyline and deprenyl prevent the neurotoxicity of 1-methyl-4-phenyl-1,2,3,6-tetrahydropyridine (MPTP) in monkeys. Eur J Pharmacol 106: 209–210

Cohen G (1988) Oxygen radicals and Parkinson's disease. In: Halliwell B (ed) Oxygen radicals and tissue injury. FASEB, Bethesda, pp 130–135

Dexter DT, Carter CJ, Wells FR, Javoy-Agid F, Agid Y, Lees A, Jenner P, Marsden CD (1989) Basal lipid peroxidation in substantia nigra is increased in Parkinson's disease. J Neurochem 52: 381–389

Donaldson J, LaBella FS, Gesser D (1981) Enhanced autoxidation of dopamine as a possible basis of manganese neurotoxicity. Neurotoxicology 2: 53–64

Fornstedt B, Brun A, Rosengren E, Carlsson A (1989) The apparent autoxidation rate of catechols in dopamine-rich regions of human brains increases with the degree of depigmentation of substantia nigra. J Neural Transm [PD Sect] 1: 279–295

Fornstedt B, Pileblad E, Carlsson A (1990) In vivo autoxidation of dopamine in guinea pig striatum increases with age. J Neurochem 55:655–659

Freisleben H-J, Lehr F, Fuchs J (1992) Lifespan of NMRI-mice is increased by deprenyl. Proceedings Int Amine Oxidase Workshop 5: 12 (Abst. #B02)

Graham DG (1978) Oxidative pathways for catecholamines in the genesis of neuromelanin and cytotoxic quinones. Mol Pharmacol 14: 633–643

Graham DG (1984) Catecholamine toxicity: a proposal for the molecular pathogenesis of manganese neurotoxicity and Parkinson's disease. Neurotoxicology 5: 83–96

Halliwell B (1989) Oxidants and the central nervous system: some fundamental questions, is oxidant damage relevant to parkinson's disease, Alzheimer's disease, traumatic injury, or stroke? Acta Neurol Scand 126: 23–33

Heikkila RE, Kindt MV, Sonsalla PK, Giovanni A, Youngster SK, McKeown KA, Singer TP (1988) Importance of monoamine oxidase A in the bioactivation of neurotoxic analogues of 1-methyl-4-phenyl-1,2,3,6-tetrahydrophyridine. Proc Natl Acad Sci USA 85: 6172–6176

Hirsch E, Graybiel AM, Agid YA (1988) Melanized dopaminergic neurons are differentially susceptible to degeneration in Parkinson's disease. Nature 334: 345–348

Jenner P, Dexter DT, Schapira AHV, Marsden CD (1990) Free radical involvement and altered iron metabolism as a cause of Parkinson's disease. In: Marsden CD, Fahn S (eds) The assessment and therapy of parkinsonism. Parthenon Publishing Group, New Jersey, pp 17–30

Knoll J (1989) The pharmacology of selegiline [(−)deprenyl]: new aspects. Acta Neurol Scand 126: 83–91

Magyar K, Tothfalusi L, Lengyel J, Gaal J (1992) Neuroprotective effect of deprenyl and p-fluoro-deprenyl against DSP-4 toxicity. Neurochem Int 21: D92 (abstract)

Markey SP, Johannessen JN, Chiueh CC, Burns RS, Herkenham MA (1984) Intraneuronal generation of a pyridinium metabolite may cause drug-induced parkinsonism. Nature 311: 464–467

Murphy DL, Sunderland T (1993) Monoamine oixdase inhibitors in neurodegenerative disorders. In: Kennedy SH (ed) Clinical advances in monoamine oxidase inhibitor therapies. American Psychiatric Press, Washington (in press)

Mytilineou C, Cohen G (1985) Deprenyl protects dopamine neurons from the neurotoxic effects of 1-methyl-4-phenylpyridinium ion. J Neurochem 5: 1951–1953

Obata T, Chiueh CC (1992) In vivo trapping of hydroxyl free radicals in the striatum utilizing intracranial microdialysis perfusion of salicylate: effects of MPTP, MPDP$^+$ and MPP$^+$. J Neural Transm [Gen Sect] 89: 139–145

Parkinson Study Group (1989) Effects of deprenyl on the progression of disability in early Parkinson's disease. N Engl J Med 321: 1364–1371

Poirier J, Donaldson J, Barbeau A (1985) The specific vulnerability of the substantia nigra to MPTP is related to the presence of transition metals. Biochem Biophys Res Commun 128: 25–33

Riederer P, Sofic E, Rausch W-D, Schmidt B, Reynolds GP, Jellinger K, Youdim MBH (1989) Transition metals, ferritin, glutathione, and ascorbic acid in parkinsonian brains. J Neurochem 52: 515–520

Rinne JO, Roytta M, Paljarvi L, Rumnukainen J, Rinne UK (1991) Selegiline (deprenyl) treatment and death of nigral neurons in Parkinson's disease. Neurology 41: 859–861

Salo PT, Tatton WG (1991) Deprenyl reduces death of motoneurons caused by axotomy. J Neurosci Res 31: 1–7

Spina MB, Cohen G (1989) Dopamine turnover and glutathione oxidation: implications for Parkinson disease. Proc Natl Acad Sci USA 86: 1398–1400

Sziraki I, Kardos V, Patthy M, Patfalusi M, Gaal J, Sloti M, Kollar E (1992) Complex mode of action of deprenyl in protection against MPTP-parkinsonism in mice. Neurochem Int 21: D69 (abstract)

Tariot PN, Sunderland T, Weingartner H, Murphy DL, Welkowitz JA, Thompson K, Coen RM (1987) Cognitive effects of l-deprenyl in Alzheimer's disease. Psychopharmacology 91: 489–495

Tatton WG, Greenwood CE (1991) Rescue of dying neurons: a new action for deprenyl in MPTP parkinsonism. J Neurosci Res 30: 666–672

Wu R-M, Chiueh CC, Murphy DL (1993) Apparent antioxidant effect of l-deprenyl on hydroxyl radical formation and nigral injury elicited by MPP^+ in vivo. Eur J Pharmacol (in press)

Authors' address: C. C. Chiueh, Ph.D., NIMH, NIH Clinical Center Bldg. 10, Rm. 3D-41, Bethesda, MD 20892, U.S.A.

J Neural Transm (1994) [Suppl] 41: 197–205

Novel toxins and Parkinson's disease: N-methylation and oxidation as metabolic bioactivation of neurotoxin

M. Naoi[1], W. Maruyama[2], T. Niwa[3], and T. Nagatsu[4]

[1] Department of Biosciences, Nagoya Institute of Technology, Nagoya,
[2] Department of Neurology, Nagoya University School of Medicine, Nagoya,
[3] Department of Internal Medicine, Nagoya University Branch Hospital, Nagoya,
[4] Division of Molecular Genetics (II) Neurochemistry, Institute for Comprehensive Medical Science, School of Medicine, Fujita Health University, Toyoake, Japan

Summary. In human brains, a series of monoamine-derived 1,2,3,4-tetrahydroisoquinolines and the 6,7-dihydroxy derivatives has been identified. A tetrahydroisoquinoline was found to cause parkinsonism in monkey, but its toxicity was not so potent as 1-methyl-4-phenyl-1,2,3,6-tetrahydropyridine. Two metabolic steps were found to increase cytotoxicity of isoquinolines. N-Methylation by a non-specific N-methyltransferase was proved by in vivo and in vitro experiments. The N-methylated compound was oxidized into N-methylisoquinolinium ion by monoamine oxidase from human brain mitochondria. The oxidation was proved by microdialysis in the rat brain. The isoquinolinium ion was more cytotoxic than the two metabolic precursors. N-Methylation of dopamine-derived 1-methyl-6,7-dihydroxy-1,2,3,4-tetrahydroisoquinolines was detected by in vivo microdialysis in the rat striatum, and their presence in the human brain was confirmed by GC-MS. The metabolic bioactivation may be a general pathway to produce neurotoxins as the pathogenic agents of Parkinson's disease.

Introduction

The major pathological finding in the brain of Parkinson's disease (PD) is the nerve degeneration and cell loss of dopamine (DA) neurons in the nigrostriatal system. The biochemical changes in the postmortem parkinsonian brains were studied by Nagatsu and Narabayashi (Nagatsu et al., 1979). Reduction of tyrosine hydroxylase [tyrosine, tetrahydropteridine: oxygen oxidoreductase (3-hydroxylating); EC 1.4.16.2, TH] and other related enzymes and of the biopterin cofactor were found. However, the biochemical basis of neurodegeneration in PD had not been well clarified until the discovery of 1-methyl-4-phenyl-1,2,3,6-tetrahydropyridine (MPTP) (Davis et al., 1979). Studies on the PD model with MPTP suggest the basic properties of neurotoxins specific for DA neurons. Dopaminergic neurotoxin should be transported into the brain through the blood-brain barrier (BBB) and oxidized by monoamine oxidase [monoamine: oxygen oxidoreductase (deaminating), EC 1.4.3.4, MAO] into a more toxic compound; MPTP is oxidized into 1-methyl-4-phenylpyridinium ion (MPP^+) (Chiba et al., 1984). The uptake of the compound into the cells by DA transport system, and the accumulation in conjugation with the intracellular components such as neuromelanin are considered as the mechanism of how the toxicity is specified to DA neurons. The results with MPTP indicate that similar biochemical processes may be underlying in the pathogenesis of idiopathic PD.

To find novel toxins that elicit parkinsonism in humans, endogenous and exogenous compounds with characteristics similar to MPTP have been intensively examined. Out of them, a series of monoamine-derived terahydroisoquinolines were proposed as candidates of dopaminergic neurotoxins. 1,2,3,4-Tetrahydroisoquinoline (TIQ) is synthesized by non-enzymatic condensation of β-phenethylamine with formaldehyde and 1-methyl-TIQ (1MeTIQ) by the reaction with acetaldehyde. TIQ is also present in food and easily transported into the brain. TIQ was identified in the human brain by gas chromatography-mass spectrography (GC-MS) (Niwa et al., 1987; Ohta et al., 1987). TIQ was found to cause parkinsonism in primates after systemic administration (Nagatsu and Yoshida, 1988). However, its toxicity was not so potent as MPTP and the symptoms were reversible. In search for more toxic compounds, chemically synthesized N-methylisoquinolinium ion ($NMIQ^+$) was proposed as a candidate, because of the similarity of the chemical structure to MPP^+. It was found to inhibit tyrosine hydroxylation in slices from rat striatum (Hirata et al., 1986), and type A monoamine oxidase (Naoi et al., 1987), as in the case of MPTP or MPP^+. It was suggested that two enzymatic processes, N-methylation and oxidation, may produce isoquinolinium ion from TIQ in the brain. Indeed by in vitro experiments using enzyme samples from the human brain, N-methylation of TIQ into N-methyl-1,2,3,4-tetrahydroisoquinoline (NMTIQ) (Naoi et al., 1989a) and its oxidation into $NMIQ^+$ were demonstrated (Naoi et al., 1989b). Occurrence of NMTIQ was also confirmed by GC-MS in the brain of monkeys administered by TIQ (Niwa et al., 1990). Cytotoxicity of $NMIQ^+$ was demonstrated by in vivo experiments with a DA cell model, clonal pheochromocytoma PC12h cells (Naoi et al., 1989c; Maruyama et al., 1992).

This paper reports N-methylation of dopamine-derived 6,7-dihydroxy-1,2,3,4-tetrahydroisoquinolines (DHTIQs), which was proved by in vitro and in vivo experiments. The presence of N-methylated alkaloids in human brain was confirmed by GC-MS. Effects on the monoamine levels in the brain and intracellular ATP levels were examined, and the results were discussed as possible mechanism of the cytotoxicity. Perturbation of monoamines by these isoquinolines is discussed in relation to pathogenesis of PD.

Materials and methods

NMTIQ and NMIQ⁺ were kindly donated by Dr. S. Matsuura; DHTIQs, such as (R) and (S)-1-methyl-6,7-dihydoxyl-1,2,3,4-tetrahydroisoquinoline (salsolinols) by Dr. P. Dostert. Standards of monoamines and their metabolites were purchased from Sigma; 6,7-dihydoxyl-1,2,3,4-tetrahydroisoquinoline (Norsal) from Janssen; TIQ from Aldrich. As reported previously, rat brain microdialysis was performed and N-methylated salsolinols and monoamines were quantitatively assayed by high-performance liquid chromatography (HPLC) with electrochemical detection (ECD) using 16 electrodes (CEAS, ESA, Bedford, MA) (Maruyama et al., 1992). Oxidation of NMTIQ was analyzed by HPLC-fluorimetric detection (FD) (Naoi et al., 1989a).

Isoquinolines and their N-methylated derivatives were identified in the human brain by GC-MS as reported previously (Niwa et al., 1987, 1992).

Results and discussion

Figure 1 shows the chemical structures and biosynthetic pathway of isoquinolines used in this article. Norsal and 1-methyl-DHTIQ (Sal) were identified in the brain, and they are synthesized by condensation of dopamine with an aldehyde or a keto acid. The later reaction is followed by decarboxylation and reduction to yield (R)-enantiomer of Sal in the human brain (Dostert et al., 1990).

As shown in Fig. 2, N-methylation of (R)Sal was demonstrated by microdialysis in the rat striatum. By perfusion of (R)Sal, production of N-methyl-(R)Sal [NM(R)Sal] was quantitatively analyzed by HPLC-ECD. The chemical structure of the product in the perfusate was confirmed as NM(R)Sal by GC-MS (Niwa et al., 1992). N-Methylated Sal and Norsal were found to occur in the human brain by GC-MS (Niwa et al., 1991). Another enantiomer of Sal, (S)Sal was also methylated in the striatum. The

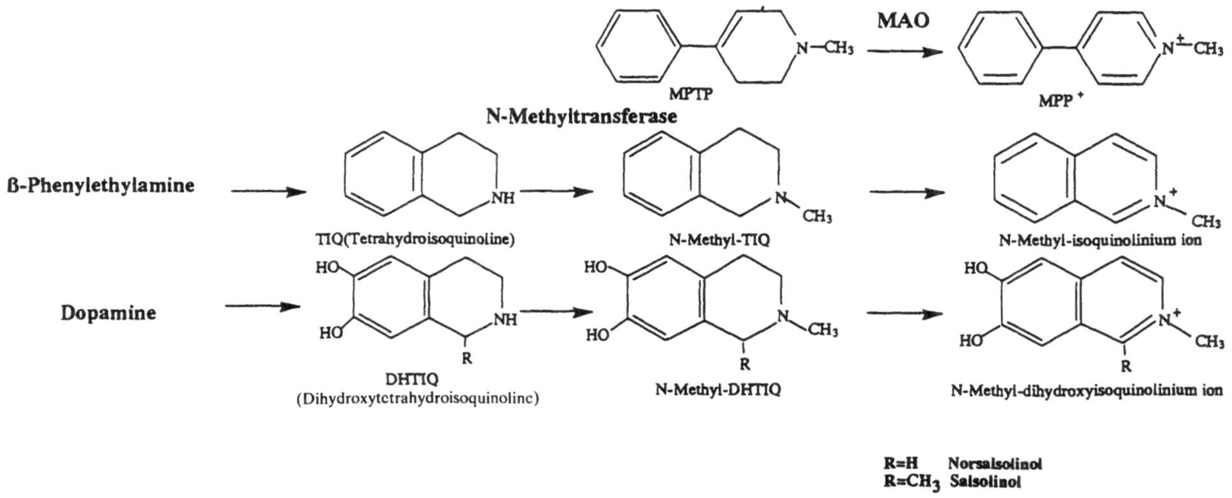

Fig. 1. Metabolic pathway and chemical structures of 1,2,3,4-tetehydroisoquinolines and 6,7-dihydroxy-1,2,3,4-tetrahydroisoquinolines. *MAO* Monoamine oxidase

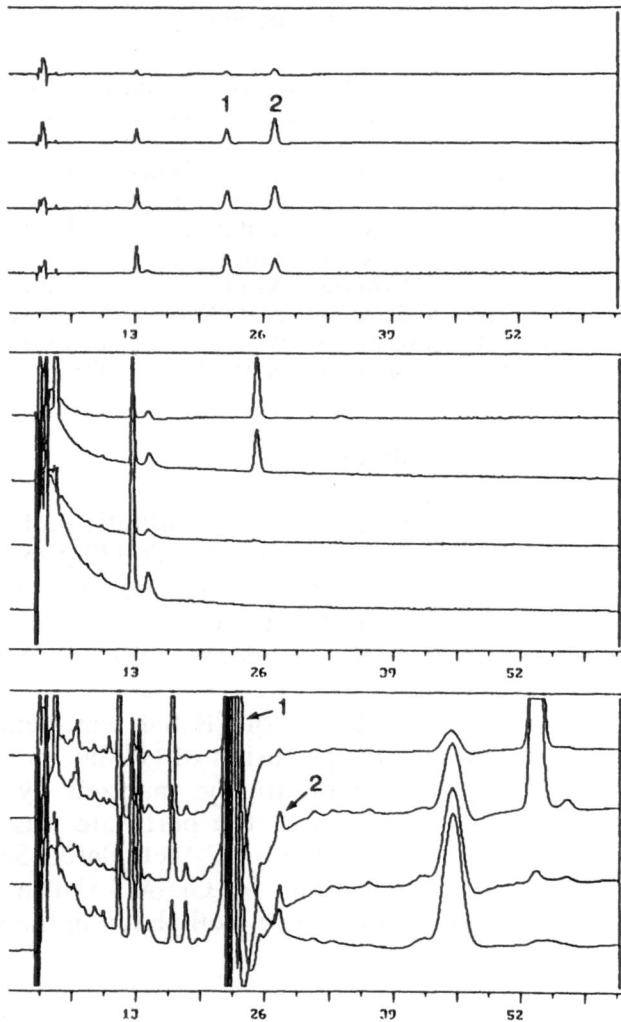

Fig. 2. HPLC chromatograms of standard and the dialysate in the rat striatum before and after (R)Sal perfusion. Rat striatum was perfused with (R)Sal by in vivo microdialysis and the dialysate was collected every 20 min. Twenty µl of sample was applied on an HPLC-ECD apparatus. Full scale of each electrode was set at 1 µA (**a**) and 150 nA (**b,c**). **a** Standard of 1 µM (R)Sal (*1*) and NM(R)Sal (*2*). **b** Dialysate before (R)Sal perfusion. **c** Dialysate after (R)Sal perfusion

N-methylation activity in rat brain regions was compared in the substantia nigra, hippocampus and hypothalamus. As summarized in Table 1, in the substantia nigra NM(R)Sal concentration was significantly higher than in other regions (p < 0.01). It suggests that Sal is N-methylated mainly in the brain region rich in DA neurons.

Oxidation of N-methylated TIQ into N-methylisoquinolinium ion was previously demonstrated with MAO sample prepared from human synaptosomes. The oxidation was confirmed also by microdialysis of NMTIQ in the

Table 1. Concentration of (R)Sal and N-methylated (R)Sal [NM(R)Sal] in the dialysate

	Striatum	Substantia nigra	Hypothalamus	Hippocampus
Concentration of (R)Sal(μM)	290.8 ± 28.9	358.2 ± 57.4	178.9	190.5
			176.6	171.4
Concentration of NM (R)Sal(μM)	289.2 ± 163.5	1,011 ± 127.4*	160.8	162.5
			199.2	144.8
Relative ratio of concentration [(R)NMSal/(R)Sal] × 10^3	0.863 ± 0.398	2.97 ± 0.536*	0.899	0.853
			1.13	0.845

The concentration of (R)Sal and NM(R)Sal in the dialysate after 80 min perfusion of (R)Sal. (R)Sal was perfused by in vivo microdialysis method. The samples were collected at 20 min intervals and flow rate was 2 μl/min. *$p < 0.01$ by Scheffe F-test

Table 2. The effects of dihydroxytetrahydroisoquinolines on extracellular levels of dopamine and serotonin

	Concentration (nM)	
	DA	5-HT
(R)Sal	303.7 ± 96.5	2,531 ± 121
(S)Sal	21.5 ± 0.99	3,692 ± 11
NM(R)Sal	84.0 ± 44.6	5.2 ± 5.2
Norsal	50.3 ± 12.8	4.3 ± 2.8
NMNorsal	102.9 ± 21.9	N.D.
Basal level in the striatal dialysate before perfusion	4.2 ± 3.6	N.D.

The concentrations of DA and 5-HT in the striatal dialysate were quantitatively analyzed using HPLC-ECD, after 120 min perfusion of DHTIQs. Each value represents mean and SE of duplicate measurements of two samples obtained by 4 [(R)Sal] or 2 [(S)Sal, NM(R)Sal, Norsal, N-methylated norsalsolinol = NMNorsal] experiments. *N.D.* Not detected

rat striatum. The amounts of NMIQ$^+$ synthesized in the brain tissues were quantitatively analyzed by HPLC-FD. The amount of NMIQ$^+$ in the brain after 240 min perfusion was estimated to be 0.16 ± 0.02 pmol/mg protein (mean and SE).

By perfusion of DHTIQs in the rat striatum, enormous amounts of serotonin (5-HT) and DA were released as summarized in Table 2. The amount of released 5-HT was the highest with Sal. Study on structure-activity relationship reveals that catechol structure is essentially required to the release of DA into extracellular fluid (ECF), since TIQs could not increase the levels in ECF (Maruyama et al., 1993a). The levels of the oxidized products of DA and 5-HT, i.e. DOPAC, HVA, and 5-HIAA in the perfusate were reduced by Sal perfusion, indicating inhibition of MAO activity (Moruyama et al., 1993b).

Table 3. The effects of isoquinolines on ATP levels in
PC12h cells

	Intracellular concentration (pmol/mg protein)	
		pre-incubated with clorgyline
Control	770.2 ± 116.6	1,142.9 ± 514.3
TIQ	658.6 ± 177.9	942.9 ± 142.9
NMTIQ	154.1 ± 98.8*	1,120.0 ± 542.9
NMIQ$^+$	644.7 ± 28.9	856.1 ± 57.1
MTPT	N.D.*	N.D.*
MPP$^+$	458.2 ± 266.0	1,028.6 ± 314.3

PC12h cells were incubated with 1 mM isoquinolines and
pyridnes for 20 min at 37°C, or after wards reacted with
100 μM clorgyline for 10 min. The intracellular ATP amounts
were quantitatively assayed using HPLC-UV. *N.D.* Not
detected. *Difference is statistically significant from the
basal value by one-sample runs test (p < 0.01)

The effect of TIQs and DHTIQs on intracellular ATP levels were
examined, using PC12h cells. As summarized in Table 3, NMTIQ was
found to reduce ATP level significantly after 20 min incubation, but TIQ
and NMIQ$^+$ did not affect ATP level. The reduction could be prevented by
pre-incubation of the cells with clorgyline, an inhibitor of type A MAO. It
suggests that oxidation of NMTIQ by MAO is directly coupled with ATP
depletion. Enzymatic oxidation of NMTIQ produces hydrogen peroxide,
and peroxide or other activated oxygen species may be involved in ATP
depletion. Sal and N-methylated Sal and N-methyl-Norsal reduced ATP
levels in PC12h cells, but Norsal did not reduce ATP level. In addition, TIQ
inhibited oxidative phosphorylation in isolated mitochondria (Suzuki et al.,
1988), as in the case with MPTP or MPP$^+$ (Mizuno et al., 1987).

To examine (sub)chronic effects of TIQs and DHTIQs, PC12h cells
were cultured in their presence of various concentrations of isoquinolines,
and the amount of the cell protein was measured as index of the living cells.
As shown in Fig. 3, DHTIQs were more potent to reduce the cell number
than TIQs. NMIQ$^+$ was found to be more cytotoxic than TIQ and NMTIQ.
Sal is produced from DA presumably in DA neurons and, furthermore,
methylated in DA cells, as indicated by high activity of N-methyltransferase
in the substantia nigra. On the other hand, TIQ is distributed diffusely in
the human brain and its specific accumulation in DA neurons has never
been confirmed. Thus, DHTIQs and their derivatives may have more possibility to be cytotoxic in DA cells. Sal is known to be increased by L-DOPA
therapy (Sandler et al., 1973) and N-methylated DHTIQs were detected in
the cerebrospinal fluid from the PD patients treated with L-DOPA (Mosel
and Koempf, 1992). These results suggest that DA itself can increase
biosynthesis of Sal and its N-methyl derivatives.

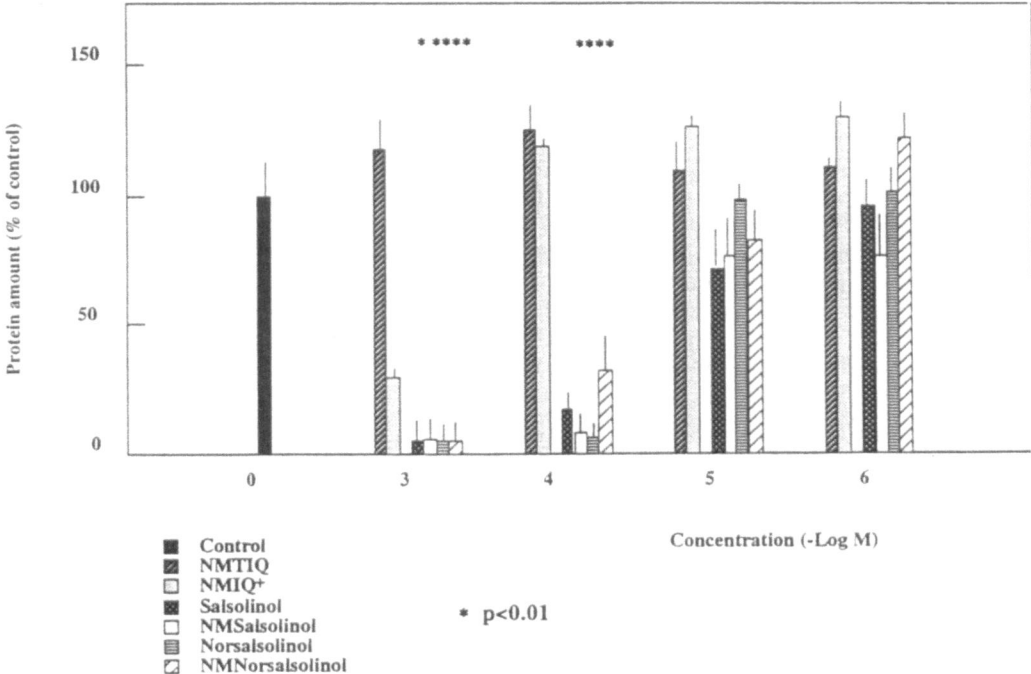

Fig. 3. Effects of TIQs and DHTIQs on the cell growth of PC12h cells. PC12h cells were cultured with various concentrations of isoquinolines for 3 days, and the cells were harvested and the cell protein was measured. The protein amount was expressed as percents of control. *Significantly different from control (p < 0.01)

Concerning the mechanism of the cell death by the toxins, the results obtained by microdialysis are suggestive. After perfusion of DHTIQs, huge amounts of DA and 5-HT were released into ECF in the striatum. Release of free DA should increase the enzymic oxidation by MAO or its auto-oxidation and may enhance formation of activated oxygen species, such as the hydroxy radical. Especially in substantia nigra the interaction of iron with free DA might enhance the radical formation and thus increase the cytotoxicity (Sofic et al., 1991).

This article presents the data that in the brain tetrahyroisoquinolines and 6,7-dihydroxytetrahydroisoquinolines are methylated by a non-specific N-methyltransferase and that N-methylated compounds become more cytotoxic and more specific to monoamine neurons. Oxidation by MAO increases further the cytotoxic potentials, as in the case of MPP^+ and $NMIQ^+$. These results seem to support a view that metabolic activation of endogenous or exogenous compounds may produce novel toxins related to pathogenesis of Parkinson's disease. At present in vivo experiments have failed to induce parkinsonism in animals with these DHTIQs. Thus, endogenous or exogenous compounds should be searched for to find compounds that can be activated by the metabolism into neurotoxins as potent and specific as MPTP.

References

Chiba K, Trevor A, Castagnoli Jr N (1984) Metabolism of the neurotoxic tertiary amine, MPTP, by brain monoamine oxidase. Biochem Biophys Res Commun 120: 574–578

Davis GCB, Williams AC, Markey SP, Ebert MH, Caine ED, Beichert CM, Kopin IJ (1979) Chronic parkinsonism secondary to intravenous injection of meperidine analogues. Psychiatry Res 1: 249–254

Dostert P, Strolin Benedetti M, Bellotti V, Allievi C, Dordain G (1990) Biosynthesis of salsolinol alkaloid, in healthy subjects. J Neural Transm 81: 215–223

Hirata Y, Sugimura H, Takei H, Nagatsu T (1986) The effects of pyridinium salts, structurally related compounds of 1-methyl-4-phenylpyridinium ion (MPP$^+$), on tyrosine hydroxylation in rat striatal tissue slices. Brain Res 397: 341–344

Maruyama W, Nakahara D, Ota M, Takahashi T, Takahashi A, Nagatsu T, Naoi M (1992) N-Methylation of dopamine-derived 6,7-dihydroxy-1,2,3,4-tetrahydroisoquinoline, (R)-salsolinol, in rat brains: in vivo microdialysis study. J Neurochem 59: 395–400

Maruyama W, Nakahara F, Dostert P, Hashiguchi H, Ohta S, Hirobe M, Takahashi A, Nagatsu T, Naoi M (1993a) Selective release of serotonin by endogenous alkaloids, 1-methyl-6,7-dihydroxy-1,2,3,4-tetrahydroisoquinolines, (R)- and (S)salsolinol, in the rat striatum: in vivo microdialysis study. Neurosci Lett 149: 115–118

Maruyama W, Nakahara D, Dostert P, Takahashi A, Naoi M (1993b) Naturally-occurring isoquinolines perturb monoamine metabolism in the brain; studied by in vivo microdialysis. J Neural Transm [Gen Sect] 94: 91

Mizuno Y, Sone N, Saitoh T (1987) Effects of 1-methyl-4-phenyl-1,2,3,6-tetrahydropyridine and 1-methyl-4-phenylpyridinim ion on activities of the enzymes in the electron transport system in mouse brain. J Neurochem 48: 1767–1793

Moser A, Koempf D (1992) Presence of methyl-6,7-dihydroxy-1,2,3,4-tetrahydro soquinolines, derivatives of the neurotoxin isoquinoline, in parkinsonian lumbar CSF. Life Sci 50: 1885–1895

Nagatsu T, Kato T, Nagatsu I, Kondo Y, Inagaki S, Iizuka R, Narabayashi H (1979) Catecholamine-related enzymes in the brain of patients with parkinsonism and Wilson's disease. In: Poirier LJ, Sourkes TL, Bedard PJ (eds) Advances in neurology, vol 24. Raven Press, New York, pp 283–292

Nagatsu T, Yoshida M (1988) An endogenous substance of the brain, tetrahydroisoquinoline, produces parkinsonism in primates with decreased dopamine, tyrosine hydroxylase and biopterin in the nigrostriatal regions. Neurosci Lett 87: 178–182

Naoi M, Hirata Y, Nagatsu T (1987) Inhibition of monoamine oxidase by N-methylisoquinolinium ion. J Neurochem 48: 709–712

Naoi M, Matsuura S, Parvez H, Takahashi T, Hirata Y, Minami M, Nagatsu T (1989a) Oxidation of N-methyl-1,2,3,4-tetrahydroisoquinoline into the N-methylisoquinolinium ion by monoamine oxidase. J Neurochem 52: 653–655

Naoi M, Matsuura S, Takahashi T, Nagatsu T (1989b) An N-methyltransferase in human brain catalyses N-methylation of 1,2,3,4-tetrahydroisoquinoline into N-methyl-1,2,3,4-tetrahydroisoquinoline, a precursor of a dopaminergic neurotoxin, N-methylisoquinolinium ion. Biochem Biophys Res Commun 161: 1213–1219

Naoi M, Takahashi T, Parvez H, Kabeya R, Taguchi E, Yamaguchi K, Hirata Y, Minami M, Nagatsu T (1989c) N-Methylisoquinolinium ion as an inhibitor of tyrosine hydroxylase, aromatic L-amino acid decarboxylase and monoamine oxidase. Neurochem Int 15: 315–320

Niwa T, Takeda N, Kaneda N, Hashizume Y, Nagatsu T (1987) Presence of tetrahydroisoquinoline and 2-methyl-tetrahydroquinoline in parkinsonian and normal human brain. Biochem Biophys Res Commun 144: 1084–1089

Niwa T, Yoshizumi H, Tatematsu A, Matsuura S, Yoshida M, Kawachi M, Naoi M, Nagatsu T (1990) Endogenous synthesis of N-methyl-1,2,3,4-tetrahydroisoquinoline, a precursor of N-methylisoquinolinium ion, in the brains of primates with parkinsonism after systemic administration of 1,2,3,4-tetrahydroisoquinoline. J Chromatogr 533: 145–151

Niwa T, Takeda N, Yoshizumi H, Tatematsu A, Yoshida M, Dostert P, Naoi M, Nagatsu T (1991) Presence of 2-methyl-6,7-dihydroxy-1,2,3,4-tetrahydroisoquinoline and 1,2-dimethyl-6,7-dihydroxy-1,2,3,4-tetrahydroisoquinolines, novel endogenous amines, in parkinsonian and normal brains. Biochem Biophys Res Commun 177: 603–609

Niwa T, Maruyama W, Nakahara D, Takeda N, Yoshizumi H, Tatematsu A, Takahashi A, Dostert P, Naoi M, Nagatsu T (1992) Endogenous synthesis of N-methylsalsolinol, an analogue of MPTP, in the rat brain during in vivo microdialysis with salsolinol as demonstrated by gas chromatography/mass spectrometry. J Chromatogr 578: 109–115

Ohta S, Kohno M, Makino Y, Tachikawa O, Hirobe M (1987) Tetrahydroisoquinoline and 1-methyl-tetrahydroisoquinoline are present in the human brain: relation to Parkinson's disease. Biomed Res 8: 453–456

Sandler M, Carter SB, Hunter KR, Stern GM (1973) Tetrahydroisoquinoline alkaloids: in vivo metabolites of L-DOPA in man. Nature 241: 439–443

Sofic E, Paulus W, Jellinger K, Riederer P, Youdim MBH (1991) Selective increase of iron in substantia nigra zona compacta of parkinsonian brains. J Neurochem 56: 978–982

Suzuki K, Mizuno Y, Yoshida M (1988) Inhibition of mitochondrial NADH-ubiquinone oxidoreductase activity and ATP synthesis by tetrahydroisoquinoline. Neurosci Lett 86: 105–108

Authors' address: Dr. M. Naoi, Department of Biosciences, Nagoya Institute of Technology, Showa-ku, Gokiso-cho, Nagoya 466, Japan.

J Neural Transm (1994) [Suppl] 41: 207–219

Amphetamine-metabolites of deprenyl involved in protection against neurotoxicity induced by MPTP and 2'-methyl-MPTP*

I. Sziráki[1], V. Kardos[1], M. Patthy[1], M. Pátfalusi[1], J. Gaál[2], M. Solti[2], E. Kollár[1], and J. Singer[1]

[1] Institute for Drug Research, and
[2] Chinoin Pharmaceutical and Chemical Works, Budapest, Hungary

Summary. The ability of l-deprenyl to protect against the parkinsonian effects of 1-methyl-4-phenyl-1,2,3,6-tetrahydropyridine (MPTP) has been attributed to the inhibition of conversion of MPTP to MPP^+ (1-methyl-4-phenylpyridinium) catalyzed by MAO-B. We report here that deprenyl-treatment in mice has an additional neuroprotective element associated with the rapid metabolization of l-deprenyl to l-methamphetamine and l-amphetamine. l-Methamphetamine and l-amphetamine inhibit MPP^+-uptake into striatal synaptosomes prepared from rats. Post-treatment by l-deprenyl, l-methamphetamine, l-amphetamine (at times when MPTP is no longer present in the striatum of mice) protects against neurotoxicity in C57BL mice by blocking the uptake of MPP^+ into dopaminergic neurons, and even against the neurotoxicity induced by $2'CH_3$-MPTP, which is partly bioactivated by MAO-A. These findings may have clinical implications since deprenyl has recently been found to delay the progression of Parkinson's disease.

Introduction

MPTP induces parkinsonian symptoms in man (Davis et al., 1979; Langston et al., 1983) and the degeneration of nigrostriatal dopaminergic neurons in animals, including mice (Heikkila et al., 1984; Hallman et al., 1984). For the neurotoxicity of MPTP, its oxidation to MPP^+ by MAO-B (Chiba et al., 1984; Markey et al., 1984; Hallman et al., 1984) in glial cells (Brooks et al., 1989) and the subsequent uptake of MPP^+ by the dopamine (DA) reuptake system (Javitch et al., 1985) are required. Pretreatment of animals with nonselective MAO-inhibitors (e.g. pargyline), the selective MAO-B inhibitor l-deprenyl (Heikkila et al., 1984; Cohen et al., 1984; Langston et al., 1984) or with dopamine reuptake blockers such as nomifensine and mazindol (Javitch et al., 1985; Mayer et al., 1986) prevents MPTP neurotoxicity

---

* This paper is dedicated to the memory of Prof. R.E. Heikkila

owing to the inhibition of conversion of MPTP to MPP$^+$ and the inhibition of the uptake of MPP$^+$ into dopaminergic neurons, respectively. In accordance with the sequence of steps required for MPTP-neurotoxicity, temporal differences were found between the antagonizing effect of pargyline and nomifensine against the neurotoxicity induced by a single dose of MPTP (Sundström and Jonsson, 1986). Pretreatment with either of these drugs completely antagonized the depletion of DA, the most characteristic neurochemical alteration in the striatum of mice after MPTP-treatment. Posttreatment with the MAO-inhibitor pargyline was effective only 5 min after MPTP whereas the catecholamine uptake blocker nomifensine was still protective 1–2 h after the MPTP treatment. Our recent results show that mazindol provides protection against DA-ergic neurotoxicity in mice even when administered as late as 4 h after MPTP-treatment suggesting that the length of time while MPP$^+$ is present in the extradopaminergic compartment is relatively long (Sziráki et al., 1990). We hypothesized that deprenyl, in contrast with pargyline, nomifensine or mazindol, has a complex mode of action against MPTP-neurotoxicity: its neuroprotective action involves both the inhibition of conversion of MPTP to MPP$^+$ and the inhibition of uptake of MPP$^+$ by l-methamphetamine and l-amphetamine which are generated (Reynolds, 1978) from l-deprenyl.

Material and methods

To evaluate deprenyl and its metabolites for their capacity to block MPP$^+$-uptake in vitro, synaptosomes were prepared from striata of CFY male rats by a standard procedure (Schaht and Heptner, 1974), and the uptake of MPP$^+$ was determined using the slightly modified method of Javitch and Snyder (1985). l-Deprenyl (invented and developed in Chinoin Pharmaceutical and Chemical Co Ltd, Budapest) and its metabolites were synthesized at Chinoin. All other chemicals and reagents were from standard sources. Synaptosomes were preincubated either with deprenyl or its metabolites or with mazindol for 3 min at 37°C, then incubated in the presence of tritiated and nonradioactive MPP$^+$ (final conc. 10^{-7}M) for 5 min. To correct for passive diffusion, control samples were incubated at 4°C.

   To determine the neuroprotective effect of deprenyl and its metabolites, C57 black male mice (24–32 g) were injected intraperitoneally with MPTP-hydrochloride (30 or 40 mg/kg; free base) either alone or in combination with MAO-inhibitors, with l-methamphetamine-/l-amphetamine-hydrochloride or in combination with mazindol according to the treatment schedules indicated in the figures. Other mice received 1-methyl-4-(2'methylphenyl)-1,2,3,6-tetrahydropyridine (2'CH$_3$-MPTP) hydrochloride (15 mg/kg; free base) either alone or in combination with substances listed above. The drugs were dissolved in saline with the exception of mazindol, which was suspended in 1% Tween 80. Three to 4 days later the mice were killed by cervical dislocation. The brains were removed and the corpora striata were dissected out and kept at −80°C until assayed. At processing the tissue samples the paired striata were weighed and then sonicated in 200 µl distilled water for 15 sec. Aliquots of water homogenates were resonicated in HPLC-mobile phase containing α-methyl dopamine as internal standard. After centrifugation (10,000 g, 15 min at 4°C) aliquots of supernatants were assayed for dopamine (DA) and its metabolites [3,4-dihydroxyphenylacetic acid (DOPAC) and homovanillic acid (HVA)] by HPLC-EC (Patthy and Gyenge, 1988).

For determination of MAO-B activities, aliquots of the water homogenates were resonicated with equal volumes of 10 mM phosphate buffer (pH = 7.0) containing 0.1% Triton × 100. The samples were centrifuged and MAO activities from aliquots of supernatants were determined by the slightly modified radioenzymatic method of Wurtman and Axelrod (Wurtman and Axelrod, 1963) using 2-phenylethylamine (PEA) as specific substrate.

For statistical analysis one-way ANOVA was used with Duncan's multiple range test.

Results

To evaluate l-deprenyl, l-desmethyldeprenyl, l-methamphetamine (l-METH), l-amphetamine (l-AMPH), the metabolites of deprenyl in man (Reynolds et al., 1978), for their capacity to block MPP^+-uptake, synaptosomes prepared from striata of rats were used. Synaptosomes were preincubated either with deprenyl or its metabolites or with mazindol for 3 min at 37°C, then incubated in the presence of MPP^+ (final concentration: $10^{-7}M$) for 5 min. l-Deprenyl and l-desmethyldeprenyl had almost no effect on MPP^+ uptake into striatal synaptosomes (less than 20% at 5 µM concentration). l-METH and l-AMPH, however, effectively blocked the accumulation of MPP^+ with a potency similar to that of mazindol. IC_{50} values (µM) for l-METH, l-AMPH and mazindol were 0.49, 0.54 and 0.17, respectively.

As shown in Table 1, l-METH (5 mg/kg, i.p.) or mazindol administered 30 min before each of the 3 MPTP-treatments completely prevents the depletion of striatal DA and DA-metabolites (DOPAC, HVA) induced by repeated administration of MPTP (30 mg/kg, free base, i.p.) injection. The ability of l-AMPH (5 mg/kg) to prevent MPTP-neurotoxicity was investigated with a different treatment schedule using a single dose of MPTP. In C57BL mice one injection of MPTP (40 mg/kg) caused a 40–70% decrease in the striatal DA content measured 3 days after the treatment. l-AMPH, when administered 30 min prior to the MPTP-treatment, significantly attenuated the striatal DA-depletion (Fig. 1). Striatal levels of DA and DA-metabolites were close to control values in mice treated with l-METH or l-AMPH alone

Table 1. Effects of l-methamphetamine or mazindol on the MPTP-induced decrease of the striatal levels of DA and its metabolites in mice

Pretreatment	Treatment	DA (µg/g)	% of control	DOPAC (µg/g)	% of control	HVA (µg/g)	% of control
Saline	Saline	14.1 ± 0.3	100	1.0 ± 0.05	100	1.0 ± 0.05	100
Saline	MPTP	7.1 ± 0.4*	50	0.6 ± 0.05*	60	0.4 ± 0.04*	40
l-METH	MPTP	14.0 ± 0.5**	99	1.1 ± 0.06**	110	1.1 ± 0.10**	110
Mazindol	MPTP	14.3 ± 0.6**	101	1.1 ± 0.06**	110	1.1 ± 0.07**	110

Mice received intraperitoneal (i.p.) injections of MPTP · HCl (30 mg free base per kg) daily for three consecutive days. l-METH · HCl or mazindol was administered i.p. (5 mg/kg; free base) 30 min before each MPTP-treatment and the animals were killed 4 days after the last injection. Values represent the means ± SEM from one of the two separate experiments with similar results. Numbers of mice varied from 5 to 7 per group. * Statistically different (P < 0.01) from saline controls. ** Statistically different (P < 0.01) from MPTP-only group

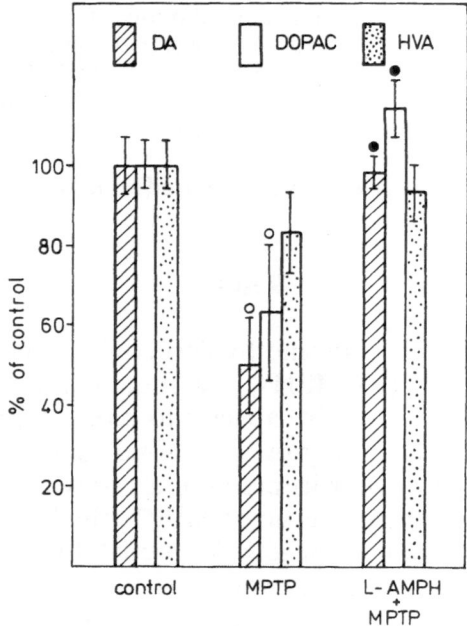

Fig. 1. Effects of l-amphetamine on MPTP-induced decrease of striatal levels of DA and its metabolites. Mice (n = 5–6) received l-AMPH (5 mg/kg; i.p.) 30 min prior to MPTP administration and were killed 3 days after the treatments. Striatal levels of *DA*, *DOPAC* and *HVA* (μg/g) in control group were 15.81 ± 1.85, 1.06 ± 0.07 and 1.06 ± 0.09, respectively. ○ Statistically significant difference (P < 0.01) compared to control group; ● statistically significant difference (P < 0.01) compared to MPTP-only group

(5 mg/kg), regardless whether the latter were applied 3 times (24 h apart) or only once. These results clearly demonstrate that pretreatment with each of these deprenyl-metabolites that inhibit MPP$^+$-uptake in vitro prevents MPTP-neurotoxicity in vivo.

To investigate whether l-METH and l-AMPH generated after deprenyl treatment have neuroprotective effects, a treatment schedule has to be applied by which the postulated metabolite-mediated element of neuroprotection can be distinguished from neuroprotection owing to the inhibition of conversion of MPTP to MPP$^+$.

Groups of mice were treated i.p. with deprenyl or pargyline (10–10 mg/kg) prior to, or after a single dose of MPTP (40 mg/kg), as indicated in Fig. 2. No protection was afforded by pargyline administered 30 or 60 min after a single dose of MPTP. In contrast, a 30-min post-treatment by deprenyl was completely, and a 60-min post-treatment was partially protective against MPTP-neurotoxicity. Since MPTP is not present in appreciable amount in the striatum of C57BL mice (Jonsson et al., 1986) 30 min after the treatment, the striking difference between pargyline and deprenyl concerning the time course of their protection seems to be unrelated to the inhibition of the conversion of MPTP to MPP$^+$.

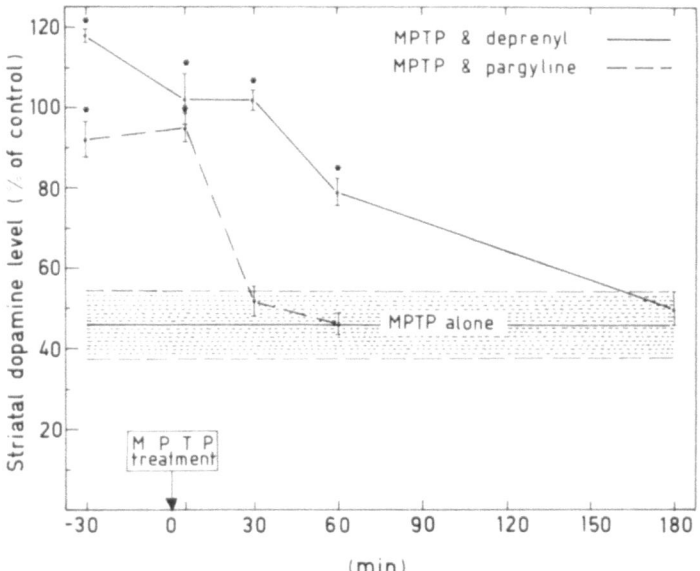

Fig. 2. Time course of protection against neurotoxicity induced by a single dose of MPTP in mice. DA levels were determined 3 days after the treatments. The abscissa is the time elapsed between MPTP-treatment and pre- (negative value) or post-treatment (positive values) with l-deprenyl or pargyline. l-Deprenyl or pargyline (10 mg/kg) was administered i.p. 30 min before or at various time points after the MPTP-treatment (40 mg/kg, i.p.). The animals were killed 3 days after the treatments. Each point represents the means ± SEM, expressed as percent of the respective mean control from four separate experiments (numbers of mice per group varied from 4–10 in each experiment). When calculating the effect of MPTP alone, data of all the MPTP-only treated mice (n = 29) were pooled. *Statistically significant differences (P < 0.05) between mice receiving MPTP alone and those receiving l-deprenyl or pargyline plus MPTP. For statistical analysis the data were expressed in µg per g tissue in each separate experiment

To examine this question more directly we, at a discriminatory time point, determined MAO-B activity in the striatal tissue homogenate of mice treated with pargyline or l-deprenyl either alone or in combination with MPTP. Two groups of mice received pargyline or deprenyl (10 mg/kg) and 30 min later they were killed. Other groups first received MPTP (40 mg/kg) and 30 min later they were post-treated with pargyline or deprenyl, then killed 30 min after the second treatment.

Figure 3 shows that 30 min after the treatment both pargyline and l-deprenyl caused a more than 80% inhibition of MAO-B when applied alone and a more than 90% inhibition when applied in combination with MPTP. If l-METH and l-AMPH generated from l-deprenyl are responsible for the neuroprotective effect of deprenyl-posttreatment, l-METH and l-AMPH would be expected to be protective when the conversion of MPTP to MPP$^+$ is already completed. We therefore investigated whether l-METH and l-AMPH could antagonize MPTP-induced striatal DA-depletion 60 min and 180 min after MPTP-treatment. (At time points when l-deprenyl was still

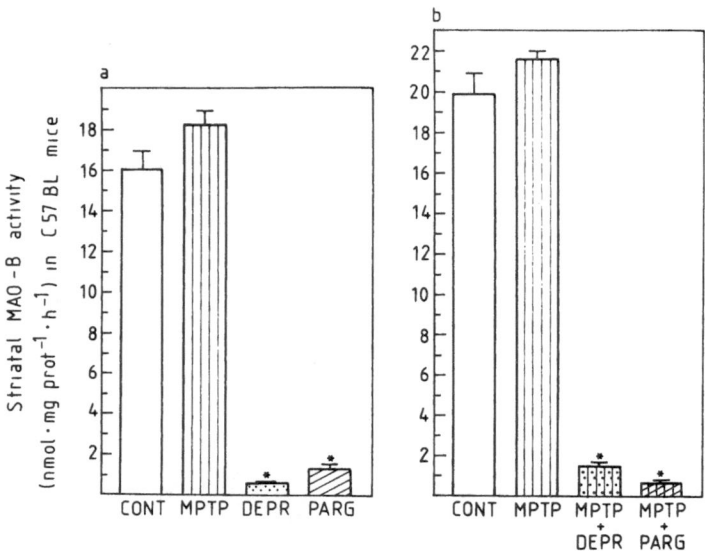

Fig. 3. Inhibition of striatal MAO-B activity after administration of l-deprenyl or pargyline alone (**a**) or in combination with MPTP (**b**). Mice (n = 3–7) were treated i.p. with *MPTP* (40 mg/kg) l-deprenyl (*DEPR*) or pargyline (*PARG*) (10 mg/kg) or with MPTP plus one of the two MAO inhibitors. Control mice (*CONT*) received saline. In groups treated with MAO-inhibitor alone or in combination with MPTP, the animals were killed 30 min after the treatment with l-deprenyl or with pargyline. The animals in control groups and in MPTP-only groups were killed 30 min (**a**) or 60 min (**b**) after the injection. The bars represent the means ± SEM (nmol product · mg prot^{-1} · h^{-1}) of MAO-B activity in homogenates of striatal tissue at a substrate concentration of 10^{-6} M. *P < 0.01 vs controls

protective and was no longer protective, respectively). Figure 4 shows that l-METH and l-AMPH (5 mg/kg, i.p.) protect against MPTP-neurotoxicity when administered 60 min after MPTP-treatment. (Their protective effect was more pronounced when they were applied 30 min after MPTP-treatment; data not shown). Like l-deprenyl, l-METH or l-AMPH was ineffective when administered 180 min after MPTP.

To gain further evidence to support our finding that deprenyl-treatment has a neuroprotective element unrelated to the inhibition of MAO-B, we investigated whether l-deprenyl and its metabolites also protect against neurotoxicity of a methylated analog of MPTP (2'CH$_3$-MPTP), which is partially bioactivated by MAO-A (Sonsalla et al., 1987). A 30 min post-treatment with l-deprenyl (10 mg/kg, i.p.) or mazindol (5 mg/kg) protected against neurotoxicity induced by a single dose of 2'CH$_3$-MPTP (15 mg/kg i.p.), whereas pargyline and the MAO-A inhibitor clorgyline were ineffective (Table 2, Experiment 1). As was the case with MPTP, l-METH and l-AMPH were also protective, regardless whether they were administered 30 min prior to, or 30 min after the 2'CH$_3$-MPTP-treatment (Table 2, Experiment 2). l-Deprenyl also was protective when administered 30 min prior to a single dose of 2'CH$_3$-MPTP.

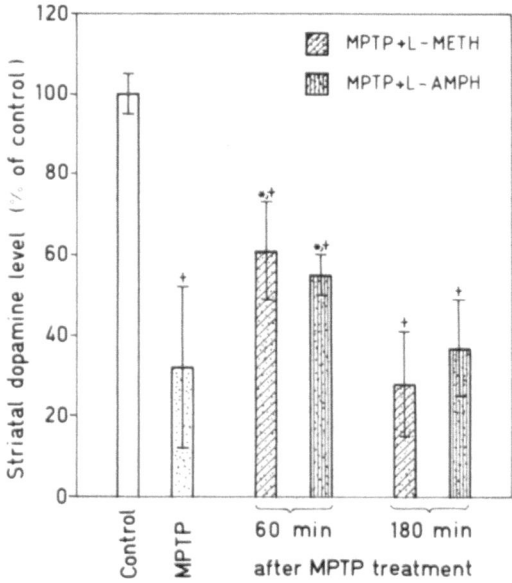

Fig. 4. The effects of l-methamphetamine or l-amphetamine post-treatment on MPTP-induced striatal dopamine depletion. Mice (n = 6) received l-METH or l-AMPH (5 mg/kg, i.p.) at either 60 or 180 min after MPTP-treatment (40 mg/kg, i.p.) and were killed 4 days after the treatments. The bars represent the means ± SEM. Striatal DA level in control group was $14.12 \pm 0.66 \mu g/g$. †Statistically significant (P < 0.01) difference compared to control group. *Statistically significant (P < 0.01) difference compared to MPTP-only treated group

Table 2. Protection against $2'CH_3$-MPTP-induced striatal DA-depletion by l-deprenyl and its metabolites in mice

Experiment 1				Experiment 2			
treatment 1	treatment 2	DA (µg/g)	% of control	treatment 1	treatment 2	DA (µg/g)	% of control
Saline	saline	12.6 ± 0.7	100	saline	saline	16.1 ± 0.9	100
$2'CH_3$-MPTP	saline	$5.9 \pm 0.5^*$	47	saline	$2'CH_3$-MPTP	$4.3 \pm 0.3^*$	27
$2'CH_3$-MPTP	pargyline	4.0 ± 0.2	32	l-METH	$2'CH_3$-MPTP	$15.3 \pm 0.3^\#$	95
$2'CH_3$-MPTP	clorgyline	5.1 ± 0.6	41	l-AMPH	$2'CH_3$-MPTP	$13.8 \pm 0.6^\#$	86
$2'CH_3$-MPTP	l-deprenyl	$12.2 \pm 0.5^\#$	97	mazindol	$2'CH_3$-MPTP	$9.8 \pm 0.8^\#$	61
$2'CH_3$-MPTP	mazindol	$11.2 \pm 0.4^\#$	89	l-deprenyl	$2'CH_3$-MPTP	$12.0 \pm 0.8^\#$	78
				$2'CH_3$-MPTP	l-METH	$13.8 \pm 0.6^\#$	86
				$2'CH_3$-MPTP	l-AMPH	$15.9 \pm 0.4^\#$	99
				$2'CH_3$-MPTP	mazindol	$16.2 \pm 0.7^\#$	101

In experiment 1 the mice received a single dose of $2'CH_3$-MPTP · HCl (15 mg free base per kg, i.p.) and 30 min later the animals were treated i.p. either with MAO inhibitors (10 mg/kg) or with mazindol (5 mg/kg). In experiment 2 l-methamphetamine · HCl or l-amphetamine · HCl (5 mg/kg; free base) was administered either before or after the $2'$-CH_3-MPTP-treatment. One group of mice in this experiment received l-deprenyl 30 min before the $2'CH_3$-MPTP. In both experiments the mice (6–7 per group) were killed 3 days after the treatments. Results are the means ± SEM. *Statistically different (P < 0.01) from saline controls. $^\#$ Statistically different (P < 0.01) from $2'$-CH_3-MPTP-only group

Discussion

In this report we provide direct evidence that l-deprenyl-treatment has a protective element against MPTP-neurotoxicity which is associated with its metabolites. In testing the hypothesis that metabolism of l-deprenyl to l-METH and l-AMPH is involved in the protection against MPTP-neurotoxicity we first investigated whether metabolites of l-deprenyl inhibit MPP^+-uptake into striatal synaptosomes and, subsequently, whether they prevent MPTP-induced striatal DA-depletion. Here, we are the first to report that l-METH and l-AMPH inhibit MPP^+-uptake in vitro with an inhibitory potency similar to that of mazindol, while l-deprenyl and its nor-metabolite have almost no effect at concentrations of $5 \, \mu M$.

Most of the drugs with the capacity to inhibit DA- and MPP^+-accumulation in rat or mouse striatal tissue slices were found to prevent striatal DA-depletion induced by repeated treatments with MPTP (Mayer et al., 1986). d-Amphetamine with DA-releasing and DA-reuptake inhibiting properties (Horn, 1978) has been found to potentiate (Sershen et al., 1986) or to antagonize (Chiueh et al., 1986; Sershen et al., 1986) the MPTP-neurotoxicity depending upon the treatment conditions. Our recent results demonstrate for the first time (Sziráki et al., submitted) that pretreatment with d-methamphetamine at 5 mg per kg protects against MPTP-neurotoxicity in C57BL mice. Our present data show that pretreatment with the l-enantiomer is also protective against MPTP-neurotoxicity (Table 1, Fig. 1) indicating that metabolism of l-deprenyl might be involved in its neuroprotective effect.

To separate the MAO-B inhibiting element of neuroprotection by l-deprenyl from that owing to the in vivo MPP^+-uptake inhibitory potency of the metabolites generated after treatment, l-deprenyl was applied prior to, or after the conversion of MPTP to MPP^+ and the time course of protection of l-deprenyl treatment was compared with that of the treatment with pargyline (Fig. 2).

In mice treated with a single intravenous injection of MPTP the striatal level of MPTP was negligible 30 min after the treatment and the formation of MPP^+ reached a maximum level about 30–60 min after MPTP administration (Jonsson et al., 1986). These data indicate that the period within which MPTP-neurotoxicity can be antagonized by blocking the MAO-B enzyme is relatively short. Indeed, post-MPTP administration of pargyline was effective only 5 min after MPTP (Sundström and Jonsson, 1986), whereas nomifensine and mazindol were still protective 1–2 h (Sundström and Jonsson, 1986) and 4 h (Sziráki et al., 1990), respectively, after the MPTP treatment. Post-MPTP protection by the uptake inhibitors nomifensine and mazindol reveals the extraneuronal presence of MPP^+ 2–4 h after the maximum level of MPP^+ in the striatum. In agreement with previous reports, pretreatment (Heikkila et al., 1984; Cohen et al., 1984; Langston et al., 1984) with pargyline or l-deprenyl antagonized the MPTP-induced striatal DA-depletion monitored 3 days after the treatments. Consistent with the findings of Sundström and Jonsson (1986), 5-min post-treatment with pargyline was also protective, but no protection was found when pargyline was given 30 or 60 min after the MPTP-treatment. In contrast, post-treatment with deprenyl 30 or 60 min after the MPTP-treatment was protective similarly to post-treatment with nomifensine or mazindol. Complete or partial protection by l-deprenyl at times when the level of nonconverted MPTP in the striatum of mice is negligible (Jonsson et al., 1986) indicates that l-deprenyl functions as a MPP^+-uptake inhibitor.

To further investigate the finding that factors other than inhibition of the conversion of MPTP to MPP^+ are responsible for the difference between time-courses of protection by l-deprenyl and by pargyline, we examined whether or not l-deprenyl and pargyline inhibit striatal MAO-B activity 60 min after the MPTP-treatment (Fig. 3). At this time point, post-treatment by l-deprenyl, but not by pargyline, is still protective against MPTP-induced DA-depletion. The near complete inhibition of striatal MAO-B activity at this time by pargyline or l-deprenyl indicates that neither the protection by post-treatment with l-deprenyl nor the lack of protection by pargyline can be associated with the inhibition of conversion of MPTP to MPP^+. Thus, the protection by post-treatment with l-deprenyl seems to be associated with the blockade of MPP^+-uptake, since the MAO-B inhibitor was applied too late to inhibit the oxidation of MPTP. l-Deprenyl itself does not affect the uptake of MPP^+ in vitro, therefore metabolite(s) with MPP^+-uptake inhibitory potency generated from l-deprenyl are likely to be present in striatum when MPP^+ is still outside the dopaminergic neurons. In fact, a relatively high amount (Philips, 1981) of amphetamine, which is probably derived from the demethylation of methamphetamine (Reynolds et al., 1978) was detected in the brain of mice treated with l-deprenyl (1 mg/kg i.p.) as early as 15 min after the treatment (84.4 ng/g fresh tissue, a value which represents a more than thousand-fold increase as compared to control level). Between 15 min to 60 min after the deprenyl-treatment, the tissue concentration of amphetamine calculated from the data of Philips (1981) was the same order of magnitude (0.89 μM at 15 min, 0.8 μM at 30 min and 0.45 μM at 60 min after the treatment) as that we found for 50% inhibition of MPP^+-uptake in vitro. Tissue concentration of amphetamine decreased to less than 20% at 120 min and to 5% at 240 min of the concentration observed at 15 min after the treatment.

If the protection by post-treatment with l-deprenyl against MPTP-neurotoxicity is associated with the MPP^+-uptake inhibitory potency of amphetamine-metabolites generated from l-deprenyl, l-METH and l-AMPH would be expected to be protective when applied not only before, but also after MPTP-treatment. Indeed, both metabolites of l-deprenyl were protective when applied 30 min (data not shown) or 60 min (Fig. 4) after the MPTP-treatment.

The dual mode of action of l-deprenyl in the mouse model of MPTP-parkinsonism is in harmony with its complex pharmacology which might include blocking effect on the uptake of DA (Knoll et al., 1978). Based on indirect evidence, it has been suggested that the ability of l-deprenyl to protect against xylamine- and DSP-4-induced neurotoxicity (Dudley, 1988;

Finnegan et al., 1990) does not depend on its MAO-B inhibiting properties and may be associated with its metabolites. Recent results demonstrate that the neuroprotective/neuronal rescue effects of l-deprenyl are even more complex. It can reduce MPTP-induced cell death when given 3 days after MPTP-treatment (Tatton and Greenwood, 1991) which indicates that the increase of neuronal survival is mediated by a novel site of action. Stimulation of aromatic l-amino acid decarboxylase by l-deprenyl (Li et al., 1992) may also be relevant to its antiparkinsonian effect providing another mechanism for the potentiation of l-DOPA therapy, in addition to the inhibition of DA-metabolism (Birkmayer et al., 1977).

l-Deprenyl, which blocks the biological effects of MPTP in various animals including monkeys (Heikkila et al., 1984; Cohen et al., 1984; Langston et al., 1984), had earlier been found to have beneficial effects as an "add on" therapy for Parkinson's disease (Birkmayer et al., 1977, 1985) and recently it was found to delay the progression of the disease (Tetrud and Langston, 1989; Parkinson Study Group, 1989), applied either alone or with α-tocopherol. It is a matter of controversy whether the metabolism of l-deprenyl to l-METH and l-AMPH may contribute to the therapeutic action of l-deprenyl in parkinsonism (Tetrud and Langston, 1989; Parkinson Study Group, 1989; Parkes et al., 1975; Karoum et al., 1982). In addition to the protective effect of l-deprenyl against MPTP-neurotoxicity in vivo, l-deprenyl was found to protect dopamine neurons in explant cultures of rat embryonic midbrain from the neurotoxic effect of MPTP and even from that of MPP^+ (Mytilineou and Cohen, 1984, 1985). Although the authors of the paper on the protection by l-deprenyl against the neurotoxic effect of MPP^+ in cultures of dopaminergic neurons arrived at a different conclusion (Mytilineou and Cohen, 1985), it is likely that the protective effect described in their work was associated with the blockade of uptake of MPP^+ by metabolites generated from l-deprenyl.

To further support the concept that metabolites of l-deprenyl are involved in protection from the effects of parkinsonogenic agents in mice, we investigated whether l-deprenyl and its metabolites could protect against neurotoxicity induced by a methylated analog of MPTP. 1-Methyl-4-(2'-methylphenyl)-1,2,3,6-tetrahydropyridine ($2'CH_3$-MPTP) is an approximately 2-fold more potent dopaminergic neurotoxin than MPTP in C57BL mice (Youngster et al., 1986), and its bioactivation is mediated not only by MAO-B but also by MAO-A (Sonsalla et al., 1987). As was the case with MPTP, pretreatment of mice with the DA-uptake inhibitor mazindol protected against $2'CH_3$-MPTP-neurotoxicity. l-Deprenyl, however, even at a dose of 40 mg/kg did not prevent the $2'CH_3$-MPTP-induced neurotoxicity. l-Deprenyl was administered 18 h before $2'CH_3$-MPTP (2 injections of 17.5 mg/kg, i.p. at 6 h intervals). Using the same treatment schedule, protection against $2'CH_3$-MPTP-induced neurotoxicity was observed at a nonselective dose (Sonsalla et al., 1987) of MDL 72,145, a MAO inhibitor, or with a combination of selective doses of clorgyline and l-deprenyl (Heikkila and Sonsalla, 1987).

We report here that 30 min pre- or post-treatment with l-deprenyl or mazindol protects against 2'CH$_3$-MPTP-induced neurotoxicity. l-METH and l-AMPH were also protective regardless whether they were applied 30 min prior to or 30 min after the 2'CH$_3$-MPTP-treatment (Table 2). Heikkila and his coworkers found no protection by deprenyl (Sonsalla et al., 1987; Heikkila and Sonsalla, 1987). In their experiments, however, l-deprenyl was administered 18h and 24h before the first and the second 2'CH$_3$-MPTP-treatment, respectively. In line with their results, a 18h-pretreatment with l-deprenyl provided no protection against striatal depletion induced by a single dose of 2'CH$_3$-MPTP (manuscript in preparation).

Since 2'CH$_3$-MPTP was a better substrate for MAO than MPTP in mitochondrial preparations from whole brains of C57BL mice (Sonsalla et al., 1987), it is likely that 2'CH$_3$-MPTP is converted to its pyridinium form at least as rapidly as MPTP to MPP$^+$. In our experiments the protection afforded by a 30 min pre- or post-treatment by l-deprenyl against 2'CH$_3$-MPTP-neurotoxicity is therefore most probably associated with the blockade of 2'CH$_3$-MPP$^+$-uptake due to the presence of l-METH and/or l-AMPH generated from l-deprenyl.

The conclusion of other investigators (Heikkila and Sonsalla, 1987) concerning the lack of protection by l-deprenyl against 2'CH$_3$-MPTP-neurotoxicity should be revised in the light of the present experiments. Our results suggest that even if idiopathic parkinsonism were caused by an MPTP-like substance bioactivated partly by MAO-A, l-deprenyl, owing to its complex mode of action, could be protective.

Acknowledgements

We thank I. Berekhelyi, G. Jancsó, E. Blazsek, I. Deák, M. Haraszti for technical assistance and E. Tóth, J. Vámosi for secretarial assistance. We also thank Dr. J. L. Neumayer, Research Biochemicals Inc., Natick, MA, and Dr. C. Vadász, Nathan Kline Institute, Orangeburg, NY, for their gifts of 2'CH$_3$-MPTP.

References

Birkmayer W, Knoll J, Riederer P, Youdim MBH, Hars V, Marton J (1985) Increased life expectancy resulting from addition of l-deprenyl to Madopar^R treatment in Parkinson's disease: a longterm study. J Neural Transm 64: 113–127
Birkmayer W, Riederer P, Ambrozi L, Youdim MBH (1977) Implications of combined treatment with Madopar and L-deprenyl in Parkinson's disease. Lancet i: 439–443
Brooks WJ, Jarvis MF, Wagner GC (1989) Astrocytes as a primary locus for the conversion of MPTP into MPP$^+$. J Neural Transm 76: 1–12
Chiba K, Trevor A, Castagnoli N (1984) Metabolism of the neurotoxic tertiary amine, MPTP, by brain monoamine oxidase. Biochem Biophys Res Commun 120: 574–578
Chiueh CC, Johannessen JN, Sun JL, Bacon JP, Markey SP (1986) Reversible neurotoxicity of MPTP in the nigrostriatal dopaminergic system of mice. In: Markey SP, Castagnoli N Jr, Trevor AJ, Kopin IJ (eds) MPTP — a neurotoxin producing a parkinsonian syndrome. Academic Press, New York, pp 473–479

Cohen G, Pasik P, Cohen B, Leist A, Mytilineneou C, Yahr MD (1984) Pargyline and deprenyl prevent neurotoxicity of 1-methyl-4-phenyl-1,2,3,4-tetrahydropiridine (MPTP) in monkeys. Eur J Pharmacol 106: 209–210

Davis GC, Williams AC, Markey SP, Ebert MH, Caine ED, Reichert CM, Kopin IJ (1979) Chronic Parkinsonism secondary to intravenous injection of meperidine analogs. Psychiatry Res 1: 249–254

Dudley MW (1988) The depletion of rat cortical norepinephrine and the inhibition of [^{3}H]-norepinephrine uptake by xylamine does not require monoamine oxidase activity. Life Sci 43: 1871–1877

Finnegan KT, Skratt JJ, Irwin I, DeLanney LE, Langston JW (1990) Protection against DSP-4-induced neurotoxicity by deprenyl is not related to its inhibition of MAO-B. Eur J Pharmacol 184: 119–126

Hallman H, Olsen L, Jonsson G (1984) Neurotoxicity of the meperidine analogue N-methyl-4-phenyl-1,2,3,6-tetrahydropyridine on brain catecholamine neurons in the mouse. Eur J Pharmacol 97: 133–136

Heikkila RE, Hess A, Duvoisin RC (1984) Dopaminergic neurotoxicity of 1-methyl-4-phenyl-1,2,3,4-tetrahydropiridine in mice. Science 224: 1451–1453

Heikkila RE, Manzino L, Cabbat FS, Duvoisin RC (1984) Protection against the dopaminergic neurotoxicity of 1-methyl-4-phenyl-1,2,5,6-tetrahydropyridine by monoamine oxidase inhibitors. Nature 311: 467–469

Horn AS (1978) Characteristics of neuronal dopamine uptake. In: Roberts PJ (ed) Advances in biochemical psychopharmacology, vol 19. Raven Press, New York, pp 25–34

Heikkila RE, Sonsalla PK (1987) The use of the MPTP-treated mouse as an animal model of parkinsonism. Can J Neurol Sci 14: 436–440

Javitch JA, D'Amato RJ, Strittmatter SM, Snyder SH (1985) Parkinsonism-inducing neurotoxin, N-methyl-4-phenyl-1,2,3,6-tetrahydropyridine: uptake of the metabolite N-methyl-4-phenylpyridine by dopamine neurons explains selective toxicity. Proc Natl Acad Sci USA 82: 2173–2177

Javitch JA, Snyder SH (1985) Uptake of MPP$^+$ by dopamine neurons explains selectivity of parkinsonism inducing neurotoxin, MPTP. Eur J Pharmacol 106: 455–456

Jonsson G, Sundström E, Nwanze E, Hallman H, Luthman J (1986) Mode of action of MPTP on catecholaminergic neurons in the mouse. In: Markey SP, Castagnoli N Jr, Trevor AJ, Kopin IJ (eds) MPTP — a neurotoxin producing a parkinsonian syndrome. Academic Press, New York, pp 253–272

Karoum F, Chuang LW, Eisler T, Calne DB, Liebowitz MR, Quitkin FM, Klein DF, Wyatt RJ (1982) Metabolism of (–)-deprenyl to amphetamine and metamphetamine may be responsible for deprenyl's therapeutic benefit: a biochemical assessment. Neurology 32: 503–509

Knoll J (1978) The possible mechanism of action of (–)-deprenyl in Parkinson's disease. J Neural Transm 43: 177–198

Langston JW, Ballard P, Tetrud JW, Irwin I (1983) Chronic parkinsonism in humans due to a product of meperidine-analog synthesis. Science 219: 979–980

Langston JW, Irwin I, Langston EB, Forno LS (1984) Pargyline prevents MPTP-induced parkinsonism in primates. Science 225: 1480–1482

Li X-M, Juorio AV, Paterson IA, Meng-Yang Z, Boulton AA (1992) Specific irreversible monoamine oxidase B inhibitors stimulate gene expression of aromatic l-amino acid decarboxylase in PC12 cells. J Neurochem 59: 2324–2327

Markey SP, Johannessen JN, Chiueh CC, Burns RS, Herkenham MA (1984) Intra-neuronal generation of a pyridinium metabolite may cause drug-induced parkinsonism. Nature 311: 464–467

Mayer RA, Kindt MV, Heikkila RE (1986) Prevention of the nigrostriatal toxicity of 1-methyl-4-phenyl-1,2,3,6-tetrahydropyridine by inhibitors of 3,4-dihydroxy-phenylethyl-amine transport. J Neurochem 47: 1073–1079

Mytilineou C, Cohen G (1984) 1-methyl-4-phenyl-1,2,3,6-tetrahydropyridine destroys dopamine neurons in explants of rat embryo mesencephalon. Science 225: 529–531

Mytilineou C, Cohen G (1985) Deprenyl protects dopamine neurons from the neurotoxic effect of 1-methyl-4-phenyl pyridinium ion. J Neurochem 45: 1951–1953

Parkes JD, Tarsy D, Marsden CD, Bovill KT, Phipps JA, Rose P, Asselman P (1975) Amphetamines in the treatment of Parkinson's disease. J Neurol Neurosurg Psychiatry 38: 232–237

Parkinson Study Group (1989) Effect of deprenyl on the progression of disability in early Parkinson's disease. N Engl J Med 321: 1364–1371

Patthy M, Gyenge R (1988) Perfluorinated acids as ion-pairing agent in the determination of monoamine transmitters and some prominent metabolites in rat brain by high-performance liquid chromatography with amperometic detection. J Chromatogr 449: 191–205

Philips RS (1981) Amphetamine, parahydroxy-amphetamine and beta-phenylethylamine in mouse brain and urine after (−)-and (+)-deprenyl administration. J Pharm Pharmacol 33: 739–741

Reynolds GP, Elsworth JD, Blau K, Sandler M, Lees AJ, Stern GM (1978) Deprenyl is metabolized to methamphetamine and amphetamine in man. Br J Clin Pharmacol 6: 542–544

Schacht U, Heptner W (1974) Effect of nomiphensine (HOE 984), a new antidepressant on uptake of noradrenaline and serotonin and on release of noradrenaline in rat brain synaptosomes. Biochem Pharmacol 23: 3413–3422

Sershen H, Mason MF, Reith MEA, Hashim A, Lajtha A (1986) Effect of amphetamine on 1-methyl-4-phenyl-1,2,3,6-tetrahydropyridine (MPTP) neurotoxicity in mice. Neuropharmacology 25: 927–930

Sershen H, Mason MF, Reith MEA, Hashim A, Lajtha A (1986) Effect of nicotine and amphetamine N-methyl-4-phenyl-1,2,3,6-tetrahydropyridine (MPTP) neurotoxicity in mice. Neuropharmacology 25: 1231–1234

Sonsalla DK, Yougster SK, Kindt MV, Heikkila RE (1987) Characteristics of 1-methyl-4-(2'methylphenyl)-1,2,3,6-tetrahydropyridine-induced neurotoxicity in the mouse. J Pharmacol Exp Ther 242: 850–857

Sundström E, Jonsson G (1986) Differential time course of protection by monoamine oxidase inhibition and uptake inhibition against MPTP neurotoxicity on central catecholamine neurons in mice. Eur J Pharmacol 122: 275–278

Sziráki I, Andrási F, Berzsenyi P, Horváth K, Kardos V, Patthy M, Pátfalusi M, Szabó G, Szabó H (1990) Pharmacological and neurochemical properties of GYKI-52 895, a new selective dopamine uptake inhibitor. In: Abstracts of Dopamine '90. Satellite Meeting of the XIth International Congress of Pharmacology, Como, Italy, July 8–11, 1990, p 36

Sziráki I, Kardos V, Patthy M, Pátfalusi M, Budai Gy (1993) Methamphetamine protects against MPTP-neurotoxicity in mice C57BL. Eur J Pharmacol (submitted)

Tatton WG, Greenwood CE (1991) Rescue of dying neurons: a new action for deprenyl in MPTP parkinsonism. J Neurosci Res 30: 666–672

Tetrud JW, Langston JW (1989) The effect of deprenyl (selegiline) on the natural history of Parkinson's disease. Science 245: 519–522

Wurtman RJ, Axelrod J (1963) A sensitive and specific assay for the estimation of monoamine oxidase. Biochem Pharmacol 12: 1439–1440

Youngster SK, Duvoisin RC, Hess A, Sonsalla PK, Kindt MV, Heillika RE (1986) 1-methyl-4-(2'-methylphenyl)-1,2,3,6-tetrahydropyridine ($2'$-CH_3-MPTP) is a more potent dopaminergic neurotoxin than MPTP in mice. Eur J Pharmacol 122: 283–287

Authors' address: I. Sziráki, Institute for Drug Research, Department of Biochemistry II, POB 82, Budapest, H-1325, Hungary.

J Neural Transm (1994) [Suppl] 41: 221–229

Is brain superoxide dismutase activity increased following chronic treatment with l-deprenyl?

C. T. Lai, D. M. Zuo, and **P. H. Yu**

Neuropsychiatric Research Unit, Department of Psychiatry, University of Saskatchewan Saskatoon, Saskatchewan, Canada

Summary. L-deprenyl, a specific MAO-B inhibitor, has been proposed to possess a neuroprotective effect. The mechanism of such an effect is unclear. L-Deprenyl has been found to increase rat striatal superoxide dismutase (SOD) activity, which inactivates singlet oxygen. It would be very interesting to know how such activation occurs and whether or not other MAO inhibitors also have such an effect. We have analyzed rat striatal SOD activity using a very sensitive nitrite method and an immunological procedure. The effect of different doses and time of treatment with l-deprenyl and M-2-PP (2-pentyl-N-methyl-propargylamine), a new highly potent, selective and non-amphetamine-like MAO-B inhibitor, on the rat brain has been investigated. We were unable to detect any increase of SOD activity in the rat striata and cerebral cortex nor any increase in the concentration of immunoreactive SOD antibody in the cortex following chronic treatment with l-deprenyl and M-2-PP. It remains to be substantiated as to whether or not l-deprenyl can enhance SOD levels.

Introduction

Free radicals are known to be highly cytotoxic. Most free radicals are derived from molecular oxygen (O_2) in cells, e.g. superoxide radical ($O_2^{\cdot-}$) and hydroxyl radical ($OH^{\cdot}$). Free radicals can be produced via different physiological processes (Colton et al., 1987; Curnutte et al., 1987). Several mechanisms are responsible for the protection of the neuronal cells from potential cytotoxic damage caused by free radicals. Superoxide dismutase (SOD) catalyzes the dismutation of $O_2^{\cdot-}$ to O_2 and hydrogen peroxide (H_2O_2). There are two types of SOD in mammals, i.e. copper, zinc-dependent superoxide dismutase (Cu, Zn SOD) and manganese-dependent superoxide dismutase (Mn SOD). It was thought that SOD in the central nervous system may take part in the neuroprotective mechanisms of neurons.

It has been reported that l-deprenyl, a selective monoamine oxidase type B (MAO-B) inhibitor, delays the onset of disability associated with early, otherwise untreated cases of Parkinson's disease (The Parkinson Study Group, 1989; Tetrud and Langston, 1989). Also, it has been reported that chronic treatment with deprenyl can prolong the lifespan in aged male rats (Knoll et al., 1989; Milgram et al., 1990). It has been, therefore, proposed that l-deprenyl may possess a neuroprotective effect, but the mechanism of effect is not yet understood.

Chronic treatment with l-deprenyl has been claimed to increase SOD activity as much as 10 fold in the rat striata (Knoll, 1988). This result has recently been confirmed, although to a much lesser extent (Carrillo et al., 1991; Clow et al., 1991). SOD catalyzes the dismutation of O_2^- and acts as a free radical scavenger. An increase in brain SOD activity, therefore, might be useful in preventing neuronal degeneration. If SOD activity is indeed increased following chronic treatment with l-deprenyl, it would be very interesting to know the underlying mechanism. It is also interesting to know whether or not such an effect on brain SOD activity is unique for l-deprenyl. We have compared the effect of chronic treatment with l-deprenyl and M-2-PP, a new highly potent, selective and non-amphetamine-like MAO-B inhibitor, on rat brain SOD. In addition to measuring SOD activity the enzyme levels were also determined by an immunoassay.

Materials and methods

Treatment of animals

Male Wistar rats (200 g) were used. MAO-B inhibitors, l-deprenyl and M-2-PP, were dissolved in isotonic saline. Doses of the inhibitors are indicated in the tables. Both compounds were administered subcutaneously every second day for three weeks. The control rats received a volume of saline equivalent to MAO-B inhibitor received by treated rats. Twenty-four hours after the last injection the rats were sacrificed and the brains were excised immediately. Brains were dissected on an ice-chilled Petri dish. Tissues were homogenized in 1 ml of ice-cold buffer solution ($Na_2B_4O_7$ 40 mM, KH_2PO_4 50 mM, EDTA 0.25 mM). The homogenates were then centrifuged at 10,000 g for 30 min. Protein concentrations were determined by the method of Bradford (Bio-Rad protein assay reagent) using BSA as standard.

Enzymatic method for SOD assay

SOD activity was assayed by a modified nitrite method of Oyanagui (1984). The assay is based on the reduction of nitrite formation that is generated from hydroxylamine in the presence of O_2^- generators (xanthine/xanthine oxidase). Samples were assayed in the 96-well plates and the optical density was read using a UVmax kinetic microplate reader (Molecular Devices Corp. Menlo Park, CA. U.S.A.).

The incubation mixture contained buffer solution ($Na_2B_4O_7$ 40 mM, KH_2PO_4 50 mM, EDTA 0.25 mM), pH 8.2; supernatant of rat brain samples; substrate reagent (hydroxylamine 2 mM; xanthine 0.2 mM); deionized water or sodium cyanide (NaCN) 1 mM; xanthine oxidase (XO) 4×10^{-3} unit/ml in a total volume of 150 μL. The reaction was started by adding the XO. The mixture was incubated at 37°C for 10 min and then 100 μl of coloring reagent (sulfanilic acid 300 μg/ml; N-1-naphthylethylenediamine 5 µg/ml; acetic acid 16.7%) was added and mixed. The final mixtures were allowed to stand at room temperature for 60 min for completion of the color reaction. The mixtures were measured at 550 nm with the UVmax kinetic microplate reader. Cu, Zn SOD activity can be selectively inhibited by NaCN (1 mM), therefore, this has been applied to differentiate Cu, Zn SOD and Mn SOD. Cu, Zn SOD activity was estimated from the difference between total activity (Cu, Zn SOD plus Mn SOD) and NaCN insensitive SOD activity (Mn SOD).

One unit of SOD activity is defined as the amount of SOD required to inhibit the formation of nitrite by 50%. The enzyme activity was calculated according to the inhibition titration curves, which were obtained using ten different increasing amounts of each brain extract. IC_{50} value was estimated from the inhibition curve by double reciprocal plot according to Wilkinson (1961).

Immunological estimation of rat brain Cu, Zn SOD

An enzyme-linked immunosorbent assay (ELISA) for SOD has been developed. The supernatant of the rat brain homogenates and the purified bovine erythrocyte SOD (Sigma, St. Louis, U.S.A.) in 0.05 M carbonate buffer (pH 9.6) were incubated at 4°C overnight in 96-well immunoplates. The plates were then saturated with 10% fetal bovine serum (FBS) in phosphate-buffer saline plus 0.05% Tween-20 (PBS-T). After washing with PBS-T, sheep anti-SOD antibody (1:500, Serotec, Kidlington, England) was added to each well and incubated at 37°C for 1 h. The plates were subsequently washed with PBS-T and then anti-sheep IgG peroxidase conjugate (1:3,000, Sigma, St. Louis, U.S.A.) was added to each well and further incubated at 37°C for 1 h. Peroxidase activity was detected by addition of o-phenylenediamine dihydrochloride (OPD) as substrate in 0.05 M citrate-phosphate buffer (pH 5.0) and 0.05% H_2O_2. The optical density (OD) was measured at 490 nm with the UVmax kinetic microplate reader. The amount of SOD in the rat brain samples was estimated according to the standard curves of the purified SOD that were included in each brain sample.

Statistics

Data were analyzed by one-way analysis of variance (ANOVA). Analysis of variance was performed on a Macintosh Microcomputer using the CLR ANOVA program (Clearlake Research, Houston, Texas, U.S.A.). Values are expressed as mean ± SE.

Results

As can be seen in Fig. 1, MAO-B inhibitors l-deprenyl and M-2-PP did not affect the rat striatal SOD activity in vitro. Both compounds were also unable to change the purified bovine erythrocyte SOD activity in vitro (results not shown).

The total SOD activities in rat striata were slightly higher than in the cortex (Table 1). A higher proportion of Cu, Zn SOD activity over Mn SOD activity in the cortex than in the striata were observed. Neither the total SOD nor the Cu, Zn SOD or Mn SOD activities was found to be significantly different between the control and the experimental animals following s.c. injection with l-deprenyl and M-2-PP for 1, 10 and 21 days.

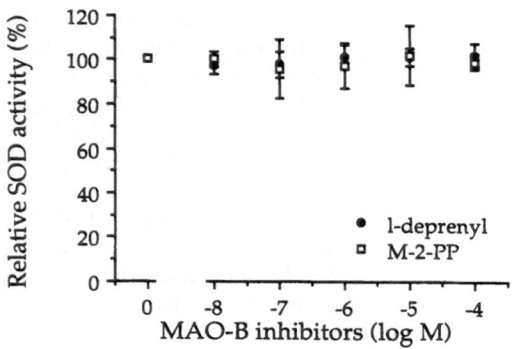

Fig. 1. Effect of l-deprenyl and M-2-PP on rat striatal SOD activity in vitro. The rat striatal enzyme preparations were pre-incubated with different concentrations of l-deprenyl and M-2-PP at 37°C for 30 min and then the total SOD activities were assessed according to the nitrite method. The striatal SOD (592 unit/mg protein) at a concentration (1:25) that would induce a 50% inhibition was applied in the assay

Table 1. Effect of MAO-B inhibitors l-deprenyl and M-2-P on superoxide dismutase activity (unit/mg protein) in the striatum and the lateral cortex of rat brain at different time period following treatment

Treatment[a] (days)	Striatum			Cortex		
	saline	l-deprenyl	M-2-PP	saline	l-deprenyl	M-2-PP
	total SOD activity					
1	594 ± 21	590 ± 21	589 ± 33	419 ± 16	429 ± 65	415 ± 32
10	593 ± 24	580 ± 33	594 ± 30	414 ± 25	414 ± 66	419 ± 18
21	591 ± 11	598 ± 17	585 ± 20	422 ± 16	420 ± 21	413 ± 15
	Cu, Zn SOD activity					
1	318 ± 25	320 ± 35	301 ± 34	287 ± 13	298 ± 45	286 ± 22
10	309 ± 19	292 ± 40	299 ± 26	274 ± 20	281 ± 50	304 ± 17
21	317 ± 17	339 ± 22	305 ± 21	300 ± 17	301 ± 20	295 ± 17
	Mn SOD activity[b]					
1	277 ± 6	271 ± 24	288 ± 31	132 ± 11	131 ± 22	130 ± 19
10	284 ± 6	288 ± 8	295 ± 13	139 ± 14	133 ± 21	115 ± 9
21	274 ± 10	260 ± 30	280 ± 18	122 ± 14	119 ± 7	118 ± 10

[a] L-deprenyl and M-2-PP (0.25 mg/Kg) were administered subcutaneously every second day for the designed time of treatment. Brain regions were dissected 24 h after the last injection of the compounds
[b] Cu, Zn SOD activity is obtained as the total SOD activity subtract the Mn SOD activity, which is insensitive towards NaCN (1 mM). Values are expressed as mean ± SE (n = 5). Statistical significance was determined by one-way ANOVA. $p < 0.05$, compared with the saline group. The enzyme activity of standard SOD from bovine erythrocytes was estimated to be 71,429 unit/mg protein using the present nitrite method, which is equivalent to 3,570 unit/mg protein using the cytochrome C reduction method according to McCord and Fridorich (1969)

Table 2. Chronic effect of different doses of MAO-B inhibitors l-deprenyl and M-2-PP on superoxide dismutase activity (unit/mg protein) in the striatum and lateral cortex of brains

Dose[a] (mg/kg)	Striatum			Cortex		
	saline	l-deprenyl	M-2-PP	saline	l-deprenyl	M-2-PP
			total SOD activity			
	596 ± 37			324 ± 29		
0.25		586 ± 46	586 ± 76		311 ± 32	332 ± 44
0.5		576 ± 46	578 ± 46		322 ± 23	306 ± 22
1		593 ± 54	567 ± 15		322 ± 33	334 ± 34
2		587 ± 42	584 ± 58		326 ± 11	302 ± 20
			Cu, Zn SOD activity			
	320 ± 70			221 ± 33		
0.25		300 ± 60	306 ± 79		209 ± 23	231 ± 37
0.5		268 ± 42	338 ± 53		211 ± 17	214 ± 24
1		288 ± 71	301 ± 51		222 ± 21	234 ± 25
2		305 ± 51	308 ± 38		236 ± 5	209 ± 9
			Mn SOD activity			
	275 ± 60			104 ± 10		
0.25		286 ± 63	280 ± 26		102 ± 9	101 ± 9
0.5		308 ± 28	240 ± 58		111 ± 8	94 ± 6
1		305 ± 46	267 ± 46		100 ± 13	99 ± 12
2		282 ± 75	267 ± 39		90 ± 9	94 ± 12

[a] Different doses of l-deprenyl and M-2-PP were administered subcutaneously every second day for 21 days. Brain regions were dissected 24 h after the last injection of the compounds. SOD activities were determined as described in Table 1

Table 3. Chronic effect of different doses of MAO-B inhibitors l-deprenyl and M-2-PP on the concentrations of immunoreactive Cu, Zn superoxide dismutase (μg/mg protein) in rat cortex

Dose[a] (mg/kg)	Saline	l-Deprenyl	M-2-PP
	3.58 ± 0.58		
0.25		3.56 ± 0.14	3.56 ± 0.26
0.5		3.55 ± 0.32	3.75 ± 0.41
1		3.77 ± 0.23	4.06 ± 0.44
2		3.93 ± 0.51	4.13 ± 0.47

[a] Different doses of l-deprenyl and M-2-PP were administered subcutaneously every second day for 21 days. Brain regions were dissected 24 h after the last injection of the compounds. ELISA fo SOD is described in Materials and methods

Table 2 shows the effect of chronic treatment (three weeks) with different doses of l-deprenyl and M-2-PP. No increase or decrease of SOD activities in rat striata and cortices was detected.

Table 3 shows the results of immunoassay of rat brain SOD levels. According to an ELISA method, the concentrations of the immunoreactive Cu, Zn SOD in the rat cortex of the untreated control rats were also found to be not significantly different from those animals following chronic treatment with different doses of l-deprenyl and M-2-PP.

Discussion

L-Deprenyl monotherapy can delay the onset of disability in Parkinson's disease, which is an illness with the nigro-striatal dopamine neurons degenerated progressively. Several mechanisms of such neuroprotective effect have been proposed (see review by Yu et al., 1992). The basal lipid peroxidation was found to be increased in the substantia nigra of post-mortem parkinsonian brains (Dexter et al., 1986, 1989a), suggesting that oxidative stress (free radical formation) is enhanced by some unknown mechanisms in the parkinsonian brain. H_2O_2 is produced in the oxidative deamination of dopamine. H_2O_2 in the presence of ferrous ion (Fe^{2+}) would be converted to hydroxyl radicals ($OH^{\cdot}$) via the Fenton reaction. $OH^{\cdot}$ is extremely reactive and can initiate a lipid peroxidation chain reaction, such as in the neuronal membrane and eventually causes neuronal degeneration. This hypothesis is supported by the reports that iron content is increased in the substantia nigra of the parkinsonian brains (Hirsch et al., 1991; Riederer et al., 1989; Dexter et al., 1989b; Sofic et al., 1988, 1991).

It has been suggested that an imbalance between the production of reduced oxygen species, such as superoxide, or hydroxyl free radicals and hydrogen peroxide, and the intracellular protective mechanism is involved in the loss of dopamine neurons in Parkinson's disease (Cohen, 1983). SOD activity has been reported to be increased in the substantia nigra of the parkinsonian post-mortem brains (Ceballos et al., 1990; Saggu et al., 1989; Marttila et al., 1988, 1990). The increase in SOD activity may be useful in the removal of superoxide anions ($O_2^{\cdot-}$), but in the SOD-catalyzed dismutation of superoxide H_2O_2 is produced, which could be subsequently converted to hydroxyl free radicals. It is, therefore, unclear whether the increase of SOD activity in the parkinsonian brain is related to the enhancement of oxidative stress or if it is a feedback mechanism to scavenge superoxide anions.

The observation that l-deprenyl can induce striatal SOD levels is very interesting (Knoll, 1988; Carrillo et al., 1991; Clow et al., 1991). SOD activity is known to increase following a hypoxia/anoxia episode (Kramer et al., 1987; Frank, 1982). It is considered to be an adaptive response to protect against the eventual oxidative stresses. There is no evidence that l-deprenyl is capable of enhancing oxidative stress. As a matter of fact l-deprenyl inhibits the oxidative deamination and the associated production of H_2O_2, which would increase oxidative stress. SOD activity is not affected by l-deprenyl in vitro. The increase of rat caudate SOD activity by l-deprenyl, if there is any, ought to be induced by some unknown mechanism.

We were unable to detect any increase of SOD activity after treatment with l-deprenyl for different lengths of time, with different doses, and in different tissues. Another MAO-B inhibitor, M-2-PP, was also found to be ineffective towards the rat brain SOD. It is presently unclear why such a discrepancy occurrs between earlier reported SOD activity increases and our results. There are a number of methods for the determination of SOD activity (Oyanagui, 1984). Basically, they are based on several indirect sequential reactions, such as (a) superoxide generation (e.g. via xanthine oxidase); (b) a reaction that would be induced by superoxide (e.g. hydroxylamine to nitrite); (c) the removal of the superoxide by SOD thereby reducing the rate of reaction (b). The assessment of such complicated non-linear responses therefore requires an inhibition curve with respect to each individual sample and an accurate estimation of the IC_{50} values could be quite crucial. Recently it was reported (Carrillo et al., 1992) that l-deprenyl induced an increase of SOD level or activity in the caudate and a decrease in other brain regions (i.e. cerebellum and hippocampus). This has further complicated the interpretation of the in vivo effects of l-deprenyl.

We have also applied an immunoassay to assess the SOD levels in the brain of the control and MAO-B inhibitor treated rats. SOD levels obtained from either measurements of enzyme activity or immunoassayss using the purified bovine erythrocyte SOD as an external standard were found to be quite similar. The results of the immunoassay are also consistent with the finding of enzyme activity, namely, the rat cortex SOD is unchanged by chronic peripheral administration of l-deprenyl or a new aliphatic propargylamine MAO-B inhibitor M-2-PP. Whether or not the brain SOD activity is induced by l-deprenyl remains to be established.

Acknowledgements

We thank the Canadian Medical Research Council and Saskatchewan Health for their continuing financial support.

References

Carrillo MC, Kanai S, Nokubo, Kitani K (1991) (−)-Deprenyl induces activities of both superoxide dismutase and catalase but not of glutathione peroxidase in the striatum of young male rats. Life Sci 48: 517–521

Carrillo MC, Kitani K, Kanai S, Sato Y, Ivy GO (1992) The ability of (−)-deprenyl to increase superoxide dismutase activities in the rat tissue and brain region selective. Life Sci 50: 1985–1992

Ceballos I, Lafon M, Agid J, Hirsch F, Nicole A, Sinet PM, Agid Y (1990) Superoxide dismutase and Parkinson's disease. Lancet i: 1035–1036

Clow A, Hussain T, Glover V, Sandler M, Dexter DT, Walker M (1990) (−)-Deprenyl can induce soluble superoxide dismutase in rat striata. J Neural Transm [Gen Sect] 86: 77–80

Cohen G (1983) The pathobiolobgy of Parkinson's disease: biochemical aspects of dopamine neuron senescence. J Neural Transm [Suppl] 19: 98–103

Colton CA, Gilbert DL (1987) Production of superoxide anions by a CNS macrophage, the microglia. FEBS Lett 223: 284–288

Curnutte JT, Babior BM (1987) Chronic granulomatous disease. Adv Hum Genet 16: 229–297

Dexter DT, Carter CJ, Agid F, Agid Y, Lees AJ, Jenner P, Marsden CD (1986) Lipid peroxidation as cause of nigral cell death in Parkinson's disease. Lancet: 639–640

Dexter DT, Carter CJ, Wells FR, Agid F, Agid Y, Lees AJ, Jenner P, Marsden CD (1989a) Basal lipid peroxidation in substantia nigra is increased in Parkinson's disease. J Neurochem 52: 381–389

Dexter DT, Wells FR, Lees AJ, Agid F, Agid Y, Jenner P, Marsden CD (1989b) Increased nigra iron content and alternations in other metal ions occuring in brain in Parkinson's disease. J Neurochem 52: 1830–1836

Frank L (1982) Protection from O_2 toxicity by pre-exposure to hypoxia: lung anti-oxidant enzyme role. J Appl Physiol 53: 475–482

Hirsch EC, Brandel J-P, Galle P, Javoy-Agid F, Agid Y (1991) Iron and aluminum increase in the substantia nigra of patients with Parkinson's disease: an X-ray microanalysis. J Neurochem 56: 446–451

Knoll J (1988) The striatal dopamine dependency of lifespan in male rats. Longevity study with (−)deprenyl. Mech Ageing Dev 46: 237–262

Knoll J, Dallo J, Yen TT (1989) Striatal dopamine, sexual activity and lifespan. Longevity of rats treated with (−)deprenyl. Life Sci 45: 525–531

Kramer K, Voss HP, Grimbergen JA, Timmerman H, Bast A (1987) The effect of ischemia and recirculation, hypoxia and recovery on anti-oxidant factors and β-adrenoceptor density. Biochem Biophys Res Commun 149: 568–575

Marttila RJ, Lorenz H, Rinne UK (1988) Oxygen toxicity protecting enzymes in Parkinson's disease. Increase of superoxide dismutase-like activity in the substantia nigra and basal nucleus. J Neurol Sci 86: 321–331

Marttila RJ, Viljanen M, Toivonen E, Lorentz H, Rinne UK (1990) Superoxide dismutase-like activity in the Parkinson's disease brain. Adv Neurol 53: 141–144

McCord JM, Fridorich I (1969) Superoxide dismutase. An enzymatic function for erythrocuprein (hemocuprein). J Biol Chem 244: 6049–6055

Milgram NW, Racine RJ, Nellis P, Mendonca A, Ivy GO (1990) Maintenance on L-deprenyl prolongs life in aged male rats. Life Sci 47: 415–420

Oyanagui Y (1984) Reevaluation of assay methods and establishment of kit for superoxide dismutase activity. Anal Biochem 142: 290–296

The Parkinson Study Group (1989) Effect of deprenyl on the progression of disability in early Parkinson's disease. N Engl J Med 321: 1364–1371

Riederer P, Sofic E, Rausch W-D, Schmidt B, Reynolds GP, Jellinger K, Youdim MBH (1989) Transition metals, ferritin, glutathione, and ascorbic acid in parkinsonian brains. J Neurochem 52: 515–520

Saggu H, Cooksey J, Dexter D, Wells FR, Lees A, Jenner P, Marsden CD (1989) A selective increase in particulate superoxide dismutase activity in parkinsonian substantia nigra. J Neurochem 53: 692–697

Sofic E, Riederer P, Heinsen H, Beckmann H, Reynolds GP, Hebenstreit G, Youdim MBH (1988) Increased iron (III) and total iron content in post mortem substantia nigra of parkinsonian brain. J Neural Transm 74: 199–205

Sofic E, Paulus W, Jellinger K, Riederer P, Youdim MBH (1991) Selective increase of iron in substantia nigra zona compacta of parkinsonian brains. J Neurochem 56: 978–982

Tetrud JW, Langston JW (1989) The effect of deprenyl (Selegiline) on the natural history of Parkinson's disease. Science 245: 519–522

Wilkinson GN (1961) Statistical estimation in enzyme kinetics. Biochem J 80: 324–332

Yu PH, Davis BA, Boulton AA (1992) Neuronal and astroglial monoamine oxidase: pharmacological implications of specific MAO-B inhibitors. Prog Brain Res 94: 309–315

Authors' address: Dr. P. H. Yu, Neuropsychiatric Research Unit, Department of Psychiatry, University of Saskatchewan, Saskatoon, Saskatchewan S7N 0W0, Canada.

J Neural Transm (1994) [Suppl] 41: 231–236

Lifespan of immunosuppressed NMRI-mice is increased by deprenyl

H.-J. Freisleben[1], F. Lehr[2], and J. Fuchs[2]

[1] Gustav-Embden-Zentrum der Biologischen Chemie, and
[2] Zentrum der Dermatologie und Venerologie, Klinikum der Johann Wolfgang
Goethe-Universität, Frankfurt am Main, Federal Republic of Germany

Summary. Immunosuppressed NMRI-mice (nu/nu) were raised and kept under germ-reduced conditions and fed with a germ-reduced diet (14 animals = controls). For another 14 mice 4 mg of selegiline were admixed to 10 kg of the diet. The 50% survival rate of the latter group was 160% from birth or 220% from the beginning of the study. The survival rate in weeks finally reached 350%, and the aerea under the curve 250%. The last mouse in the control group died at the age of 5 months, 2.5 months after the study was started; the last mouse in the selegiline group died at the age of 14.5 months, 1 year after the beginning of the study.

Introduction

The purpose of this study was to investigate the influence of selegiline HCL on the longevity of immunosuppressed nude mice. Athymic NMRI-mice lack mature T-cells. Even the B-cells do not exhibit normal capacity for production of specific IgG antibodies. Hence, cross-reacting IgM antibodies accumulate during the lives of immunosuppressed mice, because of the permanent interference of the immuno system with primarily apathogenic or only facultatively pathogenic microoogranisms mainly in the GI tract. These IgM antibodies may cause auto-imunno reactions, which appear to be the main reason of premature aging in immunosuppressed mice with characteristic alterations of the connective tissue as an accompanying symptom (Fortmeyer, 1981).

Selegiline is described in literature as a selective MAO-B-inhibitor and used in the therapy of Parkinson's disease. The compound especially ameliorates the sensitivity of the patients for L-dopa. Birkmeyer et al. (1985) reported on simultaneously increased life expectancy of selegiline-treated patients. Selegiline extended lifespan of male rats in experiments with 24 months-old animals (Knoll, 1989). According to Kornhuber et al. (1989) striatial MAO-B activity increases considerably with aging. In parallel, degeneration of dopaminergic neurons occurs. Halliwell (1989) suggested a possible reason to be the generation of free radicals via the reaction of H_2O_2 produced by MAO-B with transition metals, primarily iron

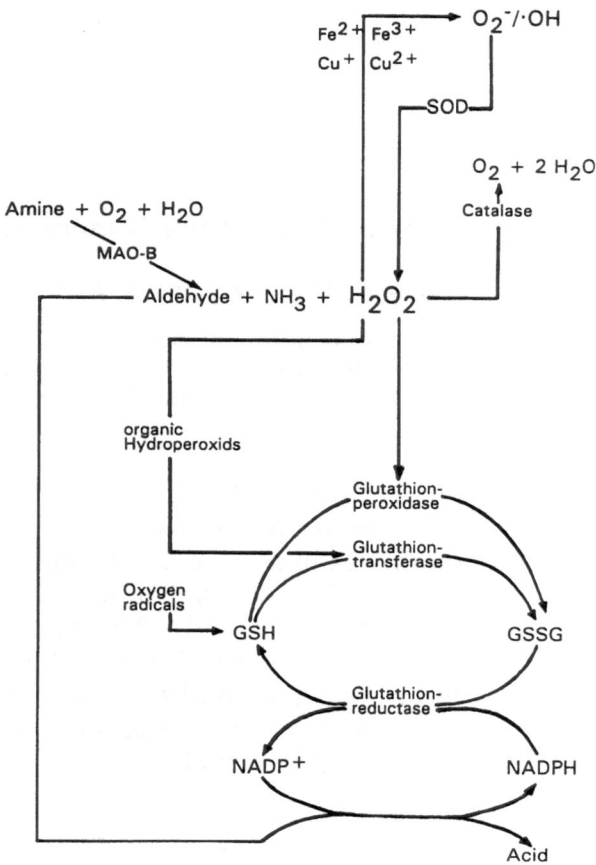

Fig. 1. Monoamine-dependent free radical generation and detoxification

(Fig. 1). A similar mechanism was discussed by Youdim et al. (1989) for premature aging of the striatum in Parkinson's disease and also for the development of Alzheimer's disease. Oldfield (1992) reported that copper ions may possibly be released from copper enzymes (i.e. other oxidases) and also copper might be responsible for the generation of free radicals from H_2O_2. Inhibition of MAO-B by selegiline explains the protection of striatal dopaminergic neurons. Furthermore, selegiline protects against their destruction by 1-methyl-4-phenyl-1,2,3,6-tetrahydropyridine (MPTP). The extension of lifespan by selegiline in aged rats, can only partially be explained via the above mechanisms.

Now, we demonstrate that selegiline extends life expectancy not only in patients suffering from Parkinsonism or in animal experiments with aged rats but also in immunosuppressed mice.

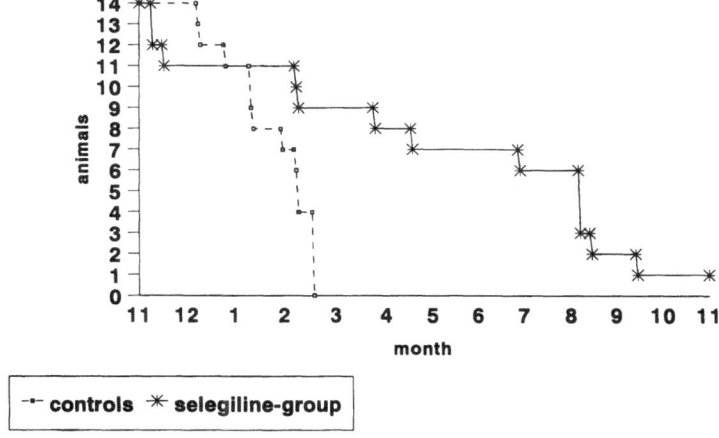

Fig. 2. Survival rate of immunosuppressed NMRI-mice

Material and methods

Animals

NMRI-mice are raised and kept in the Tierversuchsanlage of the University Clinics Frankfurt (Head: H. P. Fortmeyer). The cages are placed in "laminar flow" shelves in order to reduce air-borne infections. Feeding is accomplished by a germ-poor diet Altromin C 1003 (4,000 germs/g diet as compared to 80,000–200,000 germs/g normal mouse nutrition). Further immunoprotection is achieved by 2 g chloramphenicol per 1 l drinking water.

At the beginning of the study the mice were 10 weeks old. In the deprenyl group 4 mg of selegiline HCl was admixed to 10 kg of normal diet. An average consumption of 5.7 g per mouse and day correlates to 2.3 µg selegiline HCl per mouse and day or 70 µg per kg body weight, respectively.

Results

10 weeks after the beginning of feeding, in the control group 7 mice out of 14 were still alive (= 50%), in the selegiline-treated group 11 animals were still living (= 79%).

The last animal of the control group died on February 14th, 1992 (at the age of 5 months, or 2.5 months after the beginning of the study), when 9 mice (= 64%) were still alive in the selegiline-treated group.

On 28.04.92, 50% of the animals in the selegiline-treated group were still living (Fig. 2 and Table 1), the area under the curve was 190% as compared to the control (= 100%).

In the further course of the study it reached 250% and the survival rate in weeks exceeded 350%. The last mouse in the selegiline group died at the end of November 1992, at the age of 14.5 months, exactly 1 year after the study was started.

Table 1. Survival rate of controls vs selegiline groups

50% Survival rate	Control group	Selegiline group
Until	04.02.92	28.04.92
Weeks from start of study	10	22
Survival rate (%)	100	220
Weeks from birth	20	32
Survival rate (%)	100	160
Area under the curve (%)	100	190

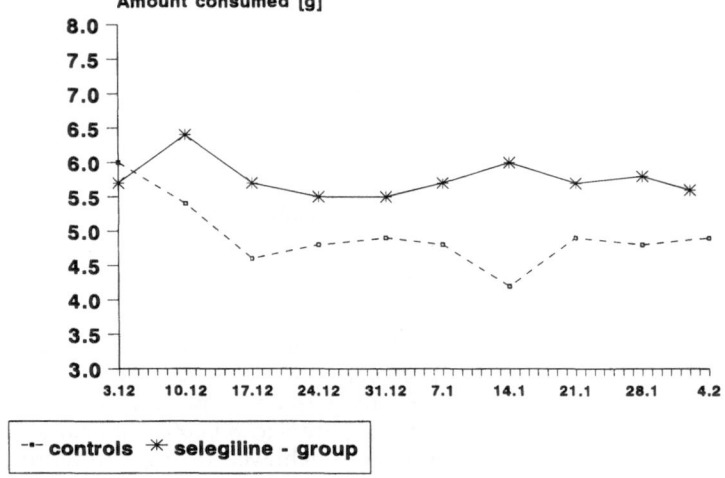

Fig. 3. Diet consumption per mouse and day

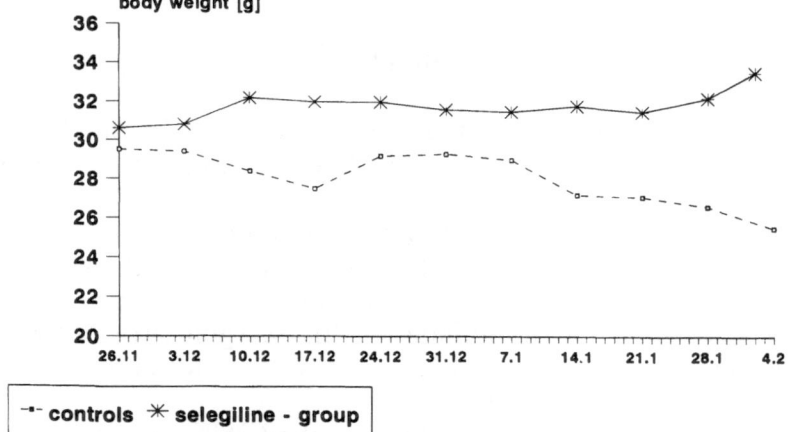

Fig. 4. Development of the mean body weight

Feeding behaviour (Fig. 3) and hence, the development of the body weight (Fig. 4) were increased in the selegiline group versus controls.

Discussion

The rapid, premature aging of immunosuppressed mice may help to understand the efficacy of selegiline in this study. So far only two conclusions are feasible: either selegiline does not only act specifically and selectively in the striatum, or the accumulation of IgM in immuno-suppressed mice causes direct or indirect effects in the striatial region and to MAO-B.

In the striatum, selegiline inhibits MAO-B and activates SOD. It protects dopaminergic neurons not only from destruction, but also rescues these neurons from cell-death after destruction by MPP+ (the active form of MPTP; Tatton et al., 1991) or after axotomy (Salo et al., 1991). Whereas the former protection is regorded as inhibition of MAO-B activity, the latter rescue appears to occur at the level of oxydative phosphorylation. MPP+ was suggested to be an inhibitor of complex I of the electron transport chain, blocking the transfer of electrons from NADH. Selegiline may counteract this inhibition and thus restitute mitochondrial ATP synthesis. However, we do not know whether or not it is the same selegiline concentration interfering with MAO-B and the electron transport chain. Our dosage of 70 µg per kg body weight is in the same magnitude as that reported by Tatton to be effective at a site obviously different from MAO-B.

References

Birkmayer W, Knoll J, Riederer P, Youdim MB, Hars V, Marton J (1985) Increased life expectancy resulting from addition of L-deprenyl to Madopar treatment in Parkinson's disease: a longterm study. J Neural Transm 64: 113–127

Fortmeyer HP (1981) Thymusaplastische Maus (nu/nu), Thymusaplastische Ratte (rnu/rnu): Haltung, Zucht, Versuchsmodelle. Paul Parey, Berlin Hamburg

Halliwell B (1989) Oxidants and the central nervous system: some fundamental questions. Acta Neurol Scand 126: 23–33

Knoll J (1989) The pharmacology of selegiline ((−)deprenyl). New aspects. Acta Neurol Scand 126: 83–91

Kornhuber J, Konradi C, Mack-Burkhard F, Riederer P, Heinsen H, Beckmann H (1989) Ontogenesis of monoamine oxidase-A and -B in the human brain frontal cortex. Brain Res 499: 81–86

Oldfield FF (1992) Free radicals, dopamine, levodopa and Parkinson's disease. Free Radic Res Commun 16 [Suppl 1] Abstract 10.3

Salo PT, Tatton WG (1991) Deprenyl reduces death of motoneurons caused by axotomy. J Neurosci Res 31: 1–7

Tatton WG, Greenwood CE (1991) Rescue of dying neurons: a new action of deprenyl in MPTP parkinsonism. J Neurosci Res 30: 666–672

Youdim MBH, Ben-Shachar D, Riederer P (1989) Is Parkinson's disease a progressive siderosis of substantia nigra resulting in iron and melanin induced neurodegeneration? Acta Neurol Scand 126: 47–54

Authors' address: Dr. H.-J. Freisleben, Gustav-Embden-Zentrum der Biologischen Chemie, Klinikum der Johann Wolfgang Goethe-Universität, Theodor Stern-Kai 7, D-60590 Frankfurt/Main, Federal Republic of Germany.

J Neural Transm (1994) [Suppl] 41: 237–241

Double blind cross over trial with deprenyl in amyotrophic lateral sclerosis

S. S. Jossan[1], J. Ekblom[2], O. Gudjonsson[2], K.-E. Hagbarth[2], and
S.-M. Aquilonius[2]

[1] Department of Medical Pharmacology, and [2] Neurology, Uppsala University,
Uppsala, Sweden

Summary. In this paper we present results from a double blind cross over trial with deprenyl, a selective and irreversible monoamine oxidase-B (MAO-B) inhibitor, in 10 patients suffering from amyotrophic lateral sclerosis. The patients were randomised in such a way that half of the patients started with the active drug and half with the placebo treatment. Each patient was given 10 mg deprenyl (eldepryl, 10 mg tablets) per day for 12 weeks and then placebo for the same length of time. There was a drug free period of 12 weeks between the courses. The neurological status of the patients were evaluated every six weeks by using Norris, spinal and bulbar scores and it was observed that all cases deteriorated in their clinical status during the 36 weeks of the controlled study. MAO-B activity in blood platelets was completely inhibited during treatment with deprenyl. In the preliminary analysis performed so far, no obvious retardation in the progress of the disease could be observed with deprenyl treatment.

Introduction

Amyotrophic lateral sclerosis (ALS) is a fatal neurodegenerative disease of unknown etiology. The disease is characterised by a progressive degeneration of the upper motor neurons in the cerebral cortex and the alpha-motor neurons of the spinal cord (Bonduelle, 1975). The prevalence of ALS is in the range of 5–10 per 10,000 inhabitants in the Western world with a predominantly adult onset. Monoamine oxidase-B (MAO-B) is a mitochondrial enzyme and is localised in various cell types, including neural and glial cells of the human nervous system (Konradi et al., 1988). This enzyme is involved in the deamination of both endogenous and exogenous amines. The conversion of MPTP to the parkinsonian inducing neurotoxin MPP^+ is also brought about by MAO-B (Chiba et al., 1984). Moreover, L-deprenyl, a MAO-B inhibitor, has been used in the treatment of patients with Parkinson's disease and this drug has been reported to retard the progress of this disease (Tetrud and Langston, 1989; Parkinson Study

Group, 1989). One of the mechanisms involved may be the inhibition of MAO-B, however, other mechanisms can not be ruled out since low concentrations of deprenyl, which do not have any inhibitory effect on MAO-B, are reported to have neuroprotective effect on experimentally induced lesions in the rat (Salo and Tatton, 1991). As there is no therapy available in this disease, we found it motivated to investigate if deprenyl have some beneficial effects in patients with ALS. In the present report, we describe results from a double blind cross-over trial with L-deprenyl in ALS patients.

Materials and methods

Patients

Ten patients with ALS were included in this study, and the diagnosis was based on clinical and electro-myographical (EMG) findings according to the criteria of the World Federation of Neurology Subcommittee on Motor Neuron Disease (1990). The study was approved by the Ethics Committee of the Faculty of Medicine, Uppsala University. The patient group consisted of 8 males and 2 females and the mean age at start in the whole group was 50 ± 3 years. The mean duration of symptoms was 19 ± 5 months. In clinical examinations 3 patients showed only lower motor neuron symptoms and 4 both lower and upper motor neuron symptoms at the start of the trial. Three patients showed bulbar as well as pseudobulbar symptoms at the start.

Drug treatment

Deprenyl (eldepryl, 10 mg) and placebo tablets were manufactured by Orion-Farmos. The effects of daily oral doses for 12 weeks were studied. The patients were randomised in two groups, one group started with the active drug and the other with placebo. After a "wash-out" period of 12 weeks between the courses, the group receiving deprenyl in the first course then started with placebo and vice versa.

Clinical parameters

The patients were evaluated every sixth week, immediately before the start, in the middle and at the termination of each treatment period, and the neurological rating included Norris score, bulbar and spinal scores (Plaitakiss et al., 1988).

MAO-B estimations in platelets

MAO-B activity was estimated in blood platelets taken at the time of each clinical evaluation. The activity estimation was used as a marker for the degree of the enzyme inhibition in blood and thus functioned as a measure of compliance.

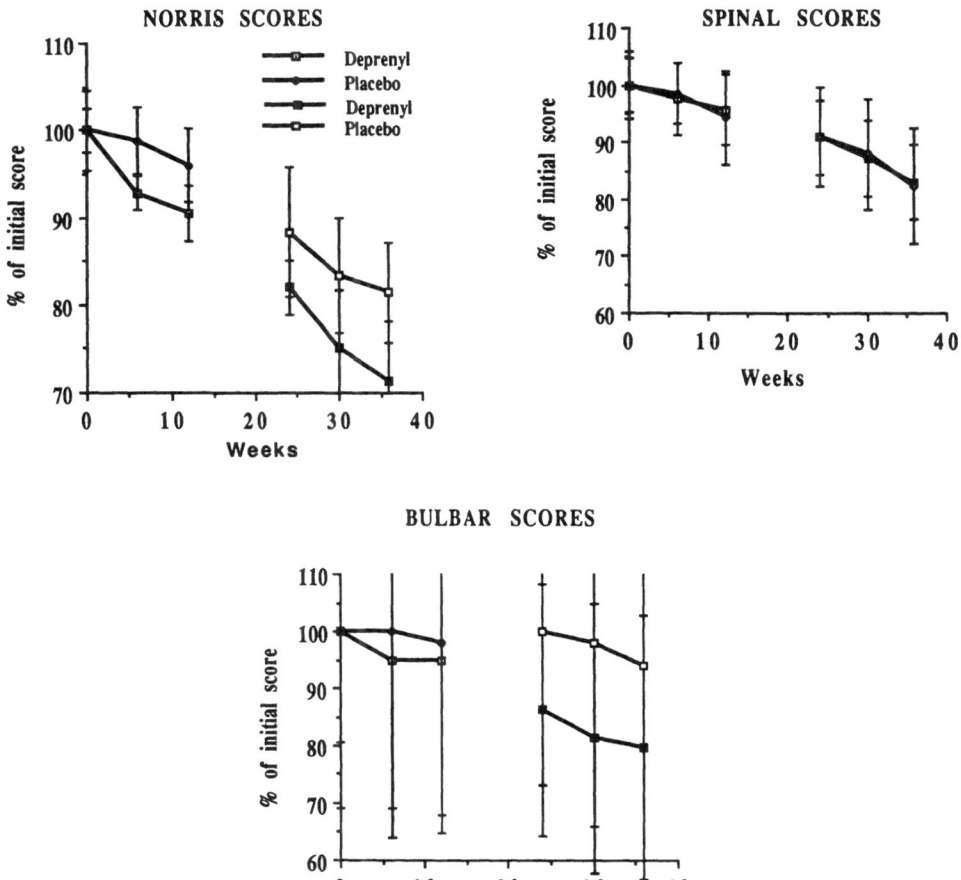

Fig. 1. The neurological status of the two groups (active drug group and placebo group) during the two treatment periods as determined each sixth week by Norris (**a**), Spinal (**b**) and Bulbar (**c**) scores. The numbers are given as mean values + S.E.M.

Statistical analysis

The statistical analysis was carried out according to the method of Hills and Armitage and the achieved cross-over differences were compared for statistical significance with a two tailed t-test.

Results

In this trial patients showed a decrease in spinal scores and Norris scores during medication with deprenyl and placebo (Fig. 1). A slight decline was observed for the bulbar score as well but the decline in this score was not as pronounced as for the spinal and Norris scores (Fig. 1). When the group treated with deprenyl was compared with the placebo group, no statistically

significant differences were observed with spinal, Norris and bulbar scores
at the two treatment periods. The estimations of MAO-B activity in platelets
from the patients showed that the enzyme activity before and after the
treatment were similar for the two groups. Furthermore, the MAO-B activity was completely inhibited during treatment with active drug (13.74 $\pm$
1.71 and 0.51 $\pm$ 0.16 nmoles/min/10^{10} platelets before and after treatment
with deprenyl, respectively).

Discussion

The three scales used to evaluate the patients showed that all cases deteriorated in their clinical status during the 36 weeks of the controlled study. The
spinal score seems to be the most sensitive scale for evaluating the disease
progress since this scale could detect deteriorations of ALS cases earlier
than the other two scales.

Already 24 hours after the first dose of deprenyl most of the MAO-B
activity in blood platelets was inhibited. Subsequent estimations of MAO-B
in blood samples from patients treated with deprenyl showed that the
enzyme activity was totally inhibited, thus, giving hold for compliance
during the trial.

For ethical reasons we found it necessary to design a cross-over study
with relatively short placebo/drug periods (12 weeks). However, in our
opinion an extended treatment period would not have led to a different
result as there were no positive effects in the following open trial. All
10 patients continued to deteriorate slowly even after completion of the
controlled study in spite of continuous treatment with 10 mg deprenyl daily.

The possibility that the dose used in this trial was too low cannot be
ignored. However, it has been demonstrated that daily 10 mg doses of
deprenyl is sufficient to inhibit intracerebral MAO-B activity (Riederer et
al., 1986). Therefore, the lack of therapeutic effect in this trial is not likely
to be due to an incomplete inhibition of MAO activity. With reference to
the findings of (Salo and Tatoon, 1991) very low doses of deprenyl, showing
no MAO-B inhibitory effect, were able to rescue motor neurons of the
facial nerve after axotomy. Moreover neurodegeneration induced by the
selective noradrenergic toxin, DSP4, was prevented by treatment with
deprenyl and it seems that mechanisms other than the MAO-B inhibition
are involved in this neuroprotective effect. However, the degeneration of
motor neurons in ALS may represent other mechanisms of cell death than
in experimental lesions and such a difference might explain the lack of
therapeutic effect of deprenyl in ALS.

In conclusion, in the preliminary evaluation of the data performed so far
no obvious improvement during the period of treatment with active drug
was observed in any of the 10 patients included in the trial.

References

Bonduelle M (1975) Amyotrophic lateral sclerosis. In: Vinken PJ, Bruyn GW (eds) Handbook of neurology, vol 22. Elsevier, Amsterdam, pp 281–338

Chiba K, Trevor A, Castagnoli N Jr (1984) Metabolism of the neurotoxic tertiary amine, MPTP, by brain monoamine oxidase. Biochem Biophys Res Commun 120: 574–578

Criteria for diagnosis of amyotrophic lateral sclerosis (1990) World Neurology 15: 12

Konradi C, Svoma E, Jellinger K, Riederer P, Denny R, Thibaults J (1988) Topographic immunocytochemical mapping of monoamine oxidase-A, monoamine oxidase B and tyrosinhydroxylase in human post mortem brain stem. Neuroscience 26: 791–801

Plaitakis A, Mandeli J, Yahr MD (1988) Pilot trial of branched chain amino acids in amyotrophic lateral sclerosis. Lancet i: 1015–1018

Parkinson Study Group (1989) Effect of deprenyl on the progression of disability in early Parkinson's disease. N Engl J Med 321: 1364–1371

Riederer P, Youdim M (1986) Monoamine oxidase activity and monoamine metabolism in brains of Parkinsonian patients treated with deprenyl. J Neurochem 46: 1359–1365

Salo PT, Tatton WG (1991) Deprenyl reduces the death of motorneurons caused by axotomy. J Neurosci Res 31: 394–400

Tetrud JW, Langston JW (1989) The effect of deprenyl (selegiline) on the natural history of Parkinson's disease. Science 245: 519–522

Authors' address: Dr. S. S. Jossan, Department of Medical Pharmacology, Uppsala University, P.O.Box 593, S-751 24 Uppsala, Sweden.

J Neural Transm (1994) [Suppl] 41: 243–248

Monoamine oxidase-B in motor cortex and spinal cord in amyotrophic lateral sclerosis studied by quantitative autoradiography

S. S. Jossan[1], J. Ekblom[2], S.-M. Aquilonius[2], and L. Oreland[1]

[1] Department of Medical Pharmacology and [2] Neurology, Uppsala University,
Uppsala, Sweden

Summary. The distribution of MAO-B was studied by using an in vitro quantitative autoradiographical method in the post-mortem spinal cord and motor cortex from control and ALS cases. ^{3}H-L-deprenyl was used as a radiotracer. Sections stained with thionine were used to count glial cells.

In both control and ALS spinal cords, high density of ^{3}H-L deprenyl binding was observed around the central canal, in the substantia gelatinosa and other grey matter regions. In the ALS cases a pronounced and statistically significant increase of MAO-B was observed in the corticospinal tract, the motor neuron areas and in the ventral white matter. An increase in the number of glial cells in spinal cords from ALS cases was also evident. Moreover, the concentration of MAO-B was highly correlated with glial cell counts in thionine stained sections in various regions of the spinal cord, both in controls and ALS cases.

An elevated level of ^{3}H-L-deprenyl binding, in ALS cases, was observed in all the individual laminae of the pre- and post-central gyri of the cerebral cortex. There was no difference in MAO-B concentration between the two groups in the occipital cortex. A substantial increase in the concentration of MAO-B was observed in the white matter of ALS cases.

Reactive gliosis has been shown to be associated with neurodegerative disorders and experimental lesions in animals. The most likely explanation for the increase of MAO-B in ALS and in other neurodegenerative disorders seems to be that the increase is a consequence of the reactive gliosis associated with these disorders.

Introduction

Amyotrophic lateral sclerosis (ALS) is a progressive neurodegenerative disorder of unknown etiology. In this disease, there is a preferential loss of the upper motor neurons in the cortex and the lower motor neurons in the brain stem and the spinal cord (Bonduelle, 1975). A marked proliferation of astrocytes in the spinal cord and in the subcortical white matter of ALS brain has also been shown (Kushner et al., 1991).

Monoamine oxidase A and B, (MAO; E.C. 1.4.3.4) are two enzymes involved in the oxidative deamination of endogenous and exogenous monoamines. The two enzymes are encoded by two different genes and show a marked difference in their primary amino acid structures (Bach et al., 1988). They also show differences in their substrate and inhibitor specificity (Johnston, 1968; Knoll and Magyar, 1972). The occurrence of the two forms varies among species and strains and also among different regions of the central nervous system (Fowler et al., 1980; Stenström et al., 1987).

The MAO-B activity increases with age in the human brain and this increase is due to the increase in the number of enzyme molecules without any change in the affinity of the enzyme towards its substrates (Fowler et al., 1980). This increase is further accelerated in neurodegenerative disorders such as Alzheimer's disease, Huntington's chorea and Parkinson's disease (Jossan et al., 1991b; review by Strolin-Benedetti and Dostert, 1989). However, no published data is available on brain or spinal cord MAO-B activity in ALS cases. In this report we have investigated, by using an in vitro quantitative autoradiographic method, the localisation of MAO-B in spinal cord in control and ALS cases. We also present data about the laminar localisation of MAO-B in three different cortical regions in control and ALS cases.

Material and methods

Spinal cords from four ALS-cases and four controls and brain-hemispheres from three control and three ALS cases were obtained at autopsy. The tissue was cut into 5-mm-thick slices and the slices were frozen on pre-cooled copper-plates in liquid nitrogen. Eighty and ten micrometer thick sections were prepared from brain and spinal cord samples respectively.

The experimental procedure was similar to that of Jossan et al. (1991a). Sections were pre-incubated for 40 min (2 × 20 min) at 20° in 0.05 M Na-K-phosphate buffer, pH 7.4. Tissue sections were then incubated for 1 h at 20° in the buffer containing 10 nM of ^{3}H-L-deprenyl (specific activity 55 mCi/µmol).

To estimate non-specific binding consecutive sections were incubated in the presence of 1 µM unlabelled L-deprenyl. After drying, the sections and a radioactive standard scale (Amersham U.K.) containing defined concentrations of tritium, were exposed at 4° for 14 days.

The regions of interest in the autoradioraphs were analysed by measurement of optical densities using a video camera system, Philips LDH 400, connected to a Cromemco (SDD) computer.

Results

Figures 1A and B show that the area around the central canal (laminae X), substantia gelatinosa (lamina-II) and the ventral horn (lamina IX) had high ^{3}H-L-deprenyl binding in the controls as well as in the ALS cases. The distribution pattern of ^{3}H-L-deprenyl binding was similar at the different

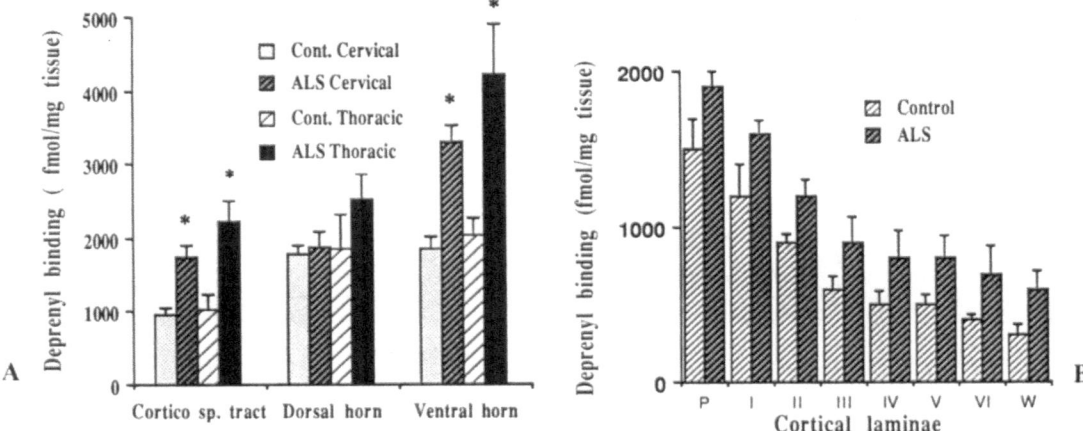

Fig. 1. Specific ³H-L-deprenyl binding in various regions of the cervical spinal cord (**A**) and the pre-central gyrus (**B**) of the control and ALS cases. Each figure indicates the mean specific deprenyl binding (fmol/mg tissue) ± S.E.M.

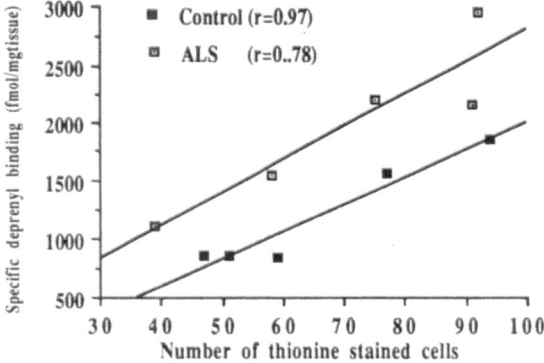

Fig. 2. Correlation between the glial cells counts and the specific deprenyl binding in the corresponding area of cervical spinal sections from control and ALS cases

segmental levels in the controls and ALS cases (Figs. 1A and B). When the mean values for ³H-L-deprenyl binding in various regions from controls were compared with those of ALS, a significant increase in the binding in the corticospinal tracts, in the motor neuron region and in the ventral white matter was observed in the ALS cases (Fig. 2A). There was also a significant increase in the number of glial cells in the corticospinal tract and the ventral horn regions of the ALS cases.

The pre- and the post central gyri and occipital cortex exhibited a clear lamination of ³H-L-deprenyl binding where pia mater and lamina I showed the highest and the white matter the lowest binding (Figs. 1C and D). There was a successive decrease in the binding from lamina I to the lamina VI. This distribution pattern was similar in the control and ALS cases. A significant increase in ³H-L-deprenyl binding in all the laminae of the pre-

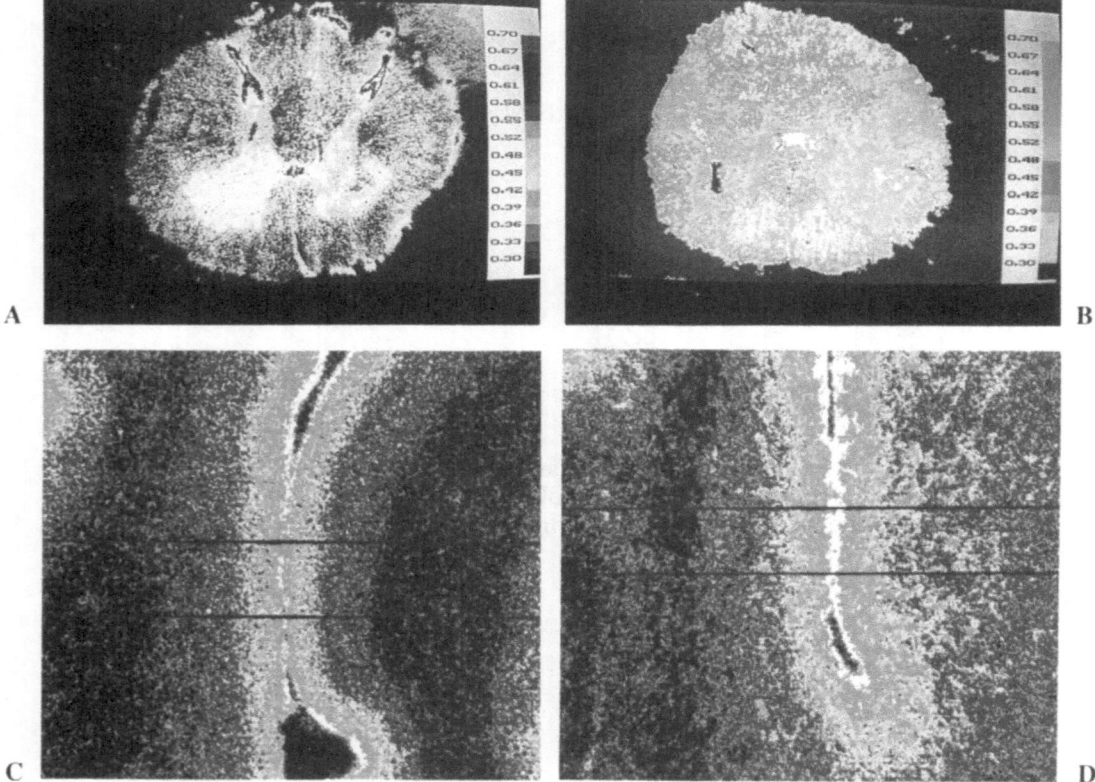

Fig. 3. Autoradiographs showing total deprenyl binding in spinal cord and cerebral cortex. **A** and **B** demonstrate the total deprenyl binding in cervical sections from control and ALS spinal cord respectively. **C** and **D** show total deprenyl binding in pre- and post-central gyri from control and ALS brain respectively

and post central gyri in the ALS cases was observed, however, the increase was more pronounced in pre-central gyrus (Fig. 2B). With regard to the ^{3}H-L-deprenyl binding in the occipital cortex, there was no difference between the two groups. White matter from ALS cases showed the highest increase in ^{3}H-L-deprenyl binding.

Discussion

The main results of this investigation are that the concentration of ^{3}H-L-deprenyl binding in various regions of the ALS spinal cord and brain is significantly increased. The increase in the binding is specific for those regions which are known to be deeply affected in this neurodegenerative disease. L-deprenyl is an irreversible MAO-B inhibitor and binds quantitatively to the enzyme protein. We have previously shown that ^{3}H-L-deprenyl have very low non-specific binding and that there is a very high correlation between the ^{3}H-L-deprenyl binding and the MAO-B activity estimated with biochemical methods (Jossan et al., 1991a). Therefore, it is reasonable to assume that the binding of ^{3}H-L-deprenyl, as measured by in vitro quantitative autoradiography, reflects the MAO-B enzyme protein.

The high correlation obtained between the mean values of glia cell counts in control and ALS spinal cord and the corresponding regional ^{3}H-L-deprenyl binding (Fig. 3) indicates that this enzyme is mainly located in the glial cells. No such correlation was observed between the motor neurons of ventral horn and the ^{3}H-L-deprenyl binding in this area. Thus the motor neurons of the human spinal cord lack or have very little MAO-B enzyme. The parallel increase in the ^{3}H-L-deprenyl binding and the number of glial cells in the corticospinal tracts, ventral horn and the ventral white matter regions of the ALS spinal cord further support the notion that the increase in MAO-B in ALS spinal cord occurs in the reactive glial cells (Fig. 3).

The increase in MAO-B in ALS spinal cord (corticospinal tracts and ventral horn) and brain (pre- and post central gyri) was found in the regions which are known to be the sites of neuronal degeneration in this particular disease. Furthermore dorsal horn and the occipital cortex, regions relatively spared in ALS, did not show any change in MAO-B concentration.

On the basis of the above results about increased MAO-B in the ALS spinal cord and brain regions together with an increase in the glial cell counts in the ALS spinal cord, we conclude that this increase mainly occurs in the reactive glial cells. Although the increase in MAO-B seems secondary to neurodegenerative process, there are theoretical possibilities that the increased turnover of monoamines might contribute to the deterioration of nerve cells by the action of neurotoxoc metabolites.

Acknowledgements

This work was supported by grants from the Swedish Medical Research Council grant no. 4145 and 4373.

References

Bach AW, Lan NC, Johnson DL, Abell CW, Bembenek ME, Kwan S-W, Seeburg PH, Shih JC (1988) c-DNA cloning of human liver monoamine oxidase A and B: molecular basis of differences in enzymatic properties. Proc Natl Acad Sci USA 85: 4934–4938

Bonduelle M (1975) Amyotrophic lateral sclerosis. In: Vinken PJ, Bruyn GW (eds) Handbook of neurology, vol 22. Elsevier, Amsterdam, pp 281–338

Fowler CJ, Oreland L, Marcusson J, Winblad B (1980) The effects of age on the activity and molecular properties of human brain monoamine oxidase. J Neural Transm 49: 1–20

Johnston JP (1968) Some observations upon a new inhibitor of monoamine oxidase. Biochem Pharmacol 17: 1285–1297

Jossan SS, Gillberg PG, d'Argy R, Aquilonius SM, Långström B, Halldin C, Oreland L (1991a) Quantitative localisation of human brain monoamine oxidase B by large section autoradiography using ^{3}H-L-deprenyl. Brain Res 547: 69–76

Jossan SS, Gillberg PG, Gottfries CG, Karlsson I, Oreland L (1991b) Monoamine oxidase B in brains from patients with Alzheimer's disease: a biochemical and autoradiographical study. Neuroscience 45: 1–12

Knoll J, Magyar K (1972) Some puzzling pharmacological effects of monoamine oxidase inhibition. Adv Biochem Psychopharmacol 5: 393–408

Kushner PD, Stephenson DT, Wright S (1991) Reactive astrogliosis is widespread in the subcortical white matter of amyotrophic lateral sclerosis brain. Exp Neurol 50: 263–267

Stenström A, Hardy J, Oreland L (1987) Intra- and extra-dopamine-synaptosomal localisation of monoamine oxidase in striatal homogenates from four species. Biochem Pharmacol 18: 2931–2935

Strolin-Benedetti M, Dostert P (1989) Monoamine oxidase, brain aging and degenerative diseases. Biochem Pharmacol 38: 555–561

Authors' address: Dr. S. S. Jossan, Department of Medical Pharmacology, Uppsala University, P.O.Box 593, S-751 24 Uppsala, Sweden.

J Neural Transm (1994) [Suppl] 41: 249–252

The effect of 6-months l-deprenyl administration on pineal MAO-A and MAO-B activity and on the content of melatonin and related indoles in aged female Fisher 344N rats

G. F. Oxenkrug[1], P. J. Requintina[1], R. M. Correa[1], and A. Yuwiler[2]

[1] Pineal Research Laboratory, Psychiatry Service, VAMC, and Department of Psychiatry and Human Behavior, Brown University School of Medicine, Providence, Rhode Island, and [2] Neurochemistry Laboratory, West Los Angeles Brentwood VAMC and Department of Psychiatry and Biobehavioral Sciences, UCLA, Los Angeles, California, U.S.A.

Summary.   Six months of administration of the selective MAO-B inhibitor, selegiline (l-deprenyl 0.25 mg/kg, s.c.) to aged female Fisher 344N rats suppressed MAO-A as well as MAO-B activity and increased serotonin (substrate for melatonin biosynthesis) and N-acetylserotonin (immediate melatonin precursor) levels in pineal glands taken from the animals during the night. Daytime values were unchanged by the treatment. The data suggest that stimulation of pineal melatonin biosynthesis might be one of the consequences of MAO-A inhibition contributing to life span prolongation induced by chronic selegiline treatment.

Introduction

It was reported that chronic administration of selegiline (l-deprenyl), an irreversible selective MAO-B inhibitor, increased longevity in *parkinsonian* patients and rats (Birkmayer et al., 1985; Knoll et al., 1989; Milgram et al., 1990). This effect was ascribed to selegiline's selective inhibition of MAO-B activity. However, selegiline (administered up to 21 days) is only selective for MAO-B at low doses; at high doses it also inhibits MAO-A (Ekstedt et al., 1979; Heikkila et al., 1990). One of the results of MAO-A inhibition by deprenyl (25 mg/kg, acute administration) is the stimulation of pineal melatonin biosynthesis (Oxenkrug et al., 1984; see for rev. Oxenkrug, 1991). Melatonin has been implicated in aging (Dilman et al., 1979; Pierpaoli and Maestroni, 1987). This raised the question of whether chronic (more than 21 days) treatment with very small doses of selegiline (0.25 mg.kg) might inhibit MAO A as well MAO B, leading to increased melatonin, and whether melatonin, so produced, might underlie selegiline facilitated longevity.

In the present study we investigated the effect of 6 month administration of 0.25 mg/kg of selegiline on pineal MAO-A and MAO-B activity and melatonin biosynthesis.

Materials and methods

Fisher 344N female rats (12 months old, body weight 200 g) (retired breeding stock) were housed two to a cage in a light-controlled (L:D 10:14, lights on at 24.00, lights off at 10.00 hrs) and temperature-regulated room with the free access to food and water.

Selegiline (0.25 mg/kg) was injected s.c. every other day in accordance with the protocol used by Knoll et al. (1989) and Milgram et al. (1990) for examining effects on life span. Controls were injected with vehicle solution. All injections were done at 08.00 (two hours before lights off).

After six months of treatment and two hours after the last injection one group of animals was decapitated at 9.55 (5 min before lights off), another group at 13.00 (after three hours of darkness) and the third at 15.00 hrs (after five hours of darkness). All manipulations during the dark phase were performed under light of 1 lux intensity. The pineal glands were quickly removed, frozen on dry ice and stored at −70°C.

Pineal glands were sonicated in 200 μl of the phosphate buffer, pH 7.4. An aliquot of the supernate was used for the determination of serotonin (5-HT), N-acetylserotonin (NAS), melatonin, 5-hydroxyindolacetic acid (5-HIAA) and 5-hydroxytryptophol (5-HTOL) content by HPLC-fluorimetric system as described elsewhere (see Oxenkrug et al., 1991). Results are expressed as mean ± S.D. (ng of the indole/pineal). MAO activity was evaluated in aliquots of supernate by the method of Wurtman and Axelrod (1963). Results are expressed as mean ± S.E.M. (nmoles/hr/pineal).

Group differences were analyzed by one-way analysis of variance and Student's t-test.

Results

Melatonin and related indoles (5-HT, NAS, 5-HIAA and 5-HTOL) levels were not different in the pineals of control and selegiline treated rats decapitated after 14 h of light (just before lights off) (Table 1). Pineal 5-HT, NAS and 5-HIAA levels were almost twice as high in deprenyl-treated rats during the dark phase (after 3 h in darkness) than in control group. Only 5-HT levels remained higher after 5 h of darkness (Table 1).

MAO-A and MAO-B activity were not different in the pineals of control and selegiline-treated rats in the light phase (just before the light off) (Table 1). MAO-B activity was significantly inhibited after 3 and 5 h of darkness, while MAO-A activity was significantly inhibited only after 5 h of darkness (Table 1).

Discussion

Our results indicate that chronically administered selegiline can lose its selectivity for MAO B. Although a low dose of 0.25 mg/kg had no effect on

Table 1. Effect of 6 month selegiline administration on rat pineal MAO, melatonin and related indoles

	MAO-A	MAO-B	Melatonin	NAS (ng/pineal)	5-HT	5-HIAA
	nmole/h/pineal					
9.55:						
Saline	1.04 ± 0.16	1.34 ± 0.03	0.13 ± 0.02	0.15 ± 0.01	60.0 ± 6.4	5.74 ± 0.62
Depr.	0.90 ± 0.15	0.84 ± 0.28	0.18 ± 0.01	0.15 ± 0.01	63.5 ± 5.2	5.00 ± 0.72
13.00:						
Saline	2.54 ± 0.53	2.43 ± 0.34	0.97 ± 0.18	0.47 ± 0.14	25.6 ± 5.16	1.87 ± 0.51
Depr.	1.55 ± 0.31	0.83 ± 0.06**	0.94 ± 0.16	0.75 ± 0.13*	49.0 ± 5.29*	2.98 ± 0.45**
15.00:						
Saline	3.87 ± 0.29	3.30 ± 0.13	1.30 ± 0.35	1.78 ± 1.24	9.60 ± 5.20	1.23 ± 0.53
Depr.	1.97 ± 0.37*	0.94 ± 0.26*	1.38 ± 0.16	2.80 ± 1.32	17.2 ± 3.12**	1.60 ± 0.63**

* $p < 0.001$ vs saline; ** $p < 0.02$

MAO A after 21 days of chronic administration (Ekstedt et al., 1979; Heikkila et al., 1990), it did inhibit MAO-A after 6 months of treatment in this present study. This inhibition was functionally significant and doubled 5-HT concentrations. Since noradrenaline is also a substrate for MAO-A it would not be surprising if its concentration also increased after chronic selegiline. These results then suggest that the consequences of MAO-A inhibition must also be considered in accounting for any selegiline-induced prolongation of life. One such consequence may be an increased in melatonin formation since although 5-HT is the primary precursor for *Melatonin biosynthesis* noradrenaline controls the rate-limiting step of its N-acetylation in melatonin biosynthesis. Although we did not observe a direct increase in pineal melatonin concentration in selegiline-treated animals, there was a two-fold increase in its immediate precursor, NAS, 3 h into the dark phase supporting the possibility of stimulated melatonin biosynthesis.

Further detailed evaluations of pineal melatonin at multiple times along the circadian cycle of animals chronically given melatonin may help determine whether chronic selegiline can counteract or reverse the age-associated decline in melatonin biosynthesis and whether melatonin itself contributes to longevity.

References

Birkmayer W, Knoll J, Riederer P, Youdim MBH, Hars V, Marton J (1985) Increased life expectancy resuting from addition of l-deprenyl to madopar treatment in Parkinson's disease: a longterm study. J Neural Transm 64: 113–127

Dilman VM, Lapin IP, Oxendrug GF (1979) Serotonin and aging. In: Essman W (ed) Serotonin in health and disease, vol 5. Spectrum Press, London, pp 113–123

Ekstedt B, Magyar K, Knoll J (1979) Does the B form of selective nonoamine oxidase inhibitor lose selectivity by long term treatment? Biochem Pharmacol 28: 919–923

Heikkila RE, Terleckyi I, Sieber BA (1990) Monoamine oxidase and the bioactivation of MPTP and related neurotoxins: relevance to "DATATOP". J Neural Transm [Suppl] 32: 217–228

Knoll, J, Dallo J, Yen T (1989) Striatal dopamine, sexual activity and lifespan. Longevity of rats treated with (−)deprenyl. Life Sci 145: 525–531

Knoll J, Eczery Z, Kelemen K, Nievel J, Knoll B (1965) Phenylisopropyl-methypropinylamine (E-250), a new spectrum psychic energizer. Arch Int Pharmacodyn 155: 154–164

Milgram NW, Racine RJ, Nellis P, Mendonca A, Ivy GO (1990) Maintenance on L-deprenyl prolongs life in aged male rats. Life Sci 47: 415–420

Oxenkrug GF, McCauley RM, McIntyre IM, Filipowicz C (1984) Effect of clorgyline and deprenyl on rat pineal melatonin. J Pharm Pharmacol 36: 55W

Oxenkrug GF (1991) The acute effect of monoamine oxidase inhibitors on serotonin conversion to melatonin. In: Coppen A, Sandler M, Harnett S (eds) 5-Hydroxytryptamine in psychiatry. A spectrum of ideas. Oxford University Press, Oxford, pp 99–108

Pierpaoli W, Maestroni GJM (1987) Melatonin: a principal neuroimmunoregulatory and anti-stress hormone: its anti-aging effects. Immunol Lett 16: 355–362

Wurtman RJ, Axelrod J (1963) A sensitive and specific assay for the estimation of monoamine oxidase. Biochem Pharmacol 62: 1439–1440

Authors' address: Dr. G. F. Oxenkrug, Psychiatry Service, VAMC, 830 Chalkstone Ave., Providence, RI 02908, U.S.A.

J Neural Transm (1994) [Suppl] 41: 253–258

Reactive gliosis and monoamine oxidase B

J. Ekblom[1], **S. S. Jossan**[2], **L. Oreland**[2], **E. Walum**[3], and **S.-M. Aquilonius**[1]

[1] Department of Neurology, University Hospital, Uppsala, [2] Department of Medical Pharmacology, Biomedical Centre, Uppsala, [3] Department of Neurochemistry and Neurotoxicology, University of Stockholm, Stockholm, Sweden

Summary. A double-staining method was applied to cryosections of human spinal cord from patients who died with amyotrophic lateral sclerosis (ALS) and corresponding controls in order to investigate cellular content of monoamine oxidase B (MAO-B). ^{3}H-L-Deprenyl emulsion autoradiography was used in combination with histochemical methods for the detection of astrocytes and monocytes/microglia. In the ALS spinal cords an increased number of astrocytes as well as an increased content of MAO-B in reactive species of astrocytes was demonstrated. No significant ^{3}H-L-deprenyl binding was observed in cells derived from the mesoderm, e.g. monocytes or microglia. Furthermore, a sub-population of reactive astrocytes that contained low levels of MAO-B was observed in spinal sections. These findings were further substantiated by studies performed on primary astrocyte cultures.

Introduction

The B-form of MAO (MAO; E.C. 1.4.3.4) has been reported to increase in several regions of the central nervous system (CNS) in conditions associated with gliosis (for review see Strolin Benedetti and Dostert, 1989). Several investigators suggest that the main pool of MAO-B in the normal as well as in the lesioned CNS is confined to astrocytes (Aquilonius et al., 1992; Ekblom et al., 1993; Levitt et al., 1982; Nakamura et al., 1990). In the present investigation we have studied the MAO-B content of reactive astrocytes as well as of reactive microglia/monocytes (cells with specific binding sites for the *Ricinus Communis* aggutinin-1 lectin, RCA-1) in spinal cords from cases who died with ALS. ALS is a neurodegenerative disorder with adult onset characterized by progressive pathological changes and symptoms preferentially restricted to motor neurons of the cerebral cortex and spinal cord (Bonduelle, 1975). In addition, we have studied the cellular MAO-B content in primary cultures of astrocytes seeded at low density, an in vitro model to produce large reactive-like astrocytes (Ekblom et al., 1993; Walum et al., 1990).

Materials and methods

Post mortem tissue

Spinal cords were dissected at autopsy from 4 ALS cases and 4 control cases with no history of neurological disease. The lumbar cords were cut horizontally into 5 mm thick segments, snap-frozen in liquid nitrogen and then stored at −70°C according to the procedure of Gillberg and co-workers (1986). Cryosections of 10 μm thickness were cut on a slide microtome and mounted on poly-L-lysine coated glass slides.

Cell cultures

Primary cultures of new-born rat cortical type-1-like astrocytes were prepared according to Ekblom and co-workers (1993). The cell suspension obtained from two hemispheres was diluted to a total volume of either 40 ml (for "normal" development) or 80 ml (for reactive-like cell development) of growth medium (Dulbecco's modification of Eagle's medium, DMEM, plus 10% fetal calf serum, 100 units of penicillin and 50 μg streptomycin per ml) and then dispensed to in 4 ml aliquotes to 60 mm plastic tissue culture dishes. The cultures were maintained in growth medium at 37°C in a humidified atmosphere of 4.5% CO_2 in air for 24 days.

Immunohistochemistry and emulsion autoradiography

The double-staining was performed as described by Ekblom et al. (1993). Briefly, the slides were incubated for 60 min at 20°C in 0.05 M Na-K-phosphate-buffer, pH 7.4, containing 10 nM ³H-L-deprenyl (spec. act. 50 Ci/mmol). After incubation with the radioligand the slides were either subjected to immunohistochemistry using a polyclonal antibody against glial fibrillary acidic protein (GFAP) or histochemical detection of RCA-1 binding (Mannoji et al., 1986). After histochemical staining, slides were air dried for 10–15 min and covered with film-emulsion LM-1 (Amersham, U.K.) and diluted in water (1:1). The emulsion-covered slides were exposed for 7 days, developed and fixed. Immediately after fixation the slides were counterstained with hematoxyline and mounted with coverslips in a watersoluble media. For the cell cultures the histochemical staining procedure was omitted.

Results

Double staining with ³H-L-deprenyl emulsion autoradiography and GFAP-immunohistochemistry revealed an excellent match of ³H-L-deprenyl binding with the GFAP-staining in the ventral white matter of both controls and the ALS cases (Fig. 1a). Furthermore, no ³H-L-deprenyl binding was found in cells lacking GFAP-immunoreactivity e.g. cells stained with RCA-1 did not exhibit any ³H-L-deprenyl binding (Fig. 1b). Some of the astrocytes in the ALS tissue were prominently enlarged and these reactive astrocytes were richer in ³H-L-deprenyl binding sites than astrocytes in the controls, probably due to their larger volume. The same phenomenon could also be observed in primary cultures of astrocytes. The cultures seeded at low density cultures were dominated by large multibranched GFAP-positive

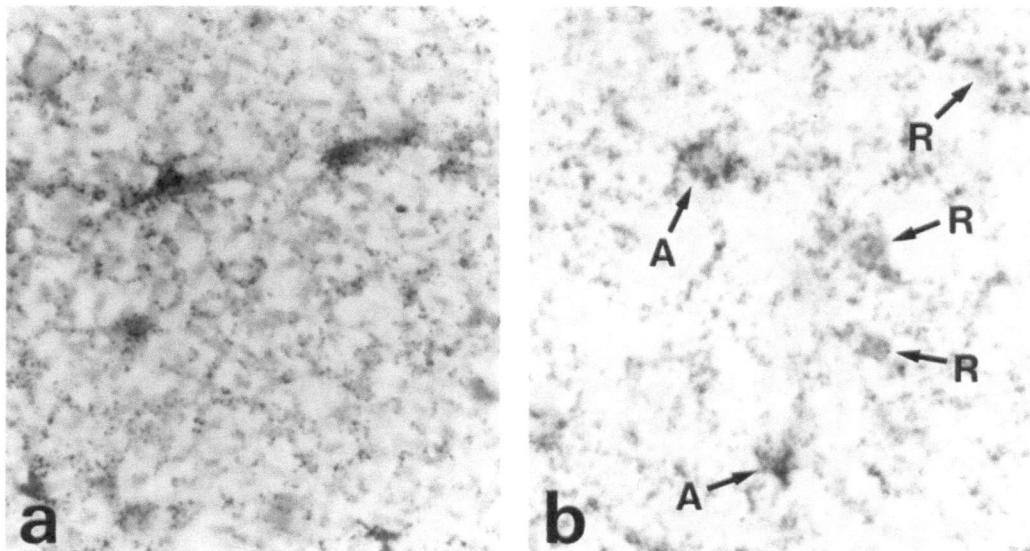

Fig. 1. Double labelling of the ventral white matter in a 10 μm spinal cryosection with histochemical staining using **a** GFAP antiserum or **b** RCA-1 lectin, and ^{3}H-L-deprenyl emulsion autoradiography (silver grains). *A* cells lacking RCA-1 staining (astrocytes) and *R* RCA-1 positive cells (microglia and/or monocytes)

reactive astrocyte-like cells similar to those described earlier (Fedoroff et al., 1984; Walum et al., 1990) and these cells were richer in MAO-B than cells with "normal" morphology.

In the white matter of the ALS spinal cords, a sub-population of enlarged reactive-like astrocytes (<10%) with relatively low deprenyl binding was observed. A heterogeneity with regard to ^{3}H-L-deprenyl binding was also evident in the primary cultures of astrocytes. The large cells in the low-density cultures exhibited a very pronounced heterogeneity with regard to the ^{3}H-L-deprenyl binding as demonstrated in Fig. 2.

Discussion

The specificity of L-deprenyl towards the B-type of MAO has been described by Magyar and Knoll (1976) and the tritiated substance has been shown to be useful as a marker of MAO-B in quantitative autoradiography (Aquilonius et al., 1992; Jossan et al., 1991). GFAP is a protein abundantly occurring in the mammalian nervous system and Bignami et al. (1972) were the first of several authors to suggest an exclusive astrocytic expression of GFAP. RCA-1 binding has been demonstrated to be a reliable marker of cells derived from the mesoderm, e.g. microglia and monocytes (Mannoji et al., 1986). Therefore. ^{3}H-L-deprenyl emulsion autoradiography, in combination with GFAP-immunohistochemistry or RCA-1 histochemistry, provides a suitable tool to selectively study MAO-B in sub-types of glial cells.

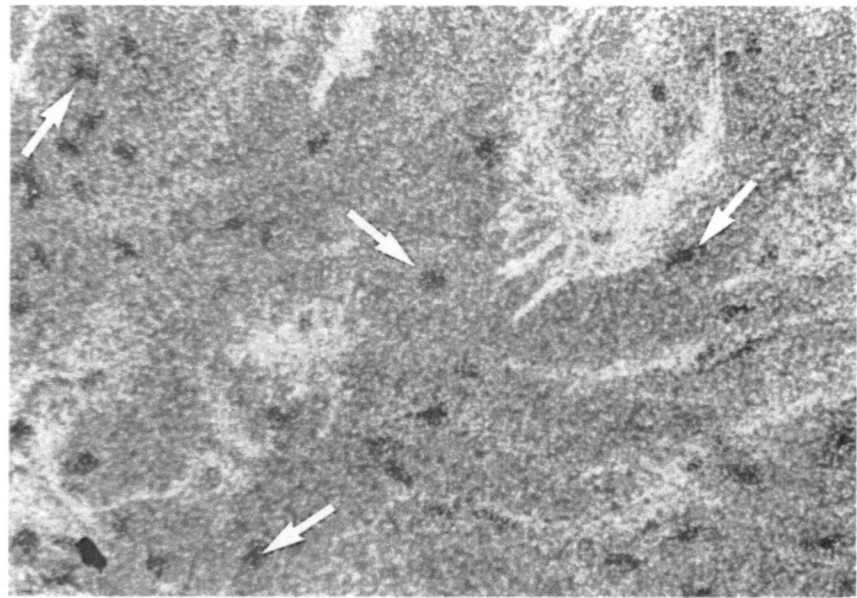

Fig. 2. Astrocytes with reactive-like morphology in low-density cultures. Arrows show reactive-like cells without detectable ^{3}H-L-deprenyl binding (bar = 20 μm)

Our results show a distinct localization of MAO-B to astrocytes in control and ALS lumbar spinal cord. This observation is in accordance with previous immunohistochemical studies on rat, monkey and human brain tissue showing MAO-B positive staining in non-neuronal cells (Levitt et al., 1982; Westlund et al., 1985, 1988). Recently, Nakamura et al. (1990) reported that reactive astrocytes in senile plaques from subjects with Alzheimer's disease were rich in MAO-B. This is in concordance with our finding that reactive-like cells exhibits relatively high MAO-B content as demonstrated both in situ and in cell cultures. Another interesting finding was that reactive astrocytes in the white matter of the spinal cord from ALS cases and cultured rat astrocytes with a morphology similar to that of reactive astroglia seemed to be heterogeneous with regard to their MAO-B content. Possible explanations for this heterogeneity include: differential modulation by factors in the local environment and/or phenotypic differences. Various enzyme systems have been shown to be altered in association with reactive astrocytosis (Diedrich et al., 1991; Schoepp and Azzaro, 1983; Tanaka et al., 1991) and such changes may lead to substantial alterations in regional metabolism. Even if such alterations most often are secondary to other processes leading to neuronal cell death, one can not exclude a possible role of the astroglial reaction as a pathogenetic mechanism in neuronal degeneration. In a recent work it has been reported that the extracellular matrix in a sub-population of reactive astrocytes have properties that inhibit neuronal regeneration (McKeon et al., 1991). It has, furthermore, been proposed that reactive astrocytes are likely to have a high synthesis rate of quinolinic acid, an endogenous NMDA receptor agonist (Schwarcz, 1992), which might cause excitotoxic effects. With regard to MAO-B, it has, for example, been suggested that the formation of H_2O_2 from the oxidative deamination reaction catalysed by MAO-B, might accelerate neurodegenerative processes (Cohen, 1983).

Acknowledgements

This work was supported by grants from The Swedish Medical Research Council (grants no 4145, 4373) and Orion-Farmos Inc., Finland.

References

Aquilonius S-M, Jossan SS, Ekblom J, Askmark H, Gillberg P-G (1992) Increased binding of L-deprenyl in spinal cords from patients with amyotrophic lateral sclerosis as demonstrated by autoradiography. J Neural Transm [Gen Sect] 89: 111–122

Bignami A, Eng LF, Dahl D, Uyeda CT (1972) Localization of the glial fibrillary acidic protein in astrocytes. Brain Res 43: 429–435

Bonduelle M (1975) Amyotrophic lateral sclerosis. In: Vinken PJ, Bruyn GW (eds) Handbook of neurology, vol 22. Elsevier, Amsterdam, pp 281–338

Cohen G (1983) The pathobiology of Parkinson's disease: biochemical aspects of neuron senescence. J Neural Transm 19: 89–103

Diedrich JF, Minnigan H, Carp RI, Whitaker JN, Race R, Freyll W, Haase AT (1991) Neuropathological changes in scrapie and Alzheimer's disease are associated with increased expression of apolipoprotein E and cathepsin D in astrocytes. J Virol 65: 4759–4768

Ekblom J, Jossan SS, Bergström M, Oreland L, Walum E, Aquilonius S-M (1993) Monoamine oxidase-B in astrocytes. Glia 8: 122–132

Fedoroff S, Neal J, Opas M, Kalnis V (1984) Astrocyte cell lineage. III. The morphology of differentiating mouse astrocytes in colony culture. J Neurocytol 13: 1–20

Gillberg P-G, Jossan SS, Askmark H, Aquilonius S-M (1986) Large-section cryomicrotomy for in vitro receptor autoradiography. J Pharmacol Toxicol Methods 15: 169–180

Jossan SS, Gillberg P-G, Gottfries C-G, Karlsson I, Oreland L (1991) Monoamine oxidase-B in brains from patients with Alzheimer's disease: a biochemical and autoradiographical study. Neuroscience 45: 1–12

Levitt P, Pintar JE, Breakfield XO (1982) Immunohistochemical demonstration of monoamine oxidase B in brain astrocytes and serotonergic neurons. Proc Natl Acad Sci USA 79: 6385–6389

Magyar K, Knoll J (1977) Selective inhibition of the B-form of monoamine oxidase. Pol J Pharmacol 29: 233–246

Mannoji H, Yeger J, Becker LF (1986) A specific histochemical marker (lectinRicinus communis agglutinin-1) for normal human microglia and application to routine histopathology. Acta Neuropathol 71: 341–343

McKeon RJ, Schreiber RC, Rudge JS, Silver J (1991) Reduction of neurite outgrowth in a model of glial scarring following CNS injury is correlated with the expression of inhibitory molecules on reactive astrocytes. J Neurosci 11: 3398–3411

Nakamura S, Kawamata T, Akiguchi I, Kameyama M, Nakamura N, Kimuara H (1990) Expression of monoamine oxidase B activity in astrocytes of senile plaques. Acta Neuropathol 86: 419–425

Schoepp DD, Azzaro AJ (1983) Effects of intrastriatal kainic acid injections on ^{3}H-dopamine metabolism in rat striatal slices: evidence for postsynaptic glial metabolism by both type A and B forms of monoamine oxidase. J Neurochem 40: 1340–1343

Schwarcz R (1992) Endogeneous excitotoxins in neurodegenerative brain diseases. Behav Pharmacol 13: 25

Strolin Benedetti M, Dostert P (1989) Monoamine oxidase, brain ageing and degenerative diseases. Biochem Pharmacol 38: 555–561

Tanaka H, Araki M, Masuzawa T (1991) Differential response of three astrocyte-specific proteins to fimbrial transection of the rat brain: immunohistochemical observations with antibodies against glial fibrillary acidic protein, glutamine synthetase and S-100 protein. Acta Histochem Cytochem 24: 11–19

Walum E, Hansson E, Harvey AL (1991) In vitro testing of neurotoxicity. ATLA 18: 153–179

Westlund KL, Denney RM, Kocherbergre LM, Rose RM, Abell CW (1985) Distinct monoamine oxidase A and B populations in primate brain. Science 230: 181–183

Westlund KL, Denney RM, Rose RM, Abell CW (1988) Localisation of distinct monoamine oxidase A and monoamine oxidase B cell populations in human brain stem. Neuroscience 25: 439–456

Authors' address: Dr. J. Ekblom, Department of Neurology, University Hospital, S-751 85 Uppsala, Sweden.

J Neural Transm (1994) [Suppl] 41: 259–266

Monoamine oxidase inhibitors, cognitive functions and neurodegenerative diseases

J. C. Delumeau[1], **D. Bentué-Ferrer**[1], **J. M. Gandon**[2], **R. Amrein**[3], **S. Belliard**[1], and **H. Allain**[1]

[1] Laboratory of Experimental and Clinical Pharmacology, University Hospital, Rennes Cedex, [2] Biotrial, Rue Jean Pecker, Technopole Rennes-Atalante, Rennes, France [3] Pharma Clinical Research CNS, F. Hoffmann-La Roche Ltd., Basel, Switzerland

Summary. Recent data obtained in animals and in humans suggest that both MAO-A and MAO-B inhibitors present cognitive enhancing properties of possible interest in the treatment of cognitive disorders. In addition, the rational for using selegiline as a neuroprotector in Parkinson's disease may also be applicable in Alzheimer's disease in which a dramatic increase in the MAO-B activity has been reported. It seems then worthwile to investigate the neuroprotective effect of MAOIs in humans and to assess, furthermore, the real therapeutical benefit of their cognitive enhancing properties.

Introduction

The monoamine oxidases (MAO) are involved in the oxidative deamination of endogenous monoamine neurotransmitters as well as exogenous monoamine, such as tyramine. This enzyme plays a major role in the deamination of biogenic amines and is present in almost all tissues in the central nervous system (CNS) as well as in peripheral tissues. Two MAO isoenzymes have been described (Johnston, 1968). The A type MAO (MAO-A) is able to catabolize dopamine (DA), noradrenaline (NA), serotonin (5-HT) and exogenous tyramine. It is selectively and irreversibly inhibited by clorgyline. The MAO activity of the gastrointestinal tract is largely of the A type (about 80%). In contrast, both A and B MAO activities are present in neurons and astrocytes, as well as in the hepatocyte in different proportions depending on the animal species. Among the monoamine neurotransmitters, DA is a substrate for MAO-A as well as MAO-B, as is tyramine, and more interestingly for the present discussion, noradrenaline (O'Caroll et al., 1986). The B isoenzyme is selectively inhibited by selegiline. Most of the brain MAO activity is located in astrocytes and, in human brain about 80% of the activity is of the B type.

Non selective MAOIs have been used for 30 years for their antidepressive properties. Unfortunately, these first generation MAOIs may be hepatotoxic and also exposed to patients the risk of a pejorative interaction with opiate and 5-HT uptake inhibitors i.e. tricyclic antidepressants (TCAs). In addition, due to their irreversible binding on their target enzymes, these drugs could precipitate hypertensive crisis in case of substantial tyramine intake. For these reasons, reversible compounds were synthesized, either specific to MAO-A such as toloxatone, moclobemide or brofaromine, or specific to the MAO-B such as Ro 19-6327 (see Da Prada et al., 1988, 1990a,b).

Since the antidepressant properties have been related to the blockade of MAO-A activity, selective MAO-A inhibitors are used as antidepressants. The recently available moclobemide, does not differ from TCAs for efficacy but is significantly better tolerated mainly due to its lack of anticholinergic effects. This property is undoubtedly an appreciable advantage for the treatment of depressive syndromes in elderly and/or deteriorated people. In addition, recent data suggest that moclobemide may have cognitive enhancing properties. Although this issue remains controversial, MAO-B inhibitors do not seem to possess antidepressive properties. However recent data obtained with selegiline strongly suggest neuroprotective properties, particularly in Parkinson's disease, and perhaps also in Alzheimer's disease. In addition, some data suggest cognitive enhancing properties.

Neurochemical basis

Although they are both of interest in the treatment of cognitive deficit in elderly and/or demented people, neuroprotective effects and cognitive enhancing properties may be related to different pathways of biological events sharing probably the same initial step: the inhibition of MAO activity. More precisely, the cognitive enhancing properties may be related to an increase of available neurotransmitters among deficient neuronal networks. Astrocyte MAO may be concerned since these cells contribute to the reuptake of monoamines. The fact that MAO inhibitors do not interfere with the catabolism of acetylcholine which plays a major role in memory and cognition is not a valid argument against such a theory since it is now obvious that the monoaminergic neuromodulatory bundles, particularly the DA ones, also play a major role in the integration of cognition. According to such a theory, inhibitors of both A and B isoenzymes may have a favourable effect; A-MAOIs by enhancing the amount of NA, DA and 5-HT available, and B-MAOIs by enhancing more selectively the amount of DA, a catecholamine of major importance in attention. Due to at least the replacement of neurons by astrocytes where it is mainly located, the MAO-B activity increases with age, and dramatically in case of Alzheimer's disease. Hence, one may hypothesize on the participation of a DA shortage, in addition to the loss of cholinergic neurons, in the pathophysiology of the age- and/or Alzheimer-related and cognitive impairment.

The oxidative deamination of monoamines by MAOs is known to induce the production of free radicals (see Riederer et al., 1989) which are able to cause cell injury (Slater, 1984). Interestingly, MAOs being located in mitochondria, the mitochondrial DNA (mtDNA) is exposed to a particularly high level of agression by free radicals. This, may explain the higher mutation rate of mtDNA compared to nuclear genome (Byrne, 1992). Actually, normal ageing has been shown to be associated with a significant deterioration of respiratory chain function (Trounce et al., 1989), leading to the theory that accumulation of mtDNA mutation, deletions in particular, may be determinant in the ageing process. For some of them, mutations are shared by several tissues whereas some others are specific to a particular cell type. There is increasing evidence that sporadic Parkinson's disease is associated with i) an accumulation of mtDNA deletions (Ikebe et al., 1990) and ii) a major reduction in respiratory chain complex activity (Schoffner et al., 1991; Mann et al., 1992). In the context of ageing but also of Alzheimer's disease in which MAO-B activity is dramatically increased, and even more in the context of Parkinson's disease, any factor leading to increasing the production of free radicals is able to produce deleterious effects on mtDNA with consequences on brain cell energy production.

Some exogenous neurotoxins that may be present in the environment or foods are able to induce CNS lesions after biotransformation. This is in particular the case of 1-methyl-4 phenyl-1,2,3,6-tetrahydroxypyridine (MPTP) which is converted by astrocytic MAO-B into MPP+, a metabolite of high toxicity for dopaminergic neurons (Tetrud and Langston, 1989). In addition preliminary data suggest that some endogenous molecules may be transformed by MAO-B following a MPTP like pathway leading to a toxic effect on respiratory chain complexes.

Even outside the neurotoxic theory for Parkinson's disease, which remains controversial, several theoretical arguments are consistent with a neuroprotective effect of MAO-B not only in Parkinson's disease but also in Alzheimer's disease and in normal ageing. In Parkinson's disease, B-MAOI's DOPA saving property by limiting the amount of DOPA administered, and therefore to be catabolised, limits the amount of free radicals to be produced in a context pre-existing neurodegenerative process. At an earlier stage, or in normal ageing, saving endogenous dopamine may also be beneficial by protecting mtDNA. The predominant location of MAO-B in astrocytes is not really an argument against a protective effect of MAO-B in neurons. By preventing the consumption of reductant resources in astrocytes, more reductant resources should be available to neurons to face oxidative stress.

Cognitive enhancing properties of MAOIs in animals

The stimulating activity of MAO-A and -B inhibitors on central adrenergic and dopaminergic systems demonstrated by Klimek et al. (1990) may explain some psychostimulating effects observed with these molecules. However conclusions should be cautious, since the substances used may have broad effects. Selegiline (l-deprenyl) for example is catabolized to l-amphetamine and l-methamphetamine, that, although levorotatory, possess mild properties related to those of the disomers category. Moreover, only a few animal studies have been devoided strictly to cognitive aspects that can be assessed exclusively by maze tests, avoidance tests, scopolamine or benzodiazepine effects antagonism, or tests in elderly animals. Globally, both A- and B-MAOIs possess facilitating properties for learning behaviours, memorization and vigilance justifying further investigations in humans in comparison to classical nootropic drugs (e.g. piracetam) or cholinesterase inhibitors (e.g. physostigmine).

Cognitive enhancing properties in humans

MAO-B

Th psychostimulating properties of selegiline were first reported in human in 1965 by Knoll who qualified it as a "psychic energizer". These effects were confirmed both in volunteer (himself!) and in patients suffering from Parkinson's disease with bradyphrenia (Lees et al., 1989). An improvement of the cognitive function was shown in patients presenting a symptomatology of senile dementia of the Alzheimer's type (SDAT) (Tariot et al., 1987a,b, 1988; Goad et al., 1991). Surprisingly, these authors showed that 10 mg but not 40 mg significantly improved the cognitive and behavioral parameters. These results have been further confirmed by Campi et al. (1990), Monteverde et al. (1990) and Piccini et al. (1990). Another study on 40 demented subjects suggested the superiority of selegiline (10 mg/d) over the classical nootropic oxiracetam (1600 mg/d), was mainly on immediate and delayed recall and in tests investigating visuo-spatial abilities (Gibson Spiral Maze) (Falsaperla et al., 1990). Since it is metabolized into two L-amphetaminic derivates, selegiline may then act through this mechanism. A mechanism involving an inhibition of the MAO-A should be also considered since the selectivity of the inhibitor tends to diminish as the concentration increases. However the evidence of a better effect with 10 mg/d rather than with 40 mg/d argues against this possibly. At present, there is no convincing argument against an involvement of the drug itself and the question of cognitive enhancing properties related to the inhibition of MAO-B itself remains open. An answer is expected from future studies using the selective inhibitor Ro 19-6327 the metabolism of which does not lead to the production of amphetaminic derivatives.

MAO-A

The administration of the anti-cholinergic scopolamine is used as a model for memory and cognitive impairment in animals as well as in humans to test the negative or positive effects of psychotropic drugs (Wesnes et al., 1988, 1989; Anand and Wesnes, 1990). Actually, moclobemide is able to antagonize scopolamine-induced amnesia in rats (Martin et al., 1989). In humans, moclobemide significantly reduces (p < 0.01) the amplitude of the scopolamine-induced cognitive impairment as judged by the Global Index. The maximal effect is observed 120 min after the administration of moclobemide (Wesnes et al., 1989). Recently, in young major depressed patients, the cognitive improvement as judged according to Global Memory Scores, Familiar Face Recognition, Critical Flicker Fusion and reaction time has been shown to be significantly superior in subjects treated with moclobemide (300 mg/d) compared to viloxazine (200 mg/d) and maprotiline (100 mg/d) (Allain et al., 1992). In addition, a recent study conducted versus placebo in elderly depressed demented people (more than 700), suggests that moclobemide may improve psychometric parameters in patients fulfilling the criteria for probable Alzheimer's dementia compared to demented subjects of various probable aetiology (Baumhackl et al., 1991; Hebenstreit et al., 1991). Hence the question of a specific effect of moclobemide in Alzheimer type dementia is raised.

Neuroprotective properties of MAOIs in degenerative diseases

In the case of Alzheimer's disease, it may be argued that the benefit observed with selegiline on cognitive abilities may be a symptomatic effect only, and not related to the neuroprotective activity expected from theoretical considerations. However, recent observations in Parkinson patients tend to support the clinical relevance of this neuroprotective property. First, Birkmayer et al. (1985) show that the administration of selegiline in Madopar-treated Parkinsonian patients, can increase their life expectancy. However, more important is the observation that selegiline significantly prolongs the delay between the diagnosis of idiopathic Parkinson's disease and the time when the substitutive dopa therapy is needed. This conclusion is based on the result of a large controlled American study involving more than 800 early Parkinsonian patients in which the association of selegiline (10 mg/d) with and without tocopherol was compared to placebo (DATATOP study, The Parkinson Study Group, 1989a,b). In this study, the time taken to reach the end point (necessity to start dopatherapy) was considered although this criterion remains unvalidated. The interim analysis of this study (which is still in progress), has shown that only 97 subjects who received selegiline reach the end point after a 12 months follow-up, compared to the 176 subjects who did not receive selegiline, the risk of reaching the end point after 12 months being decreased by 57% for the patients who had been treated with selegiline. The comparative performance of the patients after the complete wash-out period will give some further evidence on this cytoprotective hypothesis.

Conclusion

In order to be considered not only as theoretical but as clinically relevant and of therapeutic interest, the neuroprotective potential of MAOIs has to be assessed further. Studies in Parkinson's diseases, using selegiline or more selective and reversible MAO-B inhibitors would be appropriate. A clinical confirmation of the DATATOP study might support a MAO-B-dependent production of free radicals in the pathophysiological mechanism of neuro-degenerative disease, hence justifying its use in the Alzheimer's type dementia. In addition, if the cognitive enhancing properties of MAOIs were established in humans as well as in animals, they would be more widely indicated in the treatment of age-related cognitive impairment. However, the clinical relevance of this effect has to be analysed further and discussed, in regard to the actual global and social improvement MAOIs would provide.

References

Allain H, Lieury A, Brunet-Bourgin F, Mirabaud C, Trebon P, Le Coz F, Gandon J-M (1992) Antidepressants and cognition: comparative effects of moclobemide, viloxazine and maprotiline. Psychopharmacology 106: 56–61

Amrein R, Allen SR, Güntert TW, Hartmann D, Lorscheid T, Schoerlin M-P, Vranesic D (1989) The pharmacology of monoamine oxidase inhibitors. Br J Psychiatry [Suppl] 6: 66–72

Anand R, Wesnes KA (1990) Cognition-enhancing effects of moclobemide, a reversible MAO inhibitor in humans. Adv Neurol 51: 261–268

Baumhackl U, Chan-Palay V, Grüner E, Hebenstreit GF, Kasas A, Katschnig H, Krebs E, Kummer J, Martucci N, Radmayr E, Rieder L, Saletu M, Schlegel S, Lorscheid T (1991) Improvement in cognitive symptoms after treatment with moclobemide in geriatric depressed patients with dementia. Communication at the Congress of the International Psychogeriatric Association, Roma, Italy, August 18–23

Birkmayer W, Knoll J, Riederer P (1985) Increased life expectancy resulting from addition of L-deprenyl to Madopar treatment in Parkinson's disease: a long term study. J Neural Transm [Gen Sect] 64: 113–127

Byrne E (1992) New concept in respiratory chain diseases. Curr Opin Rheumatol 4: 754–793

Campi N, Todeschini GP, Scarzella L (1990) Selegiline versus L-acetylcarnitine in the treatment of Alzheimer type dementia. Clin Ther 12: 306–314

Da Prada M, Kettler R, Burkard WP, Lorez HP, Haefely W (1990a) Some basic aspects of reversible inhibitors of monoamine oxidase-A. Acta Psychiatr Scand [Suppl] 380: 7–12

Da Prada M, Kettler R, Keller HH, Burkhard WP, Muggli-Maniglio D, Haefely W (1988) Neurochemical profile of moclobemide, a short-acting and reversible inhibitor of monoamine oxidase-A. J Pharmacol Exp Ther 248: 400–413

Da Prada M, Kettler R, Keller HH, Cesura AM, Richards JG, Saura M, Muggli-Maniglio D, Wyss P-C, Kyburz E, Imhof R (1990b) From moclobemide to Ro 19-6327 and Ro 41-1049: the development of a new class of reversible, selective MAO-A and MAO-B inhibitors. J Neural Transm [Suppl] 29: 279–292

Falsaperla A, Preti AM, Oliani C (1990) Selegiline versus oxiracetam in patients with Alzheimer-Type Dementia. Clin Ther 12: 376–384

Goad DL, Davis CM, Liem P, Fuselier C, McCormack JR, Olsen KM (1991) The use of selegiline in Alzheimer's patients with behavior problems. J Clin Psychiatry 52: 342–354

Hebenstreit GF, Baumhackl U, Chan-Palay V, Grüner E, Hebenstreit, Kasas A, Katschnig H, Krebs E, Kummer J, Martucci N, Radmayr E, Rieder L, Saletu M, Schlegel S, Lorscheid T (1991) The treatment of depression in geriatric depressed and demented patients by moclobemide: results from the international multicenter double blind placebo controlled trial. Communication at the Congress of the International Psychogeriatric Association, Roma, Italy, August 18–23

Ikebe S, Tanaka M, Ohno K, Sato W, Hattori K, Kondo T, Mizuno Y, Ozawa T, (1990) Increase of deleted mitochondrial DNA in the striatum in Parkinson's disease and senescence. Biochem Biophys Res Commun 170: 1044–1048

Johnston JP (1968) Some observations upon a new inhibitor of monoamine oxidase in brain tissue. Biochem Pharmacol 17: 1285–1287

Klimek V, Nowak G, Zak J, Maj J (1990) The effect of repeated treatments with brofaromine, moclobemide and deprenyl on alpha 1-adrenergic and dopaminergic receptors in the rat brain. Neurosci Lett 108: 189–194

Knoll J (1988) The striatal dopamine dependency on life span in males rats: longevity study with selegiline. Mech Ageing Dev 46: 237–262

Lees AJ, Frankel J, Eatough V, Stern GM (1989) New approaches in the use of selegiline for the treatment of Parkinson's disease. Acta Neurol Scand 126: 139–145

Mann UM, Cooper JM, Krige D, Daniel SE, Schapira AHV, Marsden CD (1992) Brain skeletal muscle and platelet homogenate mitochondrial dysfunction in Parkinson's disease. Brain 115: 333–342

Martin JR, Schaffner R, Remennik L, Vincent GP, Sepinwall J, Lorez HP, Heafely WE (1989) Cognitive performances enhancing effects of the reversible MAO-inhibitor moclobemide in animals. In: Wurtman RJ, Corkin S, Growdon JH, Ritter-Walker E (eds) Alzheimer's disease. Center for Brain Science and Metabolism Charitable Trust, Cambridge/Mass., pp 689–694

McClelland GR, Wesnes K, Jamieson VL, Simpson PM, Christmas L (1988) A double-blind, latin square design study to investigate the effects of single oral doses of 100 and 300 mg moclobemide (Ro 11-1163) compared with placebo and trazodone (100 mg) on psychomotor performance in healthy volunteers, alone and in combination with ethanol (unpublished)

Monteverde A, Gnemmi P, Rossi F, Monteverde A (1990) Selegiline in the treatment of mild to moderate Alzheimer type dementia. Clin Ther 12: 306–314

O'Caroll AM, Bardsley ME, Tipton KF (1986) The oxydation of adrenaline and noradrenaline by the two forms of monoamine oxydase from human and rat brain. Neurochem Int 8: 493–500

Piccini GL, Finali G, Piccirilli M (1990) Neuropsychological effects of L-deprenyl in Alzheimer's dementia. Clin Pharmacol 14: 147–163

Riederer P, Konradi C, Hebenstreit G, Youdim MBH (1989) Neurochemical perspectives to the function of monoamine oxidase. Acta Neurol Scand 126: 41–45

Schoffner JM, Watts RL, Juncos JL, Torroni A, Wallace DC (1991) Mitochondrial oxidative phosphorylation defects in Parkinson's disease. Ann Neurol 30: 332–339

Slater TF (1984) Free-radical mechanisms in tissue injury. Biochem J 222: 1–15

Tariot PN, Cohen R, Sunderland T, Newhouse PA, Yount D, Mellow AM, Weingartner H, Mueller EA, Murphy DL (1987a) L-deprenyl in Alzheimer's disease. Arch Gen Psychiatry 44: 427–433

Tariot PN, Sunderland T, Weingartner H, Murphy DL, Welkowitz JA, Thompson K, Cohen RM (1987b) Cognitive effects of L-deprenyl in Alzheimer's disease. Psychopharmacology 91: 489–495

Tariot PN, Sunderland T, Cohen R, Newhaouse PA, Mueller EA, Murphy DL (1988) Tranylcypromine compared with L-deprenyl in Alzheimer's disease. J Clin Psychopharmacol 8: 23–27

Tetrud JW, Langston JW (1987) R-(−)-deprenyl as a possible protective agent in Parkinson's disease. J Neural Transm [Suppl] 25: 69–79

Tetrud JW, Langston JW (1989) The effect of deprenyl (selegiline) on the natural history of Parkinson's disease. Science 245: 519–522

The Parkinson's Study Group (1989a) Effects of deprenyl on the progression of disability in early Parkinson's disease. N Engl J Med 321: 1363–1371

The Parkinson's Study Group (1989b) DATATOP: a multicenter controlled clinical trial in early Parkinson's disease. Arch Neurol 46: 1052–1060

Trounce I, Byrne E, Marzuki S (1989) Decline in skeletal muscle mitochondrial respiratory chain function: possible factor in ageing. Lancet i:637–639

Wesnes KA, Simpson PM, Christmas L, McClelland GR, Joiner IM (1988) Acute cognitive effects of moclobemide and trazodone, alone and in combination with alcohol, in the elderly. Presentation at British Pharmacological Society, 1988 (unpublished)

Wesnes KA, Simpson PM, Christmas L, Anand R, Mc Clelland GR (1989) The effects of moclobemide on cognition. J Neural Transm [Suppl] 28: 91–102

Youdim MBH (1983) Implication of MAO-A and MAO-B inhibition for antidepressant therapy. Mod Probl Pharmacopsychiatry 19: 63–74

Zornettzer ST (1985) Catecholamine system involvement in age-related memory dysfunction. Ann NY Acad Sci 444: 242–254

Authors' address: Dr. H. Allain, Laboratory of Experimental and Clinical Pharmacology, University Hospital, F-35033 Rennes Cedex, France.

Newer monoamine oxidase inhibitors

J Neural Transm (1994) [Suppl] 41: 269–279

Can our knowledge of monoamine oxidase (MAO) help in the design of better MAO inhibitors?

P. Dostert

Farmitalia Carlo Erba, Research and Development — Erbamont Group, Milan, Italy

Summary. This paper presents a rapid overview of the mechanism by which monoamine oxidase (MAO) catalyzes the deamination of its substrates, and highlights the stereoselective nature of the active site of the enzyme. With the help of a few selected examples it is also discussed which structural factors are thought to have a preponderant influence on the affinity and selectivity of molecules towards the active site of either form of MAO.

From the currently available data on the enzyme and its inhibition, it clearly appears that new MAO inhibitors, of whatever type, could be easily designed by structural modulation of molecules already found to have MAO inhibitory properties. As to whether better MAO inhibitors could be envisaged, it is suggested that MAO inhibition might be advantageously combined with other pharmacological properties for the treatment of pathological conditions, such as stroke and epilepsy, to the occurrence of which MAO activity might contribute. The rationale of this approach is presented.

Introduction

The first generation of inhibitors of the enzyme monoamine oxidase (EC 1.4.3.4) (MAOI) introduced into clinical practice for the treatment of depression in the late 1950s and early 1960s comprised structurally heterogenous compounds. All had in common to be non-selective and irreversible MAOI. The initial period of enthusiasm for the use of MAOIs in psychiatry waned rapidly after reports of cases of hepatotoxicity and severe hypertensive crises in patients treated with these drugs. While the hepatotoxic effects were mostly seen in patients treated with hydrazinic MAOIs, the hypertensive response to the ingestion of tyramine, the so-called cheese effect, was common to all MAOIs as a consequence of their mode of action, although differences between drugs in the occurrence and severity of hypertensive crises were noted (for review see Dostert, 1984).

An important step occurred when the selective inhibition of MAO by clorgyline (Johnston, 1968) and deprenyl (Knoll and Magyar, 1972) demonstrated the presence of two forms of enzyme. The clorgyline-sensitive form was termed MAO-A and the deprenyl-sensitive form, MAO-B. The recent resurgence of interest in MAOIs is mainly based on two subsequent events: 1) the occurrence of a new generation of reversible MAO-A inhibitors, such as toloxatone, moclobemide, and brofaromine, which appear to retain the antidepressant properties of the former MAOIs without causing substantial tyramine potentiation (for review see Tipton et al., 1984; Youdim et al., 1988; Burrows and Da Prada, 1989), 2) reports that the selective and irreversible MAO-B inhibitor selegiline (l-deprenyl) can be used as an adjuvant when combined with L-dopa therapy (Chrisp et al., 1991), and, moreover, may delay the onset of disability in otherwise untreated parkinsonian patients (Lewitt and The Parkinson Study Group, 1991).

Although the occurrence of a host of new reversible and irreversible MAOIs in the last fifteen years has largely favoured a better understanding of the mechanism of action of the enzyme and of its mode of interaction with inhibitors, the improvement of the knowledge of MAO has been used only occasionally for the design of tailor-made inhibitors. The aim of the present paper is to briefly review how MAO is thought to interact with substrates and reversible MAOIs, to examine which are the factors influencing selectivity for either form of the enzyme and whether "better" inhibitors of MAO could be designed in the light of the accumulated knowledge.

Mechanistic and stereochemical aspects of the interaction of substrates and inhibitors with monoamine oxidase

MAO is a flavoenzyme in which one FAD is covalently bound to each subunit of the apoprotein (Weyler et al., 1990). The mechanism by which amines are oxidatively deaminated by MAO remains poorly understood. To date, although Silverman and colleagues have accumulated supportive evidence for the involvement of one-electron mechanisms in MAO-catalyzed amine oxidation (Silverman et al., 1980; Silverman, 1991; for review see Dostert et al., 1989; Weyler et al., 1990), the direct demonstration for a radical on the catalytic reaction coordinate of MAO is still missing. Formally the reaction catalyzed by MAO consists of two steps (Fig. 1). In the first step the amine substrate is oxidized and the FAD is reduced. After hydrolysis of the intermediate imine the final product aldehyde and $FADH_2$ are formed. In the second step $FADH_2$ is reoxidized by molecular oxygen with the formation of H_2O_2.

There is evidence that abstraction of an α-proton from the amines is the rate-limiting step of the oxidative deamination of most substrates by MAO. Using dopamine or benzylamine as substrate, Yu et al. (1986, 1988) have established that both MAO-A and MAO-B exhibit the same stereospecificity, that is, the pro-R-hydrogen (H_{Re}, Fig. 1) is abstracted by both forms of MAO.

Substitution of one α-hydrogen atom in MAO substrates, such as β-phenylethylamine, benzylamine and milacemide, by an alkyl group trans-

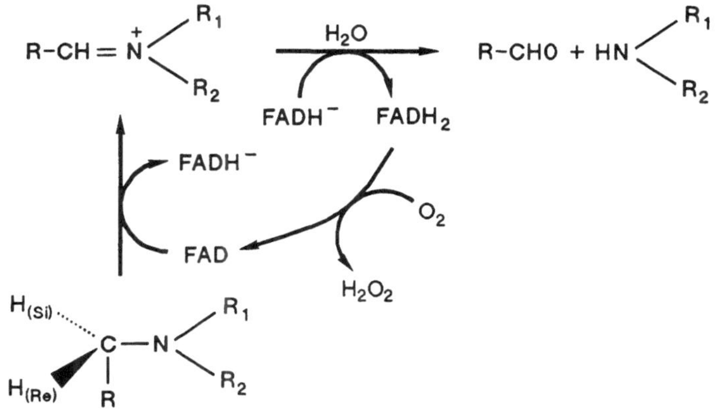

Fig. 1. Proposed mechanism for MAO-catalyzed amine oxidation

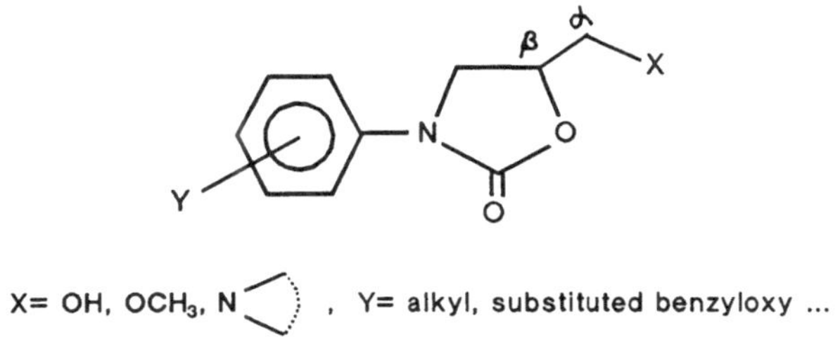

X= OH, OCH₃, N⟨⟩ , Y= alkyl, substituted benzyloxy ...

Fig. 2. General structure of oxazolidinone MAO inhibitors

forms these substrates into reversible and competitive inhibitors (Tipton et al., 1983; O'Brien et al., 1991; for review see Dostert et al., 1989). Introduction of an α-methyl residue into the selective MAO-B substrate β-phenylethylamine affords amphetamine. Using various MAO preparations, the S enantiomer of amphetamine, in which the methyl group occupies the position of the non-abstracted proton (H_{si}, Fig. 1) in MAO substrates, was repeatedly found to be a more potent MAO-A inhibitor than its R counterpart, both enantiomers having the same inhibitory potency toward MAO-B (see references quoted in Dostert et al., 1989). Conversely, Arai et al. (1986) using rat brain homogenates reported that the R enantiomer of α-methylbenzylamine, whose methyl group occupies the position of the abstracted proton (H_{Re}, Fig. 1) in benzylamine, is a more potent MAO-A and MAO-B inhibitor than the S form, in, at least partial, contradiction with Silverman (1984) who, using MAO-B purified from bovine liver, found very similar K_i values for both enantiomers of α-methylbenzylamine. Supportive evidence for the stereoselective nature of the active site of MAO-A and MAO-B has also been obtained in the series of oxazolidinone MAO inhibitors, whose general structure is given in Fig. 2 (for review see Dostert and Strolin Benedetti, 1991). Disregarding the influence of substituents X and Y (Fig. 2) on the selectivity of these inhibitors for either form of MAO, substitution of the pro-S-hydrogen in position α was shown to increase the affinity for both forms of the enzyme, whereas MAO-B might be less sensitive than MAO-A to the spatial disposition of the substituents in the β position.

Factors influencing the selective recognition of molecules by the active site of MAO-A or MAO-B

The affinity of MAO for reversible inhibitors is often higher than it is for substrates. Therefore, reversible MAOIs are more suitable than substrates for studying the relationship between the chemical structure of molecules and their recognition by the active site of MAO. With the help of a few selected examples, here is discussed which structural elements may play a determinant role in the selectivity of inhibitors for one of the MAO forms, in an attempt to illustrate the extent to which selectivity is sometimes dramatically changed as a result of very limited structural modifications.

Basically, the catalytic site of MAO is thought to consist of two distinct parts: a binding site for the amine group and a hydrophobic binding region. The 2-aminoethylcarboxamide derivatives Ro 19-6327 and Ro 41-1049 (Fig. 3) are potent and highly selective inhibitors of MAO-B and MAO-A, respectively (Da Prada et al., 1990). As shown by Cesura et al. (1989, 1990) both molecules bind exclusively to the active site of the enzyme and behave as mechanism-based reversible MAOIs. The common 2-aminoethylcarboxamide residue seems to react with a domain on MAO, which is likely to be very similar in the two forms and, most probably, not essential for substrate and inhibitor selectivity. The difference in selectivity between Ro 19-6327 and Ro 41-1049 may result from a constitutive difference between the MAO forms in the nature of the hydrophobic binding domain and/or its environment.

The influence of small structural modifications on the selectivity of reversible MAOIs has extensively been examined in oxazolidinone series (Dostert and Strolin Benedetti, 1986). For example, substitution in position meta of the phenyl ring of the benzyloxy residue present in most oxazolidinone MAOIs by a CN group or a C1 atom does substantially affect the selective recognition of otherwise identical molecules by one or the other

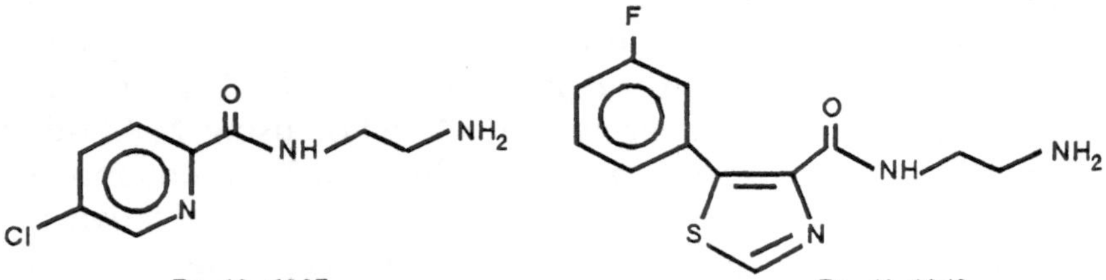

Ro 19–6327   **Ro 41–1049**

Fig. 3. Chemical structures of the selective MAO-B (Ro 10-6327) and MAO-A (Ro 41-1049) inhibitors

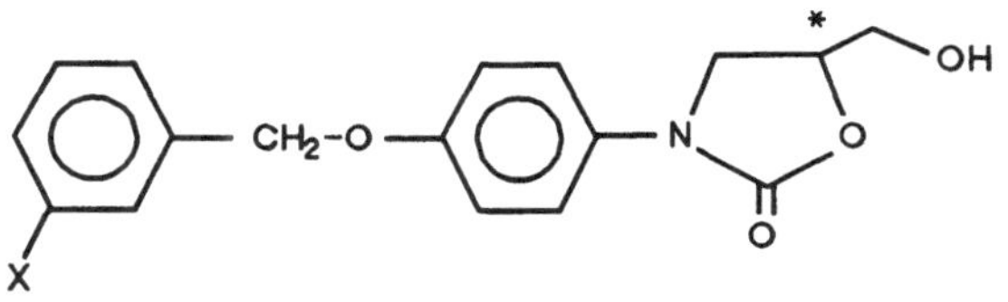

Fig. 4. *indicates the position of chiral centre

Table 1. Influence of aromatic substituents on the selective inhibition of monoamine oxidase (MAO) by oxazolidinone MAO inhibitors

X	Abs. config.	Ki (μM)		Ki (B) Ki (A)	Ki (A) Ki (B)
		MAO-A	MAO-B		
CN	R	0.006	0.36	60	
	S	0.14	0.93		0.15
Cl	R	0.022	0.013	0.6	
	S	0.66	0.027		24.4

form of MAO (Fig. 4, Table 1). Comparison of the R enantiomers shows that, while the affinity of the CN-derivative for MAO-A is about four times higher than that of the Cl-derivative, the selectivity of the former compound for that form of the enzyme is hundred times higher (60:0.6, Table 1). Conversely, regarding the S enantiomers the selectivity of the Cl-derivative for MAO-B is about 160 times higher than that of the CN-derivative (24.4:0.15). This example, together with that given previously in 2-aminoethylcarboxamide series, suggests that the selectivity of a MAOI for one or the other form of the enzyme preponderantly depends on the lipophilic part of the molecule and, to some degree, on its physicochemical characteristics, is sensitive to very subtle influences, such as distribution of the electronic charges, and, to date, can hardly be predicted.

Could new monoamine oxidase inhibitors be designed and is there a need for them?

Apart from a few exceptions concerning irreversible MAOIs, such as the series of fluoroallylamine derivatives developed by Palfreyman et al. (1986), most currently available MAOIs result from the optimization of a parent compound, whose MAO inhibitory properties were found out by serendipity. The number of natural or synthetic molecules with affinity for MAO in the micromolar range or lower is very impressive, as is also the heterogenicity of their chemical structures (for review see Zeller, 1971; Ho, 1972; Berger and Barchas, 1977; Fowler and Ross, 1984; Strolin Benedetti and

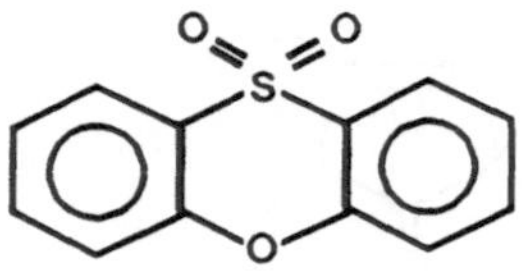

BW 1370U87

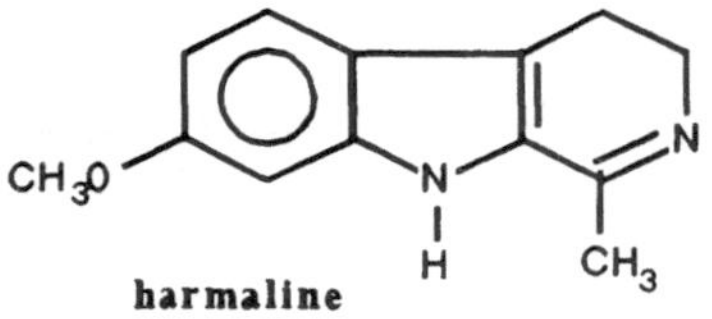

harmaline

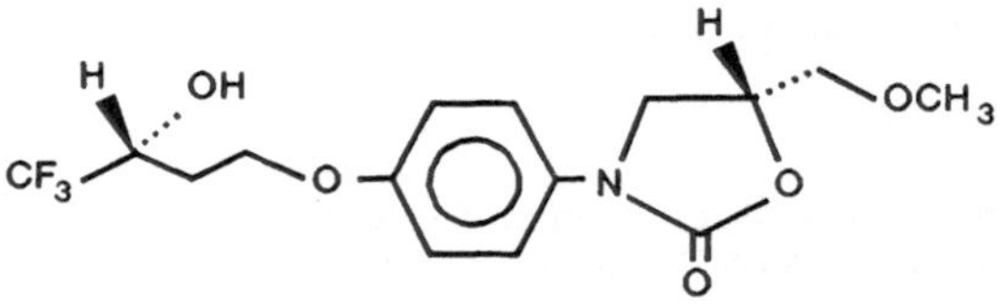

befloxatone

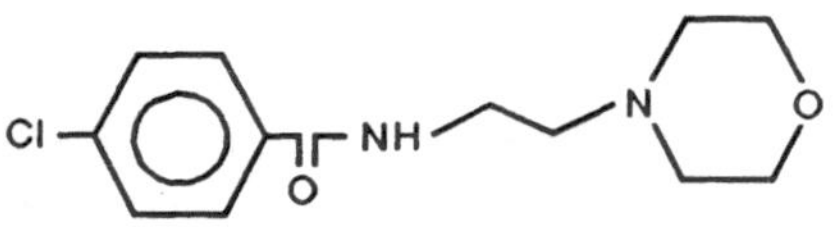

moclobemide

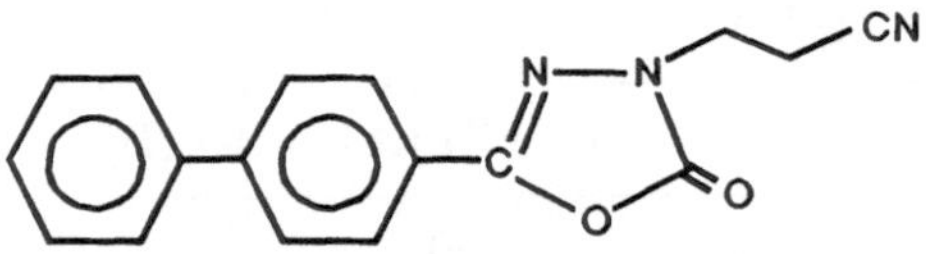

oxadiazolone derivative

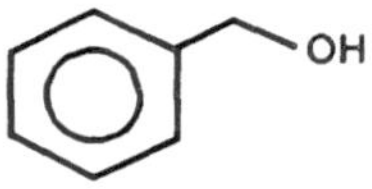

benzylalcohol

Fig. 5. Structures of some selective and reversible MAO inhibitors

Dostert, 1985, 1992; Dostert et al., 1989). For example, comparison of the structures (Fig. 5) of the competitive MAO-A inhibitors BW 1370U87 (White and Scates, 1992), harmaline (Fuller, 1968), befloxatone (Curet et al., 1992), and moclobemide (Cesura et al., 1992), and of the competitive MAO-B inhibitors Ro 19-6327 (Fig. 3), benzylalcohol (Fowler et al., 1980) and 5- (4-biphenyl)-3-(2-cyanoethyl)-1,3,4-oxadiazol-2(3H)-one (Mazouz et al., 1990), clearly indicates that a relationship between the structural features of a given molecule and its selective recognition by either form of MAO cannot a priori be established. As a fact of experience, however, it can reasonably be assumed that the affinity of any molecule casually found to interact with MAO, as also the selectivity for one of the MAO forms, can markedly be improved by chemical modulation of the basic structure. Thus, in addition to the above-mentioned examples in 2-aminoethylcarboxamide and oxazolidinone series, substitution of a methyl group for a hydrogen atom on the propargyl methylene carbon of the mechanism-based MAO inhibitor clorgyline was recently shown to markedly decrease the affinity for MAO-A, with a selective and competitive inhibition of MAO-B being found for the S enantiomer (Dostert et al., 1990).

The accumulated experience on MAO and its inhibition could undoubtedly be used for the design of new inhibitors. However, one can rightly ask whether new MAOIs are really needed, and whether the main question at the present time should not be to address the possible role of MAO in pathology and the potential benefit of MAOI treatment in the various diseases where an increase in MAO activity has been shown, or can be suspected, to be involved (for review see Strolin Benedetti and Dostert, 1992). Thus, to the question "could new MAOIs be designed from the accumulated experience on MAO and its inhibition", the answer is: yes, but no new molecules endowed with biochemical characteristics similar to those of currently available MAO-A or -B inhibitors are currently needed.

What "better" monoamine oxidase inhibitors might be

Surprisingly, in the last 10–15 years the search for new MAOIs to be used for the treatment of depression aimed at finding reversible and selective MAO-A inhibitors. While MAO-A inhibitors have undoubtedly to be reversible in nature to avoid, or at least to limit, the occurrence and severity of cheese effect, there is no evidence that MAOI antidepressants should be selective. Since both forms of MAO have been shown to be involved in the deamination of dopamine and noradrenaline in the human brain (O'Carroll et al., 1987), a molecule associating the ability to reversibly inhibit MAO-A with reversible or irreversible inhibitory properties towards MAO-B, the predominant form of the enzyme in the human brain, might prove to be more effective than selective MAO-A inhibitors for the treatment of depression. Another possible improvement of reversible MAOI for the treatment of depression would be to associate the inhibition of MAO with the ability to block the high-affinity adrenergic uptake system by which tyramine enters the nerve endings. Such a molecule would be of great interest as a new antidepressant with no, or only minimal, cheese effect, provided that the time-course of both activities is similar. For example, bifemelane [4-(o.benzylphenoxy)-N-methylbutylamine) is one of the few molecules in which both activities, reversible inhibition of MAO-A (Naoi et al., 1988) and inhibition of noradrenaline uptake (Egawa et al., 1983), although weak ($IC_{50} = 10^{-6}-10^{-7}$ M), coexist. It remains to be determined whether the structural modifications delete the simultaneous enhancement of both activities would require, are compatible.

There is convincing evidence that hydrogen peroxide generated by MAO contributes to the generation of reactive oxygen species in brain (Sinet et al., 1980; Maker et al., 1981; Seregi et al., 1982; Strolin Benedetti and Dostert, 1989). Under normal conditions, the oxidative stress generated by MAO appears to be tolerated (Cohen, 1990). However, in some pathological conditions, such as brain ischemia, an increase in oxidative stress has been associated with the release of catecholamines. Pretreatment with MAO-A inhibitors has been shown to partially prevent the consequences of transient forebrain ischemia in the rat brain (Damsma et al., 1990; Lorez et al., 1990; Kumagae and Iwata, 1990; Matsui and Kumagae, 1991), and pargyline was recently shown to decrease the sensitivity of the central nervous system to oxygen toxicity (Zhang and Piantadosi, 1991). Therefore, it appears reasonable to suggest that a molecule in which glutamate-receptor antagonism and MAO-B inhibitory properties would be combined, may be of potential interest for the treatment of stroke.

Increased monoamine neurotransmitter levels have been found in epileptic foci of the human brain (Goldstein et al., 1988; Pintor et al., 1990), and a significant higher degree of selegiline binding was recently observed in the hippocampus from patients with intractable complex partial epilepsy (Kumlien et al., 1992). In posttraumatic epilepsy, it has been suggested that treatment designed to prevent oxidative stress may be more effective than the administration of anticonvulsant drugs that mask convulsive seizures while biochemical injuries continue (Willmore and Triggs, 1984). This assumption might also be valid in other forms of epilepsy. Therefore, although changes in the catalytic properties of rat brain MAO have been reported during audiogenic epilepticform attack, possibly as a result of irreversible modifications of the enzyme induced by aldehydes formed during the oxidative deamination of MAO substrates (Medvedev and Gorkin, 1992), a drug associating anticonvulsant activity with potent MAO-B inhibitory properties appears to be of potential interest. Such molecules have recently been reported (Dostert et al., 1991). It remains to be examined whether the increase in oxidative stress that has been shown to accompany epilepticform activity (Sobaniec et al., 1989) might be, at least partially, reduced by brain MAO-B inhibition and subsequent decrease in hydrogen peroxide formation. Such a study would require the use of a species in which the composition of brain MAO is similar to that of the human brain.

References

Arai Y, Toyoshima Y, Kinemuchi H (1986) Studies of monoamine oxidase and semi-carbazide-sensitive amine oxidase. II. Inhibition by α-methylated substrate-analogue monoamines, α-methyltryptamine, α-methylbenzylamine and two enantiomers of α-methylbenzylamine. Jpn J Pharmacol 41: 191–197

Berger PA, Barchas JD (1977) Monoamine oxidase inhibitors. Psychopharmacol Ser 2: 1173–1216

Burrows GD, Da Prada M (1989) Reversible MAO-A inhibitors as antidepressants. J Neural Transm [Suppl] 28

Cesura AM, Galva MD, Imhof R, Kyburz E, Picotti GB, Da Prada M (1989) [^{3}H]Ro 19-6327: a reversible ligand and affinity labelling probe for monoamine oxidase-B. Eur J Pharmacol 162: 457–465

Cesura AM, Bös M, Galva MD, Imhof R, Da Prada M (1990) Characterization of the binding of [^{3}H]Ro 41-1049 to the active site of human monoamine oxidase-A. Mol Pharmacol 37: 358–366

Cesura AM, Kettler R, Imhof R, Da Prada M (1992) Mode of action and characteristics of monoamine oxidase-A inhibition by moclobemide. Psychopharmacology 106: S15–S16

Chrisp P, Mammen GJ, Sorkin EM (1991) Selegiline. A review of its pharmacology, symptomatic benefits and protective potential in Parkinson's disease. Drugs Aging 1: 228–248

Cohen G (1990) Monoamine oxidase and oxidative stress at dopaminergic synapses. J Neural Transm [Suppl] 32: 229–238

Curet O, Damoiseau G, Aubin N (1992) Biochemical profile of befloxatone, a new reversible MAO-A inhibitor. Clin Neuropharmacol 15 [Suppl 1]: 428B

Damsma G, Boisvert DP, Mudrick LA, Wenkstern D, Fibiger HC (1990) Effects of transient forebrain ischemia and pargyline on extracellular concentrations of dopamine, serotonin, and their metabolites in the rat striatum as determined by in vivo microdialysis. J Neurochem 54: 801–808

Da Prada M, Kettler R, Keller HH, Cesura AM, Richards JG, Saura Marti J, Muggli-Maniglio D, Wyss P-C, Kyburz E, Imhof R (1990) From moclobemide to Ro 19-6327 and Ro 41-1049: the development of a new class of reversible, selective MAO-A and MAO-B inhibitors. J Neural Transm [Suppl] 29: 279–292

Dostert P (1984) Myth and reality of the classical MAO inhibitors. Reasons for seeking a new generation. In: Tipton KF, Dostert P, Strolin Benedetti M (eds) Monoamine oxidase and disease. Prospects for therapy with reversible inhibitors. Academic Press, London, pp 9–24

Dostert P, Strolin Benedetti M (1986) Nouveaux inhibiteurs de la monoamine oxidase. Actal Chim Thér 13: 269–287

Dostert P, Strolin Benedetti M, Tipton KF (1989) Interactions of monoamine oxidase with substrates and inhibitors. Med Res Rev 9: 45–89

Dostert P, Strolin Benedetti M (1991) Structure-modulated recognition of substrates and inhibitors by monoamine oxidases A and B. Biochem Soc Trans 19: 207–211

Dostert P, Strolin Benedetti M, Tipton KF (1991) New anticonvulsants with selective MAO-B inhibitory activity. Eur Neuropsychopharmacol 1: 317–319

Dostert P, O'Brien EM, Tipton KF, Meroni M, Melloni P, Strolin Benedetti M (1992) Inhibition of monoamine oxidase by the R and S enantiomers of N[3-(2,4-dichloro-phenoxy)propyl]-N-methyl-3-butyn-2-amine. Eur J Med Chem 27: 45–52

Egawa M, Inokuchi T, Ida S, Tobe A (1983) Effects of 4-(o-benzylphenoxy)-N-methyl-butylamine hydrochloride (MCI-2016) on monoamine metabolism in the brain. Nippon Yakurigaku Zasshi 82: 351–360

Fowler CJ, Callingham BA, Mantle TJ, Tipton KF (1980) The effect of lipophilic compounds upon the activity of rat liver mitochondrial monoamine oxidase-A and -B. Biochem Pharmacol 29: 1177–1183

Fowler CJ, Ross SB (1984) Selective inhibitors of monoamine oxidase A and B: biochemical, pharmacological, and clinical properties. Med Res Rev 4: 323–358

Fuller RW (1968) Influence of substrate in the inhibition of rat liver and brain monoamine oxidase. Arch Int Pharmacodyn Ther 174: 32–36

Goldstein DS, Nadi NS, Stull R, Wyler AR, Porter RJ (1988) Levels of catechols in epileptogenic and nonepileptogenic regions of the human brain. J Neurochem 50: 225–229

Ho BT (1972) Monoamine oxidase inhibitors. J Pharm Sci 61: 821–837

Johnston JP (1968) Some observations upon a new inhibitor of monoamine oxidase in brain tissues. Biochem Pharmacol 17: 1285–1297

Knoll J, Magyar K (1972) Some puzzling pharmacological effects of monoamine oxidase inhibitors. Adv Biochem Psychopharmacol 5: 393–408

Kumagae Y, Matsui Y, Iwata N (1990) Participation of type A monoamine oxidase in the activated deamination of brain monoamines shortly after reperfusion in rats. Jpn J Pharmacol 54: 407–413

Kumlien E, Hilton-Brown P, Spännare B, Gillberg P-G (1992) In vitro quantitative autoradiography of [^{3}H]-1-deprenyl and [^{3}H]-PK 11195 binding sites in human epileptic hippocampus. Epilepsia 33: 610–617

LeWitt PA, The Parkinson Study Group (1991) Deprenyl's effect at slowing progression of parkinsonian disability: the DATATOP study. Acta Neurol Scand 84 [Suppl 136]: 79–86

Lorez HP, Harvey J, Wright L, Kollar S, Blaszat G, Thomas B, Martin JR, Kettler R, Da Prada M (1990) Moclobemide exhibits neuroprotective effects in hypoxic rat brain. In: Krieglstein J, Oberpichler H (eds) Pharmacology of cerebral ischemia. Wissenschaftliche Verlagsgesellschaft, Stuttgart, pp 477–484

Maker HS, Weiss C, Silides DJ, Cohen G (1981) Coupling of dopamine oxidation (monoamine oxidase activity) to glutathione oxidation via the generation of hydrogen peroxide in rat brain homogenates. J Neurochem 36: 589–593

Matsui Y, Kumagae Y (1991) Monoamine oxidase inhibitors prevent striatal neuronal necrosis induced by transient forebrain ischemia. Neurosci Lett 126: 175–178

Mazouz F, Lebreton L, Milcent R, Burstein C (1990) 5-Aryl-1,3,4-oxadiazol-2(3H)-one derivatives and sulfur analogues as new selective and competitive monoamine oxidase type B inhibitors. Eur J Med Chem 25: 659–671

Medvedev AE, Gorkin VZ (1992) Biogenic amines and monoamine oxidases in the regulation of activities of membrane-bound mitochondrial enzymes. Bio Amines 8: 323–337

O'Brien EM, Tipton KF, Strolin Benedetti M, Bonsignori A, Marrari P, Dostert P (1991) Is the oxidation of milacemide by monoamine oxidase a major factor in its anticonvulsant actions? Biochem Pharmacol 41: 1731–1737

O'Carroll A-M, Tipton KF, Sullivan JP, Fowler CJ, Ross SB (1987) Intra- and extras-ynaptosomal deamination of dopamine and noradrenaline by the two forms of human brain monoamine oxidase. Implications for the neurotoxicity of N-methyl-4-phenyl-1,2,3,6-tetrahydropyridine in man. Bio Amines 4: 165–178

Palfreyman MG, Mcdonald IA, Bey P, Danzin C, Zreika M, Lyles GA, Fozard JR (1986) The rational design of suicide substrates of amine oxidases. Biochem Soc Trans 14: 410–413

Pintor M, Mefford IN, Hutter I, Pocotte SL, Wyler AR, Nadi NS (1990) Levels of biogenic amines, their metabolites, and tyrosine hydroxylase activity in the human epileptic temporal cortex. Synapse 5: 152–156

Raigorodskaia DI, Medvedev AE, Gorkin VZ, Fedotova IB, Semiokhina AF (1991) Change in the catalytic properties of mitochondrial monoamine oxidase in experi-mental audiogenic epilepsy. Vopr Med Khim 37: 46–48

Seregi A, Serfözö P, Mergl Z, Schaefer A (1982) On the mechanism of the involvement of monoamine oxidase in catecholamin-estimulated prostaglandin biosynthesis in particulate fraction of rat brain homogenates: role of hydrogen peroxide. J Neurochem 38: 20–27

Sinet PM, Heikkila RE, Cohen G (1980) Hydrogen peroxide production by rat brain in vivo. J Neurochem 34: 1421–1428

Silverman RB (1984) Effect of α-methylation on inactivation of monoamine oxidase by N-cyclopropylbenzylamine. Biochemistry 23: 5206–5213

Silverman RB (1991) The use of mechanism-based inactivators to probe the mechanism of monoamine oxidase. Biochem Soc Trans 19: 201–206

Silverman RB, Hoffman SF, Catus III WB (1980) A mechanism for mitochondrial monoamine oxidase catalyzed amine oxidation. J Am Chem Soc 102: 7126–7128

Sobaniec W, Rudzinski P, Jankowicz E, Sobaniec-Lotowska M, Kulak W (1989) Cardiazol-induced seizures and the concentration of lipid peroxides in the brain of rats under the influence of valproic acid and vitamin E. Neuropatol Pol 27: 129–136

Strolin Benedetti M, Dostert P (1985) Stereochemical aspects of MAO interactions: reversible and selective inhibitors of monoamine oxidase. Trends Pharmacol Sci 6: 246–251

Strolin Benedetti M, Dostert P (1989) Effect of selective monoamine oxidase substrates and inhibitors on lipid peroxidation and their possible involvement in affective disorders. In: Lerer B, Gershon S (eds) New directions in affective disorders. Springer, Berlin Heidelberg New York Tokyo, pp 156–160

Strolin Benedetti M, Dostert P (1992) Monoamine oxidase: from physiology and pathophysiology to the design and clinical application of reversible inhibitors. Adv Drug Res 23: 65–125

Tipton KF, O'Carroll A-M, Mantle TJ, Fowler CJ (1983) Factors involved in the selective inhibition of monoamine oxidase. Mod Probl Pharmacopsychiatry 19: 15–30

Tipton KF, Dostert P, Strolin Benedetti M (1984) Monoamine oxidase and disease. Prospects for therapy with reversible inhibitors. Academic Press, London

Weyler W, Hsu Y-P P, Breakefield XO (1990) Biochemistry and genetics of monoamine oxidase. Pharmacol Ther 47: 391–417

White HL, Scates PW (1992) Mechanism of monoamine oxidase-A inhibition by BW 1370U87. Drug Dev Res 25: 191–199

Willmore LJ, Triggs WJ (1984) Effect of phenytoin and corticosteroids on seizures and lipid peroxidation in experimental posttraumatic epilepsy. J Neurosurg 60: 467–472

Youdim MBH, Da Prada M, Amrein R (1988) The cheese effect and new reversible MAO-A inhibitors. J Neural Transm [Suppl] 26

Yu PH, Bailey BA, Durden DA, Boulton AA (1986) Stereospecific deuterium substitution at the α-carbon position of dopamine and its effect on oxidative deamination catalyzed by MAO-A and MAO-B from different tissues. Biochem Pharmacol 35: 1027–1036

Yu PH, Davis BA (1988) Stereospecific deamination of benzylamine catalyzed by different amine oxidases. Int J Biochem 20: 1197–1201

Zeller EA (1971) Amine oxidases. In: Brodie BB, Gillette JR, Ackerman HS (eds) Handbook of experimental pharmacology, vol 28. Springer, Berlin Heidelberg New York, pp 518–535

Zhang J, Piantadosi CA (1991) Prevention of H_2O_2 generation by monoamine oxidase protects against CNS O_2 toxicity. J Appl Physiol 71: 1057–1061

Author's address: Dr. P. Dostert, Famitalia Carlo Erba-Research and Development, Via C. Imbonati 24, I-20159 Milan, Italy.

J Neural Transm (1994) [Suppl] 41: 281–285

Kinetic behaviour of some acetylenic indolalkylamine derivatives and their corresponding parent amines

D. Balsa[1], **V. Pérez**[1], **E. Fernández-Alvarez**[2], and **M. Unzeta**[1]

[1] Department de Bioquímica i Biologia Molecular, Facultat de Medicina, Universitat Autònoma de Barcelona, Barcelona, and
[2] Department de Química Orgánica, Consejo Superior de Investigaciones Científicas, Madrid, Spain

Summary. Different methoxy indolalkylamines based on a common structure, and differing in the side chain attached at the 2 position of the indole ring were studied as MAO A inhibitors. Some are acetylenic derivatives and consequently might behave as "suicide" MAO inhibitors (FA 42, FA 43, FA 45). The rest of compounds are the corresponding parent amines and they might behave as MAO substrates (FA 51, FA 52, FA 53, FA 54).

The kinetic behaviour of the parent amines as MAO A and MAO B inhibitors and substrates was determined. In case of acetylenic derivatives, kinetic constants defining the non-covalent adduct formation and the covalent adduct formation were also calculated for the mechanism-based inhibition.

These parameters will allow us to establish the correlation with structural features that predetermine one compound to be a good MAO substrate or a good MAO A and MAO B inhibitor.

Introduction

N-acetylenic derivatives of indolalkylamines in which the side chain is attached at the 2-position of the indole ring can be potent and selective MAO A inhibitors (Cruces et al., 1988). The aim of this work is to investigate the kinetic behaviour of these acetylenic derivatives as MAO A and MAO B inhibitors and the kinetic behaviour of their corresponding parent amines, in order to find out the structural features responsible of the conversion of a MAO substrate to good MAO A and MAO B inhibitors.

Materials and methods

The radiochemical procedures previously described (Fowler et al., 1980), using $100 \,\mu M$ 5-hydroxytryptamine (5HT) and $22.2 \,\mu M$ of phenylethylamine (PEA), as specific MAO A and MAO B substrates were used. The IC50 values were determined after 0 and 30 min preincubation of the mitochondrial MAO with each compound in the concen-

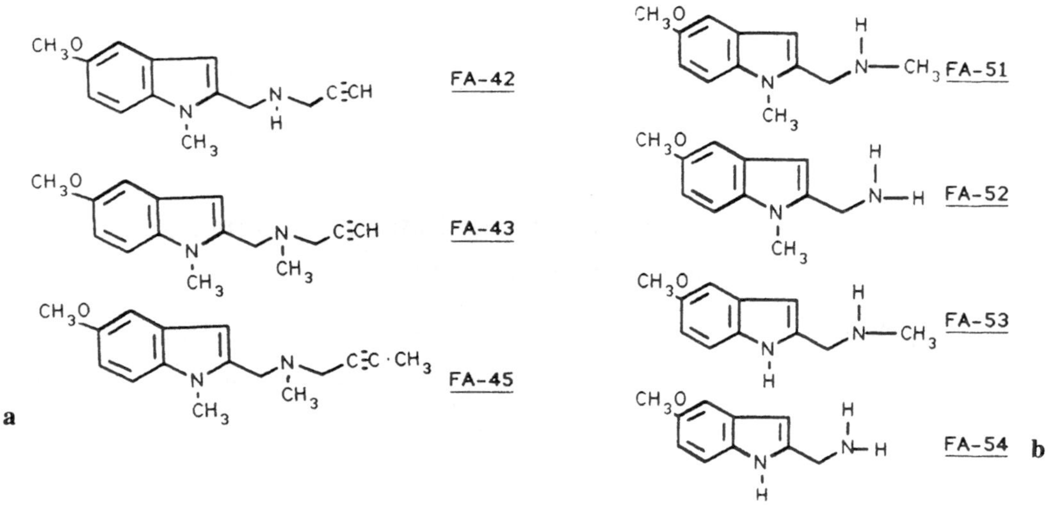

Fig. 1. Chemical structures of acetylenic derivatives of the indolealkylamines (a) and their parent amines (b)

tration range of 10^{-2}–10^{-10} M in 50 mM potassium phosphate buffer pH 7.2 at 37 C. The parent amines were assayed as MAO substrates by the aldehyde deshydrogenase coupled assay (Houslay et al., 1973), their corresponding kinetic parameters as inhibitors were determined by Lineweaver-Burk plots. The time dependence of the inhibition process was determined at inhibitor concentration, that rendered total loss of activity in the above studies after different times of enzyme-inhibitor preincubation. Reversibility of the inhibition was determined by the repeated washing and spin method (Waldemeir et al., 1983). The kinetic parameters for the mechanism-based inhibition were determined spectrophotometrically by direct analysis of reaction progress curves (Walker and Elmore et al., 1984) in the presence of varying inhibitor concentration. The method used was a modification of one previously described (Balsa et al., 1991), in which MAO A activity (400–500 μg) was assayed spectrophotometrically using kynuramine (40 μM) as substrate in 50 mM potassium phosphate buffer pH 7.2, following the increase of absorbance at 324 nm (extintion coefficient 20,000 $M^{-1}cm^{-1}$).

The methoxy derivatives of 2-[N-(2-propynyl)-aminomethyl]-1-methyl indole were synthesized by the procedure of Cruces et al. (1988).

Results and discussion

Figure 1 shows the chemical structure of the 5-methoxy derivatives of 2-[N-(2-propynyl)-aminomethyl]-1-methyl indole (FA 42, FA 43, FA 45) and their corresponding parent amines (FA 51, FA 52, FA 53, FA 54).

The inhibition of both forms of MAO by acetylenic derivatives was time-dependent, irreversible and showed a mechanism-based inhibition. The Ki values defining the affinity for the non-covalent complex formation were similar in all cases for MAO A (Fig. 2). By contrast, the kcat value corresponding to the covalent adduct formation was small in case of FA 43 compared with FA 42 and FA 45. A hydrogen attached at the N of the alkyl side chain (FA 42) has been substituted by a methyl group (FA 43) and this

MOLECULE	COMP.	MAO	Ki	kcat	kcat/Ki
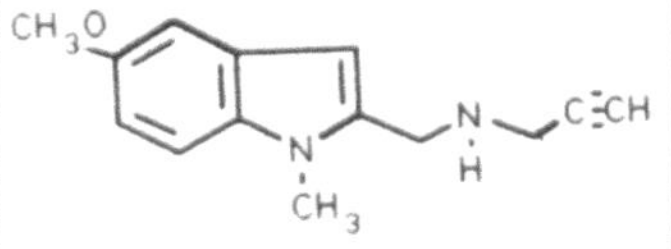 (structure)	FA-42	A	0.8	0.0062	0.007
		B	4.0	0.7	0.175
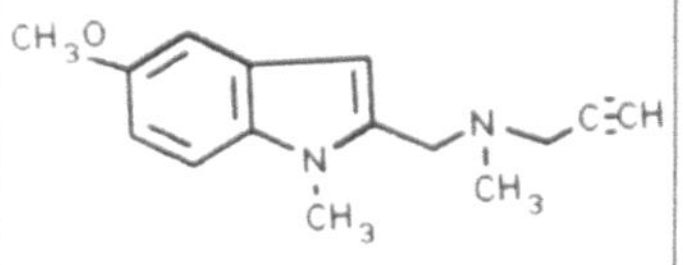 (structure)	FA-43	A	1.5	0.057	0.038
		B	30.0	0.059	0.002
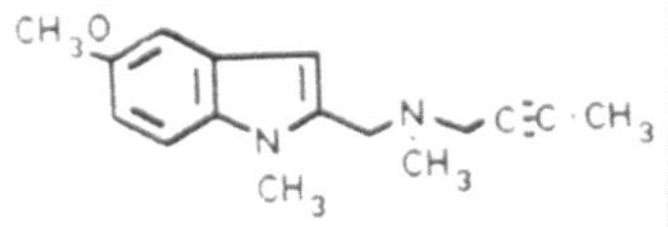 (structure)	FA-45	A	1.4	0.052	0.037
		B	370.0	0.257	0.0007

Fig. 2. Inhibition kinetic parameters of MAO A and MAO B by the acetylenic derivatives of the indolealkylamines. Ki values were expressed in nmolar and Kcat values were expressed in min^{-1}

MOLECULE	COMP.	MAO	S.Act	Km	Vmax
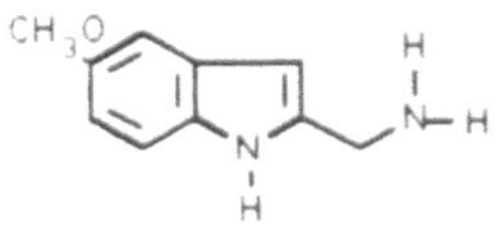 (structure)	FA-54	A	ND	ND	ND
		B	ND	ND	ND
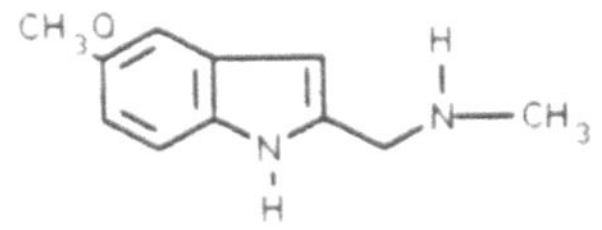 (structure)	FA-53	A	0.15	>10	ND
		B	0.40	>10	ND
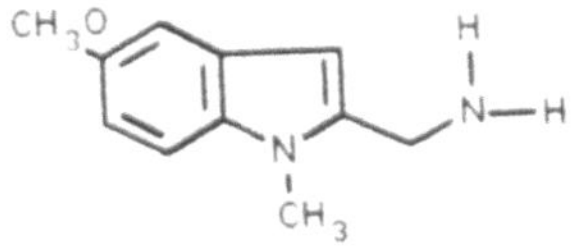 (structure)	FA-52	A	0.40	>10	ND
		B	0.64	1.74	4.91
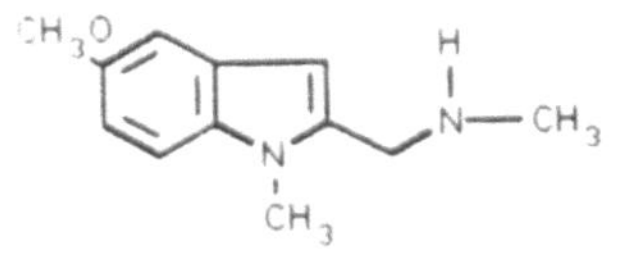 (structure)	FA-51	A	ND	ND	ND
		B	1.12	0.54	3.78

Fig. 3. Kinetic parameters of MAO A and MAO B towards the parent amines of the acetylenic derivatives as substrates. Enzymatic activities were determined by aldehyde deshydrogenase coupled assay (Houslay and Tipton, 1973). Specific activities (S. Act.) and Vmax were expressed in nmol/min mg. Km values were expressed in mM

MOLECULE	COMP.	MAO	Type of inhibition	Ki (μM)
	FA-54	A	Competitive	53.6
		B	Competitive	1410.0
	FA-53	A	Competitive	10.4
		B	Competitive	3190.0
	FA-52	A	Competitive	4.8
		B	Competitive	4.6
	FA-51	A	Competitive	7.0
		B	Competitive	55.0

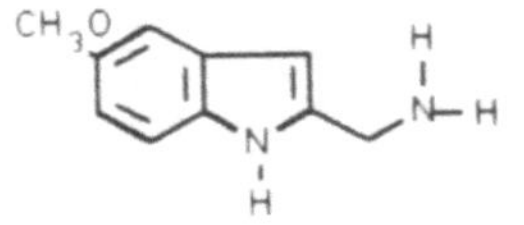

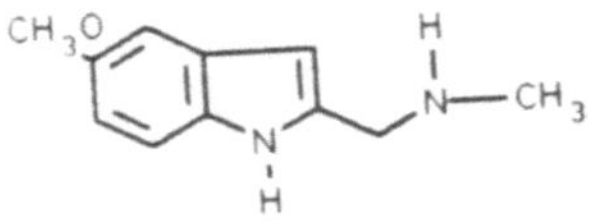

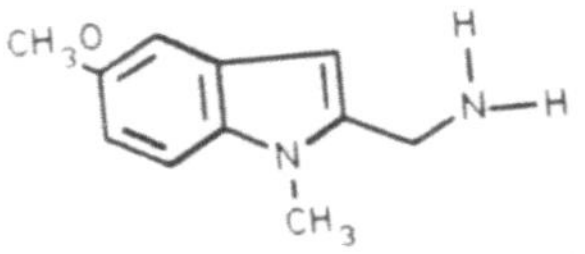

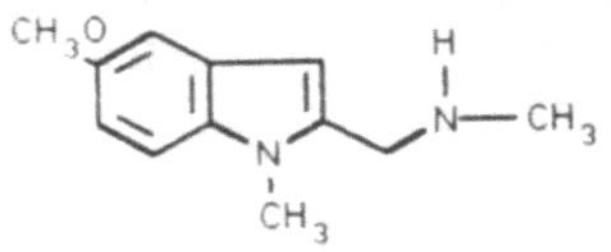

Fig. 4. Inhibition kinetic parameters of MAO A and MAO B by the parent amines of the acetylenic derivatives

group has been also attached at the acetylenic group of the side chain (FA 45). These structural changes have different effect on MAO B activity. In all cases a great decrease of the affinity was observed and consequently the catalytic efficacy diminished.

The kinetic parameters of MAO towards the parent amines derivatives (Fig. 3), showed that in general all them are bad substrates for both MAO forms, nevertheless when a methyl group is attached at the N of the indole ring (FA 52, FA 51), the affinity of MAO B increases significantly. The inhibition kinetic parameters of the parent amines (Fig. 4) showed in all cases a reversible and competitive behaviour. In case of FA 53 and FA 54 these compounds were in general bad MAO A and MAO B inhibitors, with FA 51 and FA 52 the most potent.

The substitution of an hydrogen attached at the N of the alkyl side chain (FA 52) by an —CH$_2$—C≡CH group (FA 42) involves a conversion of a bad MAO substrate into a very good "suicide" inhibitor of both MAO forms. When the hydrogen attached at the N of the alkyl side chain (FA 51) is substituted by a —CH$_2$—C≡C—CH$_3$ group (FA 45), or by a —CH$_2$—C≡CH group, (FA 43), a MAO susbtrate was converted in a very potent MAO A and MAO B suicide inhibitors.

References

Balsa D, Fernandez-Alvarez E, Tipton KF, Unzeta M (1991) Monoamine oxidase inhibitory potencies and selectivities of 2-[N-(2-propynyl) aminomethyl]-1-methyl indole derivatives. Biochem Soc Trans 19: 215–218

Cruces MA, Elorriaga C, Fernandez-Alvarez E, Nieto O (1988) Synthesis and biochemical properties of N-acetylenic and allenic derivatives of 2-amine-methylindoles as selective inhibitors of monoamine oxidase. Pharmacol Res Commun 20: 102–107

Fowler CJ, Tipton KF (1980) Concentration dependence of the oxidation of tyramine by the two forms of rat liver mitochondrial monoamine oxidase. Biochem Pharmacol 30: 3329–3332

Houslay MD, Tipton KF (1973) The nature of the electrophoretically separable multiple forms of rat liver monoamine oxidase. Biochem J 135: 173–186

Waldemeier PC, Felner AE, Tipton KF (1983) The monoamine oxidase inhibitory properties of CGP 1135A. Eur J Pharmacol 94: 73–83

Authors' address: Dr. M. Unzeta, Departament de Bioquímica i Biologia Molecular, Facultat de Medicina, Universitat Autònoma de Barcelona, E-08193 Bellaterra, Barcelona, Spain.

J Neural Transm (1994) [Suppl] 41: 287–290
© Springer-Verlag 1994

Acetylenic and allenic derivatives of 2-(5-methoxy-1-methylindolyl)alkylamines as selective inhibitors of MAO-A and MAO-B

C. Fernández García, J. L. Marco, and E. Fernández-Alvarez

Instituto de Química Orgánica, Consejo Superior de Investigaciones Científicas, Madrid, Spain

Summary. A new series of thirty derivatives of 2-(5-methoxy-1-methylindolyl)alkylamines has been synthesized and the compounds assayed as inhibitors of MAO-A and MAO-B from bovine brain mitochondria. IC_{50} values for both isoenzymes were determined using tyramine as a common substrate. Structure-activity and selectivity relationships are dicussed.
Key Words. Monoamine oxidane inhibitor design, acetylenic amine derivatives, allenic amine derivatives.

Introduction

In previous papers (Cruces et al., 1990a,b, 1991 and references cited therein) we have reported the synthesis and in vitro studies as selective inhibitors of MAO-A and MAO-B of a long series of N-acetylenic and N-allenic derivatives of 5-substituted (H; CH_3O; $C_6H_5CH_2O$; HO) 2-aminomethyl indoles. The non-acetylenic or non-allenic compounds were only mild or weak non-selective inhibitors of MAO, but the N-acetylenic and N-allenic products showed strong to very strong inhibition for both isoenzymes; in these compounds the selectivity was apparently "capricious", but, when present, MAO-A was preferently inhibited. On the other hand, this selectivity was stimulated by the presence of a methoxy (or hydroxy) group at 5 position of the heterocyclic ring, and occasionally by N-methyl substitution in the side-chain. Kinetic studies have shown that some of these N-acetylenic or N-allenic compounds (Cruces et al., 1990b) behave as substrate competitive and mechanism-based irreversible inhibitors, while the non-acetylenic or non-allenic compounds are reversible and substrate competitive inhibitors.

In this paper, as part of a search for more potent and selective agents, we report the potency of inhibition, and selectivities for MAO-A and MAO-B, of a new series of related products derived from 2-(5-methoxy-1-methylindolyl)alkylamines, in which the alkyl chain at 2 position has been elongated, shortened or modified by puting branched methyl groups in the α or β positions, with respect to the nitrogen in the side-chain. In addition, interesting data for the structure-activity relationships should be obtained.

Materials and methods

Chemistry

The synthesis of compounds 1–30 will described elsewhere. All new compounds showed correct elemental analysis and satisfactory spectroscopic data.

Enzymatic assays

Compounds 1–30 as well as clorgyline and l-deprenyl were studied as inhibitors of MAO-A and MAO-B from bovine brain mitochondria, with the common substrate tyramine, as previously described (Cruces et al., 1990a,b). Briefly, each compound was preincubated with the mitochondrial preparation for 20 min in 30 mM potassium phosphate buffer (pH 7.30) at 37°C. The residual enzyme activity was determined with ^{14}C-tyramine. IC_{50} values were obtained from plots of the residual activity (%) vs $-\log[I]$. When biphasic diagrams were obtained the different IC_{50} values for each isoenzyme were determined.

Results and discussion

In Table 1 we summarize the results obtained in this work with the previously reported data for the reference compounds (Cruces et al., 1990a). The new compounds proved to be MAO-A and MAO-B inhibitors, but only amines 6, 14, 21, 22, 38 showed significant selectivity for MAO-A. Non-acetylenic or non-allenic products (R′=H or CH$_3$) were competitive reversible inhibitors, and except for 1 or 2, they are probably substrates of both isoenzymes. For compounds with R′=H there are no important differences respect to the IC_{50} values of the reference compound A1 (except amine 1), but when R′=CH$_3$ all compounds were more potent inhibitors than the reference A2 (except amine 20). The inhibition by compounds 1 and 2 is probably related to their instability and spontaneous degradation in air to isatin derivatives, which are known as MAO inhibitors (Glover et al., 1988). Compared with the reference compounds A3–A5, the IC_{50} values for MAO-A increase in compounds 5, 7, 13–15, 25, 26, or remain similar (compound 6), while the IC_{50} values for MAO-B decrease (5, 7, 13, 25) or increase (6, 14, 15, 26). Compared with the reference compounds A6–A8, amines with R′=CH$_3$ and R′=acetylenic or allenic substituents show increased IC_{50} values for MAO-A (except in compound 10) and for MAO-B (except in 8–10, 16, 17).

In summary, we can conclude that neither the elongation and/or the α or β methyl branching in the side-chain of our previously reported compounds A1–A8 are in general favourable structural modifications for the improvement of the MAO-A selectivity.

Table 1. IC_{50} values for inhibition of MAO-A and B for compounds 1–30 and A1–A8

CH_3O— (5-methoxy-1-methylindol-2-yl)—X—N(R)(R')

- A1–A8: $X = $ —CH_2—
- 1–2: $X = (CH_2)_0$
- 3–10: $X = $ —$CH(CH_3)$—
- 11–18: $X = $ —CH_2—$CH(CH_3)$—
- 19–22: $X = $ —$CH(CH_3)CH_2$—
- 23–30: $X = $ —CH_2—CH_2—CH_2—

N[ob]	R	R'	IC_{50}-A(μM)[a]	IC_{50}-B(μM)[a]	B/A
1	H	H	1.6	1.6	1
2	H	CH_3	0.79	0.79	1
3 (A1)	H	H	14 (18)	14 (18)	1 (1)
4 (A2)	H	CH_3	80 (130)	80 (130)	1 (1)
5 (A3)	H	$CH_2C≡CH$	0.40 (0.021)	0.40 (45)	1 (2100)
6 (A4)	H	$CH_2CH=C=CH_2$	0.016 (0.022)	2.0 (0.4)	125 (180)
7 (A5)	H	$CH_2C≡C\text{-}CH_3$	2.0 (0.13)	2.0 (79)	1 (610)
8 (A6)	CH_3	$CH_2\text{-}C≡CH$	0.10 (0.0089)	0.1 (0.89)	1 (100)
9 (A7)	CH_3	$CH_2CH=C=CH_2$	0.016 (0.0054)	0.016 (0.50)	1 (93)
10 (A8)	CH_3	$CH_2C≡C\text{-}CH_3$	0.012 (0.036)	0.012 (0.036)	1 (1)
11 (A1)	H	H	25	25	1
12 (A2)	H	CH_3	63	63	1
13 (A3)	H	$CH_2C≡CH$	7.9	7.9	1
14 (A4)	H	$CH_2CH=C=CH_2$	0.079	40	506
15 (A5)	H	$CH_2C≡C\text{-}CH_3$	130	130	1
16 (A6)	CH_3	$CH_2C≡CH$	0.89	0.89	1
17 (A7)	CH_3	$CH_2CH=C=CH_2$	0.80	0.80	1
18 (A8)	CH_3	$CH_2C≡C\text{-}CH_3$	1.6	1.6	1
19 (A1)	H	H	39	39	1
20 (A2)	H	CH_3	100	100	1
21 (A6)	CH_3	$CH_2C≡CH$	1.2	320	266
22 (A7)	CH_3	$CH_2CH=C=CH_2$	0.16	63	394
23 (A1)	H	H	11	11	1
24 (A2)	H	CH_3	3.1	3.1	1
25 (A3)	H	$CH_2C≡CH$	7.9	7.9	1
26 (A4)	H	$CH_2CH=C=CH_2$	3.1	3.1	1
27	[c]	$CH_2C≡C\text{-}CH_3$	310	310	1
28 (A6)	CH_3	$CH_2C≡CH$	0.16	12	75
29 (A7)	CH_3	$CH_2CH=C=CH_2$	11	11	1
30 (A8)	CH_3	$CH_2C≡C\text{-}CH_3$	3.5	3.5	1
Clorgyline			0.0053	4.5	840
l-Deprenyl			1.3	0.01	0.0077

[a] Mean values from at least two independent and duplicated experiments
[b] Values in brackets previously reported (Cruces et al., 1990a) for compounds A1-A8
[c] For this compound $R = R' = CH_2C≡C\text{-}CH_3$

References

Cruces MA, Elorriaga C, Fernández-Alvarez E, Nieto O (1990a) Acetylenic and allenic derivatives of 2-(5-methoxyindolyl)methylamine: synthesis and evaluation as selective inhibitors of monoamine oxidases A and B. Eur J Med Chem 25: 257–265

Cruces MA, Elorriaga C, Fernández-Alvarez E (1990b) The kinetics of monoamine oxidase inhibition by three 2-indolylmethylamine derivatives. Biochem Pharmacol 40: 535–543

Cruces MA, Elorriaga C, Fernández-Alvarez E (1991) Acetylenic and allenic derivatives of 2-(5-benzyloxyindolyl) and 2-(5-hydroxyindolyl)methylamines: synthesis and in vitro evaluation as monoamine oxidase inhibitors. Eur J Med Chem 26: 33–41

Glover V, Ueki A, Goodwin BL, Watkins P, Halket J, Sandler M (1988) Isatin and its analogues as inhibitors of monoamine oxidase. Pharmacol Res Commun 20 [Suppl 4]: 117–118

Authors' address: Dr. E. Fernández-Alvarez, Instituto de Química Orgánica General (CSIC), Juan de la Cierva 3, E-28006 Madrid, Spain.

J Neural Transm (1994) [Suppl] 41: 291–293

Interactions between substituted tryptamine analogues, MAO inhibitors and cytochrome P-450

M. Valoti[2], **M. Costanzo**[1], **V. Perez**[1], **M. Unzeta**[1], and **G. P. Sgaragli**[2]

[1] Department de Bioquimica i Biologia Molecular, Universidad Autonoma
de Barcelona, Spain
[2] Istituto di Scienze Farmacologiche, Università di Siena, Italy

Summary. The effects of some MAO inhibitors, N-acetylenic analogues of tryptamine, on rat liver microsomal cytochrome P-450 (cyt P-450) have been investigated. All the compounds tested interacted with cyt P-450 with Ks values ranging between 14 and 358 µM (clorgyline Ks = 10.5 µM). Compounds with a tertiary amine and those possessing a secondary amine group in the acetylenic side chain exhibited type I and type II difference spectra, respectively. Aniline hydroxylase activity was inhibited irreversibly and in a time-dependent fashion by all compounds tested with IC_{50} ranging between 7×10^{-5} and 7×10^{-3} M (clorgyline 10^{-4} M).

Introduction

A few studies are reported in the literature on microsomal metabolism of monoamine oxidase inhibitors (MAOI). These drugs, however, present mostly chemical structures similar to compounds, which are cytochrome P-450 irreversible inhibitors, e.g. ethinyl and hydrazine derivatives. The inhibitory effects of MAOI on liver microsomal oxidising systems could help to explain some aspects of tyramine toxicity to humans associated with MAOI treatment (see Callingham et al., 1991).

Some N-acetylenic analogues of tryptamine have been recently synthetized by Cruces and coworkers (1988) and have been shown to be a new class of MAO-A and MAO-B inhibitors (see Balsa et al., 1990). In this study we have investigated the interactions of these compounds with rat liver microsomal cyt P-450.

Materials and methods

Enzyme assay

Cyt P-450 content was determined in microsomes as described by Omura and Sato (1964). Aniline hydroxylase activity was measured by recording at 37°C for 20 min the formation of 4-aminophenol according to the procedure of Nakanishi as described by Lake (1987). Rat liver microsomal proteins were estimated by the Bradford procedure (1977) using bovine serum albumin as standard.

The standard reaction mixture (1 ml, volume) contained MAOI or aniline as substrate at different concentrations, microsomal proteins (0.5 mg) and a NADPH generating system (1 mM NADPH, 4 mM glucose-6-phosphate and 1 U glucose-6-phosphate dehydrogenase).

Spectrophotometric measurements

Difference spectra were recorded at 25°C with a Shimadzu dual wavelength UV-260 spectrophotometer. The quartz cuvettes, 10 mm width, contained 2 ml of a microsomal suspension (2 mg/ml) in 0.1 M phosphate buffer, pH 7.4 and MAO inhibitors at various concentrations ($1 \times 10^{-5} - 3.5 \times 10^{-5}$ M).

Results and discussion

All the acetylenic analogues of tryptamine assayed in this study and shown in Fig. 1 interacted with cyt P-450 giving rise to the formation of characteristic differential spectra. Compounds having a secondary amine moiety in the aliphatic chain showed type II difference spectra, while those with a tertiary amine group exhibited type I spectra. The Ks values calculated from these spectra varied widely. Compounds 27, 43, 45 and 69 (type I spectra) showed the lowest affinity with Ks values ranging between 74 and 358 μM, whereas the others compounds 30 and 70 (type I spectra) and 26, 42, and 102 (type II spectra) possessed higher affinity with Ks values ranging between 14 and 37 μM. All compounds inhibited the aniline hydroxylase activity with IC$_{50}$ values which were lower for N-methylindole as compared

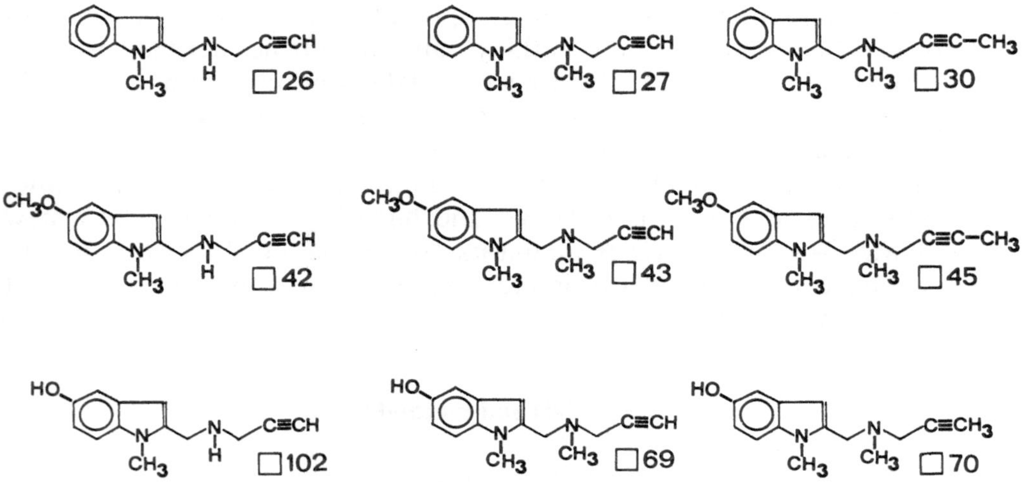

Fig. 1. Chemical structures of the N-acetylenic analogues of tryptamine used in this study

to 5-hydroxy-N-methylindole derivatives. Furthermore, it is worth emphasising that the methylethinyl derivatives exhibited a stronger inhibition when they were pre-incubated in the absence of a NADPH generating system. This indicates their inactivation by metabolism. Experiments still in progress show that these compounds are extensively N-demethylated. The inhibition of the aniline hydroxylase reaction by all the compounds studied was time dependent and was not reversed by diluting the reaction mixture, indicating irreversible inhibition.

These observations suggest that the MAO inhibitors in the present study interfere with the oxidative metabolism of xenobiotics by a dual mechanism. Firstly, by inhibiting irreversibly the aniline hydroxylase activity associated with cyt P-450 and, secondly, by inhibiting competitively N-demethylase activity. We can hypothesize that this influence takes place through direct interaction of these compounds with cyt P-450, even though further interactions with flavoproteins associated with the microsomal oxidating system cannot be discounted.

Acknowledgements

M. Costanzo was supported by an ERASMUS fellowship (ICP-91-I-0047/12).

References

Balsa D, Fernandez-Alvarez E, Tipton KF, Unzeta M (1990) Inhibition of MAO by substituted tryptamine analogues. J Neural Transm [Suppl] 32: 103–105

Bradford MM (1976) A rapid and sensitive method for the quantitation of microgram quantities of protein utilizing the principle of proteine-dye binding. Anal Biochem 72: 248–254

Callingham BA, Valoti M, Sgaragli GP (1991) Drug and enzyme interactions with moclobemide. In: Racagni G, Brunello N, Fukuda T (eds) International Congress Series 968. Elsevier Science Publisher, Amsterdam, pp 846–849

Cruces MA, Elorriaga C, Fernandez-Alvarez E, Nieto O (1988) Synthesis and biochemical properties of N-allenic derivatives of 2-amine-methylindoles as selective inhibitors of monoamine oxidase. Pharmacol Res Commun 20: 105–107

Lake B (1987) Preparation and characterisation of microsomal fractions for studies on xenobiotic metabolism. In: Snell K, Mullock B (eds) Biochemical toxicology a practical approach. IRL Press, Oxford, pp 183–215

Omura T, Sato R (1964) The carbon monoxide-binding pigment of liver microsomes II. Solubilization, purification and properties. J Biol Chem 239: 2379–2385

Authors' address: Dr. M. Valoti, Istituto di Scienze Farmacologiche, Università di Siena, via Piccolomini 170, I-53100 Siena, Italy.

J Neural Transm (1994) [Suppl] 41: 295–305

Inhibition of monoamine oxidase by clorgyline analogues

E. M. O'Brien[1], K. F. Tipton[1], M. Meroni[2], and P. Dostert[2]

[1] Department of Biochemistry, Trinity College, Dublin, Ireland
[2] Research and Development-Erbamont Group, Farmitalia Carlo Erba, Milan, Italy

Summary. N-Methyl-N-propargyl-3-(4-phenoxy)phenoxypropylamine, an analogue of the MAO-A-selective irreversible inhibitor clorgyline in which the 2,4-dichloro- substitution in clorgyline was replaced by a 2-H atom and a 4-phenoxy group, has been synthesised and assessed as an inhibitor of monoamine oxidase (MAO). This compound proved to be a time-dependent irreversible inhibitor of both MAO-A and -B. However, unlike clorgyline, it was selective towards MAO-B, both in its initial, non-covalent, binding to the enzyme and as an irreversible inhibitor. In order to assess the influence of side-chain length on inhibitory potency, analogues were synthesised in which the side-chain was reduced to 2 CH_2 units (N-methyl-N-propargyl-2-(4-phenoxy)phenoxyethylamine) or increased to 4 CH_2 units (N-methyl-N-propargyl-4-(4-phenoxy)phenoxybutylamine). Both these compounds were also time-dependent irreversible inhibitors with selectivity towards MAO-B. In the case of the initial, non-covalent, inhibition all these compounds were competitive inhibitors of MAO-A, with respect to the amine substrate, and the affinity for inhibitor binding increased with carbon chain length. In contrast the compounds were all mixed inhibitors of MAO-B. The competitive element of this inhibition (measured by K_{is}) was similar for the 2 and 3 carbon-chain compounds but decreased markedly when the chain-length was increased to 4 carbons. The uncompetitive inhibition (measured by K_{ii}) decreased as the carbon chain-length was increased from 2 to 3, but there was no significant further change when the length was increased to 4 carbons. The time-dependent irreversible inhibition (measured as the IC_{50} values after 60 min enzyme-inhibitor preincubation) showed that the potency towards MAO-A increased when the side-chain length was increased from 2 to 3 carbons but that there was no significant difference between the 3 and 4 carbon-chain compounds. In the case of MAO-B inhibition, the 2 and 3 carbon-chain compounds had similar inhibitory potencies but this increased substantially when the chain length was increased to 4 carbons. The significance of the inhibitory behaviour of these compounds is discussed in terms of the structure-activity relationships of mechanism-based irreversible MAO inhibitors.

Introduction

Clorgyline, (−)-deprenyl and pargyline act as mechanism-based ($k_{cat.}$ or "suicide") irreversible inhibitors of monoamine oxidase (MAO). The reactions can be represented by the simple mechanism:

$$E + I \underset{k_1}{\overset{k_{-1}}{\rightleftharpoons}} EI \overset{k_{+2}}{\longrightarrow} E - I$$

where E and I represent the free enzyme and inhibitor, respectively, EI represents the non-covalent enzyme inhibitor complex and $E - I$ is the covalent enzyme inhibitor adduct.

Structural changes in the clorgyline molecule — a selective inhibitor of MAO-A (Johnston, 1968), results in (−)-deprenyl — a selective inhibitor of MAO-B (Knoll and Magyar, 1972). Such differences in the selectivities of acetylenic MAO inhibitors have prompted numerous studies on the factors affecting the selectivity of these irreversible inhibitors. These studies have involved a combination of a kinetic approach and structure-potency relationships. Fowler et al. (1982) have shown that the selectivity of acetylenic inhibitors in binding to MAO is dependent on the differences in the K_i values for the non-covalent interaction and on differences in the k_2 values for the irreversible reaction. Knowledge of the structural features that confer selectivity on the acetylenic inhibitors is more limited and puzzling. Structure-activity studies have suggested the nature of the aromatic ring and the length of the carbon chain between the aromatic ring and the nitrogen moiety to be important factors in selectivity. The role of the aromatic ring in the MAO inhibitory effect of (−)-deprenyl has been examined by Knoll et al. (1978). The change of the benzene ring to a thenyl ring was found to reduce the MAO inhibitory effect, whereas a furan ring did not significantly affect the MAO inhibition. In addition, partial or total saturation of the aromatic ring resulted in extremely poor inhibition of MAO.

Williams and Lawson (1974) have shown that N-(2,4-dichlorobenzyl)-N-methyl-propargylamine is a much better inhibitor of MAO than the non-halogenated derivative, indicating substitution on the aromatic ring to be another factor affecting inhibitory potency. The influence of aromatic substituents on selectivity and potency towards MAO-A has also been examined in two series of clorgyline analogues (Williams and Walker, 1984; Ohmomo et al., 1991). Selectivity of MAO-A was found to be influenced by both the nature and position of the ring substituents with the *ortho* position having the major role in MAO-A inhibition.

With respect to the distance between the aromatic ring and the nitrogen moiety, it has been shown that introduction of one or two methylene units between the phenyl ring and the rest of the molecule in (−)-deprenyl reduces the inhibitory potency towards MAO (Knoll et al., 1978).

The aim of the present study was to examine the MAO inhibitory properties of novel clorgyline analogues (Fig. 1) in an attempt to investigate the extent to which the preferential recognition of clorgyline by MAO-A could be affected by structural modifications. The compounds investigated involved replacement of the 2,4-dichloro substitution in clorgyline by a 2-H

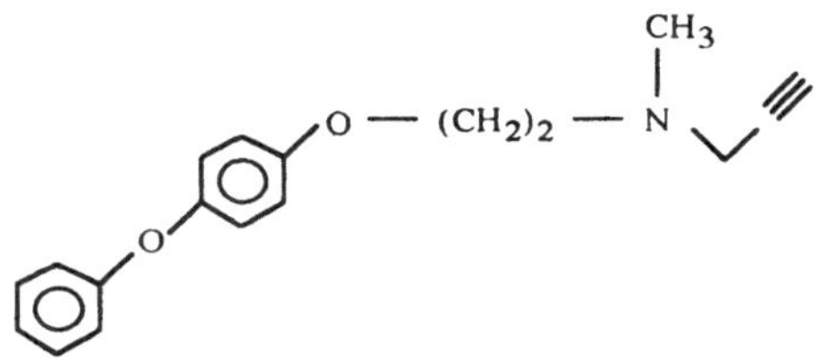

N-Methyl-N-Propargyl-2-(4-phenoxy)phenoxyethylamine (I$_a$)

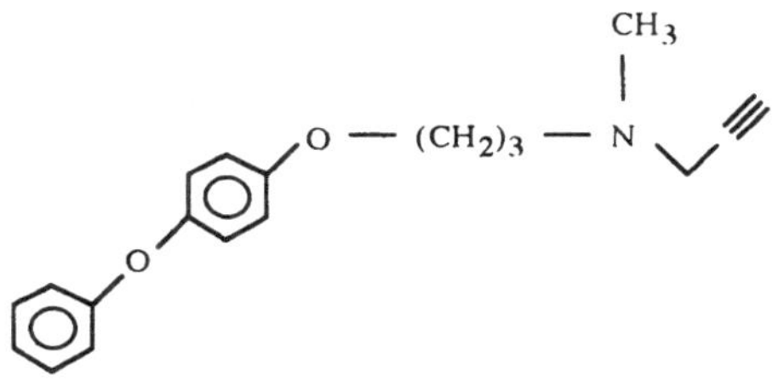

N-Methyl-N-Propargyl-3-(4-phenoxy)phenoxypropylamine (I$_b$)

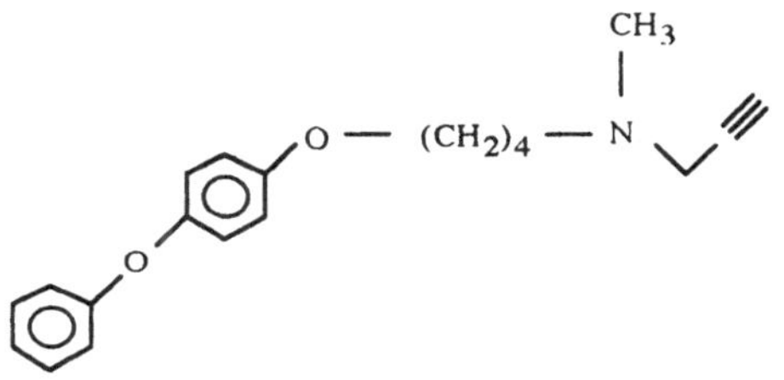

N-Methyl-N-Propargyl-4-(4-phenoxy)phenoxybutylamine (I$_c$)

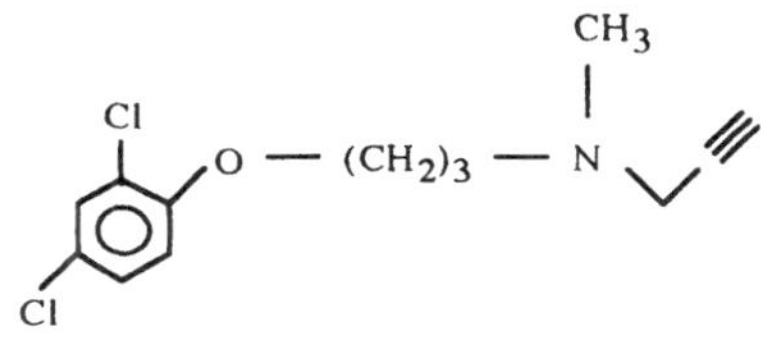

Clorgyline

Fig. 1. The structures of the N[ω-(4-phenoxyphenoxy)alkyl]-N-methyl-2-propyn-1-amines used in this study. The structure of clorgyline is also shown for comparison

atom and a 4-phenoxy group. Additionally the length of the chain between the aromatic ring and the nitrogen moiety was varied from 2–4 carbon bond units.

Materials and methods

Synthesis of N[ω-(4-phenoxyphenoxy)alkyl]-N-methyl-2-propyn-1-amines (I)

The synthetic route is summarised in Fig. 2 and is described briefly below:
1-bromo-ω-(4-phenoxyphenoxy)alkanes (2) were prepared as follows: a solution of 0.1 mole of 4-phenoxyphenol (1), 0.11 mole of 85% potassium hydroxide and 0.21

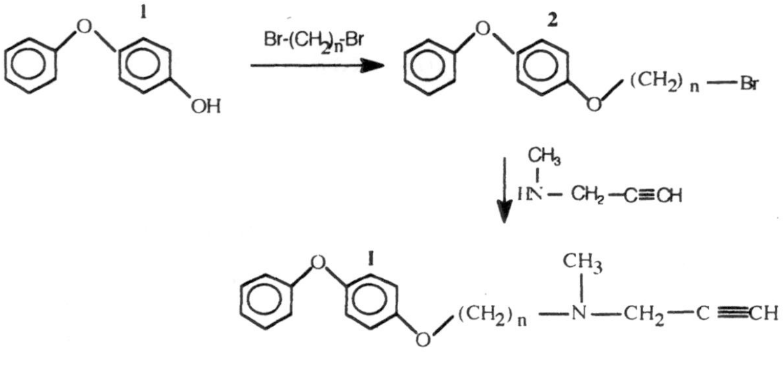

I_a (n=2); I_b (n=3); I_c (n=4)

Fig. 2. Chemical pathway for the synthesis of N[ω-(4-phenoxyphenoxy)alkyl]-N-methyl-2-propyn-1-amines (I)

mole of the corresponding dibromoalkane in 200 ml of methanol was heated to reflux for 24 h. After cooling, the solution was filtered and evaporated to dryness. The residue was dissolved in benzene and the solution was washed with water and dried over disodium sulphate. After evaporation, the residue was purified by flash chromatography using Carlo Erba RS silica gel (40–63 µm) and a mixture hexane: ethyl acetate (9:1) as eluant. Compounds "2" were obtained as oils (37%, n = 2; 52% n = 3, 4) and were used for the next step without further purification.

N[ω-(4-phenoxyphenoxy)alkyl]-N-methyl-2-propyn-1-amines (I) were prepared as follows: a solution of 0.1 mole of "2" and 0.5 mole of N-methylpropargylamine in 300 ml dimethyl formamide was heated at 80°C for 8 h in a stoppered flask. After cooling and evaporation of dimethylformamide the residue was taken up in ethyl acetate. After washing with water and drying over disodium sulphate, the solution was evaporated to dryness and the residue was dissolved in ethanol. After addition of an excess of HCl in ethanol and evaporation to dryness, the residue was triturated with anhydrous isopropyl ether and filtered to obtain the hydrochloride salt of I (Ia, 74%, mp = 123–125°C; Ib, 66%, mp = 145–146°C; Ic, 64%, mp = 87–89°C).

Preparation of solutions of the clorgyline analogues

Stock solutions of N-methyl-N-propargyl-2-(4-phenoxy) phenoxyethylamine (I_a), N-methyl-N-propargyl-3-(4-phenoxy) phenoxypropylamine (I_b) and N-methyl-N-propargyl-4-(4-phenoxy) phenoxybutylamine (I_c) were made up in dimethylsulphoxide (DMSO) and diluted as appropriate with water. The maximum concentration of DMSO which was present in any assay was 0.2% and control experiments showed that DMSO at this concentration had no effect on MAO activity.

Preparation of enzyme samples

Rat liver mitochondrial were prepared by the method of Kearney et al. (1971).

Inhibition studies

The effects of the clorgyline analogues on the activity of MAO-A and -B were determined using the radiochemical assay of Otsuka and Kobayashi (1964) as modified by Fowler and Tipton (1981), using 100 μM 5-hydroxytryptamine (5-HT) and 10 μM 2-phenethylamine (PEA) as the selective substrates, respectively.

Determination of the concentration of inhibitor giving 50% inhibition (IC$_{50}$)

IC$_{50}$ values were determined at zero-time and after 1 h preincubation of the enzyme and inhibitor at 37°C. Each value was determined in triplicate and dose-response curves were repeated at least twice in separate experiments. Calculated IC$_{50}$ values were determined using the commercially available programme KaleidaGraph version 2.1.

Reversibility studies

Reversibility was determined by repeated centrifugation and dilution (see Waldmeier et al., 1983).

Determination of K$_i$ values

The kinetic behaviour of the clorgyline analogues was determined by varying the PEA and 5-HT concentrations in the presence of several fixed concentrations of inhibitor. Short assay incubation periods of 2 min for PEA and 5 min for 5-HT were used to avoid any irreversible inhibition during the course of the assay. Each point was determined in duplicate or triplicate and inhibition plots were repeated at least twice in separate experiments. Data were analysed by non-linear regression analysis to allow the determination of the type of inhibition and the inhibitor constants.

Results

Compounds I_a–I_c were all time-dependent inhibitors of both MAO-A and -B. A comparison of the IC$_{50}$ values obtained after 0 and 1 h preincubation of enzyme and each inhibitor at 37°C is shown in Table 1. Representative graphs for the inhibition of MAO-A and -B by I_a are shown (Fig. 3).

The reversibility of the inhibition of MAO-A and -B by I_a was assessed with reference to amphetamine (Fig. 4). In contrast to amphetamine, I_a was shown to be an irreversible inhibitor of both forms of the enzyme. Similarly, both I_b and I_c were irreversible inhibitors of MAO-A and -B.

Without enzyme-inhibitor preincubation I_a, I_b and I_c were shown to be linearly competitive inhibitors of MAO-A. All three compounds displayed mixed inhibition towards MAO-B. This would suggest that the inhibitors could bind both to the free enzyme, to give a complex EI with a dissociation constant K_{is}, and either to the ES complex, to give an unreactive EIS complex, or to a free reduced form of the enzyme E^{H2} to give the complex

Table 1. IC_{50} values for the inhibition of MAO by the clorgyline analogues

Compound	Preincubation time (min)	IC_{50} (μM)	
		MAO-A (n)	MAO-B (n)
I_a	0	10.0 ± 1.500 (2)	1.60 ± 0.196 (3)
	60	0.33 ± 0.009 (2)	0.03 ± 0.002 (3)
I_b	0	1.27 ± 0.267 (3)	0.72 ± 0.020 (2)
	60	0.05 ± 0.002 (2)	0.03 ± 0.004 (2)
I_c	0	0.87 ± 0.087 (3)	0.14 ± 0.031 (3)
	60	0.05 ± 0.003 (2)	0.002 ± 0.001 (3)
Clorgyline	0	0.019 ± 0.0009 (2)	22.6 ± 1.3 (2)

Each value represents the mean ± range or S.E.M. from two or three separate determinations (n), as indicated. In each separate experiment, independent inhibition estimations were made in triplicate

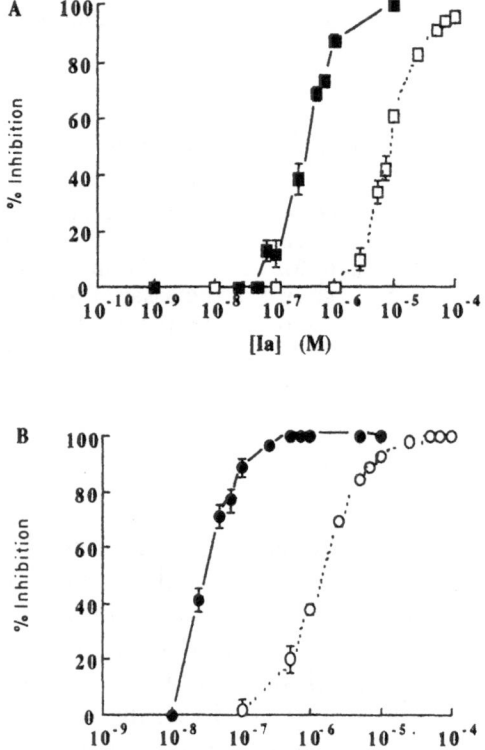

Fig. 3. The effects of N-methyl-N-propargyl-2-(4-phenoxy)phenoxyethylamine (I_a) concentration on the activities of MAO. Rat liver mitochondria ($0.2 \, \text{mg protein.ml}^{-1}$) were incubated with the indicated concentrations of inhibitor for 0-time (open symbols; dotted lines) or 1 h (closed symbols; solid lines) before activity was determined towards A $100 \, \mu\text{M}$ 5-HT (□, ■) or B $10 \, \mu\text{M}$ PEA (○, ●). Percentage inhibition was calculated with respect to samples preincubated for the same periods in the absence of inhibitor. Each point is the mean ± S.E.R. from triplicate determinations in a single experiment

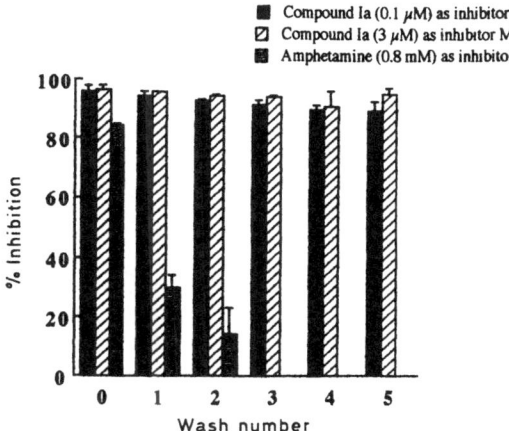

Fig. 4. Assessment of the nature of the inhibition of MAO-A and MAO-B by N-methyl-N-propargyl-2-(4-phenoxy)phenoxyethylamine (I_a). Rat liver mitochondria (2 mg protein.ml^{-1}) and compound I_a were preincubated for 1 h at 37°C before being subjected to successive centrifugation and resuspension. Amphetamine was used as a control in these studies, as it is a known reversible inhibitor of MAO-A. Activities of MAO-A and MAO-B were assayed immediately (0) and after each of 5 centrifugation-resuspension cycles (1–5). Percentage inhibition was calculated with respect to samples treated in the same way but in the absence of inhibitor. The inhibition of MAO-A by amphetamine was insignificant after the third wash. Each value is the mean ± range from triplicate determinations in two separate experiments

Table 2. K_i values for the inhibition of MAO by the clorgyline analogues

Compound	K_i (µM) MAO-A (n)	K_{is} (µM) MAO-B (n)	K_{ii} (µM) MAO-B (n)
I_a	9.30 ± 0.23 (4)	1.38 ± 0.36 (2)	17.1 ± 5.8 (2)
I_b	3.30 ± 2.20 (2)	1.47 ± 0.49 (2)	1.9 ± 0.7 (2)
I_c	1.10 ± 0.30 (2)	0.21 ± 0.05 (2)	1.4 ± 0.2 (2)

Each value represents the mean ± range or S.E.M. from a number of separate determinations (n), as indicated

$E^{H2}I$ with the dissociation constant — K_{ii}. A comparison of the inhibition constants for MAO-A and MAO-B and for each of the inhibitors is given in Table 2. For illustrative purposes representative graphs (Figs. 5, 6) for the inhibition of MAO-A and -B by the two carbon compound — I_a are shown in the form of double-reciprocal plots.

Discussion

All three compounds were time-dependent, irreversible inhibitors of both MAO-A and -B. It is reasonable to assume that they behave as mechanism-based inhibitors of MAO in a similar fashion to clorgyline and other pro-pynylamines.

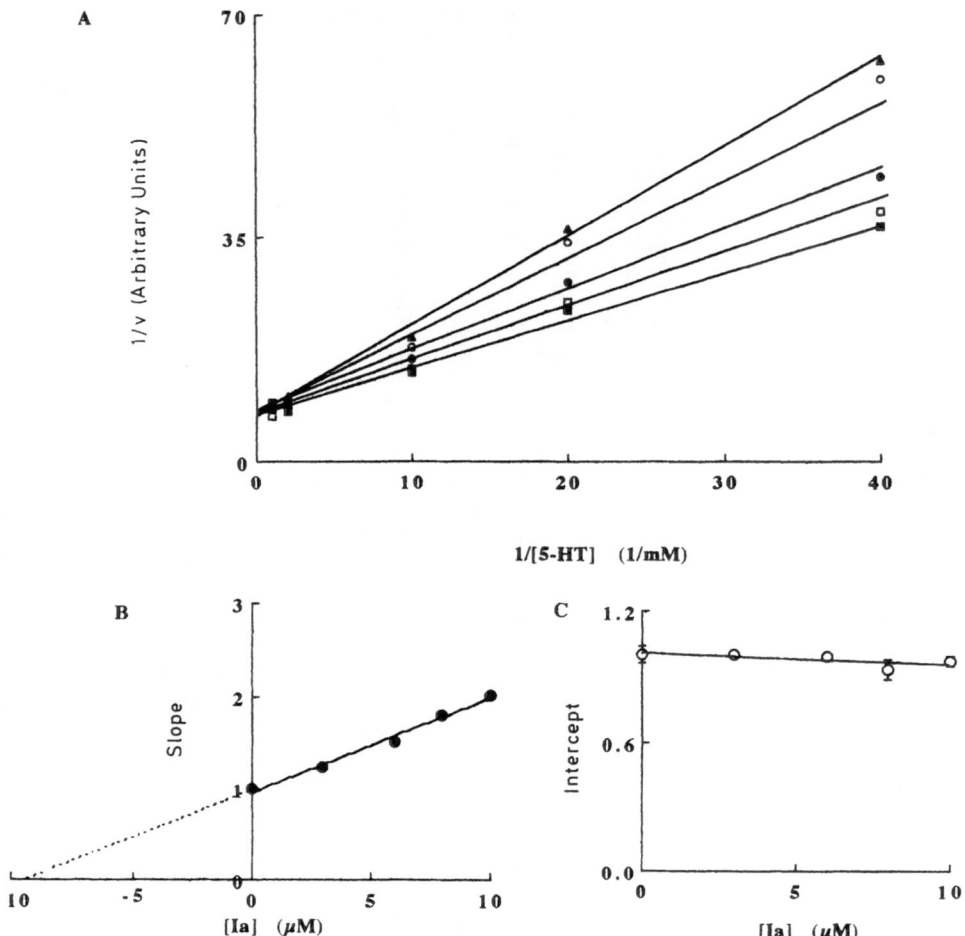

Fig. 5. Kinetics of the inhibition of MAO-A by N-methyl-N-propargyl-2-(4-phenoxy) phenoxyethylamine (I_a). A Initial rates were measured without enzyme inhibitor preincubation, in the presence of the indicated concentrations of 5-HT and in the presence of 0 (■), 3 μM (□), 6 μM (●), 8 μM (○) and 10 μM (▲) I_a. Each point is the mean of duplicate determinations in a single experiment. In all cases the errors (range) were less than 7% of the actual mean value. These error bars have been omitted for clarity. B, C The dependence of the slopes (●) and intercepts (○), respectively obtained from the double-reciprocal plot, on the I_a concentration

The three compounds were competitive inhibitors towards MAO-A but mixed inhibitors towards MAO-B. In the case of I_b the values of K_{ii} and K_{is} were similar, whereas, in the case of I_a and I_c the K_{ii} values were greater than the K_{is} values. As shown in Table 2 increasing the side-chain length from two to 4 carbon atoms resulted in a progressive increase in affinity towards MAO-A. However, this was not the case with respect to MAO-B where lengthening the side chain from 2 to 3 carbons had no significant effect on the competitive element of the inhibition (measured by K_{is}), although the uncompetitive affinity (measured by K_{ii}) was greatly increased.

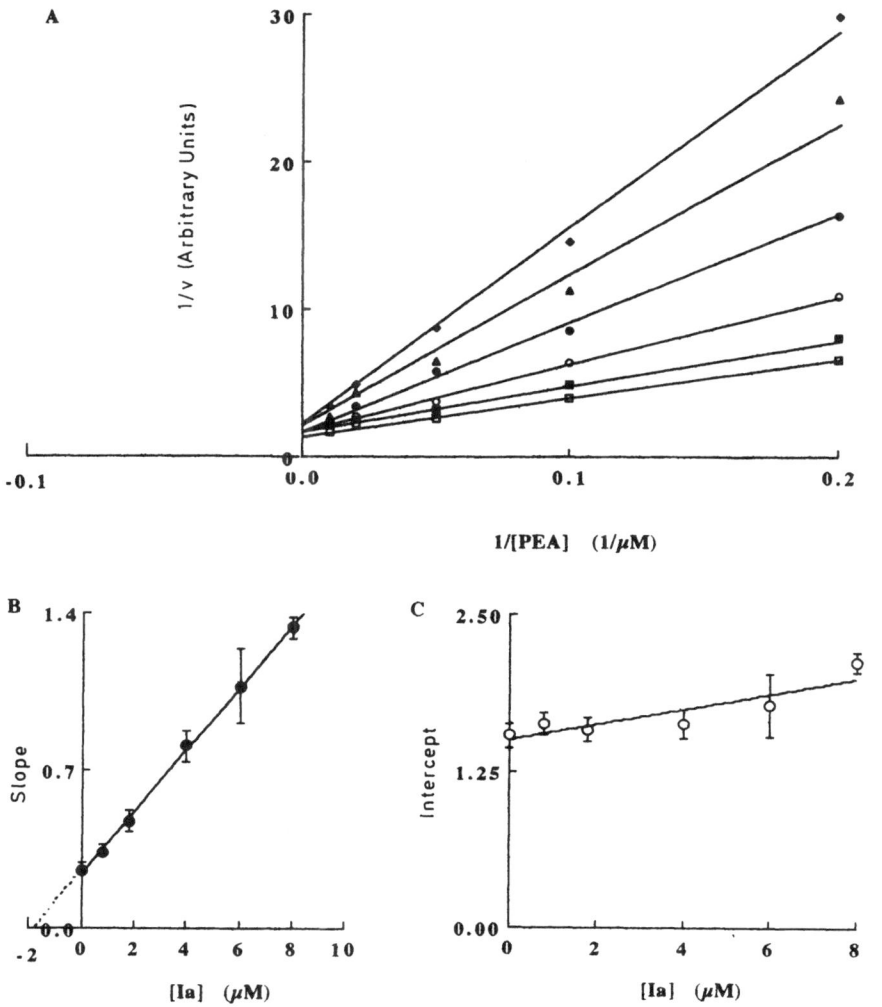

Fig. 6. Kinetics of the inhibition of MAO-B by N-methyl-N-propargyl-2-(4-phenoxy) phenoxyethylamine (I_a). A Initial rates were measured without enzyme inhibitor preincubation, in the presence of the indicated concentrations of PEA and in the presence of 0 (□), 0.8 μM (■), 1.8 μM (○), 4 μM (●), 6 μM (▲) and 8 μM (◆) I_a. Each point is the mean of duplicate determinations in a single experiment. In all cases the errors (range) were less than 5% of the actual mean value. These error bars have been omitted for clarity. B, C The dependence of the slopes (●) and intercepts (○), respectively obtained from the double-reciprocal plot, on the I_a concentration

A further lengthening of the side chain, from 3 to 4 darbons had no significant effect on K_{ii} but resulted in a large decrease in K_{is}.

Compound I_b had an identical -O-$(CH_2)_3$- linkage between the aromatic moiety and the nitrogen atom to clorgyline, but it differed from the latter compound in that the *ortho* and *para* chlorine substitutions on the aromatic ring were replaced by an *ortho*-H and a *para* phenoxy group. This change in chemical structure significantly altered the MAO inhibitory effect. In comparison to clorgyline, I_b was shown to be 67 times a less potent inhibitor of MAO-A and 31 times a more potent inhibitor of MAO-B (Table 1). Moreover, the compound showed slight selectivity towards MAO-B. These large differences in potency and selectivity between clorgyline and I_b are striking considering the structural similarity between these two compounds. The fall in selectivity and potency towards MAO-A is in agreement with the observation of Williams (1984) that substitution at the *ortho* position of clorgyline analogues tended to increase time-dependent inhibitory potency and selectivity towards MAO-A, whereas *para* substitutions diminished selectivity.

Although the 3 carbon side-chain in compound I_b was similar to that in clorgyline selectivity towards MAO-A was not retained. Therefore, the suggestion that the distance between the aromatic ring and the N-propargyl terminal is a crucial factor in designating type-A or -B inhibitory properties (Kalir et al., 1981), was investigated further. The side chain was increased to 4 carbon atoms or decreased to 2 carbon atoms (I_c and I_a, respectively). At zero-time preincubation of the enzyme and inhibitor, an increase in chain length from 2 through 4 carbon atoms was found to produce increasingly more potent inhibitors of both MAO-A and -B, with an approximate 11-fold increase in inhibitory potency for both forms of the enzyme (Table 1), although, as discussed above, the behaviour with respect to MAO-B was complicated in terms of the differential effects on the two K_i values describing the mixed inhibition.

In the absence of enzyme-inhibitor preincubation and after preincubation of the enzyme and inhibitor for 1 h all three compounds were more selective inhibitors of MAO-B (Tables 1, 2). The greatest selectivity was seen with the 4 carbon chain (IC_{50} MAO-A/IC_{50} MAO-B = 25) and the poorest selectivity was seen with the 3 carbon chain (IC_{50} MAO-A/IC_{50} MAO-B = 1.7), after 1 h enzyme inhibitor preincubation. However the responses of the two enzymes to increasing side-chain length were different. An increase from 2 to three carbons resulted in a large increase in the potency of MAO-A inhibition after a preincubation time of 1 h, but there was no further increase when the carbon chain was lengthened further. In contrast there was no significant change in the IC_{50} value towards MAO-B after 1 h preincubation when the side-chain was increased from 2 to three carbons, but there was a large increase in inhibitory potency when it was further increased to 4 carbons (Table 1). It is notable that, although a separation of 3 or 4 carbon atoms was necessary for efficient non-covalent binding to MAO-A (Table 2), it was by no means sufficient to confer selectivity towards that form of the enzyme.

The behaviour of the inhibitors presented here indicate that relatively small modifications in clorgyline structure can result in profound changes in their interactions with MAO-A and -B. This study confirms that the nature and possible position of the substituent on the aromatic ring is an important factor in conferring MAO selectivity. Indeed the "bi-benzene' nature of the aromatic moiety seems to be a dominant feature in bestowing -B selectivity in these compounds. The role of the chain-length between the aromatic ring and the nitrogen atom would appear to have an important bearing on inhibitory potency but to be less important in conferring selectivity. Clearly any attempts to predict the selectivity of MAO inhibitors simply in terms of side-chain length would be invalid. More detailed studies of structure-activity relationships and the binding sites of MAO-A and -B will be necessary before the behaviour of acetylenic inhibitors can be successfully predicted from their structures alone.

References

Fowler CJ, Tipton KF (1981) Concentration dependence of the oxidation of tyramine by the two forms of rat liver mitochondrial monoamine oxidase. Biochem Pharmacol 30: 3329–3332

Fowler CJ, Mantle TJ, Tipton KF (1982) The nature of the inhibition of rat liver monoamine oxidase types -A and -B by the acetylenic inhibitors clorgyline, l-deprenyl and pargyline. Biochem Pharmacol 31: 3555–3561

Johnston JP (1968) Some observations upon a new inhibitor of monoamine oxidase in brain tissue. Biochem Pharmacol 17: 1285–1297

Kalir AS, Sabbagh A, Youdim MBH (1981) Selective acetylenic "suicide" and reversible inhibitors of monoamine oxidase types A and B. Br J Pharmacol 73: 55–64

Kearney EB, Salach JI, Walker WH, Seng RL, Kenney W, Zeszotek E, Singer TP (1971) The covalently bound flavin of hepatic monoamine oxidase. I. Isolation and sequence of a flavin peptide and evidence for binding at the 8α position. Eur J Biochem 24: 321–327

Knoll J, Magyar K (1972) Some puzzling pharmacological effects of monoamine oxidase inhibition. Adv Biochem Psychopharmacol 5: 393–408

Knoll J, Ecsery Z, Magyar K, Sátory E (1978) Novel (−)-deprenyl-derived selective inhibitors of B-type monoamine oxidase, the relation of structure to their action. Biochem Pharmacol 27: 1739–1747

Ohmomo Y, Hirata M, Murakami K, Magata Y, Tanaka C, Yokoyama A (1991) Synthesis of fluorine and iodine analogues of clorgyline and selective inhibition of monoamine oxidase A. Chem Pharm Bull 39: 1038–1040

Otsuka S, Kobayashi Y (1964) A radioisotopic assay for monoamine oxidase determinations in human plasma. Biochem Pharmacol 13: 995–1006

Waldmeier PC, Felner AE, Tipton KF (1983) The monoamine oxidase inhibitory properties of CGP 11305 A. Eur J Pharmacol 94: 73–83

Williams CH (1984) Selective inhibitors of monoamine oxidases A and B. Biochem Pharmacol 33: 334–337

Williams CH, Lawson J (1974) Monoamine oxidase. II. Time-dependent inhibition by propargylamines. Biochem Pharmacol 23: 629–636

Williams CH, Walker B (1984) What does the binding of inhibitors reveal about the active site of monoamine oxidase? In: Tipton KF, Dostert P, Strolin Benedetti M (eds) Monoamine oxidase and disease. Prospects for therapy with reversible inhibitors. Academic Press, London, pp 41–52

Authors' address: Dr. K. F. Tipton, Department of Biochemistry, Trinity College, Dublin 2, Ireland.

J Neural Transm (1994) [Suppl] 41: 307–311

Kinetics of inhibition of MAO-B by
N-(2-aminoethyl)-p-chlorobenzamide (Ro 16-6491) and analogues

E. L. Mullan and **C. H. Williams**

Division of Biochemistry, Queen's University, Belfast, United Kingdom

Summary. Ro 16-6491 is known to be a potent reversible inhibitor of human brain MAO-B. This compound and several analogues were tested for their effect on bovine liver MAO-B. It was found that in compounds where the amide bond of Ro 16-6491 was replaced by an ester bond an increase in potency of the order of 150–200 times was obtained.

Introduction

Monoamine oxidase (MAO; EC 1.4.3.4; amine: oxygen oxidoreductase; flavin-containing) is a flavoprotein situated in the mitochondrial outer membrane of many tissues. It occurs as two separate gene products, MAO-A and MAO-B which share about 70% homology as found by sequencing of their respective cDNA's. Despite their homology they are easily distinguished by their substrate specificity and their inhibitor sensitivity. L-Deprenyl inhibits MAO-B whereas clorgyline inhibits MAO-A. MAO catalyses the oxidation of primary and secondary amines to the corresponding aldehyde according to the reaction:

$$RCH_2NH_2 + O_2 + H_2O \rightarrow RCHO + NH_3 + H_2O_2.$$

The tertiary amine 1-methyl-4-phenyl-1,2,3,6-tetrahyropyridine (MPTP) is exceptional in that it has been found not only to be oxidised by MAO-B but to be one of its best substrates. The product of oxidation is the neurotoxin, the 5,6-dihydro MPTP which gives rise to symptoms of Parkinson's disease. Hence the inhibition of MAO-B may be of importance in slowing the progression of Parkinson's disease. MAO-A is also of clinical significance because of the anti-depressant effects of many of its inhibitors. However, adverse side effects mean that MAO inhibitors are not widely prescribed anymore. Inhibitor studies are still continuing and many specific reversible and irreversible prodrug MAO inhibitors have been found which have anti-depressant properties. In recent years attention has turned to reversible inhibitors. Compounds based on aminoethyl amides of aromatic acids have been shown to be potent reversible inhibitors of MAO-B (Cesura et al., 1987). [N-(2-aminoethyl)-p-chlorobenzamide HCl] is a reversible inhibitor

Compounds examined.	K_i
(i) 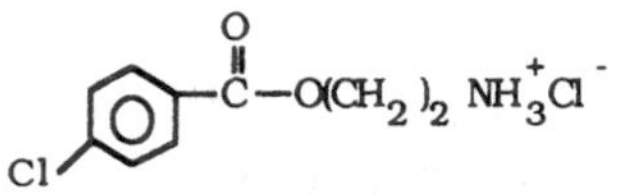 Cl-C$_6$H$_4$-C(=O)-O(CH$_2$)$_2$NH$_3^+$Cl$^-$	7.7 nm
(ii) 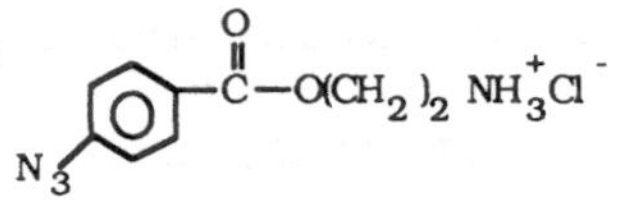 N$_3$-C$_6$H$_4$-C(=O)-O(CH$_2$)$_2$NH$_3^+$Cl$^-$	96.3 nm
(iii) 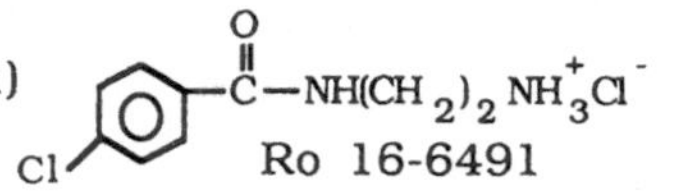 Cl-C$_6$H$_4$-C(=O)-NH(CH$_2$)$_2$NH$_3^+$Cl$^-$ Ro 16-6491	1.5 μm
(iv) 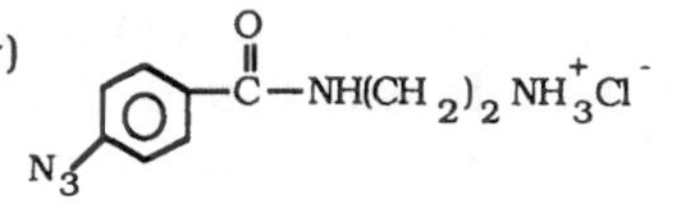 N$_3$-C$_6$H$_4$-C(=O)-NH(CH$_2$)$_2$NH$_3^+$Cl$^-$	13.2 μm
(v) 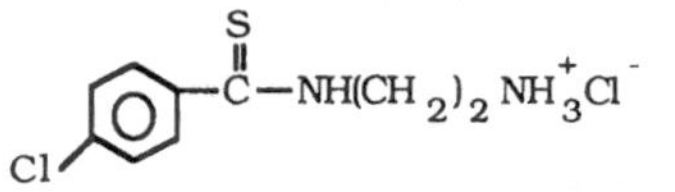 Cl-C$_6$H$_4$-C(=S)-NH(CH$_2$)$_2$NH$_3^+$Cl$^-$	14.2 μm

Fig. 1

of MAO-B which binds selectively and with high affinity to the active site of MAO-B. Studies of the effect of various concentrations of Ro 16-6491 on the oxidation of ^{14}C-PEA by rat brain homogenates with and without preincubation gave K_i values of 0.0375 μmol/l and 0.55 μmol/l respectively. K_D values for Ro 16-6491 with brain preparations and platelet membranes were 47 nM and 108 nM respectively (Cesura et al., 1987). A chloropyridine analogue of the prototype has been found to be even more potent (Cesura et al., 1988). Ro 16-6491 seems first to be oxidised by MAO-B and then deaminated to the aldehyde although with very low efficiency. It is postulated that the inhibitor behaves initially in a reversible competitive manner but then undergoes a change in structure/conformation to a form which shows enhanced affinity for the enzyme (Cesura et al., 1987). We have examined the inhibition of bovine MAO-B by Ro 16-6491 and a number of compounds related to it (Fig. 1). Effects of these inhibitors can be evaluated kinetically using the method of Williams and Morrison (1979). The effect of replacing the amide bond with a carboxylic ester was examined and it was found that the ester compounds exhibit the same inhibitory properties but with significantly increased potency. K_i values for Ro 16-6941 and the ester analogue were 1.5 μM and 7.7 nM respectively.

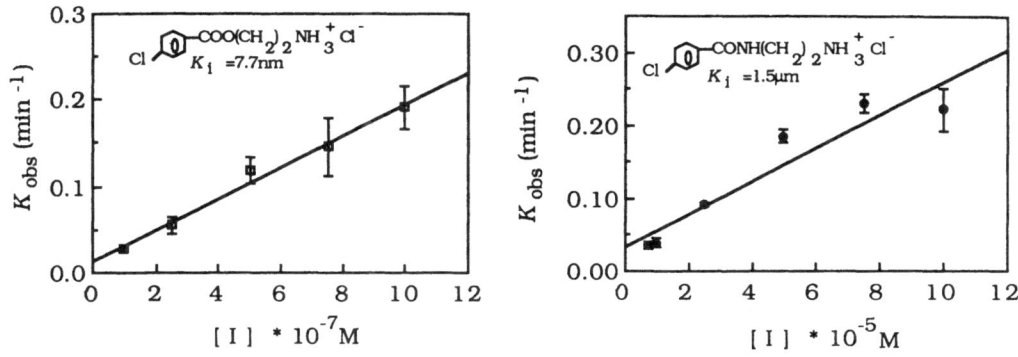

Fig. 2

Materials and methods

Inhibitors were synthesised by conventional methods and identity and purity were established by IR, NMR and TLC. Bovine liver was used as the source of MAO-B, prepared and purified by the method of Salach and Weyler (1987), excluding the sucrose gradient purification step.

Effects of inhibitors on the oxidation of benzylamine, (1 mM), were examined at a series of concentrations of inhibitor by following the increase in A_{250} due to formation of benzaldehyde (90 min, 30°C, pH 7.5). In some experiments the wavelength used was 260 nm to minimise interference from UV absorbance of the inhibitors. Correction was made for the consequent reduction in extinction value. Progress curves were fitted to the Chan Slow Tight Binding Equation shown below, using a computer based method.

$$[P] = V_s t + (V_o - V_s)(1 - e^{k_{obs}t})/k_{obs} + [P]_o$$

V_s is the final steady state velocity, V_o is the initial velocity, k_{obs} is the apparent first order rate constant for the reaction and $[P]_o$ is the concentration of product at $t = 0$. The parameters V_s, V_o and k_{obs} were evaluated for at least four inhibitor concentrations. The K_i and association rate constant for the slow tight binding process were evaluated from a plot of k_{obs} against [I].

$$k_{obs} = k_4 + k_3[I]/(1 + [S]/K_m)$$

$$K_i = k_4/k_3$$

Typical plots for N-(2-aminoethyl)-p-chlorobenzamide hydrochloride and 2-(aminoethyl)-p-chlorobenzoate hydrochloride are shown in Fig. 2.

Results

Figure 2 shows the relationship between k_{obs} and [I] for the slow tight binding inhibition of MAO-B for compounds (i) and (iii). Similar plots were generated for the other compounds and the results can be seen in Fig. 1.

Discussion

It is clear from the analysis that the inhibitors behave as slow tight binders and obey a scheme similar to the one in Fig. 3 where the ratio of the rate

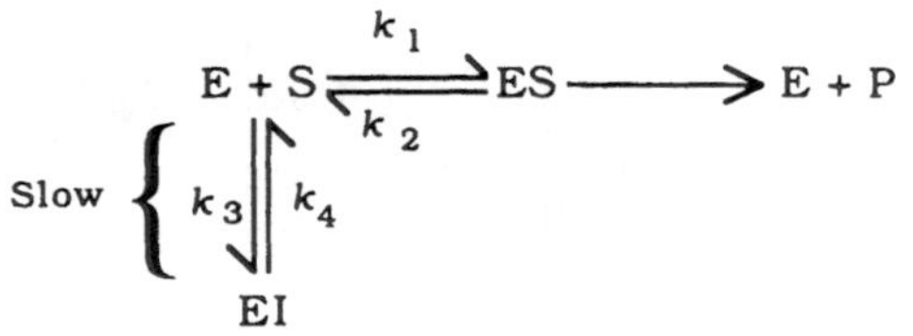

E. L. Mullan and C. H. Williams

$$E + S \underset{k_2}{\overset{k_1}{\rightleftharpoons}} ES \longrightarrow E + P$$

Slow $\Big\{\ \ k_3 \Big\Updownarrow k_4$

EI

Fig. 3

constants k_3 and k_4 gives K_i. It is also clear that the compounds incorporating a carboxylic ester in place of the amide bond show better inhibitory potencies.

There are several possible explanations for this. The most simple is that rotation of the aroyl group with respect to the remainder of the side chain is important for the binding of the inhibitor to the enzyme active site to be optimal. In the esters little or no energy barrier to such rotation exists whereas the barrier to rotation created by the presence of an amide bond is appreciable. A further conformational restraint in the amides (iii) and (iv) may be imposed by conjugation of the delocalised electrons of the amide bond with the aromatic π-system, creating a tendency to coplanarity of these two structures. The energy of activation required for rotation about the central C—N bond increases in going from X=O to X=S, for example:

$PhCXNMe_2$	X=O	X=S
E_a(kcal/mole)	7.5	15.4

Walter and Voss (1970). This is consistent with the finding that the compound (v) is less potent than the corresponding amide. The amide bond is isosteric with the C=C bond of the cinnamylamines which have been shown to be excellent substrates of MAO-B (Williams et al., 1988).

Acknowledgements

We thank Dr. M. Da Prada, Hoffmann La Roche, for a gift of Ro 16-6491 and the department of Education (N.I.) for a post graduate studentship to E. L. Mullan.

References

Cesura AM, Galva MD, Imhof R, Da Prada M (1987) Binding of [³H] Ro 16-6491, a reversible inhibitor of monoamine oxidase type B, to human brain mitochondria and platelet membranes. J Neurochem 48: 170–176

Cesura AM, Imhof R, Takacs B, Galva MD, Picotti GB, Da Prada M (1988) [³H] Ro 16-6491, a selective probe for affinity labelling of monoamine oxidase type B in human brain and platelet membranes. J Neurochem 50: 1037–1043

Salach JI, Weyler W (1987) Preparation of the flavin-containing aromatic amine oxidases of human placenta and beef liver. Methods Enzymol 42: 627–634

Tabor CW, Tabor H, Rosenthal SM (1954) Purification of amine oxidase from beef plasma. J Biol Chem 208: 645–661

Walter W, Voss J (1970) In: Zabicky J (ed) The chemistry of amides. Interscience, London, pp 383–477
Williams CH, Lawson J, Backwell FRC (1988) Oxidation of 3-amino-1-phenylprop-1-enes by monoamine oxidase and their use in a continuous assay of the enzyme. Biochem J 256: 911–915
Williams JW, Morrison JF (1979) The kinetics of reversible tight-binding inhibition. Methods Enzymol 63: 436–467

Authors' address: E. L. Mullan, Division of Biochemistry, Queen's University, Belfast, United Kingdom

J Neural Transm (1994) [Suppl] 41: 313–319

Experimental and theoretical study of reversible monoamine oxidase inhibitors: structural approach of the active site of the enzyme

J. Wouters[1], **F. Moureau**[1], **D. P. Vercauteren**[2], **G. Evrard**[1], **F. Durant**[1], **J. J. Koenig**[3], **F. Ducrey**[3], and **F. X. Jarreau**[3]

[1] Laboratoire de Chimie Moléculaire Structurale, and [2] Laboratoire de Physico-Chimie Informatique, Facultés Universitaires Notre-Dame de la Paix, Namur, Belgium
[3] Centre de Recherche Delalande, Groupe Synthélabo, Rueil-Malmaison, France

Summary. Experimental and theoretical physico-chemical methods were used to investigate the interaction between aryl-oxazolidinones and monoamine oxidase (MAO). Several arguments suggest that these compounds interact with the flavin adenine dinucleotide (FAD) cofactor of MAO. The calculation using ab initio molecular orbital methods of the electronic properties of flavin and befloxatone, a reversible inhibitor of MAO A, led to a description of the interaction between aryl-oxazolidinones and the cofactor of the enzyme. Structure activity relationship results revealed additional sites of interaction with the protein core of MAO A.

As a result of this work, a model is proposed for the reversible inhibition of MAO by oxazolidinones via long distance, reversible interactions with the FAD cofactor of the enzyme.

Introduction

Original inhibitors of monoamine oxidase (MAOIs) have been developed by Delalande Research. They belong to the aryl-oxazolidinone family, i.e. toloxatone (1) (Ferrey, 1985), marketed as a safe antidepressant (Humoryl®) since 1985 and befloxatone (2) (Rovei, 1992), presently in clinical development. They are selective for the A form of the enzyme, reversible, and competitive versus tyramine (see Fig. 1).

These second generation A-MAOI drugs retain the antidepressant efficacity and are devoid of severe food (cheese effect) and drug incompatibilities induced by first generation MAOIs since they cannot act as substrates for the enzyme.

The absence of any amino function in these drugs is essential for reversibility. According to mechanistic studies (Silverman, 1988), it has been shown that irreversible inhibitors (mechanism-based inhibitors) generate reactive species trapped by MAO A via strong covalent bonding with the flavin cofactor of the enzyme.

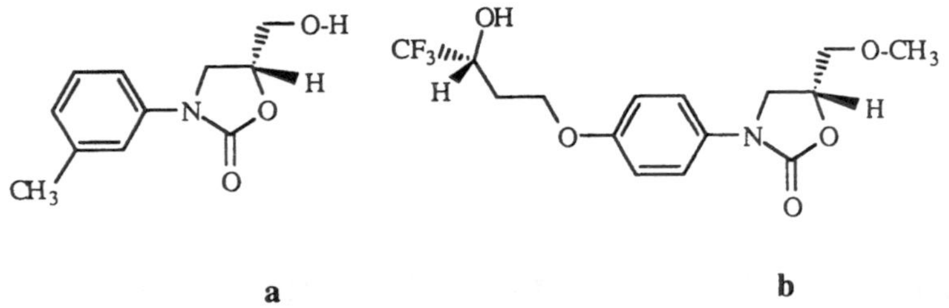

Fig. 1. Toloxatone (1) and befloxatone (2)

In order to try to understand the mechanism of reversible inhibition of monoamine oxidase by aryl-oxazolidinones and to obtain information on the active site of the enzyme, a detailed structural and electronic analysis of befloxatone (2) was undertaken. An X-ray diffraction-crystallographic study revealed the functional groups of the befloxatone which could interact with the MAO active site. Molecular orbitals calculations (ab initio RHF) and structure activity relationship results confirmed the crucial role plaid by some parts of the molecule for interaction with the protein. Finally, a model was proposed where we identify the long distance forces which induce a strong but still reversible association between substituted aryl-oxazolidinones and MAO A.

Material and methods

X-ray diffraction

3-(4-(4,4,4-trifluoro 3-hydroxy butoxy)phenyl) 5-methoxy methyl oxazolidin-2-one crystallized from an AcOEt solution at room temperature. Cell parameters were obtained by least-square refinement of twenty five medium-angle reflections. Intensities collected on an Enraf-Nonius CAD-4 diffractometer were corrected for Lorentz and polarization effects. The structure was solved by direct methods using the SHELX86 program (Sheldrick, 1986). The refinement was performed with SHELX76 (Sheldrick, 1976), by full-matrix least-squares on F. The molecular geometry analysis was carried out by the XRAY76 program (Steward, 1976). The stereoscopic view drawings of the molecular conformation and crystal packing were generated by the ORTEP program (Johnson, 1971).

Ab initio molecular orbital method

Calculations were made at the Restricted Hartree-Fock (RHF) level of electronic theory. At such a level, one considers the independent motion of a single electron in the electrostatic field of fixed nuclei and averaged Coulomb and exchange fields due to other electrons. This level of the theory results in the traditional molecular orbital (MO) language. Within this framework, calculations have been performed at the STO-3G degree of sophistication in the LCAO expansion of the molecular orbitals. The advantages of this basis set are summarized in a review paper (Pople, 1977). For qualitative interpretations, the STO-3G scheme has the great advantage of relating molecular properties to simple atomic parameters and allow for a conceptual approach common to both theoreticians and experimentalists.

The atomic coordinates of the heavy atoms considered in our calculations are obtained by the crystallographic resolution of the structures of lumiflavin (Scarbrough, 1977) and befloxatone (2); all H-atoms were located at standard positions (bond lengths, valence and torsion angles) from their carrier atoms. The generation of the electron charge density two-dimensional iso-contour maps was performed with the MOPLOT (Molecular Orbital Plot) subprogram (Hinde, 1988) available within the MOTECC (Modern Techniques in Computational Chemistry) package (Chin, 1989). All computations were carried out using the GAUSSIAN86 programs (Frisch, 1988) adapted to an IBM 9377/90-FPS M64 computer system. The 2D iso-contour maps were drawn with an in-house device-independent contouring program CPS (Contouring Plotting System) (Baudoux, 1989) developed in Fortran with the IBM graPHIGS software (Chin, 1989).

Results and discussion

The X-ray diffraction analysis of befloxatone (2) reveals that the phenyl oxazolidinone moeity adopts a nearly planar conformation. The analysis of the crystal packing (Fig. 2) shows that important Π-Π type interactions between the phenyl rings stabilize the structure of befloxatone in its solid state. The coplanarity of the phenyl oxazolidinone moeity of the molecules thus allows optimal van der Waals interactions. The mean distances, ca. 3.5 Å, between two molecules parallely stacked (Fig. 2a) is shorter than the sum of the van der Waals (vdW) radii, 3.70 Å, obtained by summing up the vdW radii of two phenyl carbon atoms. The stability of such observed dimeric structure (Fig. 2b) is moreover increased by the presence of inter-molecular hydrogen bonds: the lateral butoxy chain of one molecule folds up to form an intermolecular hydrogen bond between the hydroxyl group of one molecule and the carbonyl of the oxazolidinone moeity of the other.

This analysis of the forces responsible for the crystal packing clearly reveals the functional groups of befloxatone which could interact with the MAO active site, in particular the phenyl oxazolidinone moeity of the molecule which could interact through Π-Π type van der Waals forces and the hydroxyl group of the lateral chain which could lead to intermolecular hydrogen bonds.

Several reasons (Moureau, 1992) lead us already to postulate a privileged interaction of the planar electron-rich phenyl oxazolidinone with the flavin cofactor of the enzyme, which is also planar and known to be electron acceptor. This hypothesis is confirmed by the calculation, through ab initio MO methods, of the electronic properties of flavin and befloxatone which clearly shows the possibility of overlap between the Highest Occupied Molecular Orbital (HOMO) of the donor of electrons, befloxatone, and the Lowest Unoccupied Molecular Orbital (LUMO) of flavin (Fig. 3). This is in agreement with a charge transfer ($\lambda_{max} = 490$ nm) complex as observed in a spectrophotometric study of this MAOI in solution in presence of flavin (data not presented).

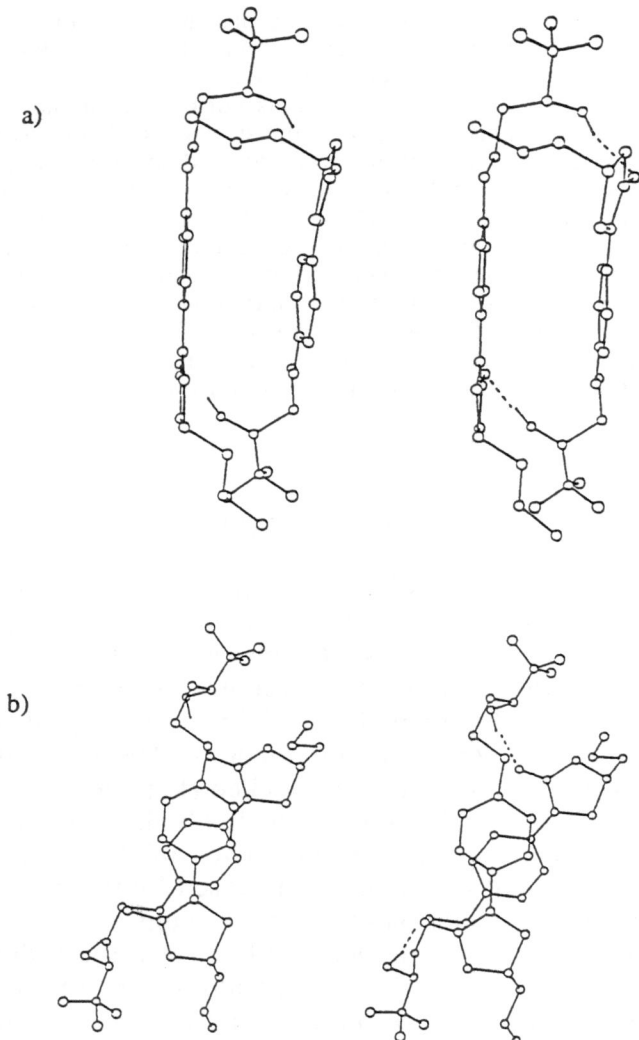

Fig. 2. Stereoview of the crystal packing of befloxatone: **a** view along the c axis; **b** view along the b axis, dotted lines represent intermolecular hydrogen bond

In addition, structure activity relationship results (Koenig, 1992) confirm the crucial role of both side chains of befloxatone. Stereochemical requirements at the oxazolidinone level (R) and for the hydroxyl function (R) argue in favour of an association with the protein core of MAO A for both functional groups of this compound.

Conclusion

As a result of this work, a model is proposed for the reversible inhibition of MAO A by aryl-oxazolidinones (Fig. 4). It is assumed that the inhibitor

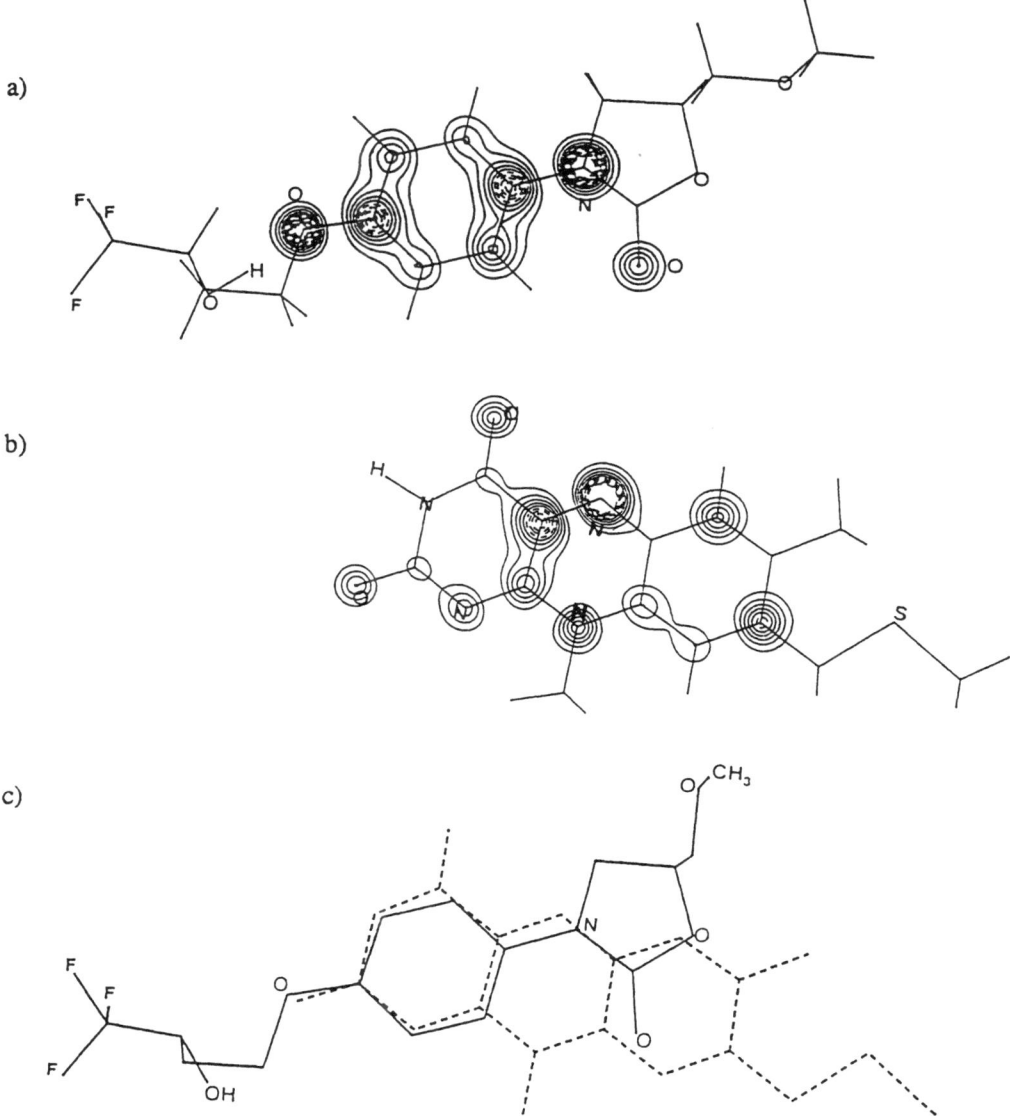

Fig. 3. Superposition of the ab initio HF STO-3G iso-electron density maps (e$^-$/Å^3) of the HOMO of befloxatone (**a**) and the LUMO of SCH$_3$ lumiflavin (**b**); in the deduced relative orientation (**c**), befloxatone is drawn in solid line and lumiflavin in dotted line

interacts via long distance interactions with the enzyme. In this model, we propose that befloxatone is engaged in a molecular association through a double attachment with MAO A. A primary binding site is the covalently bound flavin cofactor which can interact reversibly via Π-Π interactions with the phenyl oxazolidinone moeity of the inhibitor; the side chains of the inhibitor are engaged in additional interactions with the peptide core.

318   J. Wouters et al.

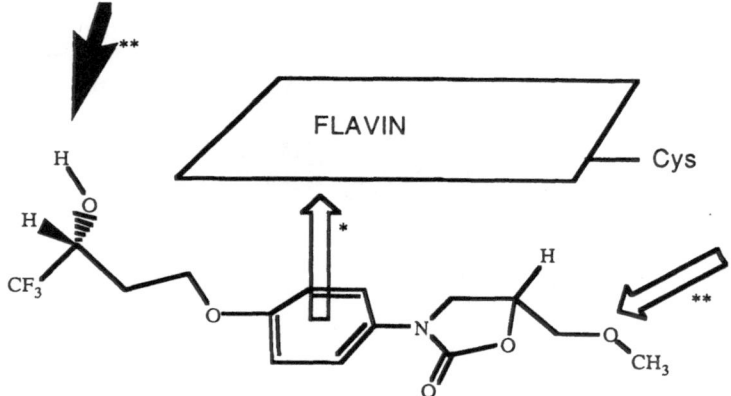

Fig. 4. Proposed model of reversible inhibition of MAO A by befloxatone. * Π-Π type interaction with the flavin cofactor of MAO; ** interactions with the peptide core of the enzyme

Acknowledgements

J. W. acknowledges the Belgian National Foundation for Scientific Research (FNRS) for his Research Assistant position. The authors thank the FNRS, IBM-Belgium, and the Facultés Universitaires Notre-Dame de la Paix for the use of the Namur Scientific Computing Facility.

References

Baudoux G, Vercauteren DP (1989) CPS, a contour plotting system. Facultés Universitaire Notre-Dame de la Paix, Namur

Chin S, Vercauteren DP, Luken WL, Re M, Scateni R, Tagliavini R, Vanderveken DJ, Baudoux G (1989) Modern techniques in computational chemistry. MOTECC 89. ESCOM, Leiden, pp 499–546

Ferrey G, Rovei V, Strolin Benedetti M, Gomeni C, Languillat J (1985) Antidepressant activity of toloxatone, a selective MAO-A inhibitor, in depressed patients. In: Borrows CD, Norman TR, Dennerstein L (eds) Clinical and pharmacological studies in psychiatric disorders. Libbey, Paris, pp 83–86

Frisch MJ, Binkley JS, Schlegel HB, Raghavachari K, Melius C, Martin R, Stewart JJ, Bobrowicz FW, Rohlfing CM, Kahn LR, Defrees DJ, Seeger R, Whiteside RA, Fox DJ, Fleuder EM, Pople JA (1988) GAUSSIAN86 Canegie-Mellon Quantum Chemistry Publishing Unit, Pittsburgh

Hinde RJ, Luken WL, Chin S (1988) IBM Kingston Technical Report KGN-141, June 20th, Kingston

Johnson CK (1971) Program ORTEP II. Oak Ridge Thermal Ellipsoid Plot, ORNL-3794 UC-4-Chemistry, Oak Ridge, Tennessee

Koenig JJ, Moureau F, Vercauteren D, Durant F, Ducrey F, Jarreau FX (1992) Befloxatone, a spontaneously reversible MAO-A inhibitor: modelisation at molecular level. Clin Neuropharmacol 15: 424B

Moureau F, Wouters J, Vercauteren DP, Collin S, Evrard G, Durant F, Ducrey F, Koenig JJ, Jarreau FX (1992) A reversible monoamine oxidase inhibitor, toloxatone: structural and electronic properties. Eur J Med Chem 27: 939–948

Pople JA (1977) In: Schaefer HF (ed) Applications of electronic structure theory. Modern technical chemistry series, vol 4. Plenum, New York, p 1

Rovei V, Thomas J, Koenig JJ, Jarreau FX (1992) Befloxatone, a new oxazolidinone reversible MAO-A inhibitor. Clin Neuropharmacol 15: 425A

Scarbrough FE, Shieh HS, Voet D (1977) The X-Ray crystal structure of the molecular complex bis(lumiflavin-2,6-diamie-9-ethylpurine)-ethanol-water. Acta Crystallogr B33: 2512–2523

Sheldrick GM (1976) SHELX76. A program for crystal structure determination. University of Cambridge, Cambridge

Sheldrick GM (1986) SHELX86. A program for crystal structure determination. Institute of Anorganic Chemistry, University of Göttingen

Silverman RB (1988) Mechanism-based enzyme inactivation. Chemistry and enzymology vol 1, 2. CRC Press, Boca Raton

Steward JM, Machin PA, Dickinson CW, Ammon HL, Heck M, Flack H (1976) XRAY76, Technical report TR-445. Computer Science Center, University of Maryland, Maryland

Authors' address: Dr. J. Wouters, Facultés Universitaires Notre-Dame de la Paix, Laboratoire CMS, 61 Rue de Bruxelles, B-5000 Namur, Belgium.

J Neural Transm (1994) [Suppl] 41: 321–325

Lazabemide (Ro 19-6327), a reversible and highly sensitive MAO-B inhibitor: preclinical and clinical findings

S. Henriot, C. Kuhn, R. Kettler, and **M. Da Prada**

Pharma Division, Preclinical Research, F. Hoffmann-La Roche Ltd, Basel, Switzerland

Summary. Ro 19-6327 (lazabemide, L), MDL 72974, selegiline, AGN 1135 and MDL 72145 were investigated for their MAO inhibitory effect in rat tissues in vitro. The selectivity of MAO-B inhibition of L, selegiline and MDL 72974 was also measured in vitro in human brain tissue as well as ex vivo in rat brain and liver after acute and subchronic administration. Of all compounds investigated L was the most selective for MAO-B inhibition under in vitro and ex vivo conditions. In volunteers, L completely but reversibly inhibited platelet MAO-B with a dose-dependent duration. Clinical trials with L are under way in both Alzheimer's and Parkinson's disease (PD).

Introduction

In several clinical studies MAO-B inhibition in blood platelets is used as a peripheral index for monitoring the degree of MAO-B inhibition in the central nervous system (CNS). The validity of this model was recently clearly demonstrated by a PET-scan study performed with Ro 19-6327 (L, lazabemide) (Bench et al., 1991). Because of the virtual absence of MAO-A activity in human platelets the selectivity of the new MAO-B inhibitors was investigated in vitro in rat and human tissues. Ex vivo studies were performed after acute and subchronic drug treatment since irreversible MAO-B inhibitors in contrast to reversible MAO-B inhibitors lose part of their MAO-B selectivity after subchronic treatments.

Materials and methods

MAO activity was tested radiochemically according to Wurtman and Axelrod with some modifications for rat and human tissues (Da Prada et al., 1989) or human platelets (Kettler and Da Prada, 1989). The substrates were 5-hydroxytryptamine (5HT, 200 μmol/l) for MAO-A, β-phenylethylamine (PEA, 20 μmol/l) for MAO-B and tyramine (200 μmol/l) for both enzyme isoforms. In vitro measurements were performed after preincubation of the homogenate with the inhibitor for 30 min at 37°C.

S. Henriot et al.

Results

The ratios of the IC_{50} values for inhibition of MAO-A and MAO-B in rat brain homogenates for L, MDL 72974, selegiline, AGN 1135 and MDL 72145 (Table 1) indicate that MAO-B inhibition induced by L in vitro is very selectively restricted to the MAO-B; our experiments show that L is about 100 times more selective than selegiline. Also under ex vivo conditions L exhibits the highest selectivity of all compounds investigated (L

Table 1. In vitro and ex vivo MAO-A and MAO-B inhibition in rat tissues (MAO activity was measured as outlined in Materials and methods)

Compound	Brain, in vitro $IC_{50} \times$ nmol/l		ratio MAO-A/MAO-B	Liver, ex vivo $ED_{50} \times$ nmol/kg p.o., 2 h		ratio MAO-A/MAO-B
	5-HT	PEA		5-HT	PEA	
Ro 19-6327	983,000	37	26,568	>1,000,000	53	18,867
MDL 72974	470	0.27	1,740	211,300	86	131
Selegiline	1,400	6	233	248,000	3,600	69
AGN 1135	3,000	30	100	64,000	1,300	49
MDL 72145	5,000	100	50	6,000	400	15

IC_{50} and ED_{50} values were calculated from at least 5 different concentrations or doses using a RS/1 computer program

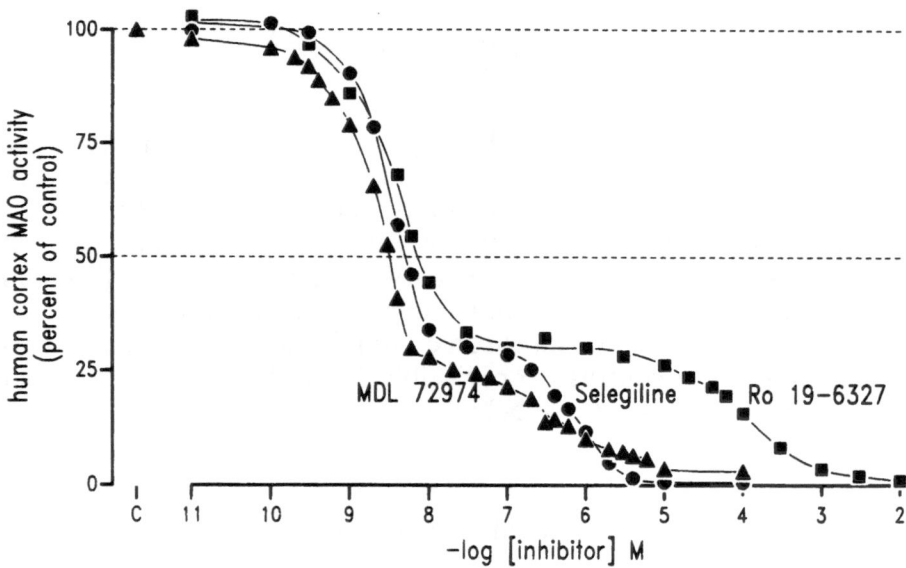

Fig. 1. Inhibition of MAO activity in human cortex homogenates by lazabemide (Ro 19-6327), selegiline or MDL 72974. MAO activity was measured with tyramine, 200 μmol/l as substrate according to Materials and methods. Values are means of two experiments performed in triplicate. The MAO activity in absence of inhibitor amounted to 20.7 nmol/h/mg tissue

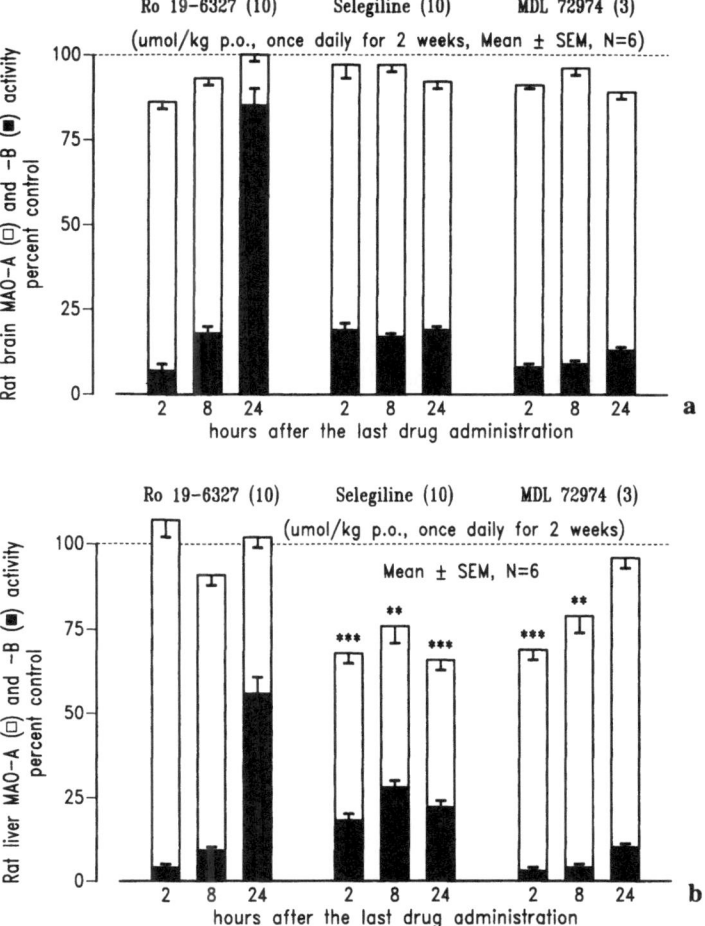

Fig. 2. Activity of MAO-A and MAO-B in brain (**a**) and liver (**b**) of rats, 2, 8 and 24 h after termination of a subchronic treatment with lazabemide (Ro 19-6327), selegiline or MDL 72974, 10, 10 or 3 µmol/kg p.o., respectively, once daily for two weeks. Control MAO-A and (MAO-B) activities (nmoles/h/mg tissue, means ± SEM, N = 5) were: brain 8.2 ± 0.3 (3.1 ± 0.2) and liver 22.1 ± 1.5 (15.5 ± 1.6)

is about 270 times more selective than selegiline) as shown by comparing the ED_{50} values for inhibition of MAO-A and MAO-B in the liver (Table 1). The higher selectivity of L for MAO-B inhibition in comparison to selegiline and MDL 72974 is also underlined by an enzyme titration experiment with human cortex homogenate using tyramine as mixed substrate for both MAO-A and -B. As shown in Fig. 1 all three compounds inhibit MAO-B with near the same potency (IC_{50} about 10 nmol/l — left part of the curve); however, half maximal inhibition of MAO-A (right part of the curve) requires about 100-fold more L than selegiline or MDL 72974.

A comparison of the irreversibly acting selegiline and MDL 72974 with the reversibly acting L for their MAO inhibition after two weeks administration once daily to rats shows the irreversible compounds selegiline and MDL 72974 to produce a significant MAO-A inhibition at doses of 10 or 3 mg/kg, respectively; at least in the case of selegiline, the dose used was not high enough to induce a complete inhibition of MAO-B in the brain (Fig. 2). In contrast, L produces a short-lasting MAO-B inhibition without affecting MAO-A activity.

Discussion

Addition of MAO-B inhibitors to the therapy is postulated to improve efficacy in PD since Birkmayer's findings with selegiline (Birkmayer et al., 1982). Now, selegiline is not specific for MAO-B inhibition (metabolism to amphetamines and limited selectivity for MAO-B inhibition). Consequently, selegiline is not free of side effects in laboratory animals as well as in human beings, e.g. central stimulation or potentiation of the cardiovascular effects of tyramine (Sunderland et al., 1985; Haefely et al., 1990). In contrast, L is free of stimulating effects. Moreover, regarding MAO-B inhibition, L in contrast to all other investigated compounds, is highly selective (Table 1, Fig. 1). L is absolutely free of tyramine potentiating liability in both rats (Haefely et al., 1990) and humans (Dingemanse et al., 1991). Therefore, L offers an unique opportunity to clarify the consequence of highly selective MAO-B inhibition on the progress of degeneration of the nigrostriatal dopaminergic neurons in Parkinson's disease and to determine a potential beneficial effect of MAO-B inhibition in Alzheimer's disease.

References

Bench CG, Prince GW, Lammertsma AA, Cremer JC, Luthra SK, Turton D, Dolan RJ, Kettler R, Dingemanse J, Da Prada M, Bizière M, McClelland GR, Jamieson VL, Wood NB, Frackowiak RSJ (1991) Measurement of human cerebral monoamine oxidase type B (MAO-B) activity with positron emission tomography (PET): a dose ranging study with the reversible inhibitior Ro 19-6327. Eur J Clin Pharmacol 40: 169–173

Birkmayer W, Riederer P, Youdim MBH (1982) (−)Deprenyl in the treatment of Parkinson's disease. Clin Neuropharmacol 5: 195–230

Da Prada M, Kettler R, Keller HH, Burkard WP, Muggli-Maniglio D, Haefely WE (1989) Neurochemical profile of moclobemide, a short-acting and reversible inhibitor of monamine oxidase type A. J Pharmacol Exp Ther 248: 400–414

Dingemanse J, Kettler R, Schmitt M, Fotteler B, Da Prada M (1991) Clinical pharmacology of selective inhibitors of MAO-B and COMT. Abstract F-19-05, 10th International Symposium on Parkinson's Disease. October 27–30, Tokyo

Haefely WE, Kettler R, Keller HH, Da Prada M (1990) Ro 19-6327, a reversible and highly selective monoamine oxidase B inhibitor: a novel tool to explore the MAO-B functions in humans. Adv Neurol 53: 505–512

Kettler R, Da Prada M (1989) Platelet MAO-B activity in humans and stump-tail monkeys: in vivo effects of the reversible MAO-B inhibitor Ro 19-6327. In: Przuntek H, Riederer P (eds) Early diagnosis and preventive therapy in Parkinson's disease. Springer, Wien New York, pp 213–219

Sunderland T, Mueller EA, Cohen RM (1985) Tyramine pressor sensitivity changes during deprenyl treatment. Psychopharmacology 9: 432–437

Authors' address: S. Henriot, Pharma Division, Preclinical Research, F. Hoffmann La-Roche Ltd, CH-4002 Basel, Switzerland.

J Neural Transm (1994) [Suppl] 41: 327–333
© Springer-Verlag 1994

Food-derived heterocyclic amines, 3-amino-1,4-dimethyl-5H-pyrido[4,3-b]indole and related amines, as inhibitors of monoamine metabolism

W. Maruyama[1], A. Ota[2], A. Takahashi[1], T. Nagatsu[3], and M. Naoi[4]

[1] Department of Neurology, Nagoya University School of Medicine, Nagoya,
[2] First Department of Physiology, School of Medicine, and [3] Division of Molecular
Genetics (II) Neurochemistry, Institute for Comprehensive Medical Science, Fujita
Health University, Toyoake, and [4] Department of Biosciences, Nagoya Institute
of Technology, Nagoya, Japan

Summary. The effects of heterocyclic amines, pyrolysis products of tryptophan, on monoamine metabolism were examined. Among these amines, 3-amino-1,4-dimethyl-5H-pyrido[4,3-b]indole (Trp-P-1) and 3-amino-1-methyl-5H-pyrido[4,3-b]indole (Trp-P-2) are potent inhibitors of the enzymes related to amine metabolism. They inhibited type A monoamine oxidase more markedly than type B. After culture of a dopamine cell model, clonal pheochromocytoma PC12h cells, with Trp-P-1 activity of tyrosine hydroxylase was decreased by reduction of its affinity to the biopterin cofactor. Trp-P-1 and Trp-P-2 inhibited tryptophan hydroxylase competitively with the substrate and non-competitively with biopterin. These results suggest that food-derived heterocyclic amines may perturb the monoamine levels in the brain through the inhibition of the biosynthesis and metabolism of biogenic amines.

Introduction

Heterocyclic amines have been isolated and identified in cooked food as products of tryptophan pyrolysis, and they are proved to be carcinogenic and mutagenic (Sugimura, 1986). Out of these carbolines, 3-amino-1,4-dimethyl-5H-pyrido[4,3-b]indole (Trp-P-1) and 3-amino-1-methyl-5H-pyrido[4,3-b]indole (Trp-P-2) were found to inhibit the dopamine metabolism by in vivo and in vitro experiments (Ichinose et al., 1988). These heterocyclic amines inhibited tyrosine hydroxylation in tissue slices of the rat striatum. A dopaminergic cell model, clonal pheochromocytoma PC12h cells, were cultured with Trp-P-1 and Trp-P-2, and the activities of tyrosine hydroxylase [L-tyrosine, tetrahydropteridine: oxygen oxidoreductase (3-hydroxylating), EC 1.14.16.2, TH] and monoamine oxidase [monoamine: oxygen oxidoreductase (deaminating), EC 1.4.3.4, MAO] were reduced

markedly (Naoi et al., 1988). However, the mechanism of the inhibition of
TH and MAO activity was not clarified.

In this article, the inhibition of MAO was studied in detail, and the
mechanism of reduction of TH activity was studied using PC12h cells cultured with the carbolines. In addition, the chemical structure and the origin
of these amines suggest that they may inhibit the serotonin metabolism
through their effects on tryptophan hydroxylase [L-tryptophan, tetrahydropteridine: oxygen oxidoreductase (5-hydroxylating), EC 1.14.16.4, TRH].
The results are discussed in relation to possible involvement of the food-
derived carbolines to biogenic amine systems in the brain.

Materials and methods

Trp-P-1 and Trp-P-2 were purchased from Wako, and other heterocyclic amines were
kindly donated by Dr. T. Sugimura. (6R)-L-erythro-5,6,7,8-tetrahydrobiopterin
[(6R)BH$_4$] was kindly donated from Dr. S. Matsuura. PC12h cells were cultured in the
presence of Trp-P-1 and TH activity was measured by HPLC-ECD as described before (Maruyama et al., 1991). MAO activity was measured fluorimetrically using
kynuramine as substrate (Kraml, 1965). TRH was prepared from murine mastocytoma
P-815 cells and the activity was measured fluorimetrically by HPLC (Naoi et al., 1991).
The value of Michaelis constant, Km, and the maximal velocity, Vmax, were obtained
according to Lineweaver and Burk or by non-linear least square method.

Results

Inhibition of MAO activity by heterocyclic amines

The chemical structures and abbreviations of the heterocyclic amines used
in this article are shown in Fig. 1. Trp-P-2 inhibited type A and B MAO
competitively to the substrate. The inhibitor constant Ki values of the
heterocyclic amines are summarized in Table 1. Trp-P-2 was found to be the
most potent inhibitor of type A, followed by Trp-P-1. These amines inhibited type A more strongly than type B. Most of the amines inhibited type
A in competition to the substrate and some of the amines inhibited type B
in a non-competitive way.

Mechanism of reduced TH activity by culture with Trp-P-1

By culture in the presence of 10 µM Trp-P-1, TH activity in PC12h cells was
reduced markedly by comparison with control. On the other hand, neither
the cell number, cell protein amount, nor the activity of a nonspecific
enzyme β-galactosidase was changed in culture with Trp-P-1 for one day.
Kinetic studies on the TH activity in the control and sample cells are
summarized in Table 2. The Km value in terms of L-tyrosine (TYR) was
not changed, but the Vmax value was significantly decreased, indicating that

Chemical structures

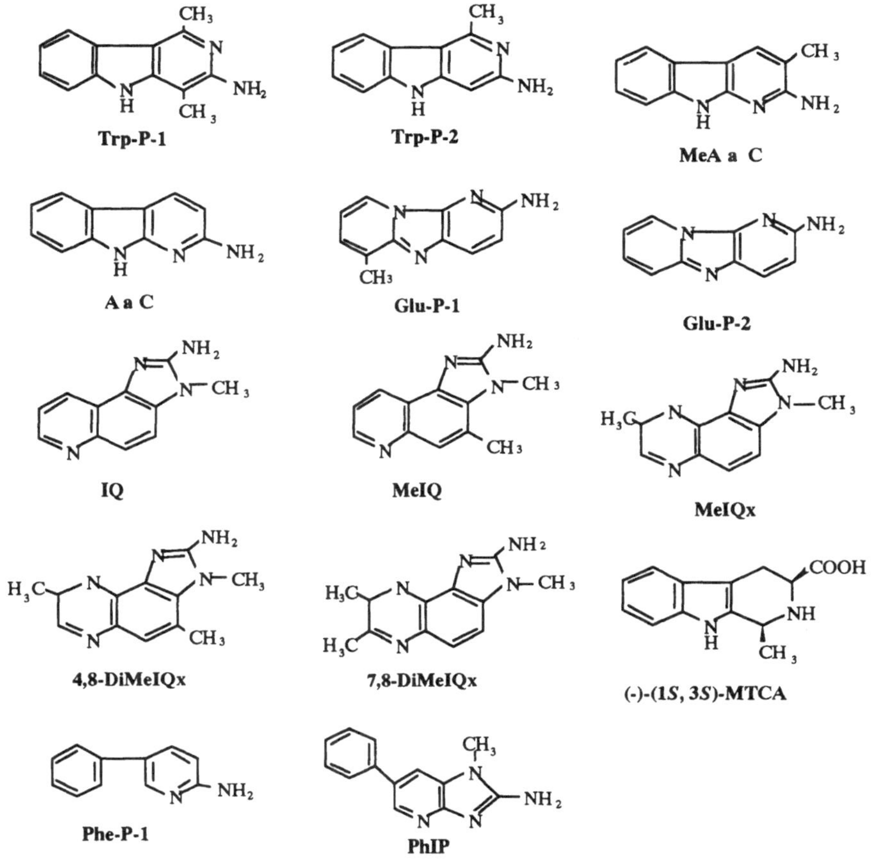

Trp-P-1

Trp-P-2

MeA a C

A a C

Glu-P-1

Glu-P-2

IQ

MeIQ

MeIQx

4,8-DiMeIQx

7,8-DiMeIQx

(-)-(1S, 3S)-MTCA

Phe-P-1

PhIP

Chemical name	Abbreviation
3-Amino-1,4-dimethyl-5H-pyrido[4,3-b]indole	Trp-P-1
3-Amino-1-methyl-5H-pyrido[4,3-b]indole	Trp-P-2
2-Amino-3-methyl-9H-pyrido[2,3-b]indole	MeAaC
2-Amino-9H-pyrido[2,3-b]indole	AaC
2-Amino-6-methyldipyrido[1,2-a:3'2'-d]imidazole	Glu-P-1
2-Aminodipyrido[1,2-a:3'2'-d]imidazole	Glu-P-2
2-Amino-3-methylimidazo[4,5-f]quinoline	IQ
2-Amino-3,4-dimethylimidazo[4,5-f]quinoline	MeIQ
2-Amino-3,8-dimethylimidazo[4,5-f]quinoxaline	MeIQx
2-Amino-3,4,8-trimethylimidazo[4,5-f]quinoxaline	4,8-DiMeIQX
2-Amino-3,7,8-trimethylimidazo[4,5-f]quinoxaline	7,8-DiMeIQX
(-)-(1S, 3S)-1-Methyl-1,2,3,4-tetrahydro-b-carboline-3-carboxylic acid	(-)-(1S, 3S)-MTCA
(-)-(1S, 3S)-1,2-Dimethy-1,2,3.4-tetrahydro-b-carboline-3-carboxylic acid	(-)-(1S, 3S)-DiMT CA
2-Amino-5-phenylpyridine	Phe-P-1
2-Amino-1-methyl-6-phenylimidazol [4,5-b] pyridine	PhIP

Fig. 1. Chemical structures and abbreviation of heterocyclic amines

Table 1. Effects of heterocyclic amines on activities of MAO-A and MAO-B

Compound	MAO-A K_i (μM)	MAO-B K_i (μM)
Trp-P-1	1.76 ± 0.42	36.9 ± 1.6
Trp-P-2	0.84 ± 0.06	137 ± 2
McAαC	27.1 ± 14.9	166 ± 18
AαC	78.7 ± 6.7	557 ± 83*
Glu-P-1	97.1 ± 22.1	107 ± 17
Glu-P-2	72.5 ± 2.3	46.5 ± 6.4
IQ	164 ± 8	454 ± 24*
MeIQ	250 ± 13	313 ± 13*
MeIQx	183 ± 10	242 ± 43*
4,8 DiMeIQx	176 ± 16	402 ± 160*
7,8-DiMeIQx	141 ± 8	386 ± 88*
($-$)-(1S,3S)-MTCA	1500 ± 240	894 ± 54*
($-$)-(1S,3S)-SiMTCA	179 ± 15	680 ± 41*
Phe-P-1	173 ± 10	143 ± 8*
K_m for kynuramine (μM)	46.4 ± 11.4	80.4 ± 8.9
V_{max} for kynuramine (nmol/min/mg protein)	0.81 ± 0.02	3.52 ± 0.963

*Inhibition was non-competitive with the substrate kynuramine, and inhibition by all other compounds was competitive with kynuramine
Each value represents the mean and SD of duplicate measurements of three experiments

Table 2. Kinetic data of TH in PC 12h cells cultured with 10μM Trp-P-1

	K_m (μM)	V_{max} (pmol/min/mg protein)
In terms of L-tyrosine		
control	5.6 ± 0.7	325 ± 15
cells cultured with 10μM Trp-P-1	5.4 ± 0.4	264 ± 6.3
In terms of (6R)BH$_4$		
control	48.7 ± 5.8	507 ± 10
cells cultured with 10μM Trp-P-1	K_{m1}: 16.6 ± 0.1 K_{m2}: 1017.9 ± 0.1	V_{max1}: 110 ± 110 V_{max2}: 590 ± 110

Each value represents the mean and S.D. obtained from duplicate measurements using 8 different concentration of subatrate or (6R)BH$_4$

the inhibition was non-competitive to the substrate. On the other hand, in terms of (6R)BH$_4$, TH in the cells cultured with Trp-P-1 was composed of two components with different kinetic properties; a low Km and Vmax value, and a high Km and Vmax value, while in control only a single component was detected.

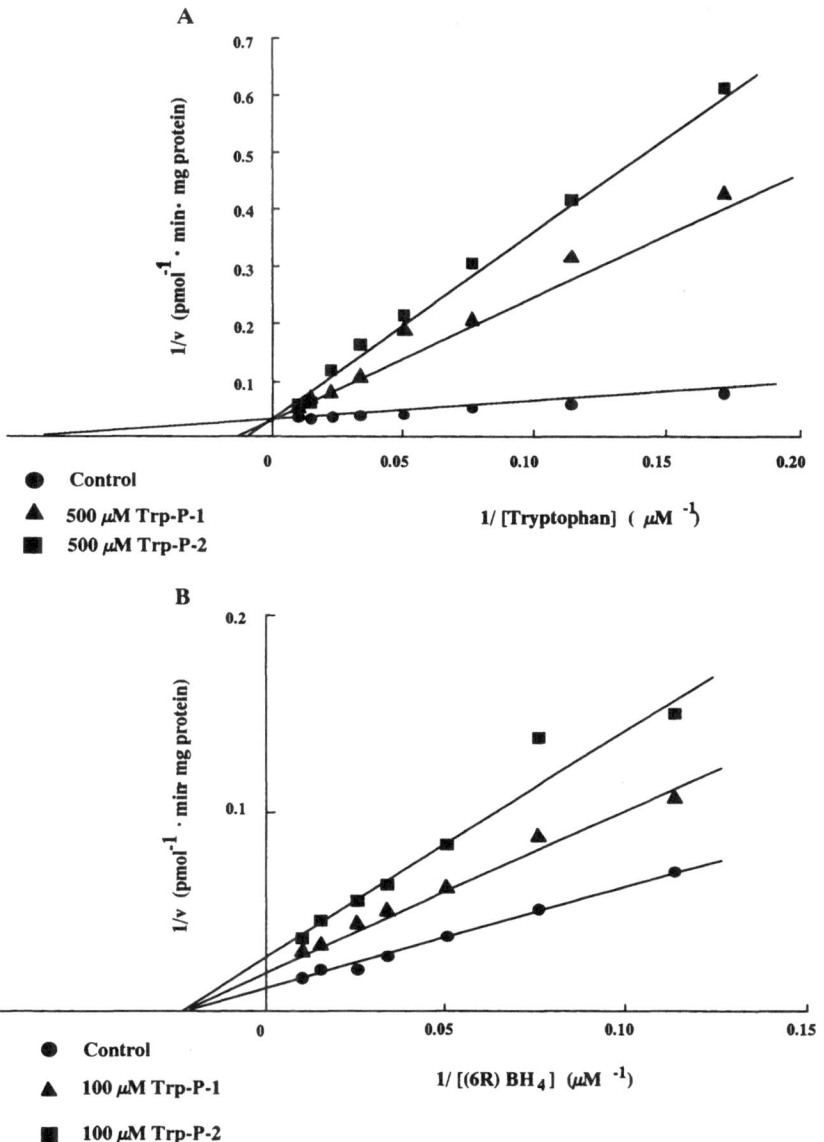

Fig. 2. Effects of Trp-P-1 and Trp-P-2 on the activity of TRH. **A** Effect of the concentration of L-tryptophan on TRH activity in the abscence and presence of Trp-P-1 and Trp-P-2. TRH activity was measured with eight different concentration of L-tryptophan and 1 mM (6R)BH$_4$. The reciprocal of the reaction velocity was plotted against that of the concentration, according to Lineweaver and Burk. **B** Effects of the concentration of (6R)BH$_4$ on TRH activity in the abscence and presence of Trp-P-1 and Trp-P-2. TRH activity was measured with seven different concentration of (6R)BH$_4$ and 100 μM L-tryptophan. The reciprocal of the reaction velocity was plotted against that of (6R)BH$_4$ concentration

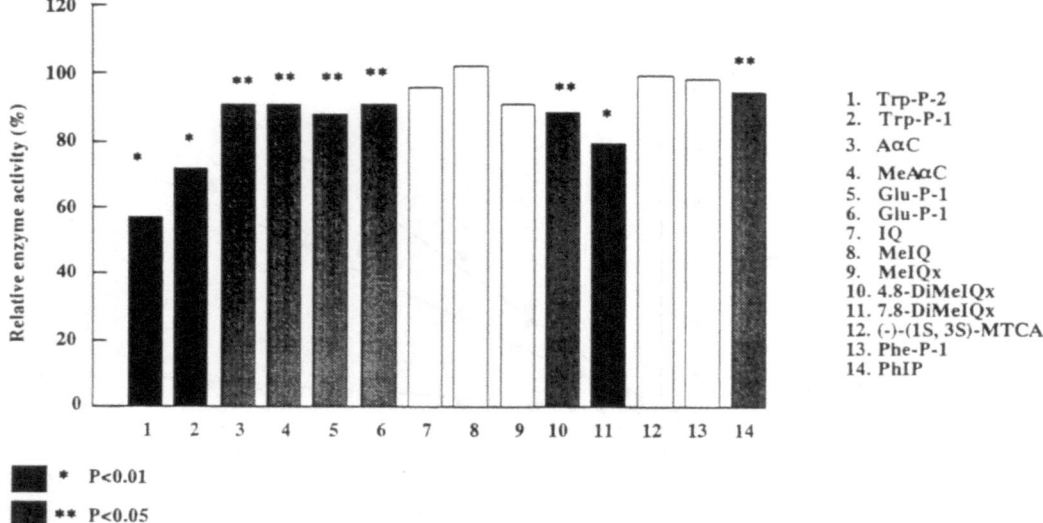

Fig. 3. Effects of heterocyclic amines on TRH activity. Enzyme activity was measured in the presence of $100\,\mu M$ heterocyclic amine, using $10\,\mu M$ L-tryptohan and $1\,mM$ (6R)BH$_4$. Each bar represents mean of triplate measurement of four experiments. *p < 0.01, **p < 0.05, by Student's t-test compared with control

Inhibition of tryptophan hydroxylase (TRH) by heterocyclic amines

Effects of these heterocyclic amines on the activity of TRH were studied. As shown in Fig. 2A, Trp-P-1 inhibited TRH activity in competition to the substrate L-tryptophan (TRY). With respect to (6R)BH$_4$ it inhibited the activity in a non-competitive way, as shown in Fig. 2B. In terms of L-TRY the inhibitor constant Ki value of Trp-P-1 and Trp-P-2 were 99.6 ± 12.9 and $41.7 \pm 5.2\,\mu M$, respectively (mean and SD), and in terms of (6R)BH$_4$, 151 ± 28 and $85.7 \pm 46\,\mu M$. The chemical structure-activity relationships, shown in Fig. 3, indicate that Trp-P-1 and Trp-P-2 are potent inhibitors of TRH.

Discussion

The results on TH activity indicate that Trp-P-1 reduces the affinity to the biopterin cofactor by induction of the conformational changes in TH protein. In the human brain the biopterin is not saturating and TH activity in dopamine cells is dependent on the concentration of biopterin. Reduction of the affinity to the cofactor decreases the activity, since the biopterin cofactor regulates TH activity by inducing the allostery in TH polymers (Minami et al., 1992). Significant reduction of TH mRNA was detected by culture with Trp-P-1 (Ota et al., 1993), which suggests that these amines reduced TH biosynthesis itself in addition to inducing the conformational changes. These results may support the view that the food-derived amines are easily transported into the brain and inhibit catecholamine biosynthesis after log-term accumulation.

Trp-P-1 and other heterocyclic amines inhibit activity of MAO and TRH, in addition to TH; the rate-limiting enzymes of monoamine metabolism in the brain. These results support a view that exogenous or endogenous compounds may perturb the biosynthesis and catabolism of biogenic amines in the brain.

References

Ichinose H, Ozaki N, Nakahara D, Naoi M, Wakabayashi K, Sugimura T, Nagatsu T (1988) Effects of heterocyclic amines in food on dopamine metabolism in nigrostriatal dopaminergic neurons. Biochem Pharmacol 37: 3289–3295

Kraml M (1965) A rapid microfluorimetric determination of monoamine oxidase. Biochem Pharmacol 14: 1683–1685

Maruyama W, Minami M, Ota A, Takahashi T, Takahashi A, Nagatsu T, Naoi M (1991) Reduction of enzymatic activity of tyrosine hydroxylase by a heterocyclic amine, 3-amino-1,4-dimethyl-5H-pyrido(4,3-b)indole (Trp-P-1), was due to reduced affinity to a cofactor of biopterin. Neurosci Lett 125: 85–88

Minami M, Takahashi T, Maruyama W, Takahashi A, Nagatsu T, Naoi M (1992) Allosteric effect of tetrahydrobiopterin cofactors on tyrosine hydroxylase activity. Life Sci 50: 15–20

Naoi M, Takahashi T, Ichinose H, Wakabayashi K, Sugimura T, Nagatsu T (1988) Reduction of enzyme activity of tyrosine hydroxylase and aromatic L-amino acid decarboxylase in clonal pheochromocytoma PC12h cells by carcinogenic heterocyclic amines. Biochem Biophys Res Commun 157: 494–499

Naoi M, Hosoda S, Ota M, Takahashi T, Nagatsu T (1991) Inhibition of tryptophan hydroxylase by food-derived carcinogenic heterocyclic amines, 3-amino-1-methyl-5H-pyrido[4,3-b]indole and 3-amino-1,4-dimethyl-5H-pyrido[4,3-b]indole. Biochem Pharmacol 41: 199–203

Ota A, Maruyama W, Takahashi T, Naoi M, Nagatsu T (1993) The effect of a heterocyclic amines, 3-amino-1,4-dimethyl-5H-pyrido[4,3-b]indole (Trp-P-1), on tyrosine hydroxylase mRNA in PC12h cells. Neurosci Lett 154: 183–186

Sugimura T (1986) Studies on environmental chemical carcinogenesis in Japan. Science 233: 312–318

Authors' address: Dr. M. Naoi, Department of Biosciences, Nagoya Institute of Technology, Gokiso-cho, Showa-ku, 466 Nagoya, Japan.

J Neural Transm (1994) [Suppl] 41: 335–337

Chronopharmacological study of moclobemide effect on the rat pineal melatonin biosynthesis

G. F. Oxenkrug[1], **P. J. Requintina**[1], **K. White**[1], and **A. Yuwiler**[2]

[1] Pineal Research Laboratory, Psychiatry Service, VAMC, and Department of Psychiatry and Human Behavior, Brown University School of Medicine, Providence, Rhode Island, and [2] Neurochemistry Laboratory, West Los Angeles Brentwood VAMC, and Department of Psychiatry and Biobehavioral Sciences, UCLA, Los Angeles, California, U.S.A.

Summary. The chronopharmacological hypothesis of the mechanism of the antidepressant effect of MAO-A inhibitors predicts that clinical efficacy depends upon the time of the day at which the drugs are administered. In the present study moclobemide (10 mg/kg, s.c.) injected at the end of the light phase advanced the onset of the nighttime increase of the rat pineal melatonin biosynthesis while the same dose of drug injected at the end of the dark phase delayed the daytime decrease of melatonin biosynthesis. The results obtained suggest that the timing of MAO-A inhibitor administration should be considered in clinical practice.

Introduction

A large body of literature indicates that depression is associated with circadian rhythm abnormalities and that normalization of the disturbed (advanced or delayed) circadian rhythms is associated with the therapeutic action of antidepressants (Wehr and Wirz-Justice, 1982). We have suggested that selective MAO-A inhibitors normalize circadian rhythms by stimulating melatonin production and that this action constitutes the pharmacological mechanism for the clinical antidepressant effect of MAO inhibitors (Oxenkrug et al., 1986; for review see Oxenkrug, 1991). Recent studies indicate that the response of the melatonin circadian rhythm to exogenously administered melatonin depends upon the time of administration (Lewy et al., 1990). One might hypothesize that the time of administration of MAO-A inhibitors might also influence their effectiveness in altering the melatonin circadian rhythm. We have been unable to find any data addressing this issue in the literature. We have previously demonstrated the stimulation of rat pineal melatonin biosynthesis by the daytime administration of moclobemide (Oxenkrug et al., 1990). In the present study we have investigated the effects of moclobemide administered at the beginning and end of the dark phase on rat pineal melatonin biosynthesis.

Materials and methods

Fisher 344N (3 month old, female) rats were maintained on a 10h dark: 14h light schedule, with lights on at 10:00 and lights off at 20:00, in a temperature-regulated room with food and water ad libitum. Moclobemide (10 mg/kg, s.c.) was injected one hour before the end of the light phase (19:00) or one hour before the end of the dark phase (09:00). Rats were decapitated at the time points indicated below (Table 1). Pineal glands were immediately removed, frozen and kept at $-70°C$ until biochemical evaluation. Pineal contents of melatonin and related indoles were determined by our HPLC-fluorescence procedure as described elsewhere (see Oxenkrug, 1991). Results were expressed as mean + S.D. (ng/pineal) (four animals in each group) and examined by one-way analysis of variance and Student's t-test.

Results

In rats treated with moclobemide one hour before the onset of dark phase (at 09:00) pineal melatonin, serotonin (5-HT) and N-acetylserotonin (NAS) content were higher than in saline-treated rats 2 h after injection (at 11:00). Pineal melatonin biosynthesis was not different between the saline and moclobemide-treated rats during the rest of the dark cycle (Table 1). In rats treated with moclobemide one hour before the end of the dark phase (at 19:00) pineal melatonin, serotonin (5-HT) and N-acetylserotonin (NAS) content were higher than in saline-treated rats within 2 hours after injection (at 21:00).

Discussion

Our data indicate that moclobemide-induced stimulation of the rat pineal melatonin biosynthesis depends on the time of moclobemide administration: moclobemide (10 mg/kg, s.c.) injected at the end of the light phase advanced the onset of the nighttime increase of rat pineal melatonin biosynthesis while the same dose of drug at the end of the dark phase delayed the daytime decrease in melatonin biosynthesis. These results suggest that MAO-A inhibitors might serve as a substitute for the environmental time cues (light/dark cycle), and, thereby, rearrange the disturbed circadian rhythms in depressed patients. Antidepressants including MAO inhibitors are routinely administered several times a day. The present study suggests that such administration might have less effect on circadian rhythm and, therefore, might be therapeutically less effective than once-daily dosing optimally timed. In this vein, a recent study showed that the time of the day of chlorimipramine administration profoundly affected the clinical efficacy of that antidepressant (Nagayama et al., 1991). This effect was not related to the blood concentrations of the antidepressant. As for the MAO inhibitors, circadian rhythmicity in the lethal toxicity of tranylcypromine has been reported (Ross et al., 1981). Clinical chronophamacological studies of MAO-A inhibitors will be of great importance.

Table 1. Moclobemide effect on the rat pineal melatonin and related indoles

	Time decapit.	Melatonin	5-HT	5-HTOL	5-HIAA	NAS
Exp. 1						
Saline	10.00	0.11 ± 0.01	53.4 ± 4.2	1.8 ± 1.07	6.1 ± 0.8	0.13 ± 0.04
M-9.00	10.00	0.46 ± 0.21*	84.1 ± 11.7*	0.4 ± 0.24*	3.0 ± 0.7*	0.31 ± 0.18*
Saline	11.00	0.11 ± 0.03	56.7 ± 6.1	3.7 ± 2.24	6.4 ± 0.8	0.14 ± 0.04
M-9.00	11.00	0.71 ± 0.31*	68.4 ± 4.6	0.3 ± 0.16*	3.7 ± 0.6*	0.41 ± 0.35*
Saline	13.00	2.15 ± 0.34	16.8 ± 3.4	0.7 ± 0.8	2.4 ± 0.9	3.12 ± 1.4
M-9.00	13.00	2.06 ± 0.54	19.7 ± 3.6	0.5 ± 0.36	1.1 ± 0.4*	5.2 ± 1.1*
Saline	16.00	2.01 ± 0.21	8.2 ± 0.78	nd	4.9 ± 0.10	3.6 ± 0.55
M-9.00	16.00	1.88 ± 0.45	9.9 ± 0.40	nd	4.4 ± 0.08	4.2 ± 0.52
Saline	19.00	1.59 ± 0.29	9.1 ± 3.09	nd	7.4 ± 2.2	3.2 ± 1.11
M-9.00	19.00	1.69 ± 0.37	10.3 ± 1.23	nd	7.5 ± 1.11	3.2 ± 0.62
Exp. 2						
Saline	20.00	0.75 ± 0.31	28.8 ± 10.2	0.35 ± 0.27	1.9 ± 0.9	0.6 ± 0.4
M-19.00	20.00	1.71 ± 0.18*	35.1 ± 12.7	0.30 ± 0.28	1.4 ± 0.2	2.8 ± 0.5*
Saline	21.00	0.31 ± 0.08	50.7 ± 4.9	0.66 ± 0.17	3.8 ± 0.2	0.2 ± 0.1
M-19.00	21.00	1.77 ± 0.24*	53.3 ± 3.9	0.21 ± 0.09*	2.1 ± 0.5*	3.1 ± 0.6*

Moclobemide (10 mg/kg, s.c.) was injected at 9.00 (M-9.00) or at 19.00 h (M-19.00); nd not detectable; mean + S.E. (ng/pineal); lights were off at 10.00 and on at 20.00 h; *p < 0.001 vs. saline

Acknowledgements

The authors wish to thank Profs. Da Prada and Haefely for the generous gift of moclobemide.

References

Lewy AJ, Sack RL, Latham JM (1991) Melatonin and the acute suppressant effect of light may help regulate circadian rhythms in humans. In: Arend J, Pevet P (eds) Advances in pineal research, vol 5. J Libbey, London Paris, pp 285–293

Nagayama H, Nagano K, Ikezaki A, Tashiro T (1991) Double-blind study of the chronopharmacology of depression. Chronobiol Int 8: 203–209

Oxenkrug GF (1991) The acute effect of monoamine oxidase inhibitors on serotonin conversion to melatonin. In: Sandler M, Coppen A, Harnett S (eds) 5-Hydroxytryptamine in psychiatry. A spectrum of ideas. Oxford University Press, New York, pp 98–109

Oxenkrug GF, Balon R, Jain AK, McIntyre IM, Appel D (1986) Single dose of tranylcypromine increases human plasma melatonin. Biol Psychiatry 21: 1085–1089

Oxenkrug GF, Requintina PJ, Yuwiler A (1990) Does moclobemide stimulate melatonin synthesis as other selective MAO-A inhibitors do? J Neural Transm [Suppl] 32: 171–175

Ross FH, Graff NL, Sermons AL (1981) Chronotoxicity of antidepressant and psychomotor stimulant in mice. In: Halberg F, Scheving LE, Powell EW (eds) XIII International conference proceedings. Il Ponte, Milano, pp 125–130

Wehr TA, Wirz-Justice A (1982) Circadian rhythm mechanisms in affective illness and in antidepressant drug action. Pharmacopsychiatry 15: 31–39

Authors' address: Dr. G. F. Oxenkrug, Psychiatry Service, VAMC, 830 Chalkstone Ave., Providence, RI 02908, U.S.A.

J Neural Transm (1994) [Suppl] 41: 339–347

Biochemical pharmacology of befloxatone (MD370503), a new potent reversible MAO-A inhibitor

V. Rovei, D. Caille, O. Curet, D. Ego, and F.-X. Jarreau

Synthelabo Recherche, Rueil Malmaison, France

Summary. In vitro and ex-vivo studies show that befloxatone, a new oxazolidinone derivative, is a potent, reversible, competitive and specific MAO-A inhibitor (KiA from 1.9 to 3.6 nM and KiB/KiA ratio between 100 and 400, in the Rat and in Man, depending on the tissue). Befloxatone possesses a marked activity in antidepressant-sensitive behavioral models in rats (from 0.03 to 0.15 mg/kg po) and mice (from 0.21 to 0.29 mg/kg po). At these doses, befloxatone does not induce a significant potentiation of oral tyramine. Befloxatone is devoid of sedative, anticholinergic and cardiovascular effects. Befloxatone is rapidly and extensively distributed in rat brain, the pharmacokinetics are linear in the rat and in man in a large range of doses. Befloxatone is well tolerated in healthy volunteers and is developed as an antidepressant.

Introduction

The new generation of reversible MAO-A inhibitors has provided a significant improvement in the treatment of affective disorders (Jarrott and Vajda, 1987; Nutt and Glue, 1989; Tyrer and Harrison-Read, 1990; Paykel, 1992). These news drugs allow a safer use with respect to the old generation of irreversible MAOIs, particularly with regards to the feared cheese effect and drug interactions (Amrein et al., 1988; Da Prada et al., 1988; Bieck and Antonin, 1989; Fitton et al., 1992). An advantage of reversible MAO-A inhibitors is the control of the intensity and the duration of the enzyme inhibition by the pharmacokinetics of the inhibitor. This, together with the competitivity vs the substrate, minimizes the interaction with tyramine absorbed from food (Rovei et al., 1990).

The specificity of these inhibitors vs MAO-A explain the lack of interactions with other drugs, at least those resulting from the inhibition of cytochromes P-450 implicated in the metabolism of xenobiotics (Callingham, 1992) or the lack of toxicity, i.e. hepatoxicity.

The chemical series of oxazolidinone derivatives has brought to the discovery of the first reversible competitive and specific MAO-A inhibitors,

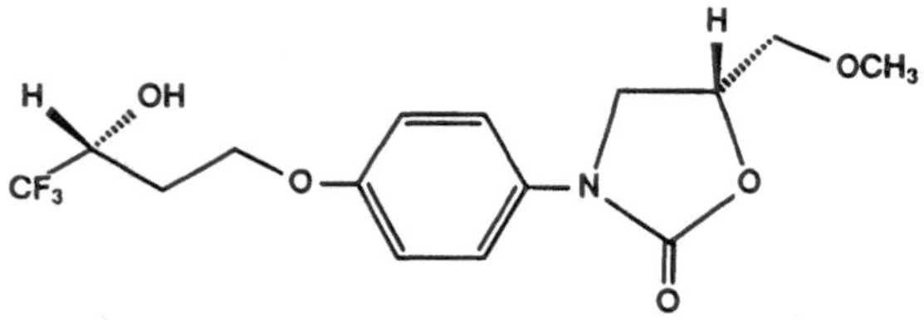

340   V. Rovei et al.

Fig. 1. Befloxatone. 3-(4,4,4-trifluoro 3(R)-hydroxy butoxy phenyl) (R)-methoxy methyl oxazolidin-2-one

the first molecule developed, toloxatone, has been introduced as an anti-depressant in the clinical practice (Ferrey et al., 1985).

This paper presents some biochemical and pharmacological preclinical studies carried out with befloxatone, a new oxazolidinone derivative, which has been selected for development as an antidepressant (see Fig. 1).

Materials and methods

Biochemistry

In vitro MAO-A and MAO-B affinity was determined from brain, intestine, liver and heart homogenates from rats (male Sprague Dawley, 130–180 g) and from human (<24 h post-mortem samples). MAO-A activity was measured with ^{14}C-5HT (spec. activ.: 2.11 GBq/mmole, final conc. 40–325 µM, 45–90 min incubation at 37°C) with concentrations of befloxatone of 2 nM, 5 nM and 10 nM (20–30 min preincubation depending on the species and the tissue). MAO-B activity was measured with ^{14}C-PEA (spec. activ.: 2.1 GBq/mmole, final conc. 2.5–20 µM, 2 min incubation) with concentrations of befloxatone of 200 nM, 600 nM and 1,400 nM. The reaction was stopped with 4N HCl and deaminated metabolites were extracted into toluene/ethyl acetate and quantified by liquid scintillation counting. The KiA and KiB values were determined from the slopes of the linear regression curves, obtained from Lineweaver-Burk plots, vs the concentration of the inhibitor.

Ex vivo MAO-A and MAO-B activity was determined in rats after single doses of befloxatone (from 0.001 to 0.75 mg/kg po). Animals were sacrificed by decapitation and brain immediately removed for determination of the enzymatic activity.

Reversibility of Befloxatone from MAO was evaluated from rat brain homogenates after dilution 1:100 (in vitro) and ex vivo, after single oral doses of the inhibitor. Competitivity of befloxatone was also determined in vitro by deamination of ^{14}C-tyramine (specific activity 1.96 GBq/mmole, final concentration 28–182 µM) in rat intestine homogenate with increasing concentrations of befloxatone (5 nM, 10 nM and 15 nM), after preincubation with 1-deprenyl (10^{-7} M, 20 min at 37°C) and by displacement of ^{3}H-befloxatone (1 nM, preincubated 45 min at 37°C) binding to rat brain membranes with tyramine (5 mM).

Brains levels of 5HT, NA, DA, 5HIAA and DOPAC were determined in rat brain homogenates by HPLC coupled to electrochemical detection (Curet et al., 1992). Plasma free DHPG concentrations (0–24 h) were determined in a group of healthy volunteers after single oral doses of befloxatone to assess the extent of MAO-A inhibition (Ansseau et al., 1992a).

The affinity of befloxatone (10^{-5} M) was evaluated on the following receptor sites: $5HT_1$, $5HT_2$, $5HT_3$, α_1, α_2, $\beta_{1,2}$, D_1, D_2, Ca^{++}, H_1, $M_{1,2}$, opiate (naloxone).

The activity of befloxatone (10^{-5} M) on monoamine uptake systems was determined in vitro in rat brain for 5HT (cortex), DA (striatum) and NA (hypothalamus). In vitro activity of befloxatone on diamine oxidase (rat intestine), semicarbazide sensitive amine oxidase (rat heart) succinate and pyruvate deshydrogenase (rat brain) was also evaluated.

Reference compounds (MAOIs and/or monoamine uptake inhibitors) were compared in each biochemical study.

Pharmacology

Potentiation of L-5-hydroxytryptophan (L-5HTP, 100 mg/kg ip) was evaluated in rats (male Sprague Dawley, 170–240 g) and in mice (male OF1, 20–30 g): effective doses (ED50 po) at 1 hr (befloxatone pretreatment) were determined.

Antagonism of ptosis and hypothermia induced by reserpine (2 mg/kg i.v.) was studied in mice (male OF1, 20–30 g): active doses (AD50 po) were determined at 1 h (ptosis) and 2 hrs (hypothermia).

Reduction of immobility in the forced swimming test was studied in rats (male Sprague Dawley, 170–220 g) after pretreatment of befloxatone (doses at 24, 5 and 1 h before the test): the minimal active dose (MAD) was determined.

Reduction of escape failures (two-way shuttle-boxes) in the learned helplessness paradigm was determined in rats (male Wistar, 170–220 g) during chronic administration of befloxatone (8 administrations): MAD was evaluated.

Potentiation of oral tyramine (12 mg/kg po) by befloxatone (from 0.1 to 2.0 mg/kg po) was studied in conscious freely moving rats (male Sprague Dawley, 280–350 g). Systolic blood pressure was measured during 60 min (catheterism of left femoral artery) after tyramine administered by gavage, mixed in an aqueous suspension prepared from milled and calibrated food pellets. Rats were pretreated with befloxatone 45 min (single dose from 0.2 to 2 mg/kg po) and 60 min (repeated doses, 7 days, 2 × 0.1 mg/kg po/day) before the administration of tyramine, in order to evaluate the interaction at plasma Cmax of befloxatone.

Other CNS and peripheral pharmacological studies of befloxatone were carried out in rats (male and female, Sprague-Dawley) and dogs (male and female Beagles).

Several reference compounds (MAOI as and/or monoamine uptake inhibitors) were systematically compared to befloxatone in the pharmacological studies.

Pharmacokinetics

The pharmacokinetics of befloxatone in plasma and in brain was studied in rats (male Sprague-Dawley, 180–200 g) following single oral doses from 0.063 to 2 mg/kg (Ego et al., 1992). Plasma pharmacokinetics of befloxatone was also investigated in healthy adult volunteers after single doses from 5 to 160 mg (Ansseau et al., 1992a).

Results and discussion

Befloxatone is a potent and selective MAO-A inhibitor, with KiA values from 1.9 to 3.6 nM in different tissues in the rat and in man (Table 1). Selectivity (KiB/KiA) varies in vitro from 100 to 400 depending on tissue and the species, and is maintained ex vivo (data not shown).

Table 1. Affinity of befloxatone for MAO-A and MAO-B
in the rat and in human. Results from different tissues

	Ki SLOPE (nM)					
	MAO-A		MAO-B		KiB/KiA	
	rat	human	rat	human	rat	human
Brain	2.5	2.3	555	269	222	117
Intestine	2.1	1.9	920	466	438	245
Liver	2.4	3.6	533	428	222	119
Heart	2.2	2.1	838	322	381	153

MAO A

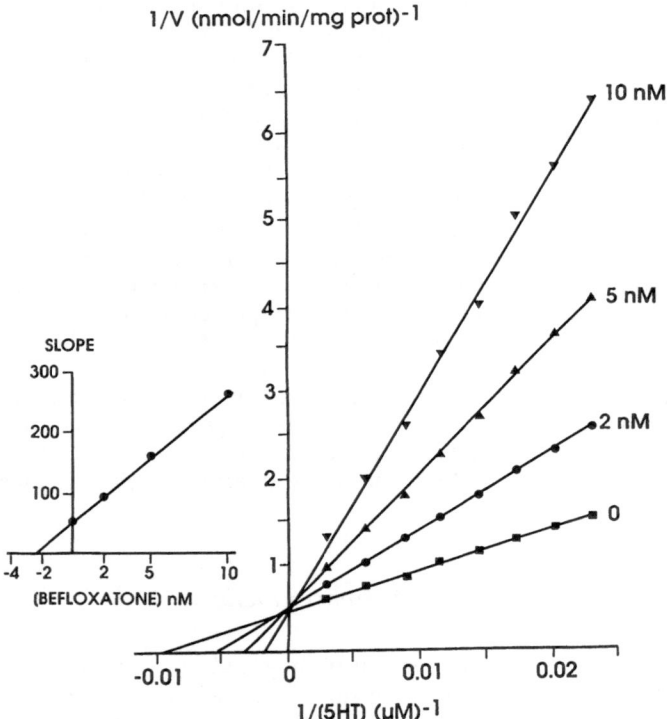

Fig. 2. Inhibition of MAO-A by befloxatone. Rat brain homogenate (0.4 mg/ml)
preincubation of 30 min

The mechanism of inhibition of MAO-A by befloxatone is competitive
(Fig. 2). ^{3}H-Befloxatone binds to brain membranes with a high affinity
(one class of sites Kd = 1.5 nM, Bmax 3.8 pmole/mg prot.); binding of
^{3}H-befloxatone to rat brain membranes is totally dissociated by tyramine
(K − 1 = 0.067 min^{-1}).

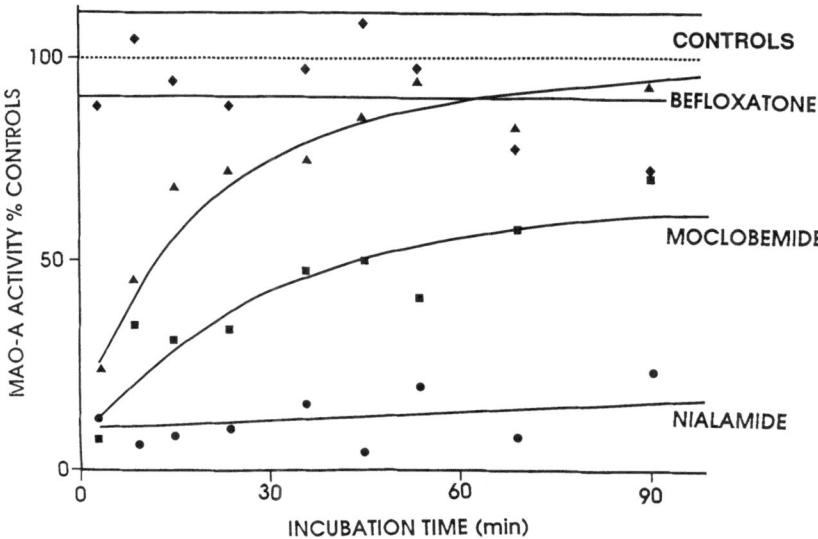

Fig. 3. In vitro reversibility of befloxatone (▲), moclobemide (■) and nialamide (●) after 1:100 dilution (v/v) of a brain homogenate preincubated with each MAOI. Values are given as the percentage of controls (♦)

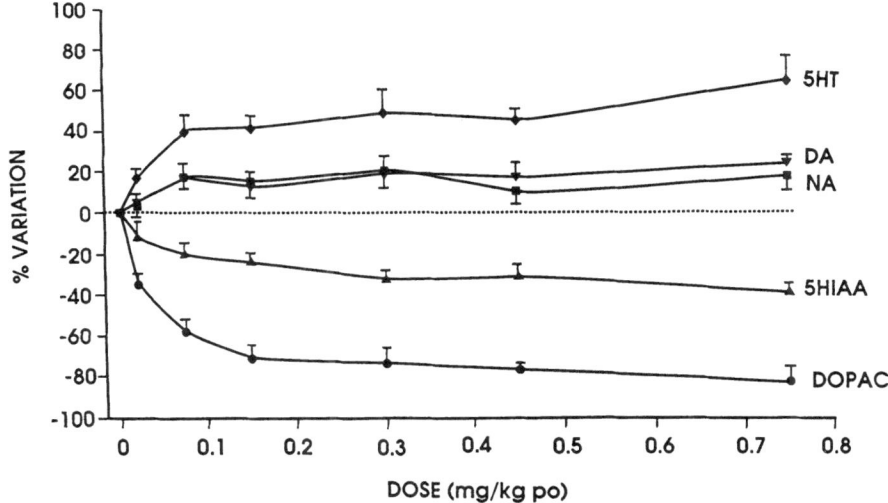

Fig. 4. Levels of monoamines and metabolites in rat brain homogenates obtained 1 h after single oral doses of befloxatone

The reversible inhibition of MAO-A by befloxatone is demonstrated both in vitro (Fig. 3) after dilution of a brain homogenate and ex vivo by a complete recovery of the enzymatic activity 24 h after a oral dose of 0.75 mg/kg. Rat brain levels of 5HT, NA and DA are significantly increased, with a decrease of 5HIAA and DOPAC at the dose of 0.075 mg/kg po of befloxatone. The dose-effect curves on brain monoamines levels and their metabolites after single increasing doses of befloxatone is shown in Fig. 4.

Table 2. Pharmacological activity of befloxatone in antidepressant related behavioral models

	Befloxatone	Moclobemide	Brofaromine	Phenelzine	Nialamide	Imipramine
L-5HTP rat (ED50, mg/kg po)	0.15	1.6	6.9	16	21	>64
L-5HTP mouse (ED50, mg/kg po)	0.21	0.73	2.1	12.3	5.2	>64
Reserpine mouse antag hypot/ptosis (AD50, mg/kg po)	0.29/0.26	1.4/1.4	2.4/1.9	7.3/8.2	6.9/6.8	>8
Forced swimming rat (MAD, mg/kg po)	0.1	7.5	10	10	6.3	50
Learned helplessness rat (MAD, mg/kg po)	0.03	5	0.2	0.6	32	16

Table 3. Interaction between befloxatone and oral tyramine (12 mg/kg po) in conscious freely moving rats

	ΔSBP max (mmHg), mean $\pm$ S.E.M. (n = 12)			
	pretreatment	day 1	day 7	day 8
Befloxatone* (2 × 0.1 mg/kg po/day)	19 ± 2	25 ± 3	27 ± 3	21 ± 3
Phenelzine* (2 × 10 mg/kg po/day)	23 ± 5	41 ± 8	58 ± 6	62 ± 10**

MAOIs were administered for 7 days (2 doses/day) at a dose equivalent to the minimal active dose in the forced swimming test in rats. The strong potentiation of the pressor effect of tyramine observed for phenelzine 24 h after the last dose (day 8) is related to the irreversible inhibition of MAO. *Each animal is its own control. **n = 10

At the highest dose of befloxatone used (0.75 mg/kg po) basal values of monoamines are recovered after 24 h.

Befloxatone is highly specific vs MAO-A since do not bind with high affinity ($>10^{-5}$ M) to a variety of receptor sites. Befloxatone has no activity on monoamine re-uptake systems, on other amine oxidases and on succinate and pyruvate deshydrogenase (FAD containing enzymes).

Befloxatone shows a potent MAOI activity (L-5HTP) in the rat and in the mouse (Table 2). Its activity on antidepressant-sensitive behavioral models is marked. In particular, the activity of befloxatone in the forced swimming test (0.1 mg/kg po) and the learned helplessness paradigm (0.03 mg/ kg po) is observed at very low doses.

The interaction studies with tyramine in conscious freely moving rats show that befloxatone after single administration (up to 0.5 mg/kg po) and during chronic administration (0.1 mg/kg po) does not potentiate (mean ΔSBP max <30 mmHg) the pressor effect of tyramine (12 mg/kg po) (Table 3). Under the same conditions, phenelzine (10 mg/kg po), an irreversible MAOI, gives a significant potentiation. The comparison of theses doses with those obtained in the antidepressant behavioral models (Table 2) suggests that befloxatone possesses a consistent safety factor with regard to cheese effect.

Other studies carried out in pharmacology show that befloxatone (up to 20 mg/kg po) is devoid of other central nervous system (in particular sedative and anticholinergic effects) or cardiovascular effects. This may be related to the specificity of befloxatone vs MAO-A.

Pharmacokinetics of befloxatone in rats are linear over a wide range of doses (Fig. 5). Befloxatone is rapidly absorbed and well distributed in brain. The biphasic curve observed in brain could be attributed both to specific (MAO-A ?) and non specific distribution. As a consequence, the brain/ plasma ratio is particularly high at the low doses, where befloxatone shows a pharmacological activity.

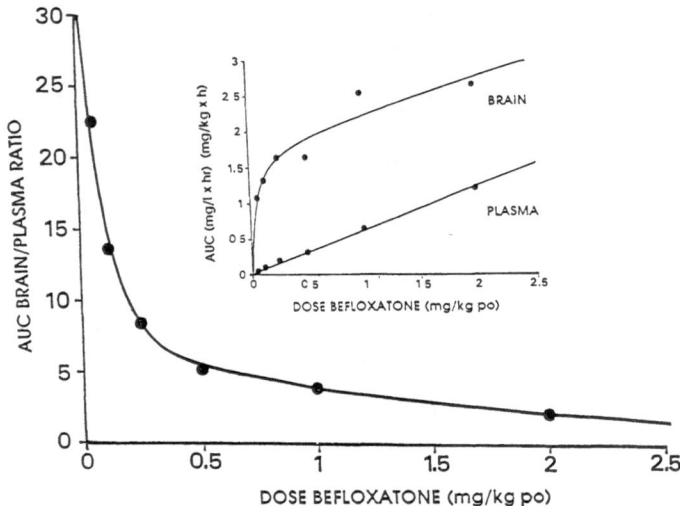

Fig. 5. Plasma and brain concentration of befloxatone, relationship between AUC values and the dose

Befloxatone is eliminated (apparent plasma t1/2 of about 2 h) mainly by metabolism to demethylated, hydroxylated and/or conjugated metabolites. Among these, only the demethyl befloxatone (MD230100) is a MAO-A inhibitor, although 5 to 10 times less active than the parent compound.

The biochemical and the pharmacokinetic properties of befloxatone may explain, at least in part, the potent pharmacological activity of befloxatone.

Phase I studies (Ansseau et al., 1992a,b) show that befloxatone is well tolerated after single (up to 160 mg) and repeated (up to 2 × 40 mg/day) oral administration to healthy adult volunteers. Plasma pharmacokinetics of befloxatone are linear with plasma Cmax obtained at 2–3 h and an apparent t1/2 of 5–6 h. Also in man, befloxatone appears a potent and reversible MAO-A inhibitor: plasma DHPG shows a dose-dependent decrease with a reduction of about 60% (2–4 h after dose) after the dose of 5 mg with a plateau reached after the dose of 10–20 mg (−70/−80%).

The overall data suggest a unique profile of befloxatone both qualitatively and quantitatively, and support further development of the drug as a promising antidepressant. Befloxatone is currently undergoing phase II studies.

Acknowledgements

The authors wish to thank Mrs. S. Haye-Gaunet for the preparation of this manuscript.

References

Amrein R, Allen SR, Vranesic D, Stabl M (1988) Antidepressant drug threrapy: associated risks. J Neural Transm [Suppl] 26: 73–86

Ansseau M, Caillé P, Derks C, Cieren-Puiseux I, Soubrane C, Wauthy J, Ego D, Curet O, Thiola A, Rovei V, Jarreau FX (1992a) Phase I study of single ascending doses of Befloxatone, a new reversible MAO-A inhibitor antidepressant. Proceedings XVIII CINP, Nice. Clin Neuropharmacol 15: 328B

Ansseau M, Caillé P, Derks C, Cieren-Puiseux I, Soubrane C, Wauthy J, Ego D, Rovei V, Jarreau FX (1992b) Phase I study of repeated doses of Befloxatone, a new reversible MAO-A inhibitor antidepressant. Proceedings XVIII CINP, Nice. Clin Neuropharmacol 15: 328B

Bieck PR, Antonin KH (1989) Tyramine potentiation during treatment with MAO inhibitors: brofaromine and moclobemide vs irreversible inhibitors. J Neural Transm [Suppl] 28: 21–31

Callingham BA (1992) Possible drug interactions with reversible MAO inhibitors. Proc XVIII CINP Nice 1992. Clin Neuropharmacol [Suppl 1] 15: 71–225

Curet O, Damoiseau G, Labaune JP, Rovei V, Jarreau FX (1992) Effects of befloxatone, a new potent reversible MAO-A inhibitor, on cortex and striatum monoamines in freely moving rats. 5th International Amine Oxidase Workshop, Galway (Ireland) (unpublished)

Da Prada M, Zurcher G, Wutrich I, Haefely WE (1988) On tyramine, food, beverages and the reversible MAO inhibitor moclobemide. J Neural Transm 26: 31–56

Ego D, Parisy D, Pattano N (1992) Pharmacokinetics of befloxatone, a new reversible inhibitor of MAO-A in rat. Proceedings XVII CINP Nice. Clin Neuropharmacol 15: 422B

Ferrey G, Rovei V, Strolin Benedetti M, Gomeni C, Languillat JM (1985) Antidepressant activity of toloxatone, a selective MAO-A inhibitor, in depressed patients. In: Burrows GD, Norman TR, Dennerstein L (eds) Clinical and pharmacological studies in psychiatric disorders. John Libbey, London Paris, pp 83–86

Fitton A, Faulds D, Goa KL (1992) Moclobemide, a review of its pharmacological properties and therapeutic use in depressive illness. Drugs 43: 561–596

Jarrott B, Vajda FJE (1987) The current status of monoamine oxidase and its inhibitors. Med J Aus 146: 634–638

Nutt D, Glue P (1989) Monoamine oxidase inhibitors: rehabilitation from recent research? Br J Psychiatry 154: 287–291

Paykel ES (1992) Role of monoamine oxidase inhibitors in the treatment of affective disorders. Proc 5th World Congress Biological Psychiatry Florence 1991. Royal Society of Medicine Services, London New York, pp 1–8

Rovei V, Ego D, Jarreau FX (1990) Assessment of the interaction between Humoryl (Toloxatone) and oral tyramine in man vol 1. XVII CINP, Kyoto, p 43

Tyrer P, Harrison-Read P (1990) New perspectives in treatment with monoamine oxidase inhibitors. Int Rev Psychiatry 2: 331–340

Authors' address: Dr. V. Rovei, Synthelabo Recherche, 10 rue des Carrières, F-92500 Rueil Malmaison, France.

J Neural Transm (1994) [Suppl] 41: 349–355

Effects of befloxatone, a new potent reversible MAO-A inhibitor, on cortex and striatum monoamines in freely moving rats

O. Curet, G. Damoiseau, J.-P. Labaune, V. Rovel, and **F.-X. Jarreau**

Synthelabo Recherche, Rueil Malmaison, France

Summary. Single administration of befloxatone (0.75 mg/kg, i.p.) in the rat increased extracellular levels of DA (+300%) in striatum. In frontal cortex, befloxatone (0.75 mg/kg, i.p.) and nialamide (100 mg/kg, i.p.) increased NA by +100% but did not modify 5HT, whereas pargyline (100 mg/kg i.p.) increased extracellular NA and 5HT by 400 and 600%, respectively. At these doses, befloxatone inhibited totally and selectively MAO-A, pargyline inhibited totally MAO-A and MAO-B. Increases of tissue and extracellular concentrations of NA and 5HT were highest after Pargyline suggesting that both monoamines may be metabolized by MAO-A and MAO-B. Befloxatone and nialamide potentiated the effects of idazoxan (20 mg/kg, i.p.) on extracellular NA in frontal cortex, which increased from 350% to 2,000 and 1,500% respectively. These results suggest that α_2-adrenoceptors play a major role in the regulation of extracellular NA in frontal cortex.

Introduction

The new generation of reversible type A monoamine oxidase (MAO) inhibitors has renewed the interest for this class of drugs, whose use in therapeutics has been restricted because of severe side effects (namely cheese effect, drug interactions) associated with the old generation of irreversible MAOIs.

Befloxatone is an oxazolidinone derivative belonging to this new generation of reversible MAO-A inhibitors (Rovei et al., 1992). In vitro and ex vivo biochemical studies demonstrate that befloxatone is a potent, reversible, selective and competitive MAO-A inhibitor in a variety of rat and human tissues (Curet et al., 1992). The specificity of befloxatone is confirmed by the absence of interaction with a variety of neurotransmitters or drug receptor sites, monoamine uptake systems and other amine oxidases. In rat brain, befloxatone increases levels of monoamines and decreases levels of deaminated metabolites in a reversible and dose-dependent manner. Furthermore, befloxatone shows a potent activity in classical anti-depressant tests such as antagonism of reserpine in mouse, and antidepressantsensitive behavioural models in rat: forced swimming and learned helplessness (Caille et al., 1992).

The aim of the present study was to evaluate in vivo the effects of befloxatone on extracellular levels of DA in striatum, 5HT and NA in frontal cortex, DOPAC, HVA and 5HIAA in both regions. In addition MAO-A and MAO-B activities and tissue levels of monoamine and their catabolites have been determined in frontal cortex following administration of befloxatone. The effects of befloxatone have been compared to those of nialamide and pargyline, two irreversible non selective MAO inhibitors. Furthermore, we investigated the involvement of α_2-adrenoceptors in the effect of befloxatone and nialamide on extracellular levels of NA in frontal cortex.

Material and methods

Intracerebral microdialysis

Male Sprague Dawley rats (250–350 g) were anesthetized with chloral hydrate (400 mg/kg, i.p.). A dialysis probe (CMA$_{10}$, Carnegie Medicin, length 4.0 mm, diameter 0.5 mm) was implanted into the frontal cortex (L: 2 mm, DV: −4 mm from the dura, AP: +3.2 mm from bregma, according to the Paxinos and Watson atlas) or into the striatum (L: 2.8 mm, DV: −8 mm, AP: +0.2 mm) and perfused with a Ringer solution (150 mM NaCl, 3 mM KCl, 1.2 mM CaCl$_2$ for NA and DA or 3.4 mM for 5HT) at a rate of 2 μl/min (for DA and 5HT) or 1.4 μl/min (for NA). Experiments were performed in freely moving rats at least 18 h after implantation of the probe. Samples were collected every 20 min (5HT, DA) or 30 min (NA) to be analyzed for monoamines and metabolites. At the end of each experiments, the position of the probe was inspected histologically. Microdialysis samples were injected into an HPLC system immediately following collection.

Assay of monoamine and metabolites

Levels of NA, 5HT, DA, HVA, DOPAC and 5HIAA were determined by HPLC coupled with electrochemical detection. Separation was achieved, at room temperature by reversed phase liquid chromatography. An ultrasphere XL-ODS 3 μM column (4.6 mm × 7 cm, Beckman) was used. For analysis of dialysates, the following mobile phases were used: 0.1 M NaH$_2$PO$_4$, 1 mM EDTA, 1.8 mM octanesulfonic acid, 8% acetonitrile, pH 3.6 for DA, HVA, DOPAC and 5HIAA or 0.1 M NaH$_2$PO$_4$, 1 mM EDTA, 2.7 mM octanesulphonic acide, 6.5% acetonitrile, pH 5.5 for NA, or 0.1 M NaH$_2$PO$_4$, 1 mM EDTA, 0.18 mM octanesulfonic acid, 8% acetonitrile, pH 4.3 for 5HT. For analysis of brain tissue extracts, the mobile phase consisted of 0.1 M NaH$_2$PO$_4$, 1 mM EDTA, 2.5 mM octanesulfonic acid, 6% acetonitrile, pH: 3.6. The flow rate was 0.9 ml/min (Waters model 510). Electrochemical detection was performed using a glassy carbon working electrode (amperometric detector Waters 460) set at +0.8 V versus Ag/AgCl reference electrode.

Determination of MAO-A and MAO-B activities ex vivo in frontal cortex

Male Sprague Dawley rats (180–200 g) were pretreated with befloxatone (0.75 mg/kg, i.p.), nialamide (100 mg/kg, i.p.) or pargyline (100 mg/kg, i.p.) and sacrified at different times after administration. MAO-A and B activities were determined by radioassay as described by Strolin Benedetti et al. (1982) with minor modifications. Frontal cortex was homogenised in 20 vol of ice-cold 0.1 M sodium phosphate buffer pH 7.4. Aliquots (0.1 ml) of the crude membrane suspension were preincubated for 5 min and incubated with a MAO-A substrate, [^{14}C]5HT (specific activity: 2.1 GBq/mmole, Amersham, final concentration 125 μM), for 5 min or with [^{14}C]PEA (specific activity: 2.1 GBq/mmole, NEN, final concentration 8 μM) for 1 min in a total volume of 0.5 ml, at 37°C. The reaction was stopped with HCl 4N and deaminated metabolites were extracted into toluene/ethyl acetate and quantified by liquid scintillation counting. For befloxatone, in order to mimimize the dissociation of the enzyme/reversible inhibitor complex and the consequent underestimation of MAO-A inhibition, frontal cortex samples were homogenated in 3 vol of buffer. Aliquots (50 μl) were preincubated 1 min at 37°C and incubated 15 sec with 10 μM [^{14}C] 5HT in a total volume of 100 μl.

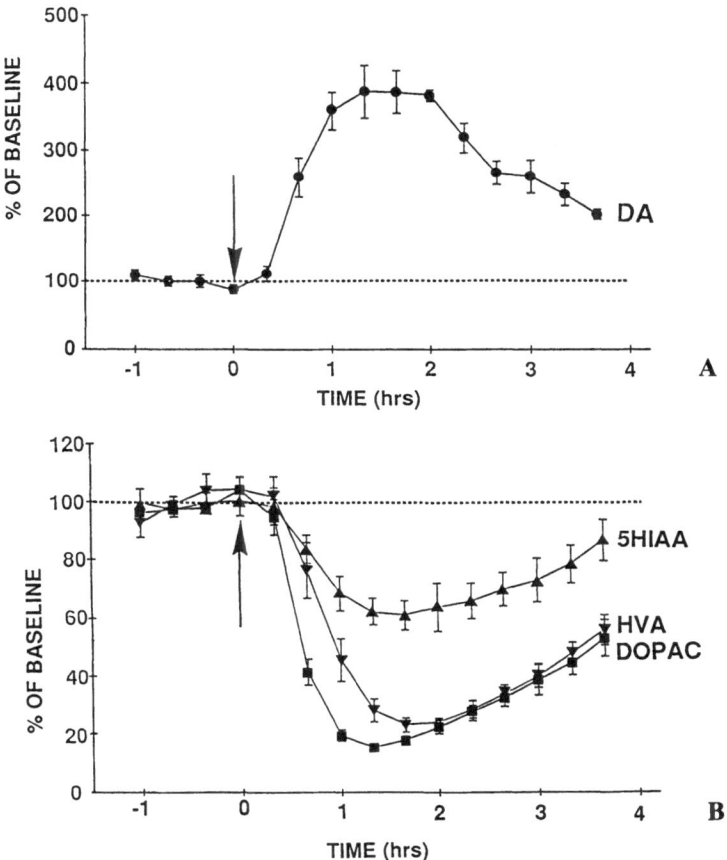

Fig. 1. Effects of befloxatone (0.75 mg/kg, i.p.) on extracellular levels of DA, (**A**) HVA, DOPAC and 5HIAA (**B**) in striatum of freely moving rats. Data are expressed as % (mean ± S.E.R., n = 5) of the mean of 4 samples taken just before drug administration. Basal values (pmoles/20 min): DA: 0.055 ± 0.005 (●), DOPAC: 60 ± 4 (■), HVA: 32 ± 2 (▼), 5HIAA: 12 ± 1 (▲).

Results and discussion

In striatum, befloxatone (0.75 mg/kg, i.p.) increased extracellular DA with a maximum of 300% over basal values, 1–2 h after injection, and decreased extracellular DOPAC, HVA and 5HIAA by 85%, 77% and 40% respectively (Fig. 1). The HVA decline was delayed compared to that of DOPAC. In frontal cortex, befloxatone (0.75 mg/kg, i.p.) did not modify extracellular 5HT (Fig. 2) but decreased DOPAC (−65%), HVA (−65%) and 5HIAA (−25%) 2 hrs after the administration (data not shown). As observed with befloxatone, nialamide (100 mg/kg, i.p.) (Fig. 2) and phenelzine (30 mg/kg, i.p., data not shown) had no effect on extracellular 5HT, whereas pargyline (100 mg/kg, i.p.) induced an increase of 5HT by 600%, 4 h after administration (Fig. 2).

Befloxatone (0.75 mg/kg, i.p.), nialamide (100 mg/kg, i.p.) and pargyline (400 mg/kg, i.p.) increased extracellular NA over basal values by 100, 100 and 400%, respectively (Fig. 2).

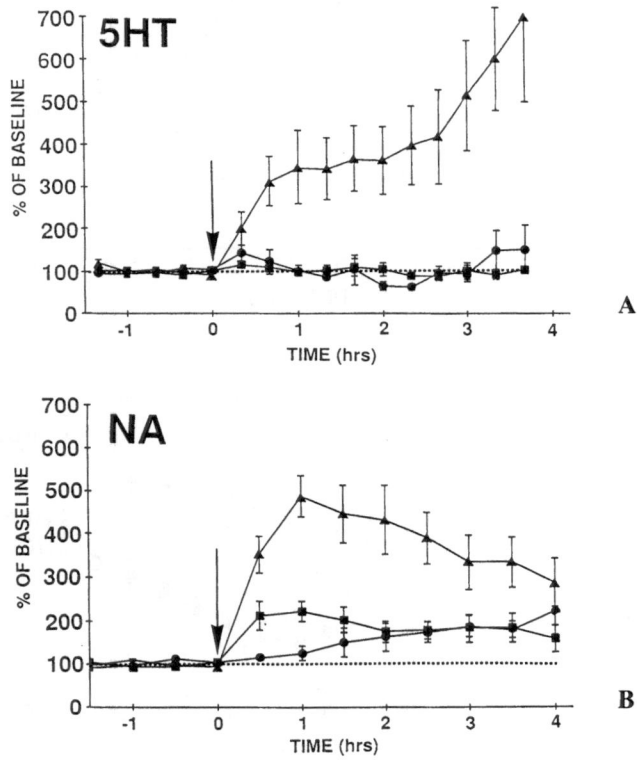

Fig. 2. Effects of befloxatone (0.75 mg/kg, i.p., ■), nialamide (100 mg/kg, i.p., ●) and pargyline (100 mg/kg, i.p., ▲) on extracellular levels of 5HT (**A**) and NA (**B**) in frontal cortex of freely moving rats. Data are expressed as % (mean ± S.E.R., n = 4–5) of the mean of 4 samples taken just before drug administration. Basal values for 5HT: 0.023 ± 0.004 pmoles/20 min and for NA: 0.031 ± 0.003 pmoles/30 min

Table 1. Time course of MAO-A and MAO-B inhibition in rat frontal cortex following acute administration of befloxatone, nialamide and pargyline

	Drugs	Dose (mg/kg ip)	% Inhibition vs time (h)					
			0.5	1	2	3	4	24
MAO A	befloxatone	0.75	97 ± 2	99 ± 1	97 ± 1	88 ± 1	88 ± 1	13 ± 5
	nialamide	100	70 ± 11	72 ± 10	99 ± 0.3	97 ± 2	98 ± 1	95 ± 1
	pargyline	100	99 ± 0.1	98 ± 1	99 ± 0.1	98 ± 0.2	98 ± 0.2	89 ± 0.2
MAO B	befloxatone	0.75	45 ± 5	39 ± 5	29 ± 7	23 ± 7	25 ± 4	8 ± 5
	nialamide	100	13 ± 6	21 ± 7	65 ± 4	83 ± 3	81 ± 2	80 ± 3
	pargyline	100	99 ± 0.1	99 ± 0.5	99 ± 0.5	99 ± 0.1	99 ± 0.1	96 ± 0.5

Control MAO-A and MAO-B activities (pmoles/min/mg tissue) were for befloxatone: MAO-A: 0.023, MAO-B: 0.038 and for nialamide and pargyline: MAO-A: 0.20, MAO-B: 0.070. Results are given as the mean ± S.E.M. of data obtained from six animals per group

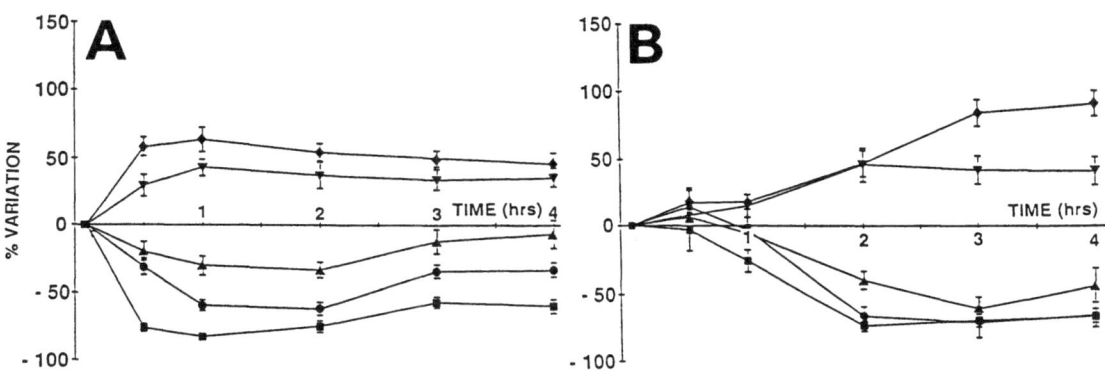

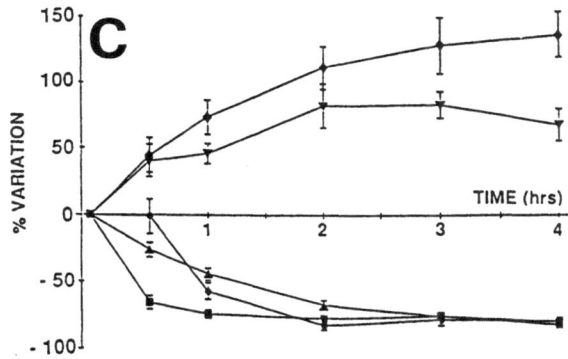

Fig. 3. Effects of a single administration of befloxatone (0.75 mg/kg, i.p.) (**A**), nialamide (100 mg/kg, i.p. (**B**) or pargyline (100 mg/kg i.p.) (**C**) on tissue levels of 5HT (◆), NA (▼), 5HIAA (▲), DOPAC (■) and HVA (●) in frontal cortex of rat. Data are expressed as % of control values (mean ± S.E.R., n = 6). Control levels (pmole/g tissue): 5HT: 3,007 ± 12, NA: 1,618 ± 68, 5HIAA: 1,104 ± 45, HVA: 492 ± 7, DOPAC: 230 ± 16

O. Curet et al.

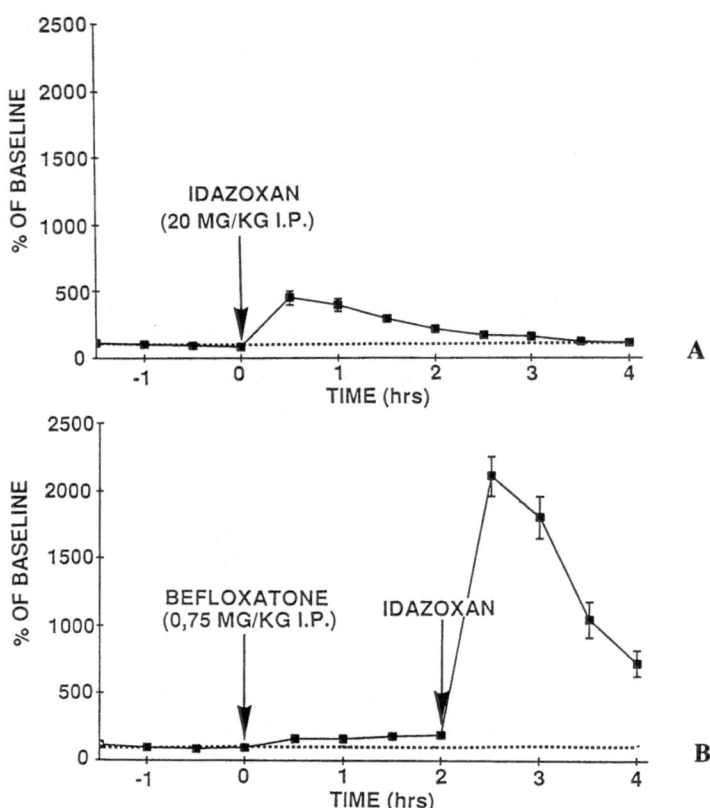

Fig. 4. Effects of befloxatone (**B**; 0.75 mg/kg, i.p.) and idazoxan (**A**; 20 mg/kg, i.p.) on extracellular levels of NA in frontal cortex of freely moving rats. Data are expressed as % (mean ± S.E.R., n = 4–5) of the mean of 4 samples taken just before drug administration. Basal values for NA: 0.023 ± 0.003 pmoles/30 min

The ex vivo MAOIs activity measurements in frontal cortex (Table 1) show that only pargyline (100 mg/kg, i.p.) induced a complete and irreversible inhibition of MAO-A (−99%) and MAO-B (−99%) 30 minutes after administration. Befloxatone (0.75 mg/kg, i.p.) decreased MAO-A in a selective and reversible manner with a maximal effect (−99%) 1 hr after administration. Nialamide (100 mg/kg, i.p.) decreased irreversibly MAO-A with a maximal effect (−98%) at time 2 h and MAO-B (−83%) at time 3 h.

As shown in Fig. 3, befloxatone (0.75 mg/kg, i.p.), nialamide (100 mg/kg, i.p.) and pargyline (100 mg/kg, i.p.) increased tissue levels of 5HT and NA, and decreased levels of DOPAC, HVA and 5HIAA in frontal cortex. Greatest variations of tissue levels of monoamines and their catabolites were observed with pargyline.

The present study shows that pargyline induced the highest increase of extracellular and tissue levels of NA and 5HT in frontal cortex, with a complete inihibition of MAO-A and MAO-B. These data support the view that 5HT and NA may be metabolized in vivo by MAO-B when MAO-A is inhibited. However, an additional effect of pargyline on release of 5HT and

NA cannot be excluded. In addition, the increase of extracellular DA following administration of befloxatone suggest that DA in rat striatum is mainly deaminated by MAO-A.

The involvement of α_2-adrenoceptors in the regulation of extracellular NA after administration of MAOIs was also investigated. As observed in a previous study (Dennis et al., 1987) idazoxan (20 mg/kg, i.p.) increased extracellular NA in frontal cortex by 350% (Fig. 4). Following single administration of befloxatone (0.75 mg/kg, i.p.) or nialamide (100 mg/kg, i.p.) idazoxan increased extracellular NA by 2,000% (Fig. 4) and 1,500% (data not shown), respectively.

Therefore, these results demonstrate that α_2-adrenoceptors play a crucial role in the regulation of extracellular NA during treatment with MAOIs such as befloxatone or nialamide.

Experiments are currently in progress to evaluate the involvement of $5HT_{1A}$, receptors on regulation of extracellular 5HT in frontal cortex following administration of befloxatone.

Acknowledgement

The authors wish to thank Mrs. S. Haye-Gaunet for the preparation of this manuscript.

References

Caille D, Adam R, Bergis OE, Fankhauser C, Gardes A, Martin P (1992) Pharmacological profile of befloxatone, a new reversible MAO-A inhibitor. Proceedings XVIII CINP, Nice. Clin Neuropharmacol 15: 423B

Curet O, Damoiseau G, Aubin N (1992) Biochemical profile of befloxatone, a new reversible MAO-A inhibitor. Proceedings XVIII CINP, Nice. Clin Neuropharmacol 15: 424B

Dennis T, L'Heureux R, Carter C, Scatton B (1987) Presynaptic alpha-2 adrenoceptors play a major role in the effects of idazoxan on cortical noradrenaline release (as measured by in vivo dialysis) in the rat. J Pharmacol Exp Ther 241: 642–649

Rovei V, Caille D, Curet O, Ego D, Jarreau FX (1994) Biochemical pharmacology of befloxatone (MD 370503), a new potent, reversible MAO-A inhibitor. J Neural Transm [Suppl] 41 (this volume)

Strolin Benedetti M, Dostert P, Boucher T, Guffroy C (1982) A new reversible selective type B monoamine oxidase inhibitor, MD780236. In: Kamijo K, Usdin E, Nagatsu T (eds) Monoamine oxidase: basic and clinical frontiers. Excerpta Medica, Amsterdam, pp 209–220

Authors' address: Dr. O. Curet, Synthelabo Recherche, 10 rue des Carrières, F-92500 Rueil Malmaison, France.

J Neural Transm (1994) [Suppl] 41: 357–363

The effects of brofaromine, a reversible MAO-A inhibitor, on extracellular serotonin in the raphe nuclei and frontal cortex of freely moving rats

P. Celada, N. Bel, and **F. Artigas**

Department of Neurochemistry, C.S.I.C., Barcelona, Spain

Summary. The effects of brofaromine, a reversible inhibitor of MAO-A, on the extracellular content of serotonin (5-hydroxytryptamine, 5-HT) and 5-hydroxyindoleacetic acid (5-HIAA) have been studied in two regions of the rat brain (midbrain raphe nuclei and frontal cortex). In both areas, locally infused brofaromine induced dose-dependent increases of 5-HT which were more marked in the raphe nuclei. Brofaromine increased extracellular 5-HT more markedly than clorgyline, suggesting that other factors (i.e. inhibition of 5-HT uptake) may be involved in its local effects. Systemic (3 mg/kg, s.c.) brofaromine did not modify extracellular 5-HT in any brain area examined. In contrast, the concurrent administration of brofaromine and deprenyl led to significant changes in the concentration of 5-HT and 5-HIAA in the brain extracellular space. The results are discussed in relation to the role of MAO-A in the control of 5-HT output.

Introduction

Brofaromine is a new antidepressant drug that inhibits MAO-A (MAO, monoamine oxidase EC 1.4.3.4) in a reversible and competitive manner (Waldmeier et al., 1983). It does not interact with other neurotransmitter receptors and transporters except with the serotonin (5-HT, 5-hydroxytryptamine) uptake site at doses higher than those inhibiting MAO-A (Waldmeier and Stocklin, 1989). Depressive patients treated repeatedly with brofaromine exhibit a reduced accumulation of 5-HT in platelets, as compared to that obtained with phenelzine (Celada et al., 1992). This suggests that inhibition of 5-HT uptake may take place in vivo at therapeutic doses in man.

The brain serotoninergic systems appear to be strongly involved in the mode of action of antidepressant drugs. As a general rule, MAO and uptake inhibitors increase the efficacy of 5-HT-mediated transmission (Blier et al., 1987). 5-HT is deaminated by the A-form of the enzyme (Yang and Neff, 1973; Fowler and Tipton, 1982) and therefore it is a clear candidate to be primarily involved in the actions of brofaromine. 5-HT uptake inhibitors such as clomipramine or fluvoxamine increase the 5-HT output in the vicinity of serotoninergic cell bodies and dendrites more markedly than in frontal cortex (Adell and Artigas, 1991; Bel and Artigas, 1992). The lesser effect in projection areas after the systemic administration of uptake inhibitors is likely to be due to the activation of 5-HT$_{1A}$ somatodendritic autoreceptors by the excess of 5-HT surrounding the cell bodies and dendrites of serotoninergic neurones and further decrease of 5-HT release, as a consequence of an impaired cell firing (Adell and Artigas, 1991). Also, irreversible MAO inhibitors increase the 5-HT output to a larger extent in the raphe nuclei (Celada and Artigas, 1993) and therefore such a negative feed-back may also occur. This mechanism may be of interest to explain the delayed onset of effect of antidepressant drugs. Therefore, we have investigated the effects of brofaromine on extracellular 5-HT and 5-HIAA (5-hydroxyindoleacetic acid) in the raphe nuclei and frontal cortex using intracerebral microdialysis in awake rats with two aims: 1) to evaluate whether brofaromine inhibits 5-HT uptake in the rat brain in vivo and 2) to examine whether brofaromine- as other antidepressant drugs- increases preferentially extracellular 5-HT in the raphe nuclei, as compared to frontal cortex.

Material and methods

Intracerebral microdialysis

Concentric dialysis probes were made as described elsewhere (Adell and Artigas, 1991). The size of the dialysis membrane (Cuprophan hollow fibers; Enka AG) was 3.5 mm long × 0.25 mm OD for both regions. Probes were implanted under sodium pentobarbitone anesthesia 20–24 h before the onset of dialysis experiments. A detailed description of the placement of the probes and the composition of the artificial CSF used to perfuse the probes is given in Adell and Artigas (1991). A flow rate of 0.5 µl/min was used. Under these conditions, mean in vitro recoveries for 5-HT and 5-HIAA through the dialysis membrane were 35% and 27%, respectively. The animals had free access to water and food pellets before and during the experiments. Dialysate fractions (15 µl) were collected every 30 min once stable values were achieved. Four basal fractions (2 h) were collected before drugs were administered. The local administration was performed by dissolving the appropriate amounts in the dialysis fluid. Animals were killed after the experiments and the placement of the cannulae was checked by perfusing methylene blue and inspecting the entire course of the probe.

Biochemical analyses and data treatment

Dialysate 5-HT and 5-HIAA were analyzed using HPLC coupled to electrochemical detection. The separation of indole compounds was performed on a Ultrasphere 3 µm ODS column (7.5 cm × 0.46 cm; Beckman). See Adell and Artigas (1991) for details.

Analysis of variance (ANOVA) has been used to examine the influence of dose and treatment factors. Differences between dose groups have been examined with a two-tail paired t-test. The areas under the curve (AUC) have also been used to examine the overall effects. Statistical significance was set at $P < 0.05$. Asterisks denote a significant effect of the treatment.

Results and discussion

Local administration

We have examined the actions of brofaromine given locally (i.e. through the dialysis probe) in two areas of the rat brain (raphe nuclei and frontal cortex) rich in serotonergic cell bodies and nerve terminals, respectively. This approach enables the study of the local effects of any given drug without the possible confounding effects due to transynaptic mechanisms. Brofaromine increased dose-dependently extracellular 5-HT in frontal cortex and the raphe nuclei (Fig. 1). As noted previously for the irreversible MAO inhibitors tranylcypromine and clorgyline, the local effects of

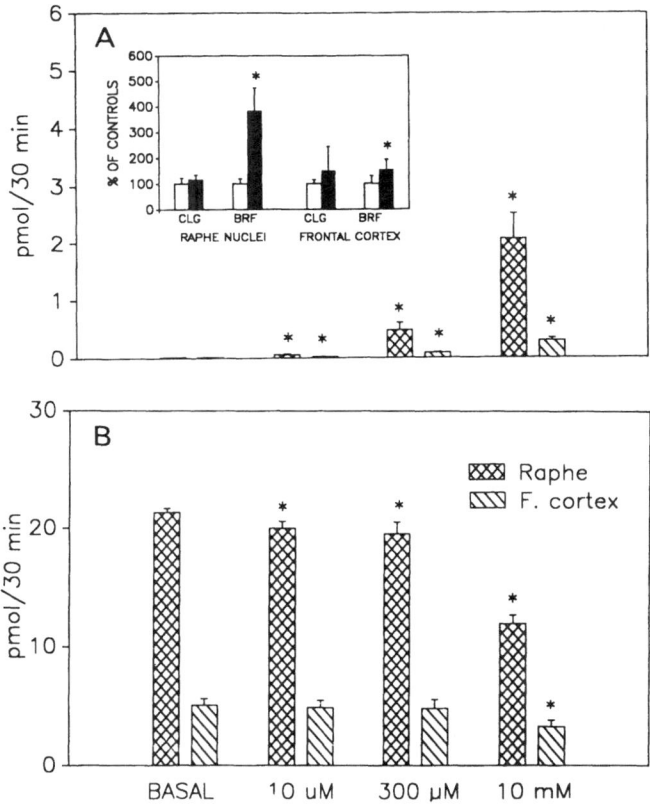

Fig. 1. Dose-related effects of brofaromine on extracellular 5-HT and 5-HIAA in raphe nuclei (cross-hatched columns) and frontal cortex (diagonal columns). Data are given as pmol/30 min-fraction corresponding to four consecutive fractions (2 h) for each animal (N = 6) (means + S.E.M.). Absolute control (basal) values were as follows: 5-HT, 13 ± 4 and 15 ± 3 fmol/fraction; 5-HIAA, 5.1 ± 0.6 and 21.4 ± 0.4 pmol/fraction in frontal cortex and raphe nuclei, respectively. Inset: comparison of the local effects of 10 µM clorgyline and brofaromine in raphe nuclei and frontal cortex. Open bars: control period; filled bars: treatment period (2 h each). Asterisks denote significant effects (P < 0.05)

brofaromine were much more marked in the raphe nuclei. This probably results from a higher 5-HT turnover in this brain area. The increase of extracellular 5-HT elicited by brofaromine was larger than that obtained after the same doses of clorgyline, a selective and irreversible inhibitor of MAO-A (see inset in Fig. 1A). It is interesting to note that clorgyline did not modify the 5-HT output at the lower dose used (10 μM for both compounds) whereas brofaromine induced a large increase of extracellular 5-HT in the raphe nuclei (ca. 380%). The effect in frontal cortex was only slightly higher for brofaromine. The lack of effect of clorgyline suggests that inhibition of MAO-A is not related to the increase elicited by this dose of brofaromine. Instead, it is likely that it can be accounted for by the uptake inhibiting properties of this compound (Waldmeier and Stöcklin, 1989). 5-HT uptake inhibitors at low (1–5 μM) concentrations induce comparable increases of extracellular 5-HT when perfused locally into the raphe using the present experimental conditions (Adell and Artigas, 1991; Bel and Artigas, unpublished). The larger effect of brofaromine in the raphe nuclei is in accordance with the fact that this brain area contains the highest density of the 5-HT transporter (Hrdina et al., 1990). Extracellular 5-HIAA was also decreased by the local application of brofaromine in a dose-dependent manner. The highest dose used (10 mM) reduced 5-HIAA to ca. 60% of control values (56% in raphe nuclei, 64% in frontal cortex). As for clorgyline (data not shown), the effects of the lower doses (10 and 300 μM) were very moderate (ca. 10%) and attained statistical significance due to the low deviation of the results.

Systemic administration

Brofaromine, at a dose known to inhibit brain MAO-A activity by about 90% (Waldmeier et al., 1983), failed to increase extracellular 5-HT in both the raphe nuclei and the frontal cortex. Figure 2A shows the effect of a s.c. dose of 3 mg/kg of brofaromine on dialysate 5-HT and 5-HIAA in the raphe nuclei. 5-HIAA was only slightly reduced in this brain area, whereas it decreased more markedly in frontal cortex (−10% in raphe nuclei, −30% in frontal cortex, AUC of post-treatment period). The unchanged extracellular 5-HT after systemic administration appears to be in overt contradiction with the increase observed after its local perfusion in both brain areas. Yet, it should be taken into account that local doses are continuously perfused for long periods of time into tiny brain areas. Hence, the amount of drug delivered to the tissue surrounding the probe is, by far, much larger than that reached after a single systemic dose and therefore unspecific drug actions may be observed. At the local doses employed (except perhaps the 10 μM dose), brofaromine is probably no longer selective for the A-form of the enzyme. In any case, it is not possible to compare these effects with previously reported in vitro data. In a separate report we have examined the actions of supramaximal doses of irreversible MAO inhbitors on extracellular 5-HT (Celada and Artigas, 1993). The full inhibition of MAO-A

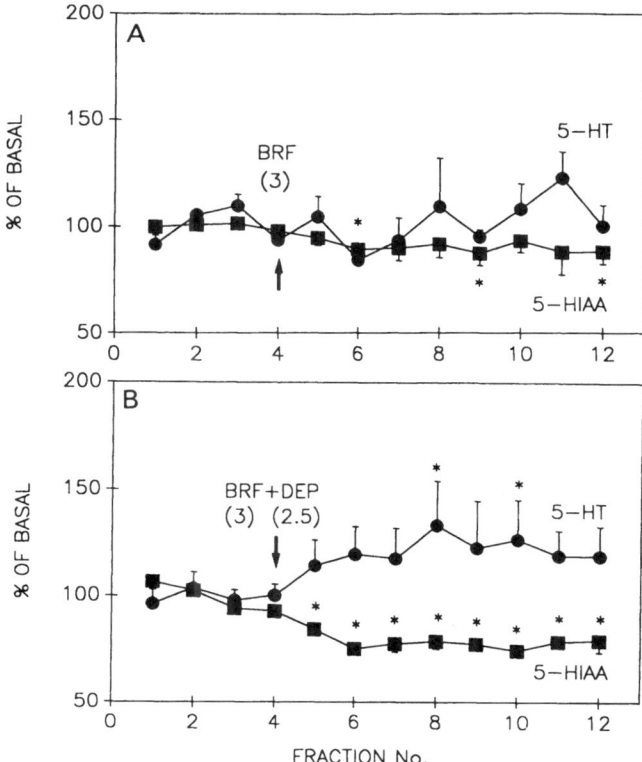

Fig. 2. A Effects of brofaromine (3 mg/kg s.c.) on dialysate 5-HT and 5-HIAA in the raphe nuclei. A very slight reduction of 5-HIAA was noted. **B** Effects of deprenyl (2.5 mg/kg) and brofaromine (3 mg/kg) given together on dialysate 5-HT and 5-HIAA in the raphe nuclei (N = 5 in all cases). Injections are marked by an arrow. Asterisks denote significant effects (P < 0.05)

by a single dose of clorgyline (10 mg/kg i.p.) induced a four-fold increase of brain tissue 5-HT but failed to increase dialysate 5-HT. In contrast, tranylcypromine or the combination of clorgyline and deprenyl- which were ineffective alone- induced very large (ca. 10-fold) increases of the 5-HT output. This suggests that following inhibition of MAO-A, 5-HT can be deaminated by MAO-B, preventing an uncontrolled intracellular rise of 5-HT. When both forms of the enzyme are inactivated, a large outflow of 5-HT occurs, probably due to the saturation of intracellular stores. In this regard, the effects of brofaromine were comparable to those of clorgyline, which did not increase significantly the 5-HT output. Also, both drugs induced a more important reduction of 5-HIAA in the frontal cortex after systemic administration, whereas it was the contrary after their local perfusion (see Fig. 1B for brofaromine, see Celada and Artigas, 1993 for clorgyline). We also examined the effects of the concurrent treatment of brofaromine and deprenyl on extracellular 5-HT and 5-HIAA (Fig. 2B). In agreement with the experiments cited above, the inhibition of MAO-A by brofaromine in combination with an irreversible blockade of MAO-B induced more marked changes in the output of 5-HT and 5-HIAA (compare Fig. 2A and 2B). This gives further support to the idea that both forms of MAO need to be inhibited to induce an acute increase of 5-HT in the brain extracellular space, although this effect was much less marked than that obtained after the combination of clorgyline and deprenyl (Celada and Artigas, 1993). Probably, this can be due to the use of a brofaromine dose that does not fully inhibit MAO-A and/or to the competitive nature of the interaction of the inhibitor with the enzyme. Further work is required to examine the effects of higher doses of brofaromine (alone and in combination with a MAO-B inhibitor) as well as after repeated treatment.

In summary, the present results indicate that the local perfusion of brofaromine increases preferentially the 5-HT output in raphe nuclei (as compared to frontal cortex). This confirms data obtained with clorgyline and tranylcypromine and suggests that the cell bodies of serotoninergic neurones (as compared to nerve terminals) may be preferentially affected by inhibitors of MAO-A. The local effects of brofaromine appear to be a combination of the inhibition of MAO-A and 5-HT uptake. However, a systemic dose that inhibits MAO-A by ca. 90% fails to modify extracellular 5-HT unless MAO-B is inhibited as well. Except for the more marked local effects, the in vivo actions of brofaromine on extracellular 5-HT and 5-HIAA in rat brain are similar to those elicited by clorgyline.

Acknowledgements

This work has been supported by grants from the Fondo de Investigación Sanitaria (89/0387 and 92/368). P. Celada (89/0387) and N. Bel (92/268) are recipients of predoctoral fellowships from the Fondo de Investigación Sanitaria. Financial support from Ciba-Geigy S. A. (Barcelona) is also acknowledged. Thanks are given to J. Torres for skillful technical assistance.

References

Adell A, Artigas F (1991) Differential effects of clomipramine given locally or systemically on extracellular 5-hydroxytryptamine in raphe nuclei and frontal cortex. An in vivo microdialysis study. Naunyn-Schmiedebergs Arch Pharmacol 343: 237–244

Bel N, Artigas F (1993) Fluvoxamine increases preferentially extracellular 5-hydroxytryptamine in the raphe nuclei: an in vivo microdialysis study. Eur J Pharmacol 229: 101–103

Blier P, De Montigny C, Chaput Y (1987) Modifications of the serotonin system by antidepressant treatments: implications for the therapeutic response in major depression. J Clin Psychopharmacol 7: 24S–35S

Celada P, Pérez J, Alvarez E, Artigas F (1992) Monoamine oxidase inhibitors phenelzine and brofaromine increase plasma serotonin and decrease 5-hydroxyindoleacetic acid in major depressive patients. Relationship to clinical improvement. J Clin Psychopharmacol 12: 309–315

Celada P, Artigas F (1993) Monoamine oxidase inhibitors increase preferentially extracellular 5-hydroxytryptamine in the midbrain raphe nuclei. A brain microdialysis study in the awake rat. Naunyn-Schmiedebergs Arch Pharmacol 347: 583–590

Fowler CJ, Tipton KF (1982) Deamination of 5-hydroxytryptamine by both forms of monoamine oxidase in the rat brain. J Neurochem 38: 733–736

Hrdina PD, Foy B, Hepner A, Summers J (1990) Antidepressant binding sites in brain: autoradiographic comparison of [^{3}H]paroxetine and [^{3}H]imipramine localization and relationship to serotonin transporter. J Pharmacol Exp Ther 252: 410–418

Waldmeier PC, Baumann PA, Delini-Stula A, Bernasconi R, Sigg K, Buech O, Feiner AE (1983) Characterization of a new, short-acting and specific inhibitor of type A monoamine oxidase. Mod Probl Pharmacopsychiatry 19: 31–52

Waldmeier PC, Stöcklin K (1989) The reversible MAO inhibitor, brofaromine, inhibits serotonin uptake in vivo. Eur J Pharmacol 169: 197–204

Yang HY, Neff NH (1973) β-Phenylethylamine: a specific substrate for type B monoamine oxidase of brain. J Pharmacol Exp Ther 187: 365–371

Authors' address: Dr. F. Artigas, Department of Neurochemistry, C.I.D., C.S.I.C., Jordi Girona 18-26, E-08034 Barcelona, Spain.

J Neural Transm (1994) [Suppl] 41: 365–370

In vivo effects of the monoamine oxidase inhibitiors Ro 41-1049 and Ro 19-6327 on the production and fate of renal dopamine

M. A. Vieira-Coelho, M. Helena Fernandes*, and **P. Soares-da-Silva**

Department of Pharmacology and Therapeutics, Faculty of Medicine,
Porto, Portugal

Summary. Ro 41-1049 (2 mg/kg), but not Ro 19-6327 (2 mg/kg), was found to reduce (70% to 93% decrease) DOPAC tissue levels in the renal cortex, outer and inner medulla and, to a similar extent, DOPAC concentrations in plasma. This inhibitory effect was, however, not accompained by an increased accumulation of newly-formed dopamine. The results presented show that type A monoamine oxidase is the predominant form of the enzyme involved in the metabolism of renal dopamine and some of the amine which escapes deamination might leave the kidney through the renal vein in the intact from.

Introduction

In kidney, the dopamine responsible for the natriuretic and diuretic effects is believed to have its origin in cortical tubular epithelial cells (Siragy et al., 1989). Renal tissues are, however, endowed with one of highest monoamine oxidase (MAO) activities in the body (Caramona and Soares-da-Silva, 1990) and deamination has been found to represent a major pathway for the inactivation of catecholamines (Kopin, 1985). MAO exists in two forms, type A and B, according to biochemical studies (Youdim et al., 1988) and results of a previous study have suggested that both forms of MAO are present in the rat kidney (Fernandes and Soares-da-Silva, 1990). In rat kidney slices loaded with L-DOPA, inhibition of type A monoamine oxidase (MAO), but not type B MAO, results in an increased accumulation of newly-formed dopamine; in both conditions, however, a decrease in the accumulation of DOPAC has been found to occur (Fernandes and Soares-da-Silva, 1990). This would suggest that type B MAO may be responsible for the deamination of renal dopamine leaving the compartment where the synthesis of dopamine has occurred (Fernandes et al., 1991). Since renal medullary elements are unable to convert L-DOPA into dopamine (Hayashi

---

* Permanent address: Faculty of Medical Dentistry, 4200 Porto, Portugal

et al., 1990; Seri et al., 1990; Soares-da-Silva and Fernandes, 1990), it could be assumed that tissue levels of dopamine in this area would reflect amine levels in the tubular filtrate.

The aim of the present work is to study, under in vivo experimental conditions, the role of type A and B MAO on the deamination of dopamine in rat renal tissues. For that purposes tropolone-treated (75 mg/kg) rats were given exogenous L-DOPA (10 and 30 mg/kg) and the influence of two selective MAO-A and MAO-B inhibitors, respectively Ro 41-1049 and Ro 19-6327 (Da Prada et al., 1989), on the accumulation of both dopamine and DOPAC in cortex and in renal areas known to reflect different phases of the tubular filtrate (outer and inner medulla) studied; levels of dopamine and DOPAC in plasma obtained from the renal vein were also measured.

Materials and methods

L-DOPA (10 and 30 mg/kg) was administered i.p. to three groups of male Wistar rats (weighing 200–280 g) 15 min before sacrifice. The group of animals in which the effect of selective type A or B MAO inhibitors was tested received either Ro 41-1049 (2 mg/kg, i.p.) or Ro 19-6327 (2 mg/kg, i.p.), respectively, 60 min before the administration of L-DOPA; the control animals were injected with the vehicle. All rats in the three experimental groups were pretreated with tropolone (75 mg/kg, i.p.) 75 min before killing in order to obtain complete inhibition of the enzyme catechol-O-methyltransferase (COMT). The rats were killed by decapitation under ether anesthesia and both kidneys removed and rinsed free from blood with saline (0.9% NaCl). The kidneys were placed on a ice cold glass plate, the kidney poles removed and fragments of the cortex, outer and inner medulla prepared with a scalpel. The renal tissues were blotted with filter paper, weighed, minced with fine scissors and placed in 2 ml of 0.2 M perchloric acid at 4°C for 24 h, before assay of catecholamines. In some experiments, blood from the renal vein was collected, the plasma obtained and plasma levels of dopamine and DOPAC determined. The assay of L-DOPA, dopamine and DOPAC in renal tissues and plasma samples was performed by means of high pressure liquid chromatography with electrochemical detection, as previously described (Fernandes and Soares-da-Silva, 1990).

Mean values ± SEM of n experiments are given. Significance of differences between one control and several experimental groups was evaluated by Tuckey-Kramer method. AP value less than 0.05 was assumed to denote a significant difference.

Dopamine hydrochloride, (−)-noradrenaline bitartrate, L-β-3,4-dihydroxyphenylalanine (L-DOPA) and 3,4-dihydroxyphenylacetic acid (DOPAC) were purchased at Sigma Chemical Company (St Louis, MO, USA). Ro 41-1049 ((N-(2-aminoethyl)-5 (m-fluoro-phenyl)-4-thiazole carboxamide hydrochloride) and Ro 19-627 ((N-(2-aminoethyl)-5-chloro-2-pyridine carboxamide hydrochloride) were kindly donated by Prof. M. Da Prada, Hoffmann-La Roche (Basel, Switzerland).

Results

The accumulation of newly-formed dopamine in renal tissues was found to be dependent on the amount of L-DOPA administered. The tissue levels (in nmol/g) of newly-formed dopamine after 10 and 30 mg/kg L-DOPA were, respectively, (n = 5 to 6): inner medulla, 128 ± 11 and 292 ± 9; outer

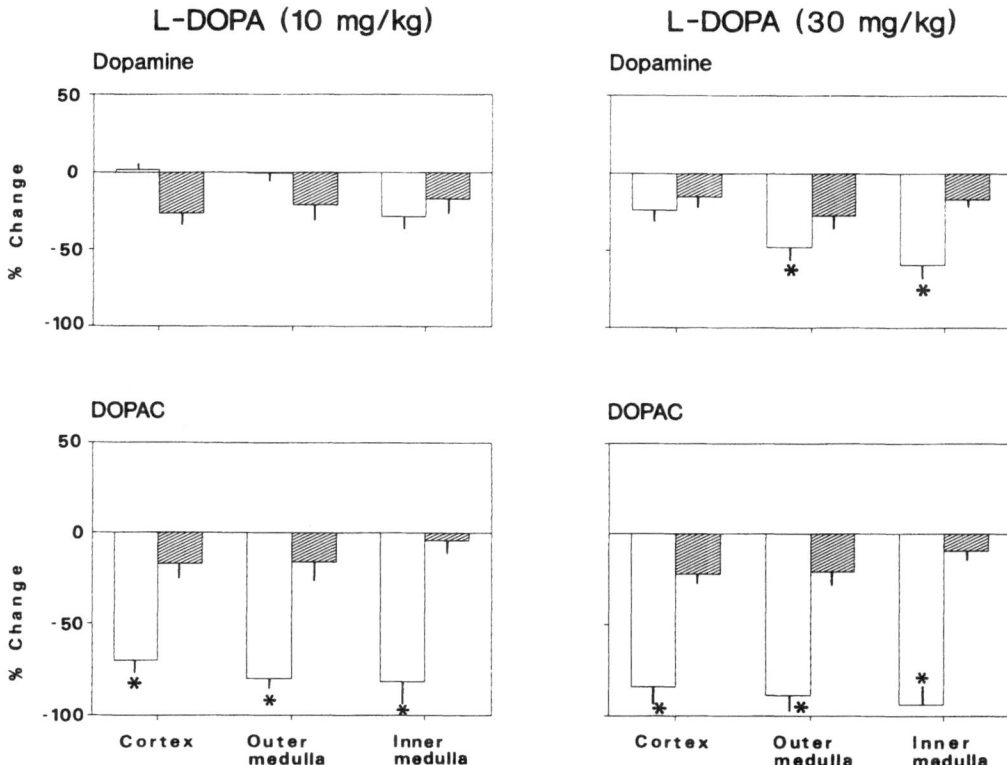

Fig. 1. Efect of Ro 41-1049 (2 mg/kg; open columns) and Ro 19-6327 (2 mg/kg; hatched columns) on dopamine and 3,4-dihydroxyphenylacetic acid (DOPAC) tissue levels in the renal cortex and outer and inner renal medulla of rats given L-DOPA (10 and 30 mg/kg, i.p.). The absolute levels of dopamine and DOPAC in controls are given in the text. Each column represent the mean of five to six experiments per group; vertical lines shower SEM.* Significantly different from corresponding control values (P < 0.02)

medulla, 42 ± 8 and 195 ± 8; cortex, 33 ± 4 and 117 ± 5. Minute amounts of L-DOPA were also found in the three renal areas studied (data not shown). The tissue levels (in nmol/g) of dopamine in non-treated rats were 0.062 ± 0.006, 0.063 ± 0.006 and 0.054 ± 0.04 (n = 6), respectively in the cortex, outer and inner medulla; DOPAC was found undetectable under these conditions. The tissue levels (in nmol/g) of DOPAC after 10 and 30 mg/kg L-DOPA were, respectively, the following (n = 5 to 6): inner medulla, 69 ± 10 and 262 ± 25; outer medulla, 37 ± 6 and 141 ± 10; cortex, 21 ± 3 and 93 ± 4. The tissue levels of noradrenaline in the cortex, outer and inner medulla of non-treated rats were, respectively, 1.2 ± 0.1, 0.22 ± 0.02 and 0.15 ± 0.01 nmol/g (n = 5). Administration of L-DOPA (10 or 30 mg/kg) did not significantly change the levels of noradrenaline in either of the kidney regions under study (data not shown).

In both groups of animals injected with 10 and 30 mg/kg L-DOPA, the administration of Ro 41-1049 (2 mg/kg) resulted in a significant decrease (70% to 93% reduction; P < 0.01) of the DOPAC levels in all kidney

Table 1. Effect of Ro 41-1049 (2 mg/kg) and Ro 19-6327 (2 mg/kg) on dopamine (DA) and 3,4-dihydroxyphenylacetic acid (DOPAC) plasma levels in the renal vein of rats given L-DOPA (30 mg/kg, i.p.). Values are mean ± SEM of five experiments per group and expressed as $p \, mol \, ml^{-1}$

	Dopamine	DOPAC
Control	1.34 ± 0.17	2.33 ± 0.25
Ro 41-1049	1.24 ± 0.15	0.54 ± 0.16*
Ro 19-6327	1.31 ± 0.14	2.23 ± 0.26

* Significantly different from corresponding control values ($P < 0.02$)

regions studied (Fig. 1). The reduction of the DOPAC levels in renal tissues of rats given L-DOPA was, however, not accompanied by an increased accumulation of its parent amine, dopamine. As a matter of fact, in rats given the highest dose of L-DOPA (30 mg/kg) tissue levels of dopamine were even found to be significantly lower than in the respective controls. The reduction of tissue levels of newly-formed dopamine in Ro 41-1049 plus L-DOPA treated rats was found to be higher in the inner medulla (59% reduction; $P < 0.02$), than in the outer medulla (48% reduction; $P < 0.02$) and the cortex (24% reduction; $P = 0.1$). As shown in Fig. 1, previous treatment of rats given 10 and 30 mg/kg L-DOPA with the type B MAO selective inhibitor Ro 19-6327 (2 mg/kg) was found slightly to decrease the levels of DOPAC (9 to 26% reduction; $P = 0.1$). The levels of dopamine in each kidney region of rats given Ro 19-6327 plus L-DOPA were found to be slightly lower than in controls (15% to 26% reduction; $P = 0.1$).

Table 1 shows the levels of dopamine and DOPAC in plasma obtained from blood collected from the renal vein of rats given L-DOPA (30 mg/kg). Pretreatment with the type A MAO inhibitior resulted in a marked reduction of DOPAC plasma levels and no change in dopamine levels. Administration of Ro 19-6327 was found to affect neither DOPAC nor dopamine plasma levels.

Discussion

Most of the dopamine appearing in urine is believed to derive from the decarboxylation of DOPA in epithelial cells of proximal convoluted tubules (Baines and Chan, 1980). This is in line with the finding that aromatic L-amino acid decarboxylase activity is mainly concentrated in the proximal segments of the nephron and almost completely absent in the renal medulla (Hayashi et al., 1990; Seri et al., 1990; Soares-da-Silva and Fernandes, 1990); for this reason, the levels of the amine in the renal medulla, namely after the administration of L-DOPA, may only reflect the presence of the amine in the tubular filtrate. In a previous study, performed under in vitro experimental conditions, we have shown that both forms of MAO are involved in the deamination of dopamine of renal origin; MAO-A was, however, found to be the predominant form of the enzyme involved in the deamination of renal dopamine. To some extent, the results reported in the present study also give support to this suggestion; in fact, only MAO-A inhibition was found to decrease the formation of DOPAC. The finding that the plasma levels of DOPAC in the renal vein were also markedly reduced by Ro 41-1049 further support this suggestion. Other evidence favouring the view that MAO-A in the rat kidney may be of greater importance in the metabolism of dopamine is that the specific activity of the enzyme, determined with the utilization of specific substrates, is almost twice that observed for MAO-B (Fernandes and Soares-da-Silva, 1992). Under in vitro experimental conditions MAO-A inhibition has been shown to result in an increased accumulation of dopamine in renal tissues (Fernandes and Soares-da-Silva, 1990); this is not the case in the present study. In fact, tissue levels of dopamine have been found to be lower after MAO-A inhibition; this might suggest that the newly-formed dopamine which escapes deamination is not accumulated in the compartment where the synthesis has occurred, in contrast with that observed in in vitro conditions. MAO-B has been suggested, however, to have a role in the deamination of the dopamine leaving the compartment where the synthesis has occurred (Fernandes et al., 1991; Pestana and Soares-da-Silva, 1992). The results presented here show that MAO-B inhibition does not result in a significant change in dopamine and DOPAC levels both in the renal tissues and plasma from the renal vein. This would suggest that the role of MAO-B in the deamination of renal dopamine may be only evident in vitro conditions.

In conclusion, the results presented show that MAO-A is the predominant form of the enzyme involved in the metabolism of renal dopamine and some of the amine which escapes deamination might leave the kidney through the renal vein in the intact from.

Acknowledgements

This study was supported by grants from the INIC (FmPl) and the JNICT (CEN 1139/92). V.-C. is recipient of a scholar fellowship from the INIC.

References

Baines AD, Chan W (1980) Production of urine free dopamine from dopa: a micropuncture study. Life Sci 26: 253–259

Caramona MM, Soares-da-Silva P (1990) Evidence for an extraneuronal location of monoamine oxidase in renal tissues. Naunyn Schmiedebergs Arch Pharmacol 341: 411–413

Da Prada M, Kettler R, Keller HH, Cesura AM, Richards JG, Marti JS, Muggli-Maniglio D, Kyburz E, Imhof R (1990) From moclobemide to Ro 19-6327 and Ro 41-1049: the development of a new class of reversible, selective MAO-A and MAO-B inhibitors. J Neural Transm [Suppl] 29: 279–292

Fernandes MH, Soares-da-Silva P (1990) Effects of MAO-A and MAO-B selective inhibitors Ro 41-1049 and Ro 19-6327 on the deamination of newly-formed dopamine in the rat kidney. J Pharmacol Exp Ther 255: 1309–1313

Fernandes MH, Pestana M, Soares-da-Silva P (1991) Deamination of newly-formed dopamine in rat renal tissues. Br J Pharmacol 102: 778–782

Fernandes MH, Soares-da-Silva P (1992) Type A and B monoamine oxidase activities in the human and rat kidney. Acta Physiol Scand 145: 363–367

Hayashi M, Yamaji Y, Kitajima W, Saruta T (1990) Aromatic L-amino acid decarboxylase activity along the rat nephron. Am J Physiol 258: F28–F33

Kopin IJ (1985) Catecholamine metabolism: basic aspects and clinical significance. Pharmacol Rev 37: 333–364

Pestana M, Soares-da-Silva P (1992) Cell outward transport of renal dopamine and the involvement of type A and B monoamine oxidase. J Am Soc Nephrol 3: 833

Seri I, Kone BC, Gullans SR, Aperia A, Brenner BA, Ballermann BJ (1990) Influence of Na^+ intake on dopamine-induced inhibition of renal cortical Na^+-K^+-ATPase. Am J Physiol 258: F52–F60

Siragy HM, Felder RA, Howell NL, Chevalier RL, Peach MJ, Carey RM (1989) Evidence that intrarenal dopamine acts as a paracrine substance at the renal tubule. Am J Physiol 257: F469–F477

Soares-da-Silva P, Fernandes MH (1990) Regulation of dopamine synthesis in the rat kidney. J Auton Pharmacol 10 [Suppl 1]: 25–30

Youdim MBH, Finberg JPM, Tipton KF (1988) Monoamine oxidase. In: Trendelenburg, U, Weiner N (eds) Catecholamines, vol I. Springer, Berlin Heidelberg New York Tokyo, pp 119-192

Authors' address: Dr. P. Soares-Da-Silva, Department of Pharmacology and Therapeutics, Faculty of Medicine, 4200 Porto, Portugal.

J Neural Transm (1994) [Suppl] 41: 371–375

Pharmacodynamics of MDL 72974A: absence of effect on the pressor response to oral tyramine

C. Hinze[1], M. Kaschube[1], and J. Hardenberg[2]

[1] Marion Merrell Dow, C.P.U., Kehl, Federal Republic of Germany
[2] Department of Clinical Research, Strasbourg, France

Summary. MDL 72974A, a new irreversible selective inhibitor of monoamine oxidase (MAO)-B which is not metabolized to amphetamine-like compounds, is currently being developed for the treatment of Parkinson's disease. In this double blind, placebo controlled randomized study 24 healthy volunteers (n = 6/dose) received single oral doses of placebo, 1, 12 or 24 mg of MDL 72974A qd over two weeks. Sensitivity to orally administered tyramine was determined under fasting conditions before and after drug administration and the doses of tyramine yielding a 30 mmHg increase of SBP (PD30) compared. The 2-fold increase of tyramine sensitivity at end of treatment seen at all MDL 72974A dose levels, however, is within the variability range of the tyramine pressor response. MDL 72974A selectively inhibits MAO-B at doses up to 24 mg orally and has a favourable safety profile.

Introduction

Monoamines are oxidized by at least two isoenzymes of mammalian monoamine oxidase (MAO), denoted MAO-A and MAO-B (Johnston, 1968). Tyramine, an indirectly-acting sympathomimetic amine present in high concentrations in certain foods, evokes its pharmacological action by displacing noradrenaline from intraneuronal stores with a subsequent increase in blood pressure (Burn et al., 1958; Davey et al., 1963). The risk of hypertensive crises during therapy with unspecific or MAO-A inhibitors is a major drawback which restricts their therapeutic use. In contrast, selective inhibitors of MAO-B have been reported to be without this effect (Elsworth et al., 1978), unless given at high doses which also inhibit MAO-A (Schulz et al., 1989). MDL 72974A (E)-4-fluoro-beta-fluoroethylene benzene butanamine a potent, selective, enzyme-activated irreversible inhibitor of MAO-B is currently in clinical development for the treatment of Parkinson's disease. Unlike L-deprenyl the prototype MAO-B inhibitor MDL 72974A is not metabolized in vivo to amphetamine, or compounds having amphetamine-like effects. The present study was conducted to investigate, in healthy volunteers, the effects of multiple oral doses of MDL 72974A on the pressor response to, and presystemic metabolism of oral tyramine.

Materials and methods

The study protocol was approved by a local ethics-committee and written informed consent was obtained from each subject before inclusion. The study followed a double-blind, placebo-controlled randomized design carried out in four groups of six subjects. During an initial single-blind placebo run-in period of 2 to 3 days oral tyramine pressor tests were performed to reduce the variability of the study population with regard to tyramine sensitivity. Subjects meeting the criteria (see below) for entry were randomized in three successive blocks of eight to receive MDL 72974A 1 mg, 12 mg or 24 mg or placebo qd for a period of 14 days.

Tyramine pressor tests were conducted under fasting conditions. Subjects rested in a supine position from at least 30 min before (equilibration period) until 2 h after oral tyramine administration. Baseline condition for blood pressure was considered to be attained when the difference between 3 consecutive readings at 5 min intervals was not greater than 5 mmHg. Tyramine capsules were ingested with 200 ml of water 1 h after administration of the study medication. Thereafter heart-rate, systolic and diastolic blood pressure were monitored (Criticon Dinamap® 1846 SX P) every 5 min for 2 h, or until blood pressure had returned to baseline. The interval between 2 consecutive blood pressure readings was reduced to 2 min when systolic blood pressure started to rise and returned to 5 min after peak pressure had been reached. Nifedipine capsules and/or intravenous doses of phentolamine were given to counteract sustained increase in blood pressure or other untoward effects of tyramine. A 3-lead ECG was monitored continuously througout. At predetermined intervals a 12-lead ECG was recorded.

On the first day of the placebo run-in subjects were administered 400 mg of tyramine. If the resulting increase in systolic blood-pressure was less than 30 mmHg a dose of 600 mg was given the next day. If the increase in systolic blood-pressure was greater than 30 mmHg the dose of tyramine was decreased to 200 mg the following day. Subjects were eligible for inclusion if they had responded with an increase in systolic blood-pressure of at least 30 mmHg following 400 or 600 mg of tyramine. They were excluded if their pressure-response to 200 mg of tyramine exceeded 30 mmHg (over-responders) or if at 600 mg the 30 mmHg were not attained (non-responders).

Starting on the morning of the fifteenth day of the randomized active treatment period, the pressor sensitivity to oral tyramine was again determined. From a uniform starting dose of 50 mg the dose of tyramine was raised from day to day in a stepwise manner at the discretion of the investigator until the dose of tyramine required to increase systolic blood-pressure by 30 mmHg (PD30) was reached or exceeded. Dose increments were at least 50 mg and the top dose was limited to 600 mg. Subjects continued to take the study medication, MDL 72974A or placebo, each day until completion of the tyramine pressor tests. Three weeks after the last administration of study medication subjects returned for a follow-up run of tyramine tests. For each subject a sequence of dose steps similar to the individual dose increments applied during the treatment phase pressor tests was used with the exception of the 50 mg dose.

On the morning following completion of the oral tyramine tests during both the placebo run-in and at end of treatment, subjects received their previously determined tyramine PD30 dose and venous blood samples were drawn for the assay of free and conjugated tyramine and HPAA. These results will be reported separately.

Checks of vital signs and routine safety laboratory tests of serum chemistry, haematology and urinalysis were conducted at regular intervals. Body weight was measured at screening, weekly during the trial and at follow-up. Subjects were ambulatory. They were informed to observer a low tyramine diet from the morning of the first placebo run-in period to completion of the last follow-up tyramine pressor test.

For estimation of the tyramine dose yielding an increase in systolic blood pressure of exactly 30 mmHg (PD30), the log tyramine dose-blood pressure relation was characterized for each subject by calculating the regression equation between the two doses of one tyramine test series which straddled the 30 mmHg increase (i.e. immediately before and after the 30 mmHg increase was surpassed). These PD30 estimates were used to calculate the ratio of the PD30 observed during the run-in period over the PD30 observed at the end of treatment. Analysis of the ratios was carried out using 1-way ANOVA. Dose group comparisons for demographic variables were made using ANOVA. For all of the remaining safety variables, measurements made across time were analyzed using ANOVA with repeated measures. Main effects were dose and time and the interaction term dose by time. Measurements made on Day 1, prior to dosing, were used as covariates. A probability of $p \leq 0.05$ was taken to indicate statistical significance.

Results

35 Subjects were screened for their tyramine sensitivity, out of which 10 were excluded. Nine subjects were non-responders with an increase in systolic blood pressure (SBP) <30 mmHg following 600 mg of tyramine. One subject was an over-responder with an increase in SBP >30 mmHg at 200 mg of tyramine. 25 subjects were entered into and 24 (mean and SD; age: 30.5, 5.7 yrs; weight: 76.2, 7.5 kg; height: 177.3, 5.3 cm) completed the second phase of the trial.

Table 1 shows the estimated PD30 and the ratios of the estimated doses for run-in/end of treatment, run-in/follow-up and follow-up/end of treatment for each subject. On three occasions subjects did not attain the threshold 30 mmHg increase in systolic blood pressure at tyramine doses of up to 600 mg. PD30 ratios calculated for run-in/end of treatment show no obvious differences between doses of MDL 72974A (mean and S.D. for n = 6; 1 mg: 2.74, 3.09; 12 mg: 2.44, 2.11; 24 mg: 2.20, 0.62). The corresponding placebo ratio (n = 5) is 1.21, 0.72. Comparison of the PD30 at run-in over that obtained at follow-up did not show any significant differences.

A significant decrease in body weight was observed in the 24 mg group as compared to the other 3 dose groups at the follow-up visit.

No clinically significant or dose-dependent changes in any of the safety laboratory parameters became apparent.

Oral tyramine led in two subjects to an episode of ventricular ectopic beats, in another to nausea and in a further to headache. One case of orthostatic collapse occurred under placebo medication. To counteract hypertensive reactions during tyramine tests nifedipine capsules (10 mg) were used 18 times and phentolamine i.v. (in steps of 5, 3 and 2 mg) in two cases.

Discussion

The large intra- and inter-subject variations in sensitivity to catecholaminergic compounds must be considered when designing protocols involv-

C. Hinze et al.

Table 1. Estimated doses of tyramine causing a 30 mmHg increase in systolic blood pressure

Subject	dose (mg)	Tyramine doses (mg)			Ratios		
		run-in	end TRT	FU	R/E	R/F	F/E
4	0	599.3	627.3	538.5	0.96	1.11	0.86
7	0	400.0	407.1	215.4	0.98	1.86	0.53
10	0	484.8	507.2	297.2	0.96	1.63	0.58
14	0	410.2	166.4	211.6	2.47	1.94	1.27
17	0	259.6	395.7	404.2	0.66	0.64	1.02
21	0	485.4	3474.6*	515.0	0.14*	0.94	0.15*
5	1	237.2	105.0	120.1	2.26	1.98	1.14
6	1	227.6	260.2	208.5	0.87	1.09	0.80
15	1	481.0	412.5	236.2	1.17	2.04	0.57
16	1	550.9	381.2	1022.5*	1.45	0.54*	2.68*
19	1	342.5	198.7	449.6	1.72	0.76	2.26
23	1	503.1	56.1	533.0	8.97	0.94	9.50
1	12	400.0	298.8	439.4	1.34	0.91	1.47
8	12	257.0	242.4	244.4	1.06	1.05	1.01
11	12	256.9	399.8	265.8	0.64	0.97	0.66
13	12	476.5	89.6	387.1	5.31	1.23	4.32
22	12	431.6	86.7	231.1	4.98	1.87	2.58
18	12	562.4	433.9	519.3	1.30	1.08	1.20
2	24	410.9	168.3	208.0	2.44	1.98	1.24
3	24	400.0	342.1	417.0	1.17	0.96	1.22
9	24	239.6	99.9	177.1	2.40	1.35	1.77
12	24	400.0	208.9	718.5*	1.91	0.56*	3.44*
20	24	273.2	91.2	318.9	3.00	0.86	3.50
24	24	273.1	118.9	356.0	2.30	0.77	2.99

* An increase of 30 mmHg was not obtained. Extrapolation of regression equation line yielded values indicated. Estimated doses (mg) of tyramine causing a 30 mmHg increase in SBP during the run-in period (R), at the end of treatment (end TRT) and at follow-up (FU). The ratios of the log doses are shown for run-in/end TRT (R/E), run-in/follow-up (R/F) and follow-up/end TRT (F/E)

ing amine pressor tests (Reimann et al., 1992). Although subjects were screened for their tyramine sensitivity and thus the population limited to tyramine responders three cases of no pressor response to the highest tyramine dose were encountered during the course of the study. For these subjects the corresponding PD30-values were extrapolated from log doses corresponding to previous actual SBP-increases. It was therefore decided to exclude these estimates from the calculation of PD30 doses and ratios per dose group. Furthermore, the protocol allowed tyramine dose increments to be different for the pressor tests at end of treatment and follow-up as compared to run-in. This may have added to the variability of the tyramine pressor response. At all doses of MDL 72974A the apparent increase in tyramine sensitivity at end of treatment compared to run-in was about two-fold and thus similar to that of the intra- and inter-subject variability.

It may be regarded as a drawback of this study that no positive control was added to demonstrate its ability to detect the onset of non-specific inhibition on MAO. However, it should be noted that the non-specific MAO-A inhibitor tranylcypromine increased tyramine sensitivity by a factor 38 and moclobemide, a MAO-A inhibitor, by a factor 5 (Berlin et al., 1989). Furthermore, this study closely followed a design described earlier and having demonstrated its ability to detect such changes in tyramine sensitivity (Schulz et al., 1989; Berlin et al., 1989; Bieck et al., 1989).

The weight loss seen in the 24 mg group was accentuated by one subject having lost 13 kg between day 1 and follow-up. It is interesting to note, that 9 kg of 13 were lost from end of treatment to follow-up. This weight loss was most likely voluntary and not seen in the other subjects of the group. Comparing mean body weight across time reveals a similar pattern from day 1 to end of treatment for all groups receiving MDL 72974A. A relationship to MDL 72974A can not be fully excluded but needs confirmation in a longer chronic administration study.

From the data presented it may be concluded that MDL 72974A over the tested dose range is a specific inhibitor of monoamine oxidase B, has a favourable safety-profile and does not potentiate the pressure response to oral tyramine. It may therefore be recommended to lift or at least alleviate dietary restrictions concerning tyramine containing food in future clinical trials in patients.

References

Berlin I, Zimmer R, Cournot A, Payan C, Pedarriosse AM, Puech AJ (1989) Determination and comparison of the pressor effect of tyramine during long-term moclobemide and tranylcypromine treatment in healthy volunteers. Clin Pharmacol Ther 46: 344–351

Bieck PR, Firkusny L, Schick C, Antonin K-H, Nilsson E, Schulz R, Schwenk M, Wollmann H (1989) Monoamine oxidase inhibition by phenelzine and brofaromine in healthy volunteers. Clin Pharmacol Ther 45: 260–269

Burn JH, Rand MJ (1958) The action of sympathomimetic amines in animals treated with reserpine. J Physiol 144: 314–346

Davey MJ, Farmer JB (1963) The mode of action of tyramine. J Pharm Pharmacol 15: 178–182

Elsworth JD, Glover V, Reynolds GP (1978) Deprenyl administration in man: a selective monoamine oxidase-B inhibitor without the "cheese effect". Psychopharmacology 57: 33–38

Johnston JP (1968) Some observations upon a new inhibitor of monoamine oxidase in brain tissue. Biochem Pharmacol 17: 1285–97

Reimann IW, Firkusny L, Antonin KH, Bieck PR (1992) Intravenous amine pressor tests in healthy volunteers. Eur J Clin Pharmacol 42: 137–141

Schulz R, Antonin K-H, Hoffmann E, Jedrychowski M, Nilsson E, Schick C, Bieck PR (1989) Tyramine kinetics and pressor sensitivity during monoamine oxidase inhibition by selegiline. Clin Pharmacol Ther 46: 528–536

Authors' address: Dr. C. Hinze, Marion Merrell Dow, Clinical Pharmacology Unit, POB 1909, D-77694 Kehl, Federal Republic of Germany.

J Neural Transm (1994) [Suppl] 41: 377–379

The acute effect of the bioprecursor of the selective brain MAO-A inhibitor, MDL 72392, on rat pineal melatonin biosynthesis

G. F. Oxenkrug[1], **P. J. Requintina**[1], **A. Yuwiler**[2], and **M. G. Palfreyman**[3]

[1] Pineal Research Laboratory, Psychiatry Service, VAMC, and Department of Psychiatry and Human Behavior, Brown University (GFO, PJR), [2] Neurochemistry Laboratory, West Los Angeles Brentwood VAMC, and Department of Psychiatry and Biobehavioral Sciences, UCLA, Los Angeles, California, and [3] Merrell Dow Research Institute, Cincinnati, Ohio, U.S.A.

Summary. The bioprecursor amino acid MDL 72394 is decarboxylated by aromatic L-amino acid decarboxylase (AADC) to liberate MDL 72392, an irreversible selective MAO-A inhibitor. Pretreatment with the AADC inhibitor carbidopa, which does not penetrate the brain-blood barrier, prevents the liberation of the MAO-A inhibitor outside the brain and results in exclusive inhibition of brain MAO-A. We found that systemic administration of MDL 72394 (0.5 mg/kg, i.p.) stimulated rat pineal melatonin biosynthesis. Carbidopa, in a dose-dependent manner, attenuated or completely prevented MDL-induced stimulation of melatonin biosynthesis in the pineal gland located outside the blood-brain-barrier.

Introduction

The pineal gland is the site of melatonin biosynthesis from serotonin. Selective MAO-A (but not MAO-B) inhibitors stimiulate pineal melatonin biosynthesis (Oxenkrug et al., 1984a; for review see Oxenkrug, 1991). The bioprecursor amino acid MDL 72394 (E-β-fluoromethylene-m-tyrosine) is decarboxylated by aromatic L-amino acid decarboxylase (AADC) to liberate MDL 72392 (R = 3-hydroxyphenyl), a potent ineversible selective MAO-A inhibitor (Palfreyman et al., 1985). Systemic administration of MDL 72394 inhibits MAO-A in both brain and peripheral tissues. However, pretreatment with the AADC inhibitor, carbidopa, which does not penetrate the brain-blood barrier, prevents the liberation of MAO-A inhibitor from its bioprecursor outside the brain. Therefore, administration of MDL 72394 to animals pretreated with carbidopa results in the selective inhibitor of MAO-A only in brain but not in the peripheral tissues (Palfreyman et al., 1985).

The first goal of the present study was to find out whether administration of MDL 72394 would stimulate melatonin biosynthesis as previously demon-

strated for MAO-A inhibitors. The second goal was to deterntive how
pretreatment with carbidopa might affect MDL 72394 effect (if any) on
melatonin biosynthesis considering that the pineal gland is located outside
the brain-blood barrier.

Materials and methods

Fisher 344N male rats (2 month old, body weight 200 g) were housed for at least 2
weeks before the experiment in a light-controlled (L:D 12:12, lights on at 06:00) and
temperature-regulated room with the free access to food and water. Carbidopa (5 or
12.5 mg/kg, i.p.) was injected in the morning (between 08:00 and 09:00). The dose
and timing of carbidopa administration were selected on the basis of preliminary
experiments to achieve maximum inhibition of AADC. MDL 72394 (0.5 mg/kg, i.p.)
was injected 60 min after carbidopa. Control animals were injected with vehicle solution or saline. Animals were decapitated 90 min after MDL injection. Each experimental group consisted of 6 animals. The pineal glands were quickly removed, frozen
on dry ice and stored at −70°C until assayed. Concentrations of melatonin and other
indoles were determined by the HPLC-fluorimetric procedure described elsewhere (see
Oxenkrug, 1991). Results are expressed as mean + S.D. (ng of the indolepineal).
Group differences were analyzed by one-way analysis of variance and Student's t-test.

Results

Carbidopa caused dose-dependent increase of 5-hydroxytryptophan (5-HTP)
and decrease of serotonin (5-HT) and, consequently, 5-hydroxyindole acetic
acid (5-HIAA) levels.

MDL 72394 administered alone elevated pineal 5-HT, N-acetylserotonin
(NAS) and melatonin levels. The increase of NAS content suggests that
elevation of melatonin levels resulted from the stimulation of melatonin
biosynthesis. Decrease of 5-HIAA levels were most likely due to inhibition
of MAO-A by MDL 72392, the product of MDL 72394 decarboxylation
(Table 1).

Pretreatment with carbidopa attenuated (5 mg/kg) or completely prevented (12.5 mg/kg) MDL 72394-induced stimulation of pineal melatonin
biosynthesis.

Discussion

Our results indicate that systemic administration of MDL 72394 stimulates
rat pineal melatonin biosynthesis most probably due to selective MAO-A
inhibition caused by MDL 72392, the product of MDL 72394 decarboxylation.
These results are in line with the previously published data on the effect
of selective MAO-A inhibitors on melatonin biosynthesis (see Oxenkrug,
1991). The carbidopa effect on daytime melatonin biosynthesis is also in
agreement with our previous report (Oxenkrug et al., 1984b).

The most interesting finding of present study is the attenuation of MDL
72394 effect on melatonin biosynthesis in animals pretreated with carbidopa,

Table 1. The effect of MDL 72,394 on the rat pineal melatonin and related indoles

Drug (mg/kg)	Melatonin**	5-HTP	5-HT	5-HIAA	NAS
Saline	0.19 ± 0.03	1.19 ± 0.04	38.13 ± 2.95	4.18 ± 0.07	0.15 ± 0.03
Carbidopa (5)	0.16 ± 0.05	4.37 ± 0.62*	29.98 ± 4.18	2.88 ± 0.54*	0.16 ± 0.02
Carbidopa (12.5)	0.04 ± 0.01	7.98 ± 0.96*	13.57 ± 0.08*	0.21 ± 0.08*	0.16 ± 0.03
MDL (0.5)	1.49 ± 0.39*	1.18 ± 0.08	52.96 ± 3.45*	2.15 ± 1.90	1.4 ± 0.06*
MDL + carbidopa (5)	0.77 ± 0.13*	2.52 ± 0.07*	41.58 ± 2.95	0.66 ± 0.09*	0.27 ± 0.07*
MDL + carbidopa (12.5)	0.18 ± 0.03	3.36 ± 1.45*	23.62 ± 12.8*	0.22 ± 0.05*	0.15 ± 0.03

nd not detectable not detectable in this table; * $p < 0.01$ vs saline; ** mean + S.E. (ng/pineal)

peripheral inhibitor of AADC. Considering that stimulation of melatonin biosynthesis requires selective MAO-A inhibition and that pineal locates outside of the brain-blood barrier, one might suggest that carbidopa prevents stimulation of melatonin biosynthesis by blocking the formation of MAO-A inhibitor from MDL 72394 outside the brain. It was suggested that stimulation of melatonin biosynthesis by MAO-A inhibitors might contribute to the antidepresant effect of these drugs (see Oxenkrug, 1991). Clinical trials of MDL 72394 alone (causing melatonin biosythesis stimulation) and in combination with carbidopa (not affecting melatonin biosynthesis) might help to clarify melatonin role in the antidepressant effect.

References

Oxenkrug GF (1991) The acute effect of monoamine oxidase inhibitors on serotonin conversion to melatonin. In: Coppen X, Sandler M, Harnett S (eds) 5-Hydroxy-tryptamine and mental illness. Oxford University Press, New York, pp 99–108

Oxenkrug GF, McCauley R, McIntyre IC, Filipowicz C (1984) Effect of clorgyline and deprenyl on rat pineal melatonin. J Pharm Pharmacol 36: 5SW

Oxenkrug GF, McIntyre IM, Novak EA, Hryhorczuk L, Froman CE (1984b) Differential effect of carbidopa on the concentration of rat pineal and hypothalamic indoleamines. J Pineal Res 1: 349–353

Palfreyman MG, McDonald IA, Fozard JR, Mely Y, Sleight AJ, Zreika M, Wagner J, Bey P, Lewis PJ (1985) Inhibition of monoamine oxidase selectively in brain mono-amine nerves using the bioprecursor (E)-β-fluoromethylene-m-tyrosine (MDL 72394), a substrate for aromatic L-amino acid decarboxylase. J Neurochem 45: 1850–1860

Authors' address: Dr. G. F. Oxenkrug, Psychiatry Service, VAMC, 830 Chelka-tone Ave., Providence, RI 02908, U.S.A.

J Neural Transm (1994) [Suppl] 41: 381–384

Chronic effect of the irreversible and reversible selective MAO-A inhibitors on rat pineal melatonin biosynthesis

G. F. Oxenkrug[1], P. J. Requintina[1], I. M. McIntyre[2], and K. White[1]

[1] Pineal Research Laboratory, Psychiatry Service, VAMC, and Department of Psychiatry and Human Behavior, Brown University School of Medicine, Providence, Rhode Island, U.S.A.
[2] Victorian Institute of Forensic Pathology, Coronial Services Centre of Victoria, South Melbourne, Australia

Summary. Acute administration of the irreversible MAO-A inhibitor, clorgyline (2.0 mg/kg, s.c.) and the reversible MAO-A inhibitor, moclobemide (10 mg/kg, s.c.), increased rat pineal melatonin and related indoles content (HPLC-fluorimetric method). Chronic (21 days) administration of clorgyline attenuated the acute effect of clorgyline on pineal melatonin biosynthesis. The acute effect of moclobemide on melatonin biosynthesis was not affected by chronic moclobemide administration. The observed difference in the chronic effects of irreversible and reversible selective MAO-A inhibitors on melatonin biosynthesis could have clinical implications.

Introduction

Dramatic stimulation of the pineal melatonin biosynthesis has been observed after acute administration of reversible and irreversible selective MAO-A inhibitors (Oxenkrug et al., 1984; for review see Oxenkrug, 1991). It is generally accepted that the clinical antidepressant effect of antidepressants is associated with their chronic (but not acute) phamacological actions (Vetulani and Sulser, 1975). Moderate elevation of blood melatonin levels in depressed patients has been reported after chronic administration of irreversible selective MAO-A and non-selective MAO inihibitors (Golden et al., 1988; Murphy et al., 1986). However, we could not find in the literature reports of studies of the chronic effects of selective MAO-A inhibitors on pineal melatonin biosynthesis.

This present study compares the acute and chronic effects of irreversible (*clorgyline*) and reversible (moclobemide) selective MAO-A inhibitors on rat pineal melatonin biosynthesis.

G. F. Oxenkrug et al.

Materials and methods

Fisher 344N (3 month old, female) rats were maintained at 12 h dark : 12 h light : schedule, with lights on/off at 10:00/22:00 h. in a temperature-regulated room with food and water ad libidum. In the acute experiment rats were injected with saline (at 09:00 hrs) daily for 20 days. Clorgyline or moclobemide (or saline) were injected on the 21st day. In the chronic experiment clorgyline (2.0 mg/kg) or moclobemide (10 mg/kg) were injected (s.c.) for 21 days.

Rats were decapitated 120 min after the last injection of drugs. Pineal glands were immediately removed, frozen and kept at $-70°C$ until assayed. Pineal melatonin and related indoles content were determined by HPLC-fluorescence procedure as described elsewhere (see Oxenkrug et al., 1991). Results were expressed as mean $\pm$ S.D. (ng/pineal) (six animals in each group) and treated by one-way analysis of variance and Student's t-test.

Results

There was no difference between the effects of acute and 21 days administration of saline on melatonin biosynthesis. Acute administration of both MAO-A inihibitors increased pineal melatonin, N-acetyl-serotonin (NAS) and serotonin (5-HT) and decreased 5-hydroxyindoleacetic acid (5-HIAA). Chronic administration of clorgyline further decreased pineal 5-HIAA levels. However, pineal melatonin and NAS levels were not increased after 21 days of clorgyline administration (Table 1). Chronic administration of moclobemide increased rat pineal melatonin content to the same extent as acute moclobemide administration.

Discussion

We have observed the remarkable difference between the chronic effects of clorgyline and moclobemide on pineal melatonin biosynthesis: chronic (and acute) moclobemide administration stimulated rat pineal *melatonin biosynthesis* while only acute (but not chronic) administration of clorgyline

Table 1. Chronic and acute effect of clorgyline and moclobemide on rat pineal melatonin

	Melatonin	NAS	5-HT	5-HIAA
Saline	0.17 ± 0.05	0.15 ± 0.02	84.50 ± 11.41	4.25 ± 0.44
Acute:				
Clorgyline	$1.25 \pm 0.07^*$	$1.19 \pm 0.29^*$	89.24 ± 8.03	$2.08 \pm 0.10^*$
Moclobemide	$0.68 \pm 0.06^*$	$0.30 \pm 0.05^*$	80.33 ± 3.95	$2.49 \pm 0.24^*$
Chronic:				
Clorgyline	0.22 ± 0.06	0.17 ± 0.03	98.41 ± 8.65	$0.95 \pm 0.01^*$
Moclobemide	$0.69 \pm 0.09^*$	$0.45 \pm 0.05^{**}$	76.32 ± 6.42	$2.97 \pm 0.41^*$

Mean $\pm$ S.D. (ng/pineal); $^* p < 0.01$ vs saline

increased pineal melatonin production. At the present time it is not clear whether such a difference is a feature of all reversible vs irreversible MAO-A inhibitors. The possible explanation of this difference might be related to the effect of MAO inhibitors on *β-adrenoceptors*. Chronic administration of clorgyline down-regulates rat brain β-adrenoceptors (see Finberg, 1987). Stimulation of the β-adrenoceptors of the pinealocytes is an essential step in melatonin biosynthesis (see Pangerl et al., 1990). Therefore, clorgyline-induced down-regulation of β-adrenoceptors might attenuate stimulation of pineal melatonin biosynthesis caused by the clorgyline administration. Although literature data indicate that moclobemide (50 mg/kg) induced 20% down-regulation of β-adrenoceptors (see Cesura and Pletcher, 1992), it is possible that at the dose level used in our study (10 mg/kg) moclobemide did not down-regulate β-adrenoceptors. Such a suggestion might be of interest since moclobemide exerts an antidepressant effect clinically and since down-regulation of the postsynaptic β-adrenoceptors has been suggested as a pharmacological mechanism of the clinical antidepressant action of inhibitors of MAO (Vetulani and Sulser, 1975).

An alternative hypothesis suggests that depression might be associated with the *circadian rhythms* abnormalities that might be normalized by the action of antidepressants (Wehr and Wirz-Justice, 1982). In this vein, we have suggested that MAO inhibitors normalize circadian rhythms via stimulation of melatonin production and that this action constitutes the pharmacological mechanism of the clinical antidepressant effect of MAO inhibitors (for rev. see Oxenkrug, 1991). Our suggestion is at variance with the hypothesis linking the clinical antidepressant effect with the pharmacological consequences of the chronic antidepressant administration (e.g. down-regulation of adrenoceptors). In fact, our hypothesis implies that down-regulation of β-adrenoceptors (and, therefore, reduction in melatonin biosynthesis) would be associated with a decrease in the clinical antidepressant action. According to our hypothesis the difference in chronic effect of clorgyline and moclobemide on melatonin biosynthesis suggests that moclobemide should be a more effective antidepressant than clorgyline. So far, we are not aware of the clinical studies comparing the antidepressant effect of clorgyline and moclobemide.

Acknowledgements

The authors wish to thank Profs. Da Prada and Haefely for the generous gift of moclobemide.

References

Cesura A, Pletcher A (1992) The new generation of monoamine oxidase inhibitors. Prog Drug Res 38: 171–297
Da Prada M, Kettler R, Keller HH, Burkard W, Muggli-Maniglio D, Haefely WE (1989) Neurochemical profile of moclobemide, a short-acting and reversible inhibitor of monoamine oxidase type A. J Pharmacol Exp Ther 248: 400–414

Finberg JPM (1987) Antidepressant drugs and down-regulation of presynaptic receptors. Biochem Pharmacol 36: 3357–3562

Golden RN, Markey SP, Risby ED, Rudorfer MV, Cowdry RW, Potter WZ (1988) Antidepressants reduce whole-body norepinephrine turnover while enhancing 6-hydroxymelatonin output. Arch Gen Psychiatry 45: 150–154

Murphy DL, Tamarkin L, Sunderland T, Garrick NA, Cohen R (1986) Human plasma melatonin is elevated during treatment with the monoamine oxidase inhibitors clorgyline and tranylcypromine but not deprenyl. Psychiatry Res 17: 119–127

Oxenkrug GF (1991) The acute effect of monoamine oxidase inhibitors on serotonin conversion to melatonin. In: Sandler M, Coppen A, Harnett S (eds) 5-Hydroxy-tryptamine in psychiatry. A spectrum of ideas. Oxford University Press, New York, pp 98–109

Oxenkrug G, McCauley R, McIntyre I, Filipowicz C (1984) Effect of clorgyline and deprenyl on rat pineal melatonin. J Pharm Pharmacol 36: 55W

Pangerl B, Pangerl A, Reiter RJ (1990) Circadian variations of adrenergic receptors in the mammalian pineal gland: a review. J Neural Transm [Gen Sect] 81: 17–30

Vetulani J, Sulser F (1975) Action of various antidepressant treatments reduced reactivity of noradrenergic cyclic AMP generating system in limbic forebrain. Nature 257: 495–496

Wehr TA, Wirz-Justice A (1982) Circadian rhythm mechanisms in affective illness and in antidepressant drug action. Pharmacopsychiatry 15: 31–39

Authors' address: Dr. G. F. Oxenkrug, Psychiatry Service, VAMC, 830 Chalkstone Ave., Providence, RI 02908, U.S.A.

Semicarbazide-sensitive amine oxidases

J Neural Transm (1994) [Suppl] 41: 387–396

Properties of mammalian tissue-bound semicarbazide-sensitive amine oxidase: possible clues to its physiological function?

G. A. Lyles

Department of Pharmacology and Clinical Pharmacology, Ninewells Hospital and Medical School, University of Dundee, Dundee, United Kingdom

Summary. Semicarbazide-sensitive amine oxidase (SSAO), occurs not only in vascular smooth muscle but also in other cell types (e.g. adipocytes, chondrocytes, odontoblasts), probably in the plasma membrane. Although certain aromatic biogenic amines (e.g. tryptamine, tyramine, β-phenylethylamine) may be endogenous substrates for SSAO in species such as the rat, the weak activity of SSAO in human tissues towards these amines makes this less likely in man. However SSAO in human and rat vascular homogenates readily converts the aliphatic biogenic amines methylamine and aminoacetone to formaldehyde and methylglyoxal, respectively. Also the xenobiotic aliphatic amine allylamine produces cardiovascular damage in experimental animals by a mechanism which involves its deamination by SSAO to acrolein. Further metabolism of these toxic aliphatic aldehydes may involve glutathione-dependent pathways. Thus, SSAO may be involved not only in the removal of physiologically-active endogenous/xenobiotic amines, but resulting metabolite (aldehyde/H_2O_2?) formation could also influence cellular function.

Introduction

Several different types of mammalian amine oxidase enzyme exhibit the property of being inhibited by semicarbazide. This inhibition is attributed to the presence in these enzymes of an organic cofactor containing one or more carbonyl groups which can react with semicarbazide. Whether or not this cofactor (pyridoxal phosphate, pyrroloquinoline quinone, 6-hydroxy dopa?) is the same in each enzyme,, with copper as a possible additional cofactor is under current investigation (Janes et al., 1990). These different amine oxidases have been named according to their postulated roles in metabolizing particular physiological amines (e.g. diamine oxidase, lysyl oxidase), or in the case of the plasma amine oxidases, indicating their localization as "soluble" enzymes within the circulation. With the latter, the polyamines spermine and spermidine have been proposed as physiological substrates in ruminant plasma, whereas in non-ruminant species, the plasma enzymes are usually most active towards the synthetic amine benzylamine, (BZ) (Blaschko, 1974).

BZ has also been described as a "preferred substrate" for a membrane-bound semicarbazide-sensitive amine oxidase (SSAO) activity found in various tissues, and with especially high activity in blood vessels (Lewinsohn, 1984). However, in the absence of convincing evidence that BZ is a biogenic amine, the physiological importance and endogenous substrate(s) of tissue-bound SSAO (and the plasma enzymes) are unclear. Although, as indicated above, inhibition by semicarbazide is not a unique characteristic of the tissue-bound "BZ oxidase", there has been an increasing tendency to use the term SSAO specifically for the latter, reflecting the current inability to define this amine oxidase on the basis of physiological function. Recent studies have suggested that SSAO may be involved not only in the metabolism of particular biogenic and xenobiotic amines, but also in the production of metabolites with physiological and/or toxicological actions. The evidence for some of these ideas is reviewed here.

Distribution of tissue-bound SSAO

Not only is SSAO activity found in a wide variety of different tissues, but evidence for its association with particular cell types within these sources is also emerging. One of the major criteria used to identify SSAO in tissue homogenate fractions is the demonstration of BZ-metabolizing activity which is sensitive to inhibition by 0.1–1 mM semicarbazide. In turn, the resistance of this activity to inhibition by similar concentrations of the acetylenic compounds clorgyline, pargyline and selegiline also distinguishes SSAO from mitochondrial monoamine oxidase (MAO), which is selectively inhibited by these drugs, and which is also capable of BZ deamination. Other drugs which have been used as SSAO inhibitors, with some selectivity for distinguishing between SSAO and MAO include various derivatives of hydrazine and allylamine (Lyles, 1984; Lyles et al., 1987).

Although the aorta of species such as the ox and rabbit were shown many years ago to contain a BZ-metabolizing amine oxidase inhibited by semicarbazide (Rucker and O'Dell, 1971; Rucker and Goettlich-Riemann, 1972), subsequent histochemical and biochemical studies of the distribution of SSAO activity in various human tissues indicated that the particularly high SSAO activity in the vasculature was associated predominantly with smooth muscle (Lewinsohn, 1984). More recently, SSAO has also been demonstrated histochemically in the smooth muscle layers of the rat aorta (Lyles and Singh, 1985), and its properties studied in rat aortic cultured smooth muscle cells (Blicharski and Lyles, 1990). Cultured vascular smooth muscle cells appear to secrete a soluble form of SSAO, perhaps indicating the origin of plasma amine oxidase (Hysmith and Boor, 1987). In addition to being in blood vessels, SSAO appears to be present in non-vascular smooth muscle also (Lewinsohn, 1984).

These findings have led to speculation that SSAO might be involved in some aspect of smooth muscle function. Although the SSAO activity found in homogenates of some whole tissues or organs probably originates at least in part from the tissue vascularity (e.g. Lyles and Archer, 1986) nevertheless it is clear that SSAO also occurs in cells other than smooth muscle. Thus, SSAO has been demonstrated in adipocytes from rat white and brown fat (Barrand et al., 1984; Raimondi et al., 1991), and in chondrocytes in rat articular cartilage (Lyles and Bertie, 1987). It was also found in various parts of the bovine eye, including the optic nerve (Fernandez de Arriba et al., 1990). Pig dental pulp SSAO appears to be associated with odontoblasts (Norqvist and Oreland, 1989). Consequently, any functional importance eventually proposed for SSAO may need to encompass its presence in a variety of different cell types.

Studies of the subcellular localization of SSAO strongly indicate that much of the enzyme activity in rat aorta (Wibo et al., 1980), brown fat (Barrand et al., 1984) and pig dental pulp (Norqvist and Oreland, 1989) is associated with the plasma membrane. However, whether or not SSAO metabolizes extracellular and/or intracellular amines is unclear, as also is the possibility that SSAO might exist to a lesser extent at other membrane loci within the cell.

Biogenic and xenobiotic aromatic amines as potential SSAO substrates

Although BZ has been the most widely-used aromatic amine substrate for studying the activity and distribution of SSAO in animal tissues, its non-physiological nature has led to investigations into the ability of SSAO to metabolize various aromatic monoamines which are formed endogenously or which may be ingested in the diet. These studies indicate that the specificity (and relative affinity) of SSAO towards various aromatic amines is often species-dependent and possibly even tissue-dependent within some species. This points strongly to the probability that SSAO represents a family of related enzymes rather than a single enzyme.

A detailed perspective of the substrate specificity of SSAO in a wide variety of tissues and species is far from complete, although the enzyme appears to act only upon primary amines. The early work of Rucker and Goettlich-Riemann (1972) indicated that rabbit aorta SSAO was active against tyramine (TYR) and tryptamine (TRYP). Later, pig aorta SSAO was found to metabolize β-phenylethylamine (PEA) (Trevethick et al., 1981), while bovine lung SSAO deaminates PEA and dopamine (DA), but apparently not TYR or TRYP (Lizcano et al., 1990). Most determinations of kinetic constants (K_m, V_{max}) for aromatic amine metabolism by SSAO have used rat tissue homogenates. Table 1 compares typical K_m values reported for BZ, with those for various biogenic amines. It is of interest that these K_m values for PEA, TYR, TRYP and DA are rather similar to K_m values for their metabolism by the A and/or B forms of MAO for which

Table 1. Km values for aromatic amine substrates of SSAO
in rat tissues

Substrate	Km (μM)
Benzylamine	2.8[a], 3.0[bc], 6.2[d]
β-Phenylethylamine	10[b], 15[ae], 44[c]
Tyramine	40[b], 52[e]
Tryptamine	54[a], 67[f]
Dopamine	130[g], 270[a], 1058[h]
Histamine	583[i]

Sources: [a] skull (Andree and Clarke, 1982), [b] brown fat (Barrand and Callingham, 1982), [c] cartilage (Lyles and Bertie, 1987), [d] aorta (Clarke et al., 1982), [e] aorta (Guffroy et al., 1985), [f] aorta (Lyles and Taneja, 1987), [g] aorta (Yu, 1988), [h] vas deferens (Lizcano et al., 1991), [i] aorta (Yu, 1990)

these amines are known to be physiological substrates (Strolin Benedetti and Dostert, 1985). Coquil et al. (1973) found no evidence for the metabolism of noradrenaline (NA) by SSAO in rat mesenteric artery, and the general consensus is that SSAO is unlikely to be involved in the breakdown of this neurotransmitter. However, this conclusion is based on limited work with this amine and caution is necessary before making general assumptions about the properties of SSAO. For example, in most tissues, 5-hydroxy tryptamine (5-HT) is not a substrate for SSAO, and yet is metabolized readily by a membrane-bound SSAO activity in pig dental pulp which also oxidizes TYR, TRYP and PEA (Norqvist et al., 1981). Furthermore, it is apparent that some biogenic aromatic amines may inhibit SSAO, in some tissues, by binding to the enzyme without being oxidized (Elliott et al., 1989a).

The possibility that SSAO in vascular smooth muscle may be capable of metabolizing circulating amines has been investigated by using perfused vascular bed preparations. Roth and Gillis (1975), studying the influence of pargyline and semicarbazide upon deamination of PEA perfused through the rabbit pulmonary vascular bed, concluded that both MAO and SSAO contributed to metabolite formation. SSAO was also found to be involved in TYR deamination in the rat perfused mesenteric arterial bed (Elliott et al., 1989b).

Evidence that SSAO can influence the vascular actions of biogenic amines has also been obtained. Contractile sensitivity of rat aortic rings to TRYP was increased by inhibition of MAO-A and further enhanced by then inhibiting SSAO, whereas inhibition of SSAO alone had no significant effect on sensitivity (Taneja and Lyles, 1988). Elliott et al. (1989c) showed that the pressor effects of TYR in the rat perfused mesenteric arterial bed were unaffected by inhibition of either MAO or SSAO alone, but were potentiated if both enzymes were inhibited. These results suggest that these enzymes may act in concert in certain tissues to regulate the pharmacological effects of particular amines.

Recent studies on the substrate specificity of SSAO in human tissues indicate clearly that the affinity of the human enzyme for aromatic amines is much lower than that of the rat, suggesting caution in using laboratory species to predict possible functions of SSAO in man. For example, K_m values (e.g. 100–300 µM) found for BZ metabolism in human vascular homogenates (Hayes et al., 1983; Suzuki and Matsumoto, 1984; Precious and Lyles, 1988) are considerably higher than in the rat (see Table 1). Suzuki and Matsumoto (1984) showed that 0.5 mM octopamine, TRYP, TYR and PEA were deaminated by SSAO in human aorta homogenates at rates of only 0.6–1.9% of that for BZ oxidation, and with no metabolism of histamine being detected. More recently, Precious and Lyles (1988) estimated K_m values of 7.9 and 10.5 mM respectively, for metabolism of PEA and TYR by SSAO in human umbilical artery. No activity towards 5-HT or TRYP was found. It has not yet been investigated fully if NA and DA are SSAO substrates in human tissues but on current evidence it would appear that the weak deaminating activity of SSAO towards aromatic biogenic amines casts doubt upon them being good physiological substrates for SSAO in man.

This section has focussed primarily upon the possible involvement of SSAO in metabolism of endogenous amines. Whether or not metabolism by SSAO might be of importance for environmental or xenobiotic aromatic amines remains to be investigated. For example, the GABA mimetic agent, kojic amine (Ferkany et al., 1981) and the MAO inhibitor, MD 220661 (Dostert et al., 1984) may be substrates in vivo for SSAO in the rat. Thus, metabolism by SSAO may be capable of influencing the pharmacological activity of various synthetic aromatic monoamines used either as experimental drugs or as potential therapeutic agents.

Biogenic and xenobiotic aliphatic amines as potential SSAO substrates

Considerable interest has been stimulated recently in the possibility that SSAO may be involved in the metabolism of various aliphatic amines. The impetus for this has come, in part, from investigations into the mechanism(s) by which the industrial unsaturated aliphatic amine, allylamine (AA), is capable of producing cardiovascular toxicity. When administered to various experimental animals (rat, dog, calf), AA causes pathological effects such as necrotic lesions of the myocardium and arterial smooth muscle (especially in the coronary circulation), smooth muscle proliferation in the intima of damaged vessels, and also lipid deposition in the vessels if animals are fed high cholesterol diets, producing lesions similar in nature to those seen in human atherosclerosis (Boor and Hysmith, 1987).

There is now strong evidence that these effects result from the ability of SSAO to convert AA to the highly toxic aldehyde, acrolein, in vascular smooth muscle (Boor et al., 1990). Cytotoxic effects of AA-exposure to pig and rat cultured aortic smooth muscle cells have also been demonstrated, and these toxic actions were reduced by pretreating cultures with SSAO inhibitors (Hysmith and Boor, 1988; Ramos et al., 1988). The latter also showed that addition of catalase to cultures also afforded some protection, suggesting that the hydrogen peroxide produced concomitantly during amine oxidation may also contribute to the cytotoxicity. Detoxification of acrolein in vivo involves its conjugation with reduced glutathione (GSH) (Boor et al., 1987), and GSH-dependent mechanisms are also involved in peroxide breakdown (DeLeve and Kaplowitz, 1991). Thus it is of interest that the cytotoxicity of AA was enhanced in cultured smooth muscle cells pretreated with buthionine sulphoximine, an inhibitor of GSH synthesis (Blicharski and Lyles, 1991).

These studies with the xenobiotic compound AA have focussed attention upon endogenously-occurring aliphatic amines as potential SSAO substrates. Methylamine (MA) can be absorbed as a result of gut bacterial degradation of dietary creatinine, lecithin and choline, or may be produced by endogenous metabolism of sarcosine, creatinine and adrenaline. MA is not a substrate for MAO, but is deaminated to formaldehyde by SSAO in homogenates of rat aorta and human umbilical artery as well as by the human plasma enzyme (Precious et al., 1988; Lyles et al., 1990). The K_m (around $800\,\mu M$) for MA metabolism by human vascular SSAO was lower than for previously-examined biogenic aromatic monoamines, with the turnover rate for MA being markedly higher than that for BZ. Species-dependent properties of SSAO were again evident in the finding that the K_m ($182\,\mu M$) for MA in the rat was much lower than that in man (Lyles et al., 1990).

Whether or not endogenous MA is of physiological or pathological significance is currently unclear (see Precious et al., 1988). In experimental studies (at concentrations greatly exceeding physiological levels), the weak base MA can partition into acidic cellular compartments to interfere with endocytotic and lysosomal function. It exhibits cytotoxicity towards cultured neurones and fibroblasts and has been proposed as a uraemic toxin since elevated plasma concentrations occur in patients with renal failure. Lyles and McDougall (1989) showed that treatment of rats with hydralazine or semicarbazide, which can inhibit SSAO, greatly enhanced the daily urinary excretion of MA, providing evidence that the enzyme is involved in the in vivo metabolism of this amine. Findings that SSAO in rat aorta homogenates can also deaminate longer chain aliphatic amines including ethylamine, n-propylamine and isoamylamine may indicate that other endogenously-occurring amines are possible physiological substrates of SSAO (Elliott et al., 1989a; Yu, 1990).

A role for SSAO in metabolizing aminoketones is also suggested by recent studies in our laboratory with aminoacetone (AMA). This aliphatic amine is a product of mitochondrial metabolism of glycine and threonine, and although previously reported to be a substrate for ruminant plasma amine oxidase, has received relatively little attention as a possible substrate for other amine oxidases. We have now demonstrated that aminoacetone is deaminated to methylglyoxal by SSAO in human umbilical artery homogenates, with a K_m (92 µM) even lower than that for BZ metabolism in this tissue (Lyles and Chalmers, 1992). More recent studies have shown that AMA is also a substrate for rat aorta (unpublished results) and bovine lung SSAO (Lizcano et al., this symposium) with K_m values of 19 and 94 µM, respectively. Whether or not AMA itself is of physiological importance in animal tissues is unclear, but currently it appears to be the highest affinity substrate yet found for human SSAO.

As well as degrading endogenous aliphatic amines, SSAO may also generate active metabolites from these compounds. Amine oxidases produce not only aldehydes but also ammonia and hydrogen peroxide. Callingham et al. (1991) have speculated that hydrogen peroxide from this source could be a physiological regulator of such processes as eicosanoid synthesis. Deamination of MA and AMA produces the potentially cytotoxic aldehydes, formaldehyde and methylglyoxal, respectively (Heck et al., 1990; Thornalley, 1990). Whether or not these could be formed by SSAO in vascular or other cells at concentrations which may affect cellular function is unknown. Detoxification of these compounds, which are also products of other metabolic pathways, involves adduct formation with GSH, before enzymatic modification. Formaldehyde dehydrogenase converts the GSH-adduct of formaldehyde to formate, which may be utilized as a single-carbon source for synthesis of purines, pyrimidines and amino acids. Methylglyoxal can be transformed (via glyoxalase I) to S-D-lactoylglutathione (LG), which is subsequently hydrolysed (by glyoxalase II) to D-lactate. Thornalley (1990) has proposed that LG may be a physiological regulator of cytoskeletal function, with possible influences upon cellular secretory activity and motility. It is intriguing to wonder if metabolism of AMA by SSAO could provide one route to this potentially interesting compound.

Conclusions

Whether the principal function of SSAO is to remove physiogically-active amines, or to produce biologically-significant products, or perhaps even both, remains to be established but recent findings suggest promising areas of investigation. Clearly species-related differences in enzyme properties must be taken into account in this respect, particularly when assessing if inhibition of SSAO may have therapeutic applications in man. Currently, certain endogenous aliphatic amines appear better substrates than aromatic amines for human vascular SSAO, and further studies of the possible effects of aliphatic amines and their metabolites upon cellular function would appear worthwhile.

References

Andree TH, Clarke DE (1982) Characteristics of rat skull benzylamine oxidase. Proc Soc Exp Biol Med 171: 298–305

Barrand MA, Callingham BA (1982) Monoamine oxidase activities in brown adipose tissues of the rat: some properties and subcellular distribution. Biochem Pharmacol 31: 2177–2184

Barrand MA, Callingham BA, Fox SA (1984) Amine oxidase activities in brown adipose tissue of the rat: identification of semicarbazide-sensitive (clorgyline-resistant activity) at the fat cell membrane. J Pharm Pharmacol 36: 652–658

Blaschko H (1974) The natural history of amine oxidases. Rev Physiol Biochem Pharmacol 70: 83–148

Blicharski JRD, Lyles GA (1990) Semicarbazide-sensitive amine oxidase activity in rat aortic cultured smooth muscle cells. J Neural Transm [Suppl] 32: 337–339

Blicharski JRD, Lyles GA (1991) D,L-buthionine sulphoximine, a glutathione-depleting agent, potentiates allylamine-induced cytotoxicity in rat aortic smooth muscle cell cultures. Br J Pharmacol 102: 184P

Boor PJ, Hysmith RM (1987) Allylamine cardiovascular toxicity. Toxicology 44: 129–145

Boor PJ, Hysmith RM, Sanduja R (1990) A role for a new vascular enzyme in the metabolism of xenobiotic amines. Circ Res 66: 249–252

Boor PJ, Sanduja R, Nelson TJ, Ansari GAS (1987) In vivo metabolism of the cardiovascular toxin, allylamine. Biochem Pharmacol 36: 4347–4353

Callingham BA, Holt A, Elliott J (1991) Properties and functions of the semicarbazide-sensitive amine oxidases. Biochem Soc Trans 19: 228–233

Clarke DE, Lyles GA, Callingham BA (1982) A comparison of cardiac and vascular clorgyline-resistant amine oxidase and monoamine oxidase. Biochem Pharmacol 31: 27–35

Coquil JF, Goridis C, Mack G, Neff NH (1973) Monoamine oxidase in rat arteries: evidence for different forms and selective localization. Br J Pharmacol 48: 590–599

DeLeve LD, Kaplowitz N (1991) Glutathione metabolism and its role in hepatotoxicity. Pharmacol Ther 52: 287–305

Dostert P, Guffroy C, Strolin Benedetti M, Boucher T (1984) Inhibition of semicarbazide-sensitive amine oxidase by monoamine oxidase B inhibitors from the oxazolidinone series. J Pharm Pharmacol 36: 782–785

Elliott J, Callingham BA, Sharman DF (1989a) Semicarbazide-sensitive amine oxidase (SSAO) of the rat aorta. Interactions with some naturally occurring amines and their structural analogues. Biochem Pharmacol 38: 1507–1515

Elliott J, Callingham BA, Sharman DF (1989b) Metabolism of amines in the isolated perfused mesenteric arterial bed of the rat. Br J Pharmacol 98: 507–514

Elliott J, Callingham BA, Sharman DF (1989c) The influence of amine metabolizing enzymes on the pharmacology of tyramine in the isolated perfused mesenteric arterial bed of the rat. Br J Pharmacol 98: 515–522

Ferkany JW, Andree TH, Clarke DE, Enna SJ (1981) Neurochemical effects of kojic amine, and its interaction with benzylamine oxidase. Neuropharmacology 20: 1177–1182

Fernandez de Arriba A, Balsa D, Tipton KF, Unzeta M (1990) Monoamine oxidase and semicarbazide-sensitive amine oxidase activities in bovine eye. J Neural Transm [Suppl] 32: 327–330

Guffroy C, Boucher T, Strolin Benedetti M (1985) Further investigations of the metabolism of two trace amines, β-phenylethylamine and p-tyramine by rat aorta semicarbazide-sensitive amine oxidase. In: Boulton AA, Maitre L, Bieck PR, Riederer P (eds) Neuropharmacology of the trace amines. Humana Press, Clifton NJ, pp 39–50

Hayes BE, Ostrow PT, Clarke DE (1983) Benzylamine oxidase in normal and athero-sclerotic aortae. Exp Mol Pathol 38: 243–254

Heck Hd'A, Casanova M, Starr TB (1990) Formaldehyde toxicity — new understanding. Crit Rev Toxicol 20: 397–426

Hysmith RM, Boor PJ (1987) In vitro expression of benzylamine oxidase activity in cultured porcine smooth muscle cells. J Cardiovasc Pharmacol 9: 668–674

Hysmith RM, Boor PJ (1988) Role of benzylamine oxidase in the cytotoxicity of allylamine toward aortic smooth muscle cells. Toxicology 51: 133–145

Janes SM, Mu D, Wemmer D, Smith AJ, Kaur S, Maltby D, Burlingame AL, Klinman JP (1990) A new redox cofactor in eukaryotic enzymes: 6-hydroxydopa at the active site of bovine serum amine oxidase. Science 248: 981–987

Lewinsohn R (1984) Mammalian monoamine-oxidizing enzymes, with special reference to benzylamine oxidase in human tissues. Brazilian J Med Biol Res 17: 223–256

Lizcano JM, Balsa D, Tipton KF, Unzeta M (1990) Amine oxidase activities in bovine lung. J Neural Transm [Suppl] 32: 341–344

Lizcano JM, Balsa D, Tipton KF, Unzeta M (1991) The oxidation of dopamine by the semicarbazide-sensitive amine oxidase (SSAO) from rat vas deferens. Biochem Pharmacol 41: 1107–1110

Lyles GA (1984) The interaction of semicarbazide-sensitive amine oxidase with MAO inhibitors. In: Tipton KF, Dostert P, Strolin Benedetti M (eds) Monoamine oxidase and disease. Academic Press, London, pp 547–556

Lyles GA, Archer DR (1986) Monoamine oxidase activity in dissociated cell fractions from rat skeletal muscle. J Pharm Pharmacol 38: 288–293

Lyles GA, Bertie KH (1987) Properties of a semicarbazide-sensitive amine oxidase in rat articular cartilage. Pharmacol Toxicol 60 [Suppl] 1: 33

Lyles GA, Chalmers J (1992) The metabolism of aminoacetone to methylglyoxal by semicarbazide-sensitive amine oxidase in human umbilical artery. Biochem Pharmacol 43: 1409–1414

Lyles GA, Holt A, Marshall CMS (1990) Further studies on the metabolism of methylamine by semicarbazide-sensitive amine oxidase activities in human plasma, umbilical artery and rat aorta. J Pharm Pharmacol 42: 332–338

Lyles GA, Marshall CMS, McDonald IA, Bey P, Palfreyman MG (1987) Inhibition of rat aorta semicarbazide-sensitive amine oxidase by 2-phenyl-3-haloallylamines and related compounds. Biochem Pharmacol 36: 2847–2853

Lyles GA, McDougall SA (1989) The enhanced daily excretion of urinary methylamine in rats treated with semicarbazide or hydralazine may be related to the inhibition of semicarbazide-sensitive amine oxidase activities. J Pharm Pharmacol 41: 97–100

Lyles GA, Singh I (1985) Vascular smooth muscle cells: a major source of the semicarbazide-sensitive amine oxidase of the rat aorta. J Pharm Pharmacol 37: 637–643

Lyles GA, Taneja DT (1987) Effects of amine oxidase inhibitors upon tryptamine metabolism and tryptamine-induced contractions of rat aorta. Br J Pharmacol 90: 16P

Norqvist A, Fowler CJ, Oreland L (1981) The deamination of monoamines by pig dental pulp. Biochem Pharmacol 30: 403–409

Norqvist A, Oreland L (1989) Localization of a semicarbazide-sensitive serotonin-oxidizing enzyme from porcine dental pulp. Biogenic Amines 6: 65–74

Precious E, Lyles GA (1988) Properties of a semicarbazide-sensitive amine oxidase in human umbilical artery. J Pharm Pharmacol 40: 627–633

Precious E, Gunn CE, Lyles GA (1988) Deamination of methylamine by semicarbazide-sensitive amine oxidase in human umbilical artery and rat aorta. Biochem Pharmacol 37: 707–713

Raimondi L, Pirisino R, Ignesti G, Capecchi S, Banchelli G, Buffoni F (1991) Semicarbazide-sensitive amine oxidase activity (SSAO) of rat epididymal white adipose tissue. Biochem Pharmacol 41: 467–470

Ramos K, Grossman SL, Cox LR (1988) Allylamine-induced vascular toxicity in vitro: prevention by semicarbazide-sensitive amine oxidase inhibitors. Toxicol Appl Pharmacol 95: 61–71

Roth JA, Gillis CN (1975) Multiple forms of amine oxidase in perfused rabbit lung. J Pharmacol Exp Ther 194: 537–544

Rucker RB, Goettlich-Riemann W (1972) Properties of rabbit aorta amine oxidases. Proc Soc Exp Biol Med 139: 286–289

Rucker RB, O'Dell BL (1971) Connective tissue amine oxidase. I. Purification of bovine aorta amine oxidase and its comparison with plasma amine oxidase. Biochim Biophys Acta 235: 32–43

Strolin Benedetti M, Dostert P (1985) Stereochemical aspects of MAO interactions: reversible and selective inhibitors of monoamine oxidase. TIPS 6: 246–251

Suzuki O, Matsumoto T (1984) Some properties of benzylamine oxidase in human aorta. Biogenic Amines 1: 249–257

Taneja DT, Lyles GA (1988) Use of an oil immersion technique to study the role of amine oxidase inhibition in potentiating tryptamine-induced contractions of rat aorta. Pharmacol Res Commun 20 [Suppl] 4: 127–128

Thornalley PJ (1990) The glyoxalase system: new developments towards functional characterization of a metabolic pathway fundamental to biological life. Biochem J 269: 1–11

Trevethick MA, Olverman HJ, Pearson JD, Gordon JL, Lyles GA, Callingham BA (1981) Monoamine oxidase activities of porcine vascular endothelial and smooth muscle cells. Biochem Pharmacol 30: 2209–2216

Wibo M, Duong AT, Godfraind T (1980) Subcellular location of semicarbazide-sensitive amine oxidase in rat aorta. Eur J Biochem 112: 87–94

Yu PH (1988) Three types of stereospecificity and the kinetic deuterium isotope effect in the oxidative deamination of dopamine as catalyzed by different amine oxidases. Biochem Cell Biol 66: 853–861

Yu PH (1990) Oxidative deamination of aliphatic amines by rat aorta semicarbazide-sensitive amine oxidase. J Pharm Pharmacol 42: 882–884

Authors' address: Dr. G. A. Lyles, Department of Pharmacology and Clinical Pharmacology, Ninewells Hospital and Medical School, University of Dundee, Dundee, DD1 9SY, United Kingdom

J Neural Transm (1994) [Suppl] 41: 397–406

Deamination of aliphatic amines by type B monoamine oxidase and semicarbazide-sensitive amine oxidase; pharmacological implications

P. H. Yu, B. A. Davis, A. A. Boulton, and **D. M. Zuo**

Neuropsychiatric Research Unit, Department of Psychiatry, University of
Saskatchewan, Saskatoon, Saskatchewan, Canada

Summary. Straight and branched chain aliphatic monoamines, which are
not normal tissue constituents, are deaminated selectively by type B monoamine oxidase (MAO-B). They exhibit a high affinity towards the active site
of MAO-B and this made them very useful pharmacologically. An anticonvulsant prodrug, Milacemide [2-(N-pentyl)glycinamide] is deaminated by
MAO-B and this facilitates a mechanism of delivering glycine into the CNS.
We have found that 2-propyl-pentylamine (2-propyl-1-aminopentane) and
N-(2-propylpentyl)glycinamide are also converted by MAO-B to valproic
acid and glycine both in vitro and in vivo; these compounds, however, cause
severe tremor. By attaching a propargylamine group the resultant series of
aliphatic propargylamine derivatives have been shown to be very potent
selective MAO-B inhibitors. They are chemically quite different from most
other MAO-B inhibitors, since they do not possess any aromatic structures.
The relatively short chain aliphatic propargylamines, i.e. N-2-pentyl-N-
methylpropargylamine and N-2-hexyl-N-methylpropargylamine, are 4 to 5
times more potent and more selective than selegiline (l-deprenyl) with
respect to the inhibition of MAO-B in brain following oral administration.

Semicarbazide-sensitive amine oxidase (SSAO) catalyzes the deamination of not only longer chain aliphatic amines but also short chain aliphatic
amines including methylamine. Formaldehyde is produced from methylamine by SSAO. Increased methylamine deamination may cause cellular
damage in some pathological conditions, such as uraemia and diabetes. We
have observed that cultured human endothelial cells are damaged by methylamine in the presence of SSAO. Inhibition of the SSAO activity completely
protects these cells from the methylamine-SSAO induced damage.

Deamination of aliphatic amines

Monoamine oxidase (MAO, E.C. 1.4.3.4) is well known for its catalytic
activity toward endogenous arylalkylamine substrates, such as neuronal
catecholamines (e.g. dopamine, noradrenaline and adrenaline), indolylethy-

lamines (e.g. serotonin) and trace amines (e.g. 2-phenylethylamine, tyramines, octopamines and tryptamine) (see review by Yu, 1986). Straight chain aliphatic amines were found to be deaminated by MAO about a half century ago. Following administration of some aliphatic amines to man, Rechenberger (1940) found that methylamine, the simplest aliphatic amine, which is normally present in the living organism, is broken down completely. About 32% of administered ethylamine, 10% of propylamine and 2% of n-butylamine were recovered in the urine. In the case of diethylamine 86% was recovered in the urine. Aliphatic amines have also been shown to be metabolized by MAO in vitro (Pugh and Quastel, 1937; Blaschko et al., 1937). Dimethylamine is not an MAO substrate and is not metabolized in vivo. It was unclear at the time, however, whether methylamine was a substrate for MAO, even though it could be metabolized totally in vivo.

Because most aliphatic amines are not endogenous substrates, their deamination has received little attention, although a few related basic studies on their deamination reaction have been reported (Symes et al., 1971; Battersby et al., 1979; Strolin-Benedetti et al., 1983; von Korff and Wolfe, 1984; Tenne et al., 1985). A systematic investigation recently conducted has revealed that straight chain aliphatic amines in the range 3 to 18 carbon atoms in length, optimum range between 5 to 10 carbons, are deaminated by MAO-B with a high affinity (Yu, 1989). Since the longer chain aliphatic amines are not endogenous constituents the deamination of these amines is physiologically unimportant except perhaps in the case of xenobiotic amines, which can then be metabolised by MAO-B.

Semicarbazide-sensitive amine oxidase (SSAO), a deaminase distinctly different from MAO (Callingham and Barrand, 1987), is also capable of deaminating aliphatic amines of various carbon chain lengths (Yu, 1990). SSAO also possesses very high affinity for aliphatic amines, yet it appears to possess a much narrower range of carbon chain lengths for optimal affinity. Nonylamine exhibited the highest affinity with a Km value even lower than that for benzylamine, which until now was thought to be the best substrate for SSAO. Unsaturated aliphatic amines, such as allylamine, can also be metabolized by SSAO but not by MAO. Acrolein, the deaminated product of allylamine, is known to cause cardiovascular damage (Boor and Nelson, 1982). SSAO is also capable of deaminating simple amines, such as methylamine and ethylamine (Yu, 1990), but not dimethylamine or ethanolamine (Yu, unpublished). Methylamine, which exists endogenously, has been thought to be a substrate for SSAO (Precious et al., 1988) and seems to be of considerable physiological interest.

Aliphatic amines as prodrugs

MAO has been found to be involved in the conversion of an antiepileptic prodrug, 2-n-pentylaminoacetamide (Milacemide) (de Varebeke et al., 1988). Milacemide, an aliphatic amine derivative, can cross the blood brain barrier, and then be oxidized by MAO to form glycinamide, which is subsequently cleaved to glycine. This mechanism for the delivery of glycine to the central nervous system is presumably related to the anticonvulsant activity of Milacemide (van Dorsser et al., 1983) and the apparent improvement in learning and memory following its administration (Handelmann et al., 1989).

We have recently discovered that the branched aliphatic amine, 2-propyl-pentylamine (2-propyl-1-aminopentane), can be deaminated by rat liver MAO and aorta SSAO; it is converted to 2-propyl-1-pentaldehyde, which is then subsequently oxidized to valproic acid (VPA) (Yu and Davis, 1991a). VPA has been widely used as an anticonvulsant and has been reported to be effective in the treatment of bipolar disorders. We were interested, therefore, to determine whether or not 2-propylpentylamine could be deaminated by MAO in vivo and thus act as a prodrug for VPA. In addition, we have also synthesized a Milacemide analogue, N-(2-propylpentyl)-glycinamide. This compound, following cleavage of the amino group by MAO, could possibly provide simultaneous delivery of both VPA and glycine to the central nervous system and this might exhibit anticonvulsant properties as well as other pharmacological effects (Yu and Davis, 1991b).

We confirmed that MAO-B was capable not only of metabolizing straight and branched chain aliphatic amines, but also branched aliphatic glycin-amides both in vitro and in vivo. VPA was detected in the brain and the serum soon after the peripheral administration of both compounds. The VPA levels in the brain following administration of these prodrugs, however, were lower than the levels necessary to produce an acute anticonvulsant effect. Furthermore we discovered that neither 2-propyl-pentylamine nor N-(2-propylpentyl)glycinamide could be employed as VPA prodrugs for the treatment of convulsion, because they induced severe tremor. In fact, they also potentiated the convulsant actions of pentylenetetrazol.

The finding that these branched aliphatic amine derivatives caused tremor is interesting although it is not clear, whether it was the amines themselves or their immediate metabolite, 2-propyl-1-pentaldehyde, that was responsible for the observed behavioral effects. The MAO-B inhibitor selegiline did not block the tremorogenic effect, suggesting that it was the amines themselves that were responsible for the tremor effect rather than the deaminated aldehyde products. This is also supported by the absence of a tremor effect induced by 2-propyl-1-pentaldehyde (unpublished results). The branched dipropyl structure is essential for the induction of the tremor, since straight chain aliphatic amides, such as n-pentylglycinamide at the same doses do not exhibit any tremorigenic effects. We have also discovered that n-decylglycinamide exerts an anticonvulsant effect but without causing any tremor (unpublished results). It is possible that the tremor induced by 2-propylpentylamine and N-(2-propylpentyl)glycinamide is related to a toxic effect of VPA; it is known that chronic administration of VPA can induce tremor in epileptic patients. The 2-PPG-induced effect may represent a useful animal model for the study of certain kinds of tremors, although at this time its mechanism of action and the nature of the neuronal systems involved remain to be elucidated.

Aliphatic N-methylpropargylamine MAO-B inhibitors

The MAO-B inhibitor selegiline has been used as an effective adjuvant to L-DOPA in the treatment of Parkinson's disease (Birkmayer et al., 1985). It was thought to potentiate dopamine's action by reducing its deamination. Recently, it has been reported that selegiline can by itself, significantly delay the onset of disability associated with early, otherwise untreated, cases of Parkinson's disease (The Parkinson Study Group, 1989; Tetrud and Langston, 1989). Selegiline, along with other MAO inhibitors, has been shown to prevent 1-methyl-4-phenyl-1,2,3,6-tetrahydropyridine (MPTP)-induced Parkinson-like neurotoxicity in animals (Heikkila et al., 1984). MPTP itself is not toxic, but is converted to its distal toxin, 1-methyl-4-phenylpyridinium ion (MPP^+), by MAO-B in the brain. The precise neuroprotective mechanism(s) by which selegiline acts is not yet understood. Perhaps it reduces oxidative stress (i.e. free radical formation) by decreasing oxidative deamination (Cohen, 1990). In addition, however, selegiline has been shown also to exhibit neuronal rescue effects, and in this mechanism of action inhibition of MAO-B activity is probably not directly involved (Tatton and Greenwood, 1991; Salo and Tatton, 1992). A prolonged life span and improved sexual activity in rodents (Knoll et al., 1989; Milgram et al., 1990) and humans (Birkmayer et al., 1983) by selegiline have also been claimed.

Selegiline, a structural analogue of amphetamine, is catabolized to produce desmethyldeprenyl, methamphetamine and amphetamine (Heinonen et al., 1989); this has caused some concern since it might cause selegiline to become a drug subject to substance abuse. It has also been asked whether or not it is the amphetamine moiety of selegiline that is related to its clinical efficacy. Different MAO-B inhibitors, not possessing the amphetamine moiety or amphetamine-like properties, should therefore be assessed (Langston, 1990). The high affinity of some aliphatic amines for the active site of MAO-B has now been applied in the design of some specific MAO-B inhibitors. We have synthesized a series of aliphatic propargylamine derivatives (Yu et al., 1992). Some of these aliphatic propargylamines, such as N-(2-hexyl)-N-methylpropargylamine (2-HxMP) and N-methyl-N-2-pentylpropargylamine (M-2-PP), are highly potent, selective and irreversible MAO-B inhibitors. MAO inhibitory activity appears to be correlated with the lipophilicity of these compounds. The carbon chain length of the N-alkyl group is not only related to the inhibitory potency, but it also affects the selectivity towards MAO-A and MAO-B. When the terminal carbon is substituted with a hydroxyl group, MAO-B inhibitory activity is markedly reduced. Compounds substituted with a carboxy or carbethoxy group at the terminal carbon also causes a considerable reduction in the MAO inhibitory activity.

Stereospecificity exists; for example, the (R) stereoisomer of N-(2-butyl)-N-methylpropargylamine is about 20 fold more potent than the (S) enantiomer in the inhibition of MAO-B activity, which is quite similar to the stereospecific effect exhibited by selegiline (Robinson, 1985).

Aliphatic propargylamines with shorter carbon chain lengths appear to be more potent than their longer chain analogues in the inhibition of brain MAO-B activity (i.e. as assessed from their ED_{50} values) following intraperitoneal administration. The shorter chain length molecules are more easily absorbed (into lipids, membranes, etc.) and/or more readily transported into the brain. In comparison to the compounds with longer carbon chain lengths, they are less lipophilic, and therefore less likely to bind or associate with peripheral lipophilic components. The effects of 2-HxMP and M-2-PP were found to be even more potent in blocking MAO-B activity in the brain following oral administration than was the case for 1-deprenyl. It is also interesting to note that aliphatic propargylamines, such as M-2-PP, are more stable in aqueous solution than is selegiline.

Chronic administration of the aliphatic propargylamine MAO inhibitors (via i.p. injection or oral administration) and their concomitant chronic effects on mouse brain MAO activity levels have been assessed. Inhibition of MAO-A and MAO-B was dependent both on the inhibitor and the dose applied. Both 2-HxMP and M-2-PP at a low dose (0.25 mg/Kg) were without effect on MAO-A and MAO-B 24 hours after a single i.p. injection. This dose was effective on MAO-B, however, after 13 days of treatment. At higher doses (2 mg/Kg) M-2-PP selectively inhibited MAO-B activity. After 10 and 21 daily treatments a greater inhibition of MAO-B was observed, but MAO-A by this time had also become slightly inhibited. At lower oral doses (i.e. 1 and 10 µg/mL in drinking water), selective inhibition of MAO-B activity was achieved, while at higher doses (100 µg/mL), MAO-A also became inhibited following three weeks of treatment. A prolonged inhibition of MAO-B activity following a single and higher dose (2 mg/Kg, i.p.) of several aliphatic propargylamines (e.g. M-2-PP) was observed, confirming that the inhibition of MAO-B by aliphatic propargylamines is like that by selegiline, i.e. irreversible in vivo, and that the synthesis of new brain mitochondrial MAO appears to be an extremely slow process.

M-2-PP was found to be capable of protecting against MPTP-induced depletion of striatal dopamine neurons as well as against DSP-4-induced depletion of noradrenaline in the hippocampus (Yu et al., unpublished results). The aliphatic propargylamine derivatives are highly potent, selective, irreversible MAO-B inhibitors. These inhibitors are apparently nontoxic and do not possess an amphetamine moiety within their structures. Some of them may be useful in the treatment of neuropsychiatric disorders such as Parkinson's and Alzheimer's diseases.

Deamination of methylamine by SSAO and vascular damage

SSAO is located in the vascular smooth muscle tissues of several mammalian species including human and it also circulates in the blood. Methylamine has recently been shown to be deaminated by this enzyme and to produce formaldehyde (Precious et al., 1988) as follows:

$$CH_3NH_2 + O_2 + H_2O \rightarrow HCHO + H_2O_2 + NH_3$$

Since formaldehyde is well known to be extremely reactive and toxic, it is important to know whether or not this reaction occurs in nature and whether it is related to certain physiological and pathological conditions.

Methylamine has been detected in many different mammalian tissues (Smith and Jepson, 1967; Nixon, 1972) and human blood (Baba et al., 1984). It is excreted in relatively large quantities by humans (Blau, 1961). It can be derived from the diet as a result of gut bacterial degradation of dietary precursors (Lowis et al., 1985; Zeisel et al., 1983) as well as from various metabolic pathways for the degradation of creatine, sarcosine (Davis and De Ropp, 1961) and adrenaline (Schayer et al., 1952). Normal circulating blood and urinary concentrations of methylamine are increased in certain physiological (pregnancy, exercise) and pathological (i.e. diabetes, uraemia) conditions (Kapeller-Adler and Toda, 1932). SSAO exhibits a reasonably high affinity for methylamine (Yu, 1990). The urinary excretion of methylamine is increased in rats treated with SSAO inhibitors. Methylamine therefore is an endogenous substrate for SSAO. Its deamination generates formaldehyde. It is possible therefore that a cytotoxic effect may occur where SSAO is located (i.e. in blood and the plasma membrane of vascular smooth muscle cells of blood vessels) (Wibo et al., 1980; Lewinsohn, 1981) and in cartilage (Lyles and Bertie, 1987).

The deamination of methylamine is indeed toxic towards an endothelial cell line (CPA 47, ACTT) and human umbilical vein primary endothelial cells (Yu and Zuo, 1992b). Methylamine alone did not significantly affect the viability and growth of these cells, neither did SSAO alone, but when SSAO and methylamine were present together a severe toxic effect on the cultured endothelial cells was observed. The SSAO inhibitors semicarbazide and MDL-72974A (Yu and Zuo, 1992a) were able to prevent these toxic effects completely. In the presence of human serum from normal subjects (which contains SSAO) and methylamine the survival of the endothelial cells was also found to be reduced. Morphological changes such as the disappearance of pseudopodia, vacuolization, and contraction of the cells were observed following exposure to methylamine in the presence of SSAO. Again the SSAO inhibitor MDL-72974A prevented these morphological changes.

In addition to formaldehyde, hydrogen peroxide is also produced during deamination of methylamine catalyzed by SSAO. Hydrogen peroxide is well known to be toxic. In our tissue culture study, however, formaldehyde was found to be significantly more toxic than the hydrogen peroxide. It is interesting to note that formaldehyde can be activated by hydrogen peroxide and becomes very reactive, such as to interact with the ε-amino group of basic amino acid lysine (Trezl et al., 1992), suggesting that the toxicity of formaldehyde can be potentiated in the presence of hydrogen peroxide. It seems astonishing, that deamination of methylamine can exhibit such destructive consequences.

Since SSAO is located in the vascular smooth muscles and circulates in the blood, it is conceivable that the SSAO-mediated conversion of methylamine to formaldehyde can occur in blood and/or in vascular tissues and thus induce vascular disorders. Blood methylamine levels in patients with chronic renal failure have been shown to be markedly elevated (Baba et al., 1984). The blood methylamine levels in these patients are quite comparable to those used in our experiments causing damage to endothelial cells. Vascular problems are known to be associated with patients exhibiting renal failure and dysfunction.

Earlier clinical reports indicated that serum amine oxidase activity was elevated in patients with diabetes mellitus (McEwen et al., 1967; Tryding et al., 1969). An increase in blood and kidney SSAO has also recently been demonstrated in streptozotocin (STZ)-induced diabetic rats (Hayes et al., 1990). None of these previous studies remarked on the possibility that SSAO activity may possess significant implications with respect to formaldehyde toxicity. Methylamine is known to be increased in the urine of diabetics (Kapeller-Adler and Toda, 1932). Vascular damage is also a major contributing factor to morbidity and mortality in diabetes (Steiner, 1981). Damage to the endothelium may therefore occur in diabetes, atherosclerosis and even in hypertension (Rau, 1991).

Retinopathy is more prevalent in diabetes than in the normal population (Klein et al., 1989). It is interesting that a relatively large amount of SSAO activity has been detected in the rat eye (Cao Danh et al., 1985). It is possible that bio-conversion of methylamine to formaldehyde may be enhanced in the microvessels of the retina or in the smooth muscles of the eye in diabetics. As a result the eye will be damaged perhaps being a cause of the blindness seen in these patients. Methanol is well known to cause blindness and its mechanism of action is known to be related to the formation of formaldehyde by enzymatic dehydrogenation. How blood methylamine and SSAO are increased in diabetes is yet to be established.

Stress induces the release of adrenaline; which can then be deaminated by monoamine oxidase to form methylamine, which can then be further oxidized by SSAO to produce formaldehyde. The cardiovascular problems that are associated with chronic stress could also be related to the chronic toxic effects of an elevated formaldehyde production. Finally, the pain in patients suffering from arthritis and gout may also, in some way, be related to the higher content of SSAO in the cartilage tissues. Deamination of methylamine in the joint tissues (Lyles and Bertie, 1987) to form formaldehyde is well known to cause severe pain and indeed it has been used to induce pain in classical pain studies.

Abnormal deamination of methylamine, catalyzed by SSAO, may be associated with certain cardiovascular and atherosclerotic conditions (i.e. diabetes, uraemia, etc.). A treatment to decrease this methylamine production or to inhibit its deamination (i.e. by SSAO inhibitor) may therefore be useful in preventing the occurrence of such conditions.

Acknowledgements

We thank Saskatchewan Health, the Medical Research Council of Canada and Diabetes Association of Canada for their continuing financial support.

References

Baba S, Watanabe Y, Gejyo F, Arakwa M (1984) High-performance liquid chromatographic determination of serum aliphatic amines in chronic renal failure. Clin Chim Acta 136: 49–56

Battersby AR, Buckley DG, Staunton J, Williams PJ (1979) Studies of enzyme-mediated reactions, part 10. Stereochemical course of the dehydrogenation of stereospecifically labelled l-amino-heptanes by the amine oxidase from rat liver mitochondria (E.C. 1.4.3.4). J Chem Soc (Perkin I) 10: 2550–2558

Birkmayer W, Knoll J, Riederer P, Hars V, Marton J (1985) Increased life expectancy resulting from addition of L-deprenyl to Madopar treatment in Parkinson's disease: a long-term study. J Neural Transm 64: 113–127

Birkmayer W, Knoll J, Riederer P, Youdim MBH (1983) (−)-Deprenyl leads to prolongation of l-Dopa efficacy in Parkinson's disease. Mod Probl Pharmacopsychiatry 19: 170–176

Blaschko H, Richter D, Schlossman H (1937) The oxidation of adrenaline and other amines. Biochem J 31: 2187–2196

Blau K (1961) Chromatographic methods for the study of amines from biological material. Biochem J 80: 193–200

Callingham BA, Barrand MA (1987) Some properties of semicarbazide-sensitive amine oxidase. J Neural Transm 23: 37–54

Cao Danh H, Strolin-Benedetti M, Doster P, Mousset A (1985) Age-related changes in monoamine oxidase and semicarbazide-sensitive amine oxidase activities of rat aorta. J Pharm Pharmacol 37: 354–357

Cohen G (1990) Monoamine oxidase and oxidative stress of dopaminergic synapses. J Neural Transm [Suppl] 32: 229–238

Davis EJ, De Ropp RS (1961) Metabolic origin of urinary methylamine in the rat. Nature 190: 636–637

De Varebeke PJ, Cavalier R, David-Remacle M, Youdim MBH (1988) Formation of the neurotransmitter glycine from the anticonvulsant milacemide is mediated by brain monoamine oxidase-B. J Neurochem 50: 1011–1016

Handelmann GE, Nevins ME, Mueller LL, Arnolde SM, Cordi AA (1989) Milacemide, a glycine prodrug, enhances performance of learning tasks in normal and amnestic rodents. Pharmacol Biochem Behav 34: 823–828

Heikkila RE, Manzino L, Cabbat FS, Duvoisin RC (1984) Protection against the dopaminergic neurotoxicity of 1-methyl-4-phenyl-1,2,5,6-tetrahydropyridine by monoamine oxidase Inhibitors. Nature 311: 467–469

Heinonen EH, Myllyla V, Sotaniemi K, Lammintausta R, Salonen JS, Anttila M, Savijarvi M, Rinne UK (1989) Pharmacokinetics and metabolism of selegiline. Acta Neurol Scand 126: 93–99

Hayes BE, Clarke DE (1990) Semicarbazide-sensitive amine oxidase activity in streptozotocin diabetic rats. Res Comm Chem Pathol Pharmacol 69: 71–83

Kapeller-Adler R, Toda K (1932) Über das Vorkommen von Monomethylamine im Harn. Biochem Z 248: 403–425

Klein R, Klein BEK, Moss SE, Davis MD, DeMets DL (1989) The Wisconsin epidemiologic study of diabetic retinopathy. IX. Four-year incidence and progression of diabetic retinopathy when age at diagnosis is less than 30 years. Arch Ophthalmol 107: 237–243

Knoll J, Dallo J, Yen TT (1989) Striatal dopamine, sexual activity and lifespan longevity of rats treated with (−)deprenyl. Life Sci 45: 525–531

Langston JW (1990) Selegiline as neuroprotective therapy in Parkinson's disease: concepts and controversies. Neurology 40: 61–66

Lewinsohn R (1981) Amine oxidase in human blood vessels and non-vascular smooth muscle. J Pharm Pharmacol 33: 569–575

Lowis S, Eastwood MA, Brydon WG (1985) The influence of creatinine, lecithin and choline feeding on aliphatic amine production and excretion in rat. Br J Nutr 54: 43–51

Lyles GA, Bertie KH (1987) Properties of a semicarbazide-sensitive amine oxidase in rat articular cartilage. Pharmacol Toxicol [Suppl] 1: 33

McEwen CMJr, Castell DO (1967) Abnormalities of serum monoamine oxidase in chronic liver disease. J Lab Clin Med 70: 36–47

Milgram NW, Racine RJ, Nellis P, Mendonca A, Ivy GO (1990) Maintenance on l-deprenyl prolongs life in aged male rats. Life Sci 47: 415–420

Nixon R (1972) Volatile amines in mouse brain: a radioassay with picogram sensitivity. Anal Biochem 48: 460–470

Precious E, Gunn CE, Lyles GA (1988) Deamination of methylamine by semicarbazide-sensitive amine oxidase in human umbilical artery and rat aorta. Biochem Pharmacol 37: 707–713

Pugh CEM, Quastel JH (1937) Oxidation of aliphatic amines by brain and other tissues. Biochem J 31: 286–291

Rau L (1991) Hypertension, endothelium, and cardiovascular risk factors. Am J Med 90 [Suppl] 2A: 13S–18S

Rechenberger J (1940) Über die flüchtigen Alkylamine im menschlichen Stoffwechsel. II. Mitteilung: Ausscheidung im Harn nach oraler Zufuhr. Hoppe-Seyl Z 256: 275–284

Robinson BJ (1985) Stereoselectivity and isozyme selectivity of monoamine oxidase inhibitors: enantiomers of amphetamine, N-methylamphetamine and deprenyl. Biochem Pharmacol 34: 4105–4108

Salo PT, Tatton WG (1991) Deprenyl reduces the death of motoneurons caused by axotomy. J Neurosci Res 31: 394–400

Schayer RW, Smiley LR, Kaplan HE (1952) The metabolism of epinephrine containing isotopic carbon. J Biol Chem 198: 545–551

Smith AD, Jepson JB (1967) Chromatography of urinary and tissue amines and amino alcohols as 2,4-dinitrophenyl derivatives prepared with 2-nitrobenzenesulfonic acid. Anal Biochem 18: 36–45

Steiner G (1981) Diabetes and atherosclerosis. Diabetes [Suppl] 30: 1–7

Strolin-Benedetti M, Sontag N, Boucher T, Kan JP (1981) Aliphatic amines as MAO substrates in the rat: the effect of selective inhibitors on the deamination of n-pentylamine. In: Usdin E, Weiner N, Youdim MBH (eds) Function and regulation of monoamine enzymes: basic and clinical aspects. MacMillan, London, pp 527–538

Symes HL, Missala K, Sourkes TL (1971) Iron and riboflavine-dependent metabolism of a monoamine in the rat in vivo. Science 174: 153–155

Tatton WG, Greenwood CE (1992) Rescue of dying neurons: a new action for deprenyl in MPTP Parkinsonism. J Neurosci Res 30: 666–672

Tenne M, Youdim MBH, Ulitzur S, Finberg JPM (1985) Deamination of aliphatic amines by monoamine oxidase A and B studied using a bioluminescence technique. J Neurochem 44: 1373–1377

The Parkinson Study Group (1989) Effect of deprenyl in the progression of disability in early Parkinson's disease. N Engl J Med 321: 1364–1371

Tetrud JW, Langston JW (1989) The effect of deprenyl (selegiline) on the natural history of Parkinson's disease. Science 245: 519–522

Trézl L, Török G, Vasvéri G, Pipek J (1992) Formation of burst chemiluminescence, excited aldehydes, and singlet oxygen in model reactions and from carcinogenic compounds in rat liver S9 fractions. In: Tyihak E (ed) Role of formaldehyde in biological systems. Hung Biochem Society, Sopron, pp 111–124

Tryding N, Nilsson SE, Tufvesson G, Berg R, Carlstrom S, Elmfors B, Nilsson JE (1969) Physiological and pathological influences on serum monoamine oxidase level. Scand J Clin Lab Invest 23: 79–84 van Dorsser W, Barris D, Cordi A, Roba J (1983) Anticonvulsant activity of milacemide. Arch Int Pharmacodyn 266: 239–249

von Korff RW, Wolfe AR (1984) Saturated amines and diamines as substrates which inhibit beef liver mitochondrial monoamine oxidase. J Bioenerg Biomem 16: 597–609

Wibo M, Duong AT, Godfraind T (1980) Subcellular location of semicarbazide-sensitive amine oxidase in rat aorta. Eur J Biochem 112: 87–94

Yu PH (1990) Oxidative deamination of aliphatic amines by rat aorta semicarbazide-sensitive amine oxidase. J Pharm Pharmacol 42: 882–884

Yu PH (1989) Deamination of aliphatic amines of different chain lengths by rat liver monoamine oxidase A and B. J Pharm Pharmacol 41: 205–208

Yu PH (1986) Monoamine oxidase. In: Boulton AA, Baker GB, Yu PH (eds) Neuromethods V: neurotransmitter enzymes. Humana Press, NJ, pp 235–272

Yu PH, Davis BA (1991a) 2-Propyl-1aminopentane, its deamination by monoamine oxidase and semicarbazide-sensitive amine oxidase, conversion to valproic acid and behavioral effects. Neuropharmacology 30: 507–515

Yu PH, Davis DA (1991b) Simultaneous delivery of valproic acid and glycine to the brain, deamination of 2-propylpentylglycinamide by monoamine oxidase B. Mol Chem Neuropathol 15: 37–49

Yu PH, Davis DA, Boulton AA (1992) Aliphatic propargylamines: potent selective irreversible monoamine oxidase B inhibitors. J Med Chem 35: 3705–3713

Yu PH, Zuo DM (1992a) (E)-4-Fluoro-beta-fluoroethylene benzene butamine (MDL-72974A) as a highly potent inhibitor for semicarbazide-sensitive amine oxidase from vascular tissues and serum of different species. Biochem Pharmacol 43: 307–312

Yu PH, Zuo DM (1992b) Methylamine, a potential endogenous toxin for vascular tissues: formation of formaldehyde via enzymatic deamination and the cytotoxic effects on endothelial cells. Diabetes 42: 594–603

Zeisel SH, Wishnok JS, Blusztjan JK (1983) Formation of methylamines from ingested choline and lecithin. J Pharmacol Exp Ther 225: 320–324

Authors' address: Dr. P. H. Yu, Neuropsychiatric Research Unit, Department of Psychiatry, University of Saskatchewan, Saskatoon, Saskatchewan, Canada, S7N OWO.

J Neural Transm (1994) [Suppl] 41: 407–414

Haloallylamine inhibitors of MAO and SSAO and their therapeutic potential

M. G. Palfreyman[1], I. A. McDonald[1], P. Bey[1], C. Danzin[2], M. Zreika[2], and G. Cremer[2]

Marion Merrell Dow Research Institute, [1] Cincinnati, Ohio, U.S.A.,
and [2] Strasbourg, France

Summary. Based on mechanistic understandings, molecular modeling and extensive quantitative structure-activity relationships, appropriately substituted haloallylamine derivatives were designed as potential mechanism-based inhibitors of MAO and/or SSAO. Potent inhibition of MAO-B and SSAO occurred with fluoroallylamines whereas chloroallylamines, such as MDL 72274A ((E)-2-phenyl-3-chloroallylamine hydrochloride), were selective and potent inhibitors of SSAO. MDL 72974A (E)-2-(4-fluorophenethyl)-3-fluoroallylamine hydrochloride is a potent ($IC_{50} = 10^{-9}$M) inhibitor of both MAO-B and SSAO, with 190-fold lower affinity for MAO-A. In clinical studies, oral doses as low as 100 µg produced substantial inhibition of platelet MAO-B. Essentially complete inhibition occurred at 1 mg with the effect lasting 6–10 days. One or 4 mg MDL 72974A given daily for 28 days to 40 Parkinson's patients treated with L-dopa produced statistically significant reductions in the Unified Parkinson's Disease Rating Scale. MAO-B inhibitors, such as MDL 72974A and L-deprenyl, offer the potential of being neuroprotective in Parkinson's Disease and other neurogenerative disorders. Concommitant inhibition of SSAO may provide additional, but as yet unproven, advantages over pure inhibitors of MAO-B.

Introduction

In the 1970's, Rando (1973) and Rando and Eigner (1977) described the weak and non-selective MAO inhibiting properties of allylamine and bromoallylamine. Walsh (1984) also noted that chloroallylamine inhibited the oxidative deamination of benzylamine. With an understanding of MAO-A and B substrate selectivities coupled with molecular modeling and extensive structure-activity studies, we have elaborated a series of very potent monoamine oxidase type B (MAO-B) and semicarbazide-sensitive amine oxidase inhibitors (Bey et al., 1984; McDonald et al., 1985, 1986; Palfreyman et al., 1984, 1986; Zreika et al., 1984, 1989; Lyles et al., 1987).

These compounds were initially evaluated as inhibitors of MAO-A and MAO-B with the goal of identifying selective MAO-B inhibitors to be used as an adjunct to L-dopa for the treatment of Parkinson's disease. Subsequently, Lyles et al. (1987) reported that some of these compounds also inhibited SSAO. The emerging evidence that L-deprenyl can actually *rescue* neurons from death after MPTP treatment, in contrast to the known ability of MAO-B inhibitors to *prevent* MPTP-induced neuronal destruction, has prompted us to reexamine some properties of these haloallylamine-based inhibitors.

All flavin- and copper-dependent amine oxidases liberate hydrogen peroxide during the normal catalytic oxidative deamination process upon cofactor reoxidation. It is conceivable that local concentrations of this powerful oxidant may play a role in chronic neurodegeneration processes. Selective inhibitors of individual amine oxidases will offer tools to examine the consequences of peroxide generation.

This report will summarize the preclinical and early clinical evaluation of MDL 72974A ((E)2-(4-fluorophenethyl)-3-fluoroallylamine hydrochloride), a potent inhibitor of both MAO-B and SSAO, in Parkinson's disease. MDL 72274A is an example of a potent and selective SSAO inhibitor and some details of its in vitro and in vivo properties will be described.

Material and methods

Male Sprague Dawley rats (Charles River) were used for biochemical studies. Monoamine oxidase activity was determined in vitro and in vivo as described in detail by Zreika et al. (1984) and Fozard et al. (1985).

SSAO activity was determined as described by Lyles et al. (1987). Human platelet MAO-B activity was determined by the methods described by Alken et al. (1984) and Harland et al. (1990).

Human volunteers were administered MDL 72974A by mouth at various single or once daily doses over the dose range 0.1–12 mg. Blood samples were withdrawn at different times to determine platelet MAO activity. Parkinson patients with a Hoehn and Yahr disease stage of 2 to 4 (Hoehn and Yahr, 1967) were treated with 1 or 4 mg MDL 72974A, added to the concurrent L-dopa regimen (mean dose = 637.5 mg/day; range = 30–2,350 mg/day) for 28 days. Efficacy of treatment was determined on days 0, 16, 29 and 50 according to the Unified Parkinson's Disease Rating Scale (UPDRS; Aasly et al., 1991).

Results

The fluoroallylamine, MDL 72974A, is a potent, mechanism-based, irreversible inhibitor of both MAO-B (Zreika et al., 1989) and SSAO (Table 1). In contrast, the chloro-containing analogue, MDL 72274A, is a potent inhibitor of SSAO, with little effect on MAO-B (Table 1). In vivo inhibition of SSAO in the rat aorta and lung occurs at submilligram/kg doses of MDL 72274A with an estimated ED_{50} of 0.3 mg/kg in the aorta and 0.75 mg/kg in the lung (Table 1 and Fig. 1A). Only slight inhibition of MAO-B was observed in the heart (Fig. 1B) at these doses.

Table 1. MAO-B and SSAO inhibiting properties of the haloallylamines, MDL 72974A and MDL 72274A

	IC$_{50}$		
	MAO-B[c]	MAO-A[e]	SSAO[a]
MDL 72,974A	3.6 nM	680 nM	5 nM[b]
MDL 72,274A	100 μM	>2 mM	8 nM[a]
	ED$_{50}$ (mg/kg)		
	MAO-B[c]	MAO-A[e]	SSAO[d]
MDL 72,974A	0.18	8	ND
MDL 72,274A	10	>50	0.3

[a] Lyles et al., 1987; [b] Yu and Zuo, 1992; [c] Rat brain [^{14}C] phenethylamine as substrate; [d] Rat aorta [^{14}C]benzylamine as substrate; [e] Rat brain [^{14}C]5-HT; *ND* not determined

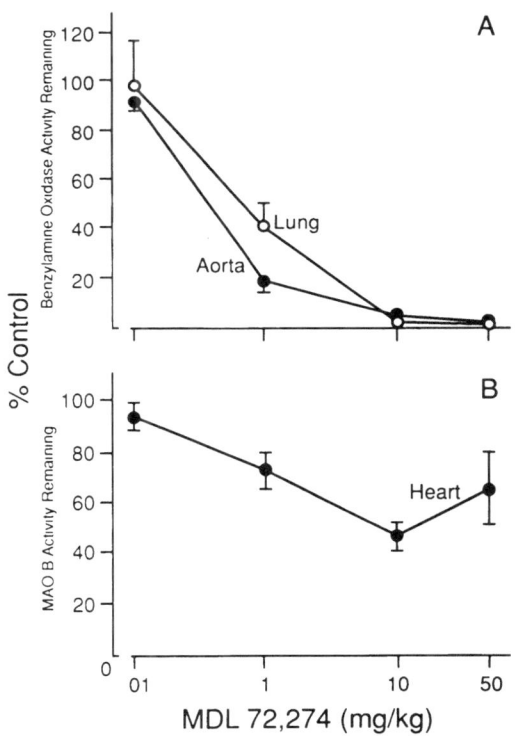

Fig. 1. Dose response effect of a single i.p. doses of MDL 72274A on SSAO (**A**) and MAO-B (**B**) activity in rat tissues 4 hours after treatment (mean ± S.E.M., n = 5)

An estimation of the half-life of SSAO can be made from the data presented in Fig. 2. In the aorta approximately half the enzyme activity had returned by 60 hours after dosing with 2 mg/kg MDL 72274A. Return of enzyme activity in the rat lung was slightly slower (Fig. 2). Inhibition of

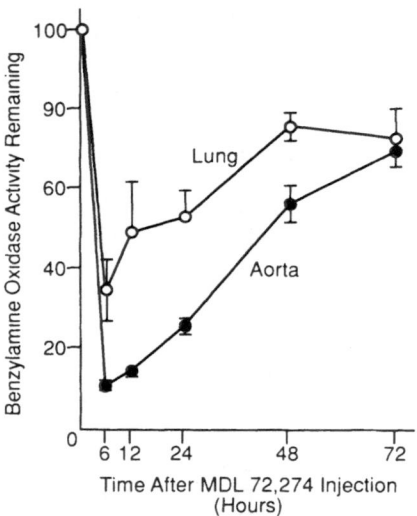

Fig. 2. Time course of effect of a single i.p. dose of MDL 72274A (2 mg/kg) on SSAO activity of rat lung and aorta (mean ± S.E.M., n = 5)

SSAO by MDL 72274A has been shown to be irreversible (Lyles et al., 1987); the recovery of enzyme activity is thus a consequence of new enzyme synthesis. It is interesting to note that the estimated half-life for SSAO is considerably shorter than most estimates for MAO turnover (6–11 days).

MDL 72974A was an extremely potent inhibitor of MAO-B in platelets from normal volunteers treated with the compound by mouth (ED$_{50}$ → 90 μg per os). These data are shown in Fig. 3 (Aasby et al., 1991; Palfreyman et al., 1993; Hinze et al., 1990; Harland et al., 1990). Almost complete inhibition of platelet MAO-B occurred at the 1 mg p.o. dose. The compound was well tolerated at several multiples of the MAO-B inhibiting dose.

In Parkinson's disease patients, addition of 1 or 4 mg MDL 72974A to the standard L-dopa treatment (L-dopa plus a decarboxylase inhibitor) produced a statistically significant improvement (p < 0.01) in the UPRDS total and motor scores on days 16 and 29 versus day 0 in both MDL 72974A treated groups, with a tendency to return to baseline values on day 50. There were no significant differences between the doses (Aasly et al., 1991; Palfreyman et al., 1992). To date, more than 250 patients have received this novel MAO-B/SSAO inhibitor with excellent tolerability being noted with doses up to 24 mg.

Discussion

Two recent events have stimulated renewed interest in MAO-B inhibitors as neuroprotective agents for the treatment of Parkinson's disease and potentially for other neurodegenerative disorders. It is now clear from the DATATOP study that L-deprenyl is able to significantly slow the rate of

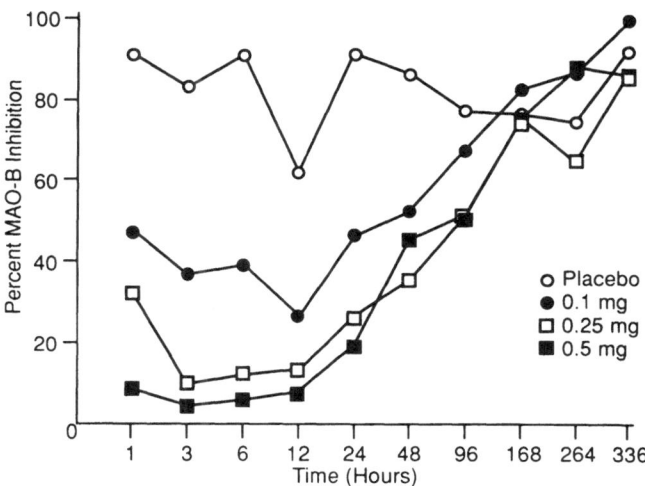

Fig. 3. Inhibition of platelet MAO-B activity after oral MDL 72974A in human volunteers. Platelet MAO-B activity was determined after 0.1, 0.25 and 0.5 mg MDL 72974A or placebo at the times shown and expressed as the percent of baseline (predrug) platelet activity. N = 5 for MDL 72974A dose groups and N = 3 for the placebo group

progression of Parkinson's disease. Equally exciting is the data from Tatton and Greenwood (1991) which demonstrates that L-deprenyl can *reverse* the neurotoxic consequences of 1-methyl-4-phenyl-1,2,5,6-tetrahydropyridine (MPTP)-induced neurodegeneration of the nigro-striatal dopamine pathway.

The neuronal rescue action of L-deprenyl is not easy to rationalize in terms of selective MAO-B inhibition. An obvious explanation would invoke previously identified neurotrophic properties of L-deprenyl. However, Nicklas (1992) recently reported that MDL 72974A is similarly able to rescue neurones from certain MPTP-induced death. It is unlikely that two compounds of quite different structures would coincidently have identical neurotrophic properties. Moreover, the MAO-A selective inhibitor, clorgyline, cannot reverse the effects of MPTP. L-Deprenyl, in contrast to MDL 72974A, does not inhibit SSAO. Until there is unambiguous evidence to the contrary, it must be assumed that MDL 72974A and L-deprenyl share some properties directly or indirectly associated with a neurotrophic effect. Conceivably, there maybe an undiscovered new amine oxidase for which both these compounds are exquisitely sensitive inhibitors and which plays a key role in the neuronal rescue process. Whether the generation of local high concentrations of hydrogen peroxide is a factor in the neuronal degeneration process is also a key question to be addressed.

As mentioned above, MDL 72974A is a potent and selective inhibitor of both MAO and SSAO. Both of these enzymes generate peroxide during the oxidative deamination process upon cofactor reoxidation. MDL 72974A is a potent neuroprotectant against the MPTP induced depletion of striatal dopamine in the mouse (Zreika et al., 1989) when administered prior to the MPTP. This effect has been rationalized on the basis of preventing the MAO-B catalyzed conversion of MPTP to MPP$^+$. The results from Nicklas' group, however, where animals were treated with MDL 72974A several hours after the complete conversion of MPTP to MPP^+ had occurred, cannot be explained in terms of MAO-B inhibition.

It will be very interesting to test the selective SSAO inhibitor, MDL 72274A, in a model of neuronal rescue. This compound is structurally related to MDL 72974A and is a potent inhibitor of SSAO (Lyles et al., 1987), but is a weak inhibitor of MAO-B. In addition in studying this compound alone, it will also be useful to ascertain whether MDL 72274A can increase the effectiveness of L-deprenyl when given in combination. However, it is important to note that the SSAO activity of neuronal tissue is very low and rather the activity lies in the vascular wall. If neuroprotection occurs with SSAO inhibitor it will suggest the involvement of another enzyme or a role for radicals generated in the vascular walls.

Although MDL 72274A has not been studied in man, clinical trials with MDL 72974A have been progressing for some years. In both normal volunteers or parkinsonian patients MDL 72974A was found to be a very potent inhibitor of platelet MAO activity after oral administration. The ED_{50} value is approximately $90 \mu g$ p.o. and daily doses of $1 mg$ produce essentially complete inhibition. One or $4 mg$ given once daily to patients afflicted with Parkinson's disease, concomitantly receiving L-dopa and a decarboxylase inhibitor (carbidopa or benserazide) led to significant ($p < 0.01$) improvement in the UPRDS total and motor scores suggesting that the inhibitor readily enters the brain and inhibits MAO-B. We have no clinical data to confirm that SSAO is also inhibited over this dose range. MDL 72974A has been evaluated in over 250 patients with excellent tolerability. Large scale, double blind studies are now underway.

In conclusion, potent and selective inhibitors of MAO-B and/or SSAO are now available. They will be very useful tools for evaluating the neuroprotective significance of the inhibition of these two enzymes both in animals and man. Whether the positive clinical effects that have been observed with MDL 72974A are due to MAO-B inhibition, a combination of MAO-B and SSAO inhibition, or an unknown neuronal rescue mechanism remains to be unambiguously established.

References

Aasly J, Green D, Hardenberg J, Harland D, Huebert N, Myllvla V, Rinne U, Seppala A, Sjaastad, Sotaniemi K (1991) An open multicenter study of the efficacy of MDL 72974A in Parkinson's disease when administered at a daily dose of 1 or 4 mg for 4 weeks as adjuvant therapy to L-DOPA. Proceedings, 10th International Symposium on Parkinson's Disease, Tokyo, Japan

Alken RG, Palfreyman MG, Brown MJ, Davies DS, Lewis PJ, Schechter PJ (1984) Selective inhibition of MAO type B in normal volunteers by MDL 72145. Br J Pharmacol 17: 615–616P

Bey P, Fozard J, Lacoste JM, McDonald IA, Zreika M, Palfreyman MG (1984) (E)-2-(3,4-Dimethoxy)-3-fluoroallylamine: a selective, enzyme activated inhibitor of type B monoamine oxidase. J Med Chem 27: 9–10

Fozard JR, Zreika M, Robin M, Palfreyman MG (1985) The functional consequences of inhibition of monoamine oxidase type B; comparison of the pharmacological properties of L-deprenyl and MDL 72145. Naunyn Schmiedebergs Arch Pharmacol 331: 186–193

Harland D, Seppala A, Hinze C, Hardenberg J, Zreika M (1990) A double-blind placebo-controlled study of the tolerability and effects on platelet MAO-B activity of single oral doses of MDL 72974A in normal volunteers. Eur J Pharmacol 183: 530

Hinze C, Harland D, Zreika M, Dulery B, Hardenberg J (1990) A double-blind, placebo-controlled study of the tolerability and effects on platelet MAO-B activity of single oral doses of MDL 72974A in normal volunteers. J Neural Transm [Suppl] 32: 203–209

Hoehn MM, Yahr MD (1967) Parkinsonism: onset, progression and mortality. Neurology 17: 427–442

Lyles GA, Marshall CMS, McDonald IA, Bey P, Palfreyman MG (1987) Inhibition of rat aorta semicarbazide-sensitive amine oxidase by 2-phenyl-3-haloallylamines and related compounds. Biochem Pharmacol 36: 2847–2853

McDonald IA, Lacoste J-M, Bey P, Palfreyman MG, Zreika M (1985) Enzyme activated irreversible inhibitors of monoamine oxidase: phenylallylamine structure-activity relationships. J Med Chem 28: 186–193

McDonald IA, Palfreyman MG, Zreika M, Bey P (1986) (Z)-2-(2,4-Dichlorophenoxy)methyl-3-fluoroallylamine (MDL 72638): a clorgyline analogue with surprising selectivity for monoamine oxidase type B. Biochem Pharmacol 35: 349–351

McDonald IA, Bey P, Zreika M, Palfreyman MG (1991) MDL 72974A: monoamine oxidase type B inhibitor, antiparkinson. Drugs of the Future 16 5: 428–431

Nicklas WJ, Terleckyj T (1992) Pharmacological aspects of chronic MAO-B inhibitor administration to rodents. 15th Annual Scientific Meeting of the Canadian College of Neuropsychopharmacology, Saskatoon (poster T-11)

Palfreyman MG, Zreika M, McDonald IA, Fozard J, Bey P (1984) MDL 72145, an irreversible inhibitor of MAO-B. In: Tipton KF, Dostert P, Strolin-Benedetti (eds) Monoamine oxidase and disease. Academic Press, London, pp 563–564

Palfreyman MG, McDonald IA, Bey, P, Danzin C, Zreika M, Lyles GA, Fozard JR (1986) The rational design of suicide substrates of amine oxidases. Biochem Soci Transact 14: 410–413

Palfreyman MG, McDonald IA, Zreika M, Dudley M, Bey P (1991) Potent and selective irreversible inhibition of MAO-B by MDL 72974A. Eur Neuropsychopharmacol 1: 319–321

Palfreyman MG, McDonald IA, Zreika M, Cremer G, Haegele K, Bey P (1993) MDL 72974A: a selective MAO-B inhibitor with potential for treatment of Parkinson's disease. J Neural Transm (in press)

Rando RR (1973) 3-Bromoallylamine induced irreversible inhibition of monoamine oxidase. J Am Chem Soc 95: 4438–4439

Rando RR, Eigner A (1977) The pseudoirreversible inhibition of monoamine oxidase by allylamine. Mol Pharmacol 13: 1005–1013

Tatton WG, Greenwood CE (1991) Rescue of dying neurons: a new action for deprenyl in MPTP parkinsonism. J Neurosci Res 30: 666–672

Walsh CT (1984) Suicide substrates, mechanism-based enzyme inactivators: recent developments. Ann Rev Biochem 53: 493–535

Yu PH, Zuo DM (1992) Inhibition of a type B monoamine oxidase inhibitor, (E)-2-(4-fluorophenethyl)-3-fluoroallylamine (MDL-72974A), on semicarbazide-sensitive amine oxidases isolated from vascular tissues and sera of different species. Biochem Pharmacol 43: 307–312

Zreika M, McDonald IA, Bey P, Palfreyman MG (1984) MDL 72145, an enzyme activated irreversible inhibitor with selectivity for monoamine oxidase type B. J Neurochem 43: 448–454

Zreika M, Fozard JR, Dudley MW, Bey P, McDonald IA, Palfreyman MG (1989) MDL 72974: a potent and selective enzyme-activated irreversible inhibitor of monoamine oxidase type B with potential for use in Parkinson's disease. J Neural Transm [P-D sect] 1: 243–254

Authors' address: Dr. M. G. Palfreyman, Marion Merrell Dow Research Institute, 2110 E. Galbraith Road, Cincinnati, OH 45215, U.S.A.

J Neural Transm (1994) [Suppl] 41: 415–420
© Springer-Verlag 1994

Several aspects on the amine oxidation by semicarbazide-sensitive amine oxidase (SSAO) from bovine lung*

J. M. Lizcano[1], A. Fernandez de Arriba[1], G. A. Lyles[2], and M. Unzeta[1]

[1] Departament de Bioquimica i Biologia Molecular, Facultat de Medicina, Universitat Autonoma de Barcelona, Barcelona, Spain
[2] Department of Pharmacology and Clinical Pharmacology, Ninewells Hospital and Medical School, University of Dundee, Dundee, United Kingdom

Summary. The lung has been shown to be potentially important in the metabolism of amines. Since SSAO has been demonstrated to be active towards some volatile short-chain aliphatic amines in other tissues, the current study determined the specificity and kinetic constants for the metabolism by bovine lung SSAO, of several aliphatic and aromatic amines some of which have been suggested to be physiological substrates (e.g. methylamine, aminoacetone and β-phenylethylamine), and others (e.g. benzylamine) which are non-physiological. In the case of benzylamine, an inhibition at high substrate concentration was observed. Kinetic assays ruled out the possibility that this inhibition was caused by products of the deamination of benzylamine, and consequently it is suggested that these results may indicate the presence of two binding sites for the interaction of benzylamine with SSAO.

Introduction

The family of amine oxidase enzyme activities contain an undefined group of enzymes named semicarbazide-sensitive amine oxidases (SSAO) (E.C. 1.4.3.6) due to their sensitivity to semicarbazide inhibition. They are widely distributed in almost all animal tissues (Lewinsohn, 1984), particularly in blood vessels (Lyles and Singh, 1985). Their presence in the lung of species such as the ox (Lizcano et al., 1990) suggests that SSAO activities could play an important physiological role in the metabolism of inhaled volatile amines.

In order to assess the contribution of SSAO in the deamination of amines, it was of interest to study the specificity of this enzyme, present in the microsomal fraction of bovine lung homogenates, towards several amines, including methylamine, aminoacetone and β-phenylethylamine

---

* J. M. Lizcano was supported by an Erasmus fellowship (ICP-91-I-0047/12)

which have been suggested might be physiological substrates (Lyles and Chalmers, 1992), and others such as benzylamine which is a non-physiological amine. In addition to determining apparent kinetic constants (K_m, V_{max}) for metabolism of these amines by SSAO in this tissue, additional studies have been carried out to investigate the mechanism responsible for the inhibition of this enzyme produced by high substrate concentrations of benzylamine.

Materials and methods

Bovine lung microsomes were obtained by a modification of the method described by Lizcano et al. (1990). After weighing, tissue was homogenized 1:10 (w/v) in 20 mM Tris-HCl buffer pH 7.2 containing 0.25 M sucrose, by the use of a Waring blender. This homogenate was centrifuged at 10,000 g for 10 min to remove nuclei, cellular debris and mitochondria. After centrifuging the supernatant at the same velocity, the pellet was discarded and 10 mM $CaCl_2$ was added to the supernatant which was stirred at 4°C for 15 min. Then this was centrifuged at 25,000 g for 30 min in order to obtain the microsomal pellet. This was resuspended in 20 mM potassium phosphate buffer pH 7.2 containing 150 mM KCl and centrifuged at 25,000 g for 30 min. The final pellets were resuspended in 20 mM potassium phosphate buffer and were stored in aliquots at −20°C until assay.

SSAO activity towards β-phenylethylamine (PEA), benzylamine (BZ) and methylamine (MA) was determined by previously-described radiochemical methods (Fowler and Tipton, 1981; Lyles et al., 1990). A spectrophotometric assay was used to study aminoacetone (AA) metabolism (Lyles and Chalmers, 1992). The contribution of SSAO to the metabolism of PEA, BZ and MA was calculated from inhibition produced by 1 mM semicarbazide, or the resistance to inhibition of amine metabolism in the presence of 1 mM clorgyline. In the case of AA, it was not possible to use clorgyline and semicarbazide due to their interference with the colorimetric assay method (see Lyles and Chalmers, 1992). Consequently, 1 mM pargyline and deprenyl were used as MAO inhibitors and 1 mM propargylamine and MDL 72145 as inhibitors of both MAO and SSAO.

Kinetic constants were determined from Lineweaver-Burk plots using substrate concentrations from 50–2,500 μM with PEA, 0.1–2 mM with MA, and 25–150 μM with AA. When AA was used as a competitive inhibitor of BZ metabolism, its concentration range was 50–200 μM.

For the studies with BZ as substrate, it was used at a concentration range of 25 μM–2 mM. The possibility that SSAO may be inhibited by benzaldehyde, benzyl alcohol and benzoic acid which are products of BZ metabolism, was examined by testing the effects of adding a range of concentrations (0.01–10 mM) of these compounds upon BZ deamination in assays. Possible effects of another metabolite, NH_4^+ were determined at concentrations between 10–100 mM. Ki values for competitive inhibitors were estimated by a non-linear Regression Analysis program using data from appropriate Lineweaver-Burk plots.

Results and discussion

The contribution of SSAO and MAO activities from bovine lung to the metabolism of different substrates was studied by determining the percentage inhibition of total deaminating activity produced by 1 mM semicarbazide and clorgyline respectively. We have previously shown with 20 μM substrate concentrations, that the relative contributions were 90% SSAO: 10% MAO-B for BZ, and 91% MAO-B: 9% SSAO for PEA (Lizcano et al., 1990). When MA was used as a substrate at either 100 μM or 1 mM, semicarbazide produced complete inhibition, whereas 1 mM clorgyline produced 20% inhibition. The metabolism of 100 μM AA was completely inhibited by 1 mM propargylamine and MDL 72145, inhibitors of both SSAO and MAO activities, whereas the MAO-selective inhibitors pargyline and deprenyl (at 1 mM) produced no inhibition. From these results it can be concluded that BZ, AA, and MA are metabolized mainly by SSAO, whereas PEA is metabolized largely by MAO but with some contribution by SSAO also.

The kinetic parameters for metabolism of these amines by SSAO, determined in the presence of 1 mM clorgyline to inhibit any MAO activity, are shown in Table 1. It can be seen that the highest affinity and catalytic efficacy are shown towards the non-physiological substrate BZ. The K_m found was intermediate between that previously reported in various studies for rat and human SSAO. Among the physiologically-occurring amines, AA had the lowest K_m, being similar to the value recently reported for human umbilical artery SSAO (Lyles and Chalmers, 1992). In separate kinetic experiments, AA was found to be a competitive inhibitor of BZ metabolism with an apparent K_i of $114 \pm 1 \mu M$ (mean $\pm$ s.e. of 3 determinations). When BZ deamination was studied, the velocity of the reaction decreased with increasing substrate concentration (Fig. 1). From the Lineweaver-Burk plots it was found that Michaelis-Menten enzyme behaviour was obeyed at relatively low concentration of substrate, whereas deviation from linearity was seen in the plots at high substrate concentration (Fig. 2). The apparent K_m was estimated by extrapolation from the linear part of this graph. The high substrate inhibition was reversible, of a mixed type and was observed at all assay incubation times studied. In order to test the possibility of product inhibition being responsible, benzaldehyde, benzoic acid, benzyl alcohol and NH_4^+ were studied as potential inhibitors of Bz metabolism. Only NH_4^+ showed inhibitory activity, which was competitive in nature with a K_i of $18 \pm 3 mM$ (mean $\pm$ s.e. of 3 experiments). These results as a whole, suggest that end-product inhibition is unlikely to account for the high substrate inhibition seen with BZ.

Table 1. Kinetic constants for amine metabolism by bovine lung microsomal SSAO

Substrate	K_m (μM)	V_{max} (pmol/min/mg prot.)	V_{max}/K_m
β-Phenylethylamine	312 ± 30	980 ± 99	3.14
Benzylamine	40 ± 3	$2,010 \pm 165$	50.25
Aminoacetone	94 ± 15	$1,416 \pm 220$	15.06
Methylamine	340 ± 73	$3,416 \pm 600$	10.04

Where appropriate microsomal fractions were preincubated with 1 mM clorgyline at 37°C for 30 min to inhibit MAO activity. Values are means $\pm$ s.e. of three different experiments

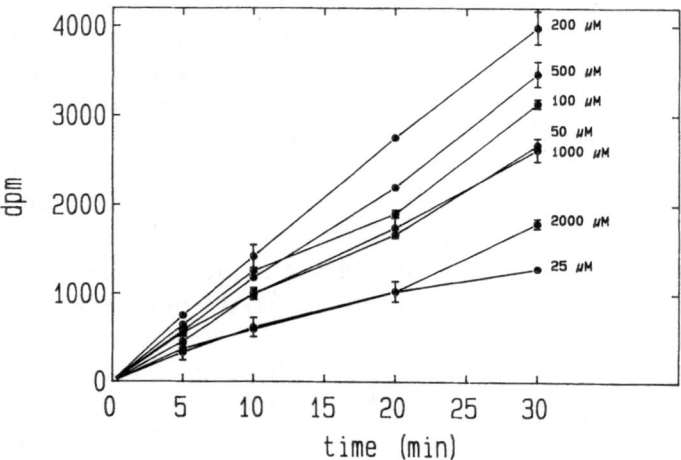

Fig. 1. Time course of product formation with benzylamine as substrate. Benzylamine was used as substrate in the concentration range 25–2,000 μM

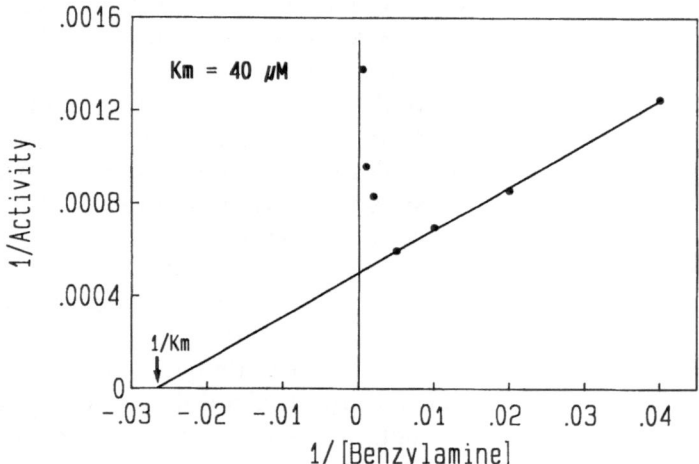

Fig. 2. Lineweaver-Burk plot of benzylamine oxidation. Benzylamine was used as substrate in the concentration range 25–2,000 μM

A possible mechanism for such substrate inhibition could be the simultaneous binding of more than one substrate molecule at the active centre of the enzyme with the subsequent formation of a relatively inactive substrate-enzyme complex. The general equation derived from Haldane and representative of such an inhibitory mechanism (Dixon et al., 1979) is:

$$\frac{V_{max}}{v} = 1 + \frac{K_m}{(S)} + \frac{(S)}{K_m'} \tag{1}$$

with v being the velocity of the reaction, V_{max} the theoretical maximal velocity, (S) the substrate concentration, K_m the dissociation constant for

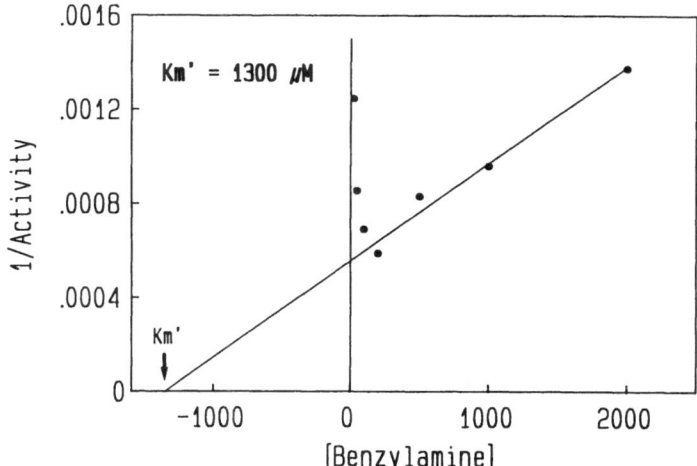

Fig. 3. Dixon Plot of benzylamine oxidation. Benzylamine was used as substrate in the concentration range 25–2,000 µM

the substrate in activable position, and K_m' the apparent dissociation constant for the same substrate as inhibitor. The corresponding plot of initial velocities of BZ oxidation against the log of substrate concentration gives a symmetrical curve, with the maximum corresponding to S_0 (data not shown), the optimum substrate concentration expressed by the derived equation (Friedenwald et al., 1954):

$$S_0 = \sqrt{K_m K_m'} \qquad (2)$$

The apparent K_m for BZ oxidation obtained from the Lineweaver-Burk plot was 40 µM, while the apparent K_m' estimated from a Dixon plot (Fig. 3) of reciprocal velocity against substrate concentration was 1,300 µM. The S_0 value calculated from Eq. 2 was 230 µM, and coincides with the value calculated graphically. The observed maximal velocity, 2,010 pmol/min mg, was in agreement with the value calculated from Eq. 1.

Since the experimental data are consistent with Eq. 1, it seems that the active centre of SSAO in microsomal fractions of bovine lung has more than one site for the interaction with BZ as substrate. The existence of two binding sites suggested by these results are in agreement with those obtained in our laboratory with the competitive inhibitor phenelzine, for which two binding sites per enzyme molecule have been found (unpublished data). Further work needs to be done to confirm these hypotheses.

References

Dixon M, Webb EC, Thorne CJR, Tipton KF (1979) Enzymes, 3rd edn. Longman, London, pp 126–136

Fowler CJ, Tipton KF (1981) Concentration dependence of the oxidation of tyramine by the two forms of rat liver mitochondrial monoamine oxidase. Biochem Pharmacol 30: 3329–3332

Friedenwald JS, Maengwyn-Davies GD, McElroy WD, Glass B (eds) (1954) The mechanism of enzyme action. The Johns Hopkins Press, Baltimore, p 180

Lewinsohn R (1984) Mammalian monoamine oxidizing enzymes with special reference to benzylamine oxidase in human tissues. Brazilian J Med Biol Res 17: 223–256

Lizcano JM, Balsa D, Tipton KF, Unzeta M (1990) Amine oxidase activities in bovine lung. J Neural Transm [Suppl] 32: 341–344

Lyles GA, Singh I (1985) Vascular smooth muscle cells: a major source of the semi-carbazide-sensitive amine oxidase of the rat aorta. J Pharm Pharmacol 37: 637–647

Lyles GA, Holt A, Marshall S (1990) Further studies on the metabolism of methyla-mine by semicarbazide-sensitive amine oxidase in human plasma, umbilical artery and rat aorta. J Pharm Pharmacol 42: 332–338

Lyles GA, Chalmers J (1992) The metabolism of aminoacetone to methylglyoxal by semicarbazide-sensitive amine oxidase in human umbilical artery. Biochem Pharmacol 43: 1409–1414

Authors' address: Dr. M. Unzeta, Departament de Bioquimica i Biologia Mol-ecular, Facultat de Medicina, Universitat Autonoma de Barcelona, E-08193 Bellaterra (Barcelona), Spain.

J Neural Transm (1994) [Suppl] 41: 421–426

Semicarbazide-sensitive amine oxidase activity of guinea pig dorsal skin

F. Buffoni, S. Cambi, G. Banchelli, G. Ignesti, R. Pirisino, and **L. Raimondi**

Department of Pharmacology, University of Florence, Italy

Summary. A semicarbazide-sensitive amine oxidase activity with a high affinity for benzylamine (Bz.SSAO) (E.C. 1.4.3.6) is present in guinea pig dorsal skin. This enzymic activity oxidized benzylamine, histamine, 1,4-methylhistamine and acetylputrescine and was inhibited by semicarbazide and by B24 (3,5-diethoxy-4-aminomethylpyridine), a selective inhibitor of Bz.SSAO enzymes. It cross reacted with the antibodies raised against pure pig plasma benzylamine oxidase. Immunohistochemistry showed that it was localized in fibroblasts.

Bz.SSAO activity of guinea pig dorsal skin increased during the process of skin healing.

A treatment of the wounds with 3 μg of b-FGF significantly accelerated the process of skin healing and the increase of Bz.SSAO activity.

Introduction

The purpose of this work was the study of the semicarbazide-sensitive amine oxidase activity (Bz.SSAO) (E.C. 1.4.3.6) of the dorsal skin of the guinea pig and the study of its variations during the process of skin regeneration after the removal of 7×7 mm pieces of epidermal and a part of dermal layer.

Methods

Male guinea pig Dunkin-Hartley of 300 g were used. Hairs were removed from the dorsal skin of guinea pig and 5 wounds (7×7 mm) were surgically induced under anesthesia totally removing the epidermal surface and a superficial part of dermal layer. The newly formed wound tissues were dissected at different times during the process and analyzed by biochemical and histological methods according to Buffoni et al. (1992).

The enzymic activity (Bz.SSAO) was determined at 37°C in air in the 12,000 g supernatant of the skin homogenate. Skin was homogenized in 1 mM Na-K phosphate buffer pH 7.4 containing 0.25 M saccharose. After 30 min of preincubation in the presence of 1 mM pargyline (to inhibit both MAO A and B) the reaction was followed either adding 80 μM 14C-benzylamine and measuring the formed oldehyole or unlabelled substrates and measuring hydrogen peroxide production according to Ignesti et al. (1992). The activity was linear for 180 min.

Enzyme kinetic was obtained in the same conditions using 2-4-8-10-20-40-80 μM 14C-benzylamine. The kinetic constants were computed according to Wilkinson (1961).

Antibodies against pure pig plasma benzylamine oxidase were raised in rabbit according to Buffoni et al. (1977).

Immunohistochemistry was carried out using the Bio-Rad immun-blot assay kit with peroxidase coupled goat antirabbit IgG.

The DNA content was determined according to Labarca and Paigen (1980), the protein content according to Lowry et al. (1951) and the hydroxyproline content according to Woessner (1961).

Results

Bz.SSAO activity in guinea pig dorsal skin

In 150 samples of dorsal skin from male guinea pigs analyzed in duplicate, using 80 μM 14C-benzylamine at pH 7.4 and 37°C in air, the following enzymic activity was found: 603 ± 30 pmoles/g of fresh tissue/min, 31 ± 2.2 pmoles/mg of protein/min (mean $\pm$ s.e.).

In 3 experiments carried out in duplicate, each on 4 samples of skin, the following apparent Michaelis Menten constants were obtained: K_m 5.3 ± 1.4 μM, V_m 15.0 ± 1.0 pmoles/mg of protein/min.

Purification, substrate specificity and inhibitor sensitivity

The skin was homogenized in 4 M urea and extracted for 24 hours at 4°C, the clear supernatant obtained at 12,000 g of centrifugation was dialysed against 0.01 M Na-K phosphate buffer pH 7.0 and applied on a DEAE-cellulose column (cm 2 × 24) previously equilibrated with the same buffer. The enzyme was eluted with 0.1 M buffer.

The pooled active fractions were collected and the enzyme was precipitated with ammonium sulphate at 0–55% of saturation. The collected precipitate was fully dialysed against 0.05 M Na-K phosphate buffer pH 7.0.

The substrate specificity of this enzyme and its sensitivity to inhibitors is reported in Table 1.

Molecular weight and immunohistochemistry

The Bz.SSAO activity of guinea pig skin cross-reacted with the antibodies raised in rabbit against the pig plasma benzylamine oxidase. Figure 1 shows the SDS electrophoresis of the partially purified skin Bz.SSAO and of the pure crystalline benzylamine oxidase (BAO) of pig plasma in reducing conditions. The pig plasma BAO gave a subunits of 97 KDa m.w. and the skin Bz.SSAO a main band at 66 Kda m.w. and a small band at 97 kDa m.w., both immunoreactive.

Table 1. Substrate specificity of partially purified Bz. SSAO of the guinea pig dorsal skin and its inhibition by semicarbazide and B24

	Activity pmoles/mg of protein/min
Benzylamine	75.2 ± 4.3
B54	43.8 ± 0.01
Histamine	6.9 ± 0.9
1,4-methylhistamine	7.3 ± 0.1
acetylputrescine	2.9 ± 0.1
putrescine	0
	% Inhibition
B24 0.00002 M	38
0.00004 M	73
0.0001 M	82.5
0.0002 M	93
Semicarbazide 0.00002 M	8
0.0001 M	40
0.0002 M	67

B24 3,5-diethoxy-4-aminomethylpyridine bichloridate*. *B54* 1,2-bis-aminomethanbenzene bichloridate. The inhibitor were preincubated for 30 min at 37°C, substrates concentration 1 mM. The activity on benzylamine is the mean of 5 determinations in duplicate, the others of 3 determinations in duplicate (mean ± s.e.). *B24 specific inhibitor of Bz SSAO enzymes (Bertini et al., 1985)

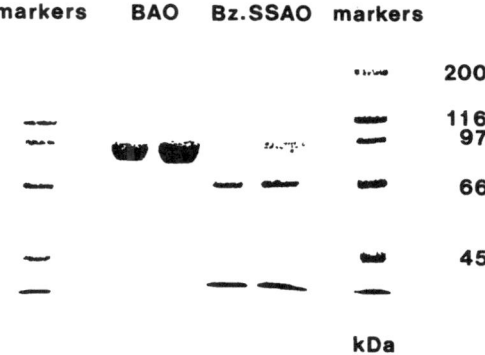

Fig. 1. SDS electrophoresis (7.5%) of the partially purified Bz.SSAO of the guinea pig dorsal skin and its comparison to the crystalline pure pig plasma benzylamine oxidase (2.4 μg and 4.8 μg of BAO and 2.8 μg and 5.6 μg of skin Bz.SSAO were deposed)

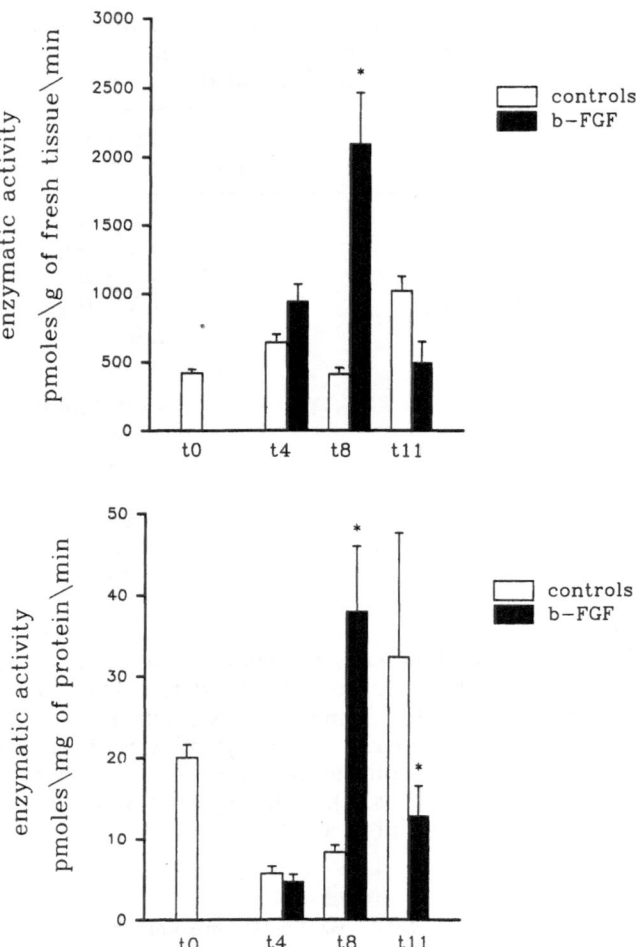

Fig. 2. Effect of b-FGF on the levels of the Bz.SSAO activity of the guinea pig dorsal skin during the process of skin healing. Mean ± s.e. *t4,t8,t11* samples obtained 4, 8, 11 days after surgery At t0 60 samples, at the other times 10 samples were analyzed in duplicate. * The difference between this value and the corresponding value in the control group is statistically significant (P < 0.05)

The immunohistochemistry showed that the enzyme was localized in the skin fibroblasts (data not shown).

Variation of Bz.SSAO activity of the dorsal skin of guinea pig during the process of skin healing

The Bz.SSAO activity of guinea pig dorsal skin significantly increased 11 days after surgery whereas its increase was very modest 4 days after surgery when there was a strong increase of the DNA and protein content of the skin.

At 4 days the increase in DNA was not only dependent on cells reclutation but also on cells reproduction as it was shown measuring the 3H-thymidine incorporation (data not shown).

The content of hydroxyproline had always greatly decreased in respect to that of normal skin.

Variation of Bz.SSAO activity of the dorsal skin of guinea pig produced by b-FGF during skin healing

Figure 2 shows that a treatment of the wound with 3 μg of b-FGF (basic fibroblast groth factor) immediately after surgery significantly increased the level of Bz.SSAO activity 8 days after surgery whereas in the controls animal this activity increased only 11 days after surgery. This increase was correlated to an increase of the fibroblasts content of the tissue.

Conclusion

In the dorsal skin of guinea pig a Bz.SSAO activity is present. It acts on benzylamine, histamine, methylhistamine and acetylputrescine with a rate (made 1 the rate of benzylamine oxidation) in the order: 1-0.09-0.10-0.04. It acts also on 1,2-bis-aminomethanbenzene but not on putrescine and it is inhibited by semicarbazide and B24.

This enzymic activity is localized in skin fibroblasts. It increases during the skin healing process 11 days after surgery and this increase is anticipated at 8 days by the stimulation of fibroblasts with a treatment of the wound with b-FGF.

Acknowledgements

We thank the National Research Council (CNR) and the MURST for financial support and V. Bertini et al. (University of Genova, Italy) for the kind gift of B24 and B54.

References

Bertini V, De Munno A, Lucchesini F, Buffoni F, Bertocci B (1985) Italian patent application n.47906- A/85 extended to Europe, U.S.A., Canada, Japan
Buffoni F, Della Corte L, Hope DB (1977) Immunofluorescence histochemistry of porcine tissues using antibodies to pig plasma oxidase. Proc Roy Soc B195: 417–423
Buffoni F, Banchelli G, Ignesti G, Pirisino R, Raimondi L (1992) Skin wound healing: some biochemical parameters in guinea pig skin. Pharmacol Res 25 [Suppl] 2: 332–333
Ignesti G, Banchelli G, Raimondi L, Pirisino R, Buffoni F (1992) Histaminase activity in rat lung and its comparison with intestinal mucosal diamine oxidase. Agents Actions 35: 192–199

Labarca C, Paigen K (1980) A simple, rapid and sensitive DNA assay procedure. Anal
    Biochem 102: 344–352
Lowry HO, Rosebrough NJ, Lewis Farr A, Randall RJ (1951) Protein measurement
    with the Folin phenol reagent. J Biol Chem 193: 265–275
Wilkinson GN (1961) Statistical estimation of enzyme kinetics. Biochem J 80: 324–332
Woessner JF (1961) The determination of hydroxyproline in tissue and protein samples
    containing small proportion of this imino acid. Arch Biochem Biophys 93: 440–447

Author's address: Prof. Dr. F. Buffoni, Dipartimento di Farmacologia preclinica e
clinica, Università di Firenze, Viale G. B. Morgagni, 65, I-50134 Firenze, Italy.

J Neural Transm (1994) [Suppl] 41: 427–432

Semicarbazide-sensitive amine oxidases in sheep plasma: interactions with some substrates and inhibitors

A. E. Crosbie and **B. A. Callingham**

Department of Pharmacology, University of Cambridge, Cambridge, United Kingdom

Summary. The present study has examined the affinities of sheep plasma semicarbazide-sensitive amine oxidase (SSAO) enzymes for a range of aliphatic amines and also the effects of two inhibitory compounds, β-aminopropionitrile (BAPN) and mexiletine. Two kinetically separable enzyme activities appeared to be responsible for the metabolism of amines containing 2–5 carbon atoms while the deamination of higher amines and methylamine and allylamine produced kinetic plots characteristic of only one enzyme activity. When benzylamine metabolism was used as an indication of enzyme activity, the two inhibitors had different effects. BAPN exhibited predominantly a mixed pattern of inhibition while the effects of low concentrations of mexiletine were largely competitive. These results present evidence confirming the presence of two kinetically separable SSAO activities in sheep plasma, although we must await the development of highly selective inhibitors before these two activities can be fully resolved.

Introduction

The substrate specificity of mammalian plasma semicarbazide-sensitive amine oxidase (SSAO) is highly species dependent and highlighted by the fact that only ruminant enzymes have activity against the polyamines, spermidine and spermine (Blaschko, 1974). This property, which was originally demonstrated in the sheep by Hirsch (1953), gave rise to the proposal that these enzymes served a protective rôle against polyamines produced by cellulose degradation in the rumen (Blaschko and Hawes, 1959). Later studies have revealed that alterations in the levels of plasma SSAO can be shown in both pregnant and diabetic ewes (Elliott et al., 1991), which may be associated with endocrine changes or the availability of the copper co-factor. Elliott and co-workers (1992) have also shown that when benzylamine is used as a substrate, two separate enzyme activities responsible for the metabolism of this amine can be distinguished kinetically. These two enzyme components have K_m values which differ approximately 100-fold. This is in contrast to the metabolism of spermidine where only one deaminating activity could be identified.

The aims of this study are to determine the affinities of these plasma enzymes for a range of aliphatic monoamines and to examine the effects of two inhibitory compounds, β-aminopropionitrile (BAPN), a known inhibitor of lysyl oxidase and the antidysrhythmic drug, mexiletine, as a preliminary to investigating the individual responses of these enzymes to the hormone and copper status of the animals. The affinity of plasma SSAO for aliphatic amines has been previously studied in the horse (Blaschko et al., 1959), rabbit (McEwen et al., 1966) and man (McEwen, 1965), although without any kinetic data being presented.

Materials and methods

Blood was obtained from mixed breed sheep by jugular venepuncture and centrifuged at 2,500 g for 10 min, to yield platelet-poor plasma. This was stored at −20°C until required. Plasma samples were shaken with manganese dioxide (20 mg ml⁻¹), to oxidise any reducing substances or any endogenous substrates for the plasma enzymes. The MnO_2 was removed by centrifugation at 2,000 g for 10 min and decantation.

Treated plasma samples (10 μl) were incubated for 30 min at 37°C with amine substrates in 0.2 M potassium phosphate buffer containing sodium azide to inhibit endogenous catalase. Amine oxidase activity was measured fluorimetrically by a method based on that described by Guilbault et al. (1968), which detected the amount of hydrogen peroxide formed during the reaction. However, in the present work, instead of homovanillic acid as the proton donor, adrenaline was used. In the presence of H_2O_2 and peroxidase, adrenaline is converted to the highly fluorescent derivative, adrenolutine by the "trihydroxyindole" reaction (Loew, 1918; Lund, 1950; Callingham and Cass, 1963), via adrenaline-quinone and adrenochrome. The adrenolutine was detected in a Locarte filter fluorimeter (Activation Filters: Corning 5113 and 3060, 405 nm; Emission Filter: Corning 3384, above 500 nm) and the fluorescence intensity recorded on a pen recorder.

Substrate concentration-velocity relations were determined by the method of Wilkinson (1961). Where two enzyme activities contributing to the metabolism of a single substrate were identified, the iterative curve stripping method of Spears et al. (1971) was used to estimate kinetic parameters for each.

In inhibitor studies, amine oxidase activity was determined using a radiochemical assay (see Lyles and Callingham, 1982) where plasma samples were incubated for 10 min at 37°C with the appropriate inhibitor concentration and [¹⁴C]-benzylamine. Enzyme activity was estimated using liquid scintillation spectrometry, following solvent extraction of the metabolites. Again the results were subjected to kinetic analysis as described above.

Results

The metabolism of a series of aliphatic amines ranging from methylamine to octylamine was studied. Figures 1 and 2 show Hanes-Woolf plots for deamination of amylamine and octylamine. Kinetic constants for each amine are shown in Table 1.

As in the case of benzylamine, it would appear that two enzyme activities are responsible for the metabolism of amines with 2–5 carbon atoms in their alkyl chain. However, when higher amines were used as substrates,

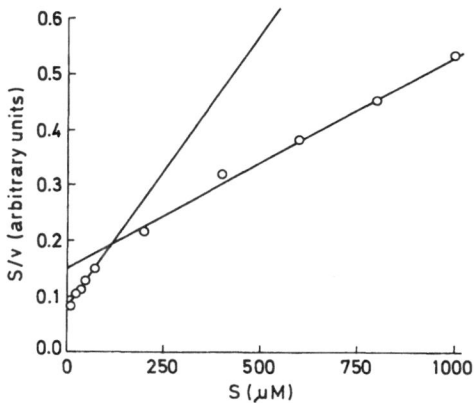

Fig. 1. The kinetics of amylamine metabolism by sheep plasma SSAO. Plasma samples were incubated for 30 min with concentrations of amylamine (10 μM – 1,000 μM). The results were analysed by the method of Wilkinson (1961) and plotted by eye on a Hanes-Woolf plot to reveal two separate catalytic activities

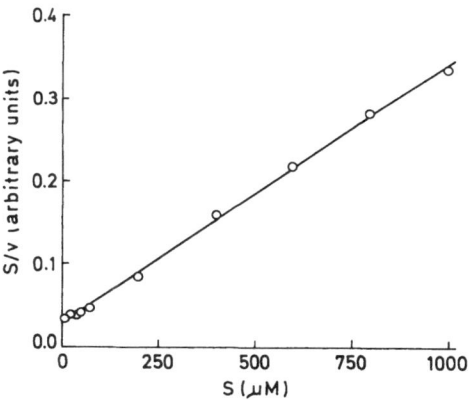

Fig. 2. The kinetics of octylamine metabolism by sheep plasma SSAO. The experimental conditions and results analysis are the same as for Fig. 1

only one enzyme activity could be detected by the Hanes-Woolf plot. This was also true for the metabolism of the simplest amine, methylamine and the unsubstituted propylamine derivative, allylamine.

When the effects of the two inhibitors on benzylamine metabolism were studied, the type of inhibition observed for each compound differed. For BAPN, a 50 μM concentration appeared to inhibit the high K_m enzyme competitively, but with a higher concentration (200 μM), a mixed inhibitory effect was observed. Only a mixed pattern of inhibition was recorded for the low K_m enzyme.

With mexiletine, the two enantiomers (100 μM) appeared to be competitive on both enzyme activities responsible for benzylamine metabolism in sheep plasma but at 500 μM the regressions deviated substantially from

Table 1. Kinetic constants for the metabolism of aliphatic amines by sheep plasma SSAO. When only one enzyme activity was measured, K_m and V_{max} values were obtained by the method of Wilkinson (1961). Where two enzyme activities were responsible for the metabolism of a single substrate, kinetic constants were derived using the method of Spears et al. (1971). The units of the V_{max} values are nmoles $H_2O_2 h^{-1}$ (ml plasma)$^{-1}$. The values for benzylamine metabolism are also shown for comparison

Amine substrate	Low K_m (μM)	V_{max} (units)	High K_m (μM)	V_{max} (units)
Benzylamine	8.9	401	813	3,252
Methylamine	—	—	390	777
Ethylamine	8.99	14.5	769	624
Propylamine	14.2	56.7	1,020	2,453
Butylamine	2.35	77.9	943	4,080
Amylamine	29.5	321	609	2,521
Octylamine	79.9	3,153	—	—
		K_m (μM)	V_{max} (units)	
Hexylamine		145	3,415	
Heptylamine		130	4,301	
Allylamine		194	5,088	

parallel. On both enzymes there was little difference in potency between the enantiomers although the (−)-isomer appeared to be slightly more effective.

Discussion

Two SSAO enzyme activities in sheep plasma, which could be separated kinetically, appeared to be responsible for the deamination of straight chain aliphatic amines containing 2–5 carbon atoms. Higher amines, such as octylamine, produced kinetic curves characteristic of a single enzyme activity.

While it is possible that as the amine carbon chain length increases, there is a progressively diminishing contribution by the "high" K_m component enzyme to the metabolism of these compounds, it is also possible that the K_m values come too close together to be resolved in this way. This may be the case with hexylamine and heptylamine, where the measured K_m values for the overall metabolism of these amines are intermediate between those seen with the other amines for the "high" and the "low" K_m activities. The metabolism of the unsaturated amine, allylamine (3-aminopropene), may be a further example, where again there is probably a contribution by two enzyme activities. As a result, the K_m is of such a value that it cannot be confidently placed in either the "low" or "high" value in Table 1. With the simplest amine in the series, methylamine, only a single regression can be found. However, the calculated K_m value is relatively high, suggesting that, kinetically, it is a high K_m enzyme activity which appears to deaminate this amine in sheep plasma.

Recently, kinetic data regarding aliphatic amine metabolism has been obtained for rat aorta SSAO (Yu, 1990) and horse plasma SSAO (Dar, 1992). These studies report that the SSAO activity towards these amines is increased with increasing carbon chain length, suggesting that the active site of these enzymes at least, binds more lipophilic amine substrates. The picture for sheep plasma is more complicated with the existence of two enzyme activities and it is difficult to make clear assumptions about substrate preferences until the properties of the two activities have been studied further.

BAPN is a natural compound which can be extracted from the sweet pea plant, Lathyrus odoratus and is a well characterised irreversible inhibitor of lysyl oxidase (Levene, 1971). It has also been reported to inhibit competitively other SSAO enzymes (see Lyles and Singh, 1985). With sheep plasma SSAO, an overall mixed pattern of inhibition was observed for this compound on both low and high K_m component enzymes responsible for the metabolism of benzylamine.

The presence of an amino grouping on the structure of BAPN suggests that this part of the molecule may interact directly with the active site of the enzymes. In addition, the BAPN molecule is characterised by the presence of a nitrile group. Cyanide ions are known to inhibit certain SSAO enzymes by chelating the copper ions which are required for activity. It is possible that part of BAPN's inhibitory effects on these sheep plasma enzymes may result from an interaction with this copper prosthetic group and account for the mixed inhibition. It seems unlikely, that either of the two plasma SSAO enzymes can be a soluble lysyl oxidase, but the conditions differ substantially from those under which lysyl oxidase is normally measured.

As in rat heart and aorta (Clarke et al., 1982), both forms of mexiletine (100 µM) competitively inhibited the two sheep plasma activities but, in contrast to the rat, with about the same potency as well as a possibly mixed interaction at 500 µM. It appears, therefore, that this agent not only competitively inhibits MAO-A (Callingham, 1977) but also interacts with both tissue-bound and plasma SSAO. However, in the case of the plasma enzymes the change in kinetics with concentration might indicate the presence of more than one binding site.

In conclusion, although evidence confirming the presence of two kinetically separable SSAO activities in sheep plasma has been presented, identification of those activities responsible for deaminating a particular substrate must await the discovery of highly selective inhibitors.

Acknowledgements

A.E.C. is a Medical Research Council Scholar. Funding was provided by the Horserace Betting Levy Board. We wish to thank Dr. R. Connan for assistance in obtaining sheep blood samples.

References

Blaschko H, Friedman PJ, Hawes R, Nilsson K (1959) The amine oxidases of mammalian plasma. J Physiol (London) 145: 384–404

Blaschko H, Hawes R (1959) Observations on spermine oxidase of mammalian plasma. J Physiol (London) 145: 124–131

Blaschko H (1974) The natural history of amine oxidases. Rev Physiol Biochem Pharmacol 70: 83–148

Callingham BA, Cass R (1963) The determination of catecholamines in biological materials. In: Varley H, Gowenlock AH (eds) The clinical chemistry of monoamines. Elsevier, Amsterdam, pp 19–30

Callingham BA (1977) Substrate selective inhibition of monoamine oxidase by mexiletine. Br J Pharmacol 61: 118–119P

Clarke DE, Lyles GA, Callingham BA (1982) A comparison of cardiac and vascular clorgyline-resistant amine oxidase and monoamine oxidase: inhibition by amphetamine, mexiletine and other drugs. Biochem Pharmacol 31 1: 27–35

Dar A (1992) An in vitro study of pulmonary vascular reactivity and amine oxidase activity in the horse. Thesis, University of Cambridge

Elliott J, Fowden AL, Callingham BA, Sharman DF, Silver M (1991) Physiological and pathological influences on sheep blood plasma amine oxidase: effect of pregnancy and experimental alloxan-induced diabetes mellitus. Res Vet Sci 50: 334–339

Elliott J, Callingham BA, Sharman DF (1992) Amine oxidase enzymes of sheep blood vessels and blood plasma: a comparison of their properties. Comp Biochem Physiol 102C 1: 83–89

Guilbault GG, Brignac PJ, Juneau M (1968) New substrates for the fluorimetric determination of oxidative enzymes. Anal Biochem 40: 1256–1263

Hirsch JG (1953) Spermine oxidase: an amine oxidase with specificity for spermine and spermidine. J Exp Med 97: 345–355

Levene CI (1971) Effect of lathyrogenic compounds on the cross-linking of collagen and elastin in vivo. In: Aldridge N (ed) A symposium on mechanisms of toxicity. Macmillan, London, pp 67–85

Loew O (1918) Über die Natur der Giftwirkung des Suprarenins. Biochem Z 85: 295–305

Lund A (1950) Simultaneous fluorimetric determinations of adrenaline and noradrenaline in blood. Acta Pharmacol 6: 137–146

Lyles GA, Callingham BA (1982) In vitro and in vivo inhibition by benserazide of clorgyline-resistant amine oxidases in rat cardiovascular tissues. Biochem Pharmacol 31 7: 1417–1424

Lyles GA, Singh I (1985) Vascular smooth muscle cells: a major source of the semicarbazide-sensitive amine oxidase of the rat aorta. J Pharm Pharmacol 37: 637–643

McEwen CM Jr (1965) Human plasma monoamine oxidase. II. Kinetic studies. J Biol Chem 240 5: 2011–2018

McEwen CM Jr, Cullen KT, Sober AJ (1966) Rabbit serum monoamine oxidase. I. Purification and characterization. J Biol Chem 241 19: 4544–4556

Spears G, Sneyd JGT, Loten EG (1971) A method for deriving constants for two enzymes acting on the same substrate. Biochem J 125: 1149–1151

Wilkinson GN (1961) Statistical estimations in enzyme kinetics. Biochem J 80: 324–332

Yu PH (1990) Oxidative deamination of aliphatic amines by rat aorta semicarbazide-sensitive amine oxidase. J Pharm Pharmacol 42: 882–884

Author's address: Dr. A. E. Crosbie, Department of Pharmacology, University of Cambridge, Tennis Court Road, Cambridge, CB2 1QJ, United Kingdom

J Neural Transm (1994) [Suppl] 41: 433–437

Location of the active site of rat vascular semicarbazide-sensitive amine oxidase

A. Holt and **B. A. Callingham**

Department of Pharmacology, University of Cambridge, Cambridge, United Kingdom

Summary. Semicarbazide-sensitive amine oxidase (SSAO) activity in rat vascular smooth muscle cells is associated extensively with the plasmalemma. To determine which side of the plasmalemma the active sites of these enzymes face, the non-permeating agent, diazotised sulphanilic acid (DSA; 4.4 mM) was perfused through the isolated mesenteric arterial bed of the rat, in an attempt to inactivate only those active sites facing extracellularly. DSA perfusion abolished the pressor responses to noradrenaline via inactivation of extracellular α_1 receptors but had no effect on cytosolic lactate dehydrogenase activity. SSAO activity, estimated by perfusing [^{14}C] benzylamine, was reduced following DSA perfusion to $55.9 \pm 4.9\%$ of that in control beds and to $52.4 \pm 6.0\%$ in homogenates of these vessels. These results suggest that almost half of SSAO active sites in rat mesenteric arteries face outwards.

Introduction

Semicarbazide-sensitive amine oxidase (SSAO) enzymes exist in tissue-bound and soluble forms in many species of plants and animals (Callingham and Barrand, 1987). Vascular smooth muscle cells (VSMCs) exhibit particularly high SSAO activity when compared to that in other mammalian tissues (Lyles and Singh, 1985).

Potent and selective SSAO inhibitors, such as the carcinostatic agent, procarbazine, have recently been described (Holt et al., 1992a,b). However, our ignorance surrounding some fundamental aspects concerning SSAO means a physiological rôle has yet to be ascribed to most of these enzymes.

In order to determine the subcellular location of the SSAO active site in VSMCs, we investigated the effect of diazotised sulphanilic acid (DSA) on benzylamine metabolism in the isolated perfused superior mesenteric arterial bed of the rat. DSA, a low molecular weight, slowly permeating reagent which can form covalent derivatives with proteins and lipids, has been used previously as a label for the outer membrane of erythrocytes (Berg, 1969) and to study ectonucleotidase activities in isolated synaptosomes (Grondal and Zimmermann, 1986).

In the present study, we have perfused DSA through the rat isolated mesenteric arterial bed in an attempt to inhibit selectively any SSAO with an extracellularly-facing active site. Remaining SSAO activity could then be quantified by perfusing with the radiolabelled, permeant substrate, [^{14}C] benzylamine.

Materials and methods

Male Wistar rats (240–300 g) were anaesthetised (sodium pentobarbitone, 60 mg kg^{-1}, i.p.) and heparinised (1500 u kg^{-1}, i.p.). Superior mesenteric arterial beds from two rats were cannulated and perfused, at 2 ml min^{-1} and 37°C, with oxygenated modified Krebs-Henseleit solution (KHS; [Ca^{2+}] 1.27 mM). The perfusion pressure was monitored with Washington PT400 pressure transducers, connected via an amplifier to an X/T pen recorder. After a 45 min equilibration period, preparations were challenged with noradrenaline (25 nmol, 200 μl; Sigma), administered through an injection port, until increases recorded in the back pressure were consistent. The control bed was then perfused with KHS for 25 min, while DSA (4.4 mM; Fluka Chemie) was perfused through the test bed for 15 min, followed by KHS for 10 min. Tissues were then rechallenged with noradrenaline.

Both tissues were subsequently perfused with the permeant substrate, [^{14}C] benzylamine (25 μM, 2 μCi μmol^{-1} in KHS; ICN Flow) for 5 min, followed by KHS alone, and eluate collected from each preparation at 15 s intervals for 15 min. After removing a 20 μl sample from each fraction to measure total radioactivity present, benzylamine oxidation was assessed by organic solvent extraction of the metabolites from the remainder of the fraction (Lyles and Callingham, 1982).

After washing through with KHS for 20 min, perfusion was stopped and beds were chilled to 4°C. Arteries were dissected and homogenised in 1 mM phosphate buffer containing 2% Triton X-100.

Lactate dehydrogenase (LDH) activity in these homogenates was measured by following the decrease in absorbance of NADH, at 340 nm and 37°C. SSAO activity was measured in homogenates by addition of [^{14}C] benzylamine and solvent extraction of metabolites.

Results

Following perfusion with 4.4 mM DSA, the contractile response to 25 nmol noradrenaline was reduced by more than 90%, when compared to that in control beds.

Profiles of total perfused radioactivity did not differ between control and DSA-treated preparations (Fig. 1a), indicating that uptake and release of benzylamine or its metabolites had not been affected by perfusion with DSA. However, the yield of metabolites had been reduced to 55.9 ± 4.9% of that from control beds (Fig. 1b). In homogenates of the perfused preparations, DSA had reduced SSAO activity to 52.4 ± 6% (n = 3) of activity in control homogenates.

Cytosolic LDH was not significantly inhibited by perfusion with 4.4 mM DSA. However, control experiments had shown that, when added to arterial homogenates, 100 μM DSA could totally inhibit LDH activity.

Results, shown as mean values ± SEM, are summarised in Table 1.

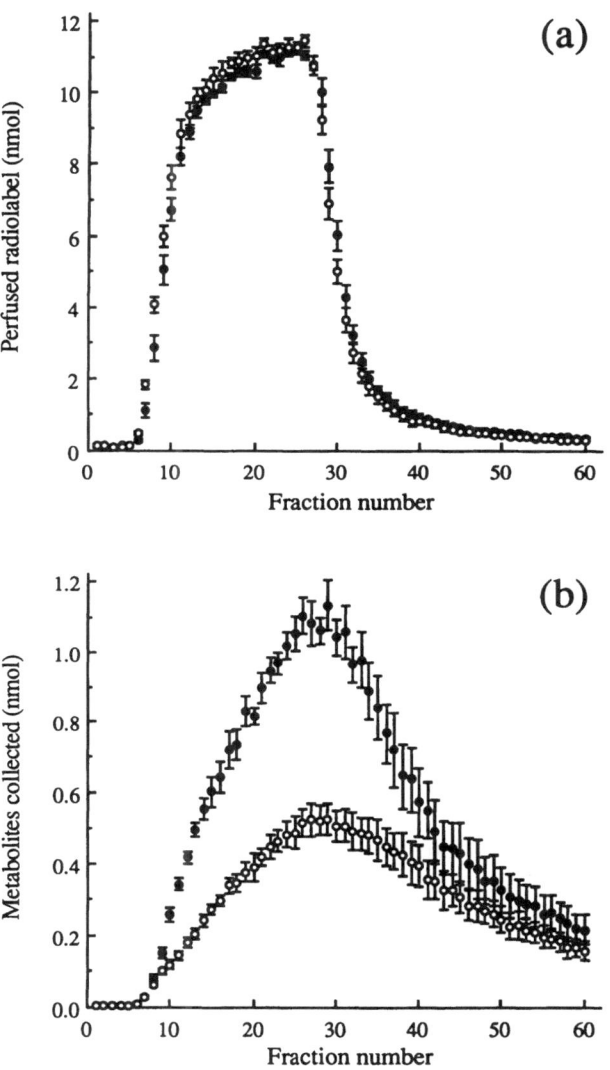

Fig. 1. Profiles of all radiolabelled species (**a**) and radiolabelled metabolites of benzy-lamine (**b**) present in the perfusate collected from isolated perfused rat mesenteries. Control mesenteries (●) were perfused with KHS for 25 min, while sample mesenteries (○) were perfused with 4.4 mM DSA for 15 min followed by KHS for 10 min. Both preparations were then perfused with 25 μM [^{14}C] benzylamine for 5 min and eluate collected from each preparation at 15 s intervals for 15 min

Discussion

A DSA perfusion schedule, which abolished almost completely the con-tractile response of rat mesenteric arteries to noradrenaline but had no effect on cytosolic LDH activity, reduced metabolism of [^{14}C] benzylamine to 55.9 ± 4.9% of that in control beds. SSAO activity in homogenates of arterial beds perfused with DSA was reduced to 52.4 ± 6.0% of con-

Table 1. The effects of perfusing DSA on the pressor response to noradrenaline and on various enzyme activities in vascular smooth muscle cells

	Response to noradrenaline (Increase over baseline pressure, mm Hg)	Area under curve: total radioactivity (arbitrary units)	Area under curve: metabolites (arbitrary units)	LDH activity in homogenates (μmol min^{-1} (mg protein)$^{-1}$)
Control Mesentery	49.5 ± 9.0 (n = 5)	532.9 ± 6.2 (n = 6)	68.3 ± 5.0 (n = 5)	0.502 ± 0.067 (n = 4)
DSA Mesentery	3.4 ± 0.8** (n = 4)	537.8 ± 5.9 (n = 6)	37.8 ± 3.4*** (n = 5)	0.387 ± 0.093 (n = 4)

** $P < 0.01$ and *** $P < 0.001$ compared to control responses and activities

trol values, confirming that all SSAO activity remaining after perfusion with DSA can be measured by perfusing the permeant substrate, [^{14}C] benzylamine. These results therefore demonstrate that DSA has inhibited membrane bound proteins but has not penetrated the muscle cell membranes, leaving intracellular enzyme activities largely unaffected. Furthermore, they imply that almost half of the SSAO active sites in VSMCs face extracellularly.

Fractionation of rat brown adipocytes reveals an equal distribution of SSAO activity between microsomal and plasma membrane vesicles (Barrand and Callingham, 1982). The present studies provide evidence that a similar distribution ratio might exist for intracellular and plasmalemmal SSAO in VSMCs. The reduction in SSAO activity which occurs following perfusion of DSA is thus possibly due to complete inhibition of an SSAO enzyme located on the plasmalemma of VSMCs and thus ideally situated to metabolise blood-borne substrates.

The results presented do not, however, provide conclusive evidence in favour of this hypothesis. We cannot be sure that the extracellular distribution of SSAO mirrors that of α_1 receptors and, therefore, that complete abolition of the pressor response to noradrenaline is indicative that DSA has reached all outer surfaces of the VSMCs. However, results (not shown) from preliminary studies of the inhibition by DSA of SSAO in cultured rat aortic SMCs suggest that at least some of the SSAO in these cells is inaccessible to DSA. Similar experiments using dissociated rat mesenteric arterial SMCs might therefore provide more conclusive evidence.

Acknowledgements

A. H. is a Medical Research Council Scholar. Funding was provided by the Horserace Betting Levy Board. We wish to thank Dr. P. J. Richardson for helpful discussion of these experiments.

References

Barrand MA, Callingham BA (1982) Monoamine oxidase activities in brown adipose tissue of the rat: some properties and subcellular distribution. Biochem Pharmacol 31: 2177–2184

Berg HC (1969) Sulfanilic acid diazonium salt: a label for the outside of the human erythrocyte membrane. Biochim Biophys Acta 183: 65–78

Callingham BA, Barrand MA (1987) Some properties of semicarbazide-sensitive amine oxidases. J Neural Transm [Suppl] 23: 37–54

Grondal EJM, Zimmermann H (1986) Ectonucleotidase activities associated with cholinergic synaptosomes isolated from *Torpedo* electric organ. J Neurochem 47: 871–881

Holt A, Sharman DF, Callingham BA, Kettler R (1992a) Characteristics of procarbazine as an inhibitor in vitro of rat semicarbazide-sensitive amine oxidase. J Pharm Pharmacol 44: 487–493

Holt A, Sharman DF, Callingham BA (1992b) Effects in vitro of procarbazine metabolites on some amine oxidase activities in the rat. J Pharm Pharmacol 44: 494–499

Lyles GA, Callingham BA (1982) In vitro and in vivo inhibition by benserazide of clorgyline-resistant amine oxidases in rat cardiovascular tissues. Biochem Pharmacol 31: 1417–1424

Lyles GA, Singh I (1985) Vascular smooth muscle cells: a major source of the semicarbazide-sensitive amine oxidase of the rat aorta. J Pharm Pharmacol 37: 637–643

Authors' address: Dr. A. Holt, Department of Pharmacology, University of Cambridge, Tennis Court Road, Cambridge, CB2 1QJ, United Kingdom

J Neural Transm (1994) [Suppl] 41: 439–443

The ex vivo effects of procarbazine and methylhydrazine on some rat amine oxidase activities

A. Holt and **B. A. Callingham**

Department of Pharmacology, University of Cambridge, Cambridge, United Kingdom

Summary. Monoamine oxidase (MAO) and semicarbazide-sensitive amine oxidase (SSAO) activities were examined in homogenates of various rat tissues following i.p. administration of procarbazine or methylhydrazine. Both compounds inhibited SSAO in a dose-dependent manner in all tissues examined, with methylhydrazine the more potent agent in this respect. Little inhibition of MAO could be detected in most cases. However, hepatic MAO-B activity was potentiated significantly in rats receiving methylhydrazine and both drugs caused a dose-dependent potentiation of MAO-A in homogenates of brown adipose tissue. The potential use of these compounds in vivo as selective SSAO inhibitors is discussed.

Introduction

Hydrazine derivatives generally rank among the most potent of semicarbazide-sensitive amine oxidase (SSAO) inhibitors and one of these, the benzylhydrazine derivative, procarbazine is employed in high doses, both in the treatment of Hodgkin's lymphoma (Bonadonna et al., 1969) and of CNS tumours in children (van Eys et al., 1985).

The ability of procarbazine to cross the blood-brain barrier means that central neurotoxicity is not uncommon and this has been attributed, at least in part, to the competitive inhibition of MAO (De Vita et al., 1965; Pfefferbaum et al., 1989). Recent in vitro studies have demonstrated that procarbazine and two of its metabolites, azoprocarbazine and monomethylhydrazine, are potent and selective inhibitors of SSAO in rat brown adipose tissue, when compared with their effects on rat hepatic MAO activities (Holt et al., 1992a,b). The present study examines whether or not these properties are retained ex vivo by estimating SSAO and MAO activities in homogenates of various tissues following administration of procarbazine or methylhydrazine to rats.

Materials and methods

Male Wistar rats (60–185 g) received i.p. injections of 0.9% w/v saline (controls), procarbazine ($0.1–100 \, mg \, kg^{-1}$; a gift from Hoffmann-La Roche) or methylhydrazine ($1 \, \mu g \, kg^{-1} – 1 \, mg \, kg^{-1}$; Aldrich) and were killed by decapitation two hours later. Aortae, interscapular brown adipose tissue (BAT), brains, hearts, livers and lungs were removed and homogenates of these tissues prepared in potassium phosphate buffer (1 mM, pH 7.8). Mitochondria in BAT from control and methylhydrazine-treated rats ($1 \, mg \, kg^{-1}$ i.p.) were examined by electron microscopy.

Amine oxidase activities in homogenates were estimated by incubating samples with [^{3}H] 5-hydroxytryptamine (250 μM, $2 \, \mu Ci \, \mu mol^{-1}$; MAO-A), [^{14}C] benzylamine (250 μM, $1 \, \mu Ci \, \mu mol^{-1}$; MAO-B) and [^{14}C] benzylamine (10 μM, $10 \, \mu Ci \, \mu mol^{-1}$; SSAO) at 37°C under oxygen followed by solvent extraction of metabolites (Lyles and Callingham, 1982).

Protein contents of homogenates were estimated by the method of Lowry (1951).

Results

Figures 1a–c are representative graphs showing the effects of drug treatment on amine oxidase activities in heart, liver and BAT, respectively. Treatment of rats with either procarbazine or methylhydrazine resulted in a dose-dependent inhibition of SSAO activities in all tissues examined. However, low doses of both drugs produced a small but significant potentiation of SSAO in lung homogenates. Little inhibition of either form of MAO occurred although MAO-A in aortic homogenates displayed some sensitivity to procarbazine.

In rats treated with methylhydrazine, significant potentiation of hepatic MAO-B occurred in a non-dose-dependent manner (Fig. 1b), but potentiation by procarbazine did not reach statistical significance. In contrast, both procarbazine and methylhydrazine caused significant, dose-dependent potentiation of MAO-A in homogenates of BAT (Fig. 1c).

Electron microscopic examination of BAT did not reveal any obvious differences in number, size or outer membrane integrity of mitochondria from control or methylhydrazine-treated rats.

Discussion

Following i.p. administration, procarbazine and a putative metabolite of this drug, methylhydrazine, inhibited selectively SSAO enzymes in

Fig. 1. Ex vivo effects of a range of doses of procarbazine (left hand panels) and methylhydrazine (right hand panels) on MAO-A (◯), MAO-B (△) and SSAO (☐) activities in homogenates of heart (**a**), liver (**b**) and BAT (**c**). Enzyme activities are expressed as percentages of those measured in homogenates of tissues from control animals, with each point the mean of five triplicate determinations. Standard errors of the ratios are shown where error bars exceed symbol size. Wilcoxon's two-sample rank test was used to compare activities from drug treated animals with control values, shown to the left of each plot. *p < 0.05 and **p < 0.01 compared to control activities

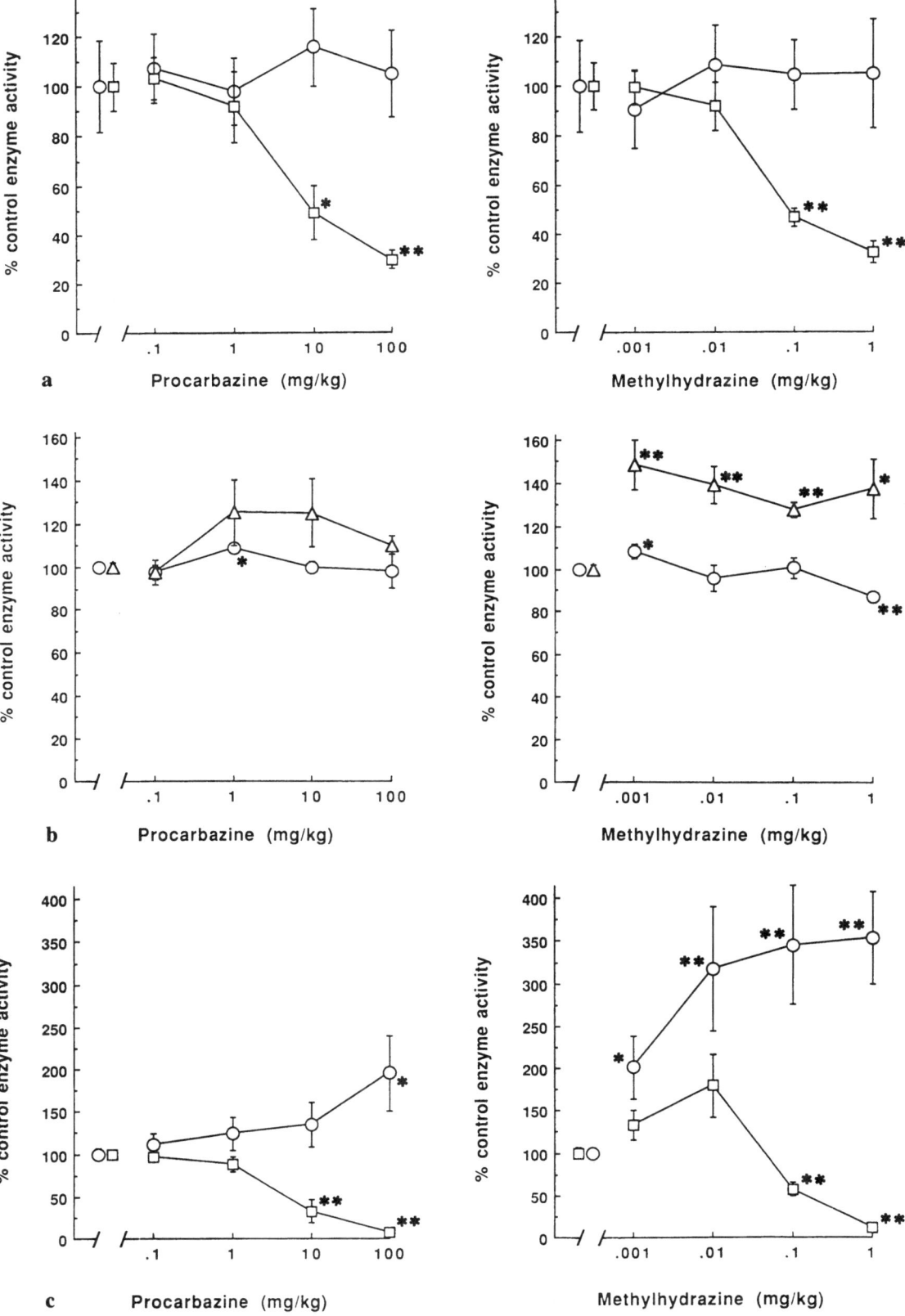

homogenates of various rat tissues when compared with their effects on MAO activities in the same tissues. Methylhydrazine was more potent than procarbazine in inhibiting SSAO, with its ID_{50} values of around $0.1\,mg\,kg^{-1}$ in all tissues examined some 100 times lower than those measured for procarbazine.

These results suggest that both compounds might be suitable for administration to rats when inhibition of SSAO with no concomitant inhibition of MAO is required. However, the possible consequences of increased MAO activity which occurred in some tissues must be considered when interpreting results from such experiments.

Procarbazine is thought to be a suicide inhibitor of SSAO, initially forming a loose association with the active site, thereafter being metabolised slowly to a reactive intermediate which binds tightly to the cofactor (Holt et al., 1992a). However, rapid hepatic metabolism of procarbazine, by MAO and P_{450}-dependent activities, to azoprocarbazine (Coomes and Prough, 1983) might mean that SSAO inhibition results rather from the action of this proximal metabolite than from that of procarbazine itself. Another putative metabolite, methylhydrazine, is more potent than either of these compounds in vitro but is much more rapidly reversible by dialysis (Holt et al., 1992b).

The turnover of rat MAO is sufficiently slow (see Strolin Benedetti and Dostert, 1992) that increased enzyme activity following administration in vivo of procarbazine or methylhydrazine most likely results rather from a direct modulatory effect on the enzyme molecule than on increased enzyme synthesis. Recently, it has been proposed by Koenig et al. (1992) that the high potency of befloxatone against MAO-A might be explained by a combination of an association between the FAD cofactor and the oxazolidinone moiety of befloxatone and an interaction between the trifluorohydroxyl moiety of the inhibitor and a secondary binding site, close to the MAO-A active site.

This second modulatory site is probably not obscured by bound substrate, perhaps allowing certain endogenous or exogenous small molecules such as methylhydrazine to bind and thus modify MAO activity. It is hoped that an examination of the effects of methylhydrazine on the kinetic constants of MAO-A in BAT will provide clues as to whether or not such an interaction might explain the dramatic potentiation of MAO which was measured in these experiments.

Acknowledgements

A. H. is a Medical Research Council Scholar. Funding was provided by the Horserace Betting Levy Board.

References

Bonadonna G, Monfardini S, Oldini C (1969) Comparative effects of vinblastine and procarbazine in advanced Hodgkin's disease. Eur J Cancer 5: 393–402

Coomes MW, Prough RA (1983) The mitochondrial metabolism of 1,2-disubstituted hydrazines, procarbazine and 1,2-dimethylhydrazine. Drug Metab Dispos 11: 550–555

De Vita VT, Hahn MA, Oliverio VT (1965) Monoamine oxidase inhibition by a new carcinostatic agent, N-isopropyl-α-(2-methylhydrazino)-p-toluamide (MIH). Proc Soc Exp Biol Med 120: 561–565

Holt A, Sharman DF, Callingham BA, Kettler R (1992a) Characteristics of procarbazine as an inhibitor in vitro of rat semicarbazide-sensitive amine oxidase. J Pharm Pharmacol 44: 487–493

Holt A, Sharman DF, Callingham BA (1992b) Effects in vitro of procarbazine metabolites on some amine oxidase activities in the rat. J Pharm Pharmacol 44: 494–499

Koenig J-J, Moureau F, Vercauteren D, Durant F, Ducrey F, Jarreau F-X (1992) Befloxatone, a spontaneously reversible MAO-A inhibitor: modelisation at molecular level. Clin Neuropharmacol 15 [Suppl] 1: 424B

Lowry OH, Rosebrough NJ, Farr AL, Randall RJ (1951) Protein measurement with the Folin phenol reagent. J Biol Chem 193: 265–275

Lyles GA, Callingham BA (1982) In vitro and in vivo inhibition by benserazide of clorgyline-resistant amine oxidases in rat cardiovascular tissues. Biochem Pharmacol 31: 1417–1424

Pffeferbaum B, Pack R, van Eys J (1989) Monoamine oxidase inhibitor toxicity. J Am Acad Child Adolesc Psychiatry 28: 954–955

Strolin Benedetti M, Dostert P (1992) Monoamine oxidase: from physiology and pathophysiology to the design and clinical application of reversible inhibitors. Adv Drug Res 23: 65–125

van Eys J, Cangir A, Coody D, Smith B (1985) MOPP regimen as primary chemotherapy for brain tumors in infants. J Neurooncol 3: 237–243

Authors' address: Dr. A. Holt, Department of Pharmacology, University of Cambridge, Tennis Court Road, Cambridge, CB2 1QJ, United Kingdom

J Neural Transm (1994) [Suppl] 41: 445–448

Histaminase activity of mesenteric artery of the rat

G. Banchelli, G. Ignesti, R. Pirisino, L. Raimondi, and **F. Buffoni**

Department of Pharmacology, University of Florence, Italy

Summary. In rat mesenteric artery homogenates histamine is oxidatively deaminated at high rate whereas putrescine is a poor substrate. The oxidation of histamine appears to be mainly catalyzed by an SSAO enzyme with high affinity for benzylamine (Bz.SSAO). Histamine oxidation is inhibited by B24 (2,5-diethoxy-4-aminomethylpyridine), a selective inhibitor of Bz.SSAO enzymes, and reduced by the presence of benzylamine.

The Bz.SSAO enzyme which is present in the rat mesenteric artery is not only able to oxidize histamine, but also methylamine, acetylputrescine and some methylated forms of histamine including 1,4-methylhistamine.

Introduction

Histamine is stored in the mast cells surrounding blood vessels on which it has important physiological and pathophysiological functions (Levi et al., 1991). The metabolism of histamine in mammalian tissue is carried out by two main enzymatic pathways: direct oxidative deamination generally considered to be catalyzed by a diamine oxidase (Maslinski and Fogel, 1991) and imidazole ring-tele-methylation catalyzed by a methyltransferase followed by oxidative deamination by the mitochondrial monoamine oxidase (MAO).

Recently it was shown that some tissue bound Bz.SSAO enzymes may have a role in the metabolism of histamine (Raimondi et al., 1991; Ignesti et al., 1992). The mesenteric arteries of the rat are very rich in Bz.SSAO activity (Coquil et al., 1973; Goridis and Neff, 1974). These observations encouraged us to attempt the characterization of the enzyme able to catalyze the oxidative deamination of histamine in the mesenteric artery of the rat.

Methods

Male Wistar rats of 180–250 g were killed by decapitation and bled. The mesenteric blood vessels were cleaned of adhering fat and frozen until needed. Samples were obtained from a pool of 3 rat mesenteric arterial beds and homogenized (1 in 20 w/v) in 10 mM phosphate buffer pH 7.4 containing 0.25 M sucrose. Homogenates were centrifuged at 600 g for 10 min to remove unbroken cells and nuclei. The enzymic activity was assayed at 37°C and pH 7.4 in air using ^{14}C-putrescine and ^{14}C-benzylamine and unlabelled substrates according to the method of Ignesti et al. (1992). Protein content was determined by the method of Lowry et al. (1951).

Results

In the rat mesenteric artery homogenates a feeble oxidation of ^{14}C-putrescine was observed, i.e., 12 ± 10 pmoles/mg of protein/min (mean $\pm$ s.e. of 5 samples in duplicate). Similar results were obtained measuring the hydrogen peroxide production in the oxidation of putrescine. This indicated that the level of the diamine oxidase activity was very low in this tissue.

Figure 1 shows the oxidative deamination of some amines and the effects of selective inhibitors. It is evident that the oxidation of all the amines, with the exception of tyramine, was strongly inhibited by B24 but not significantly by deprenyl. Therefore this enzyme appeared to be a Bz.SSAO. It was inhibited by high benzylamine concentrations, for this reason 25 μM benzylamine was used in this experiments when such inhibition was not evident.

The apparent kinetic constants for histamine computed according to Wilkinson (1961) using 6 different substrate concentrations from 0.1 to 5 mM were: K_m 0.60 ± 0.05 mM and V_{max} 1.6 ± 0.04 nmoles/mg of protein/min. In the presence of 0.02 mM α-aminoguanidine they were modified as follows: K_m 0.70 ± 0.08 mM and V_{max} 0.85 ± 0.03 nmoles/mg of protein/min.

The effect of B24 on the apparent kinetic constants of histamine oxidation (Fig. 2) is similar to its effect on other Bz.SSAO enzymes (Banchelli et al., 1990).

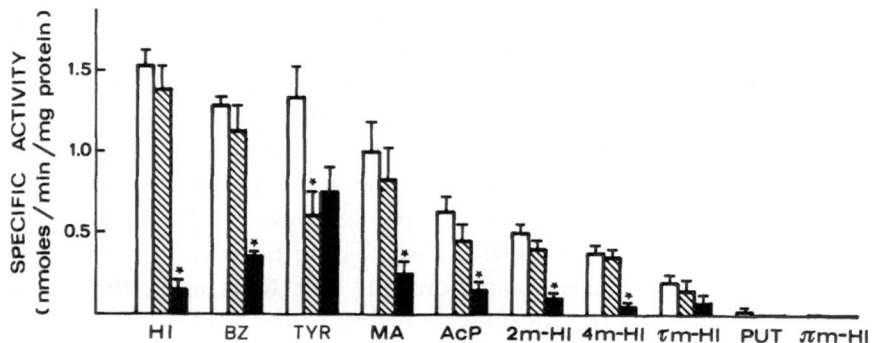

Fig. 1. The metabolism of some amines by rat mesenteric artery homogenates measured by a fluorimetric method. Homogenates were preincubated for 30 min at 37°C prior to incubation with amines. The values given for each amine in all three groups are the means $\pm$ S.E. (n = 4 in duplicate). ·*HI* histamine 1 mM, *TYR* tyramine 1 mM, *BZ* benzylamine 25 μM, *MA* methylamine 1 mM, *AcP* acetylputrescine 1 mM, *2m-HI* 2-methylhistamine 1 mM, *4m-HI* 4-methylhistamine 1 mM, *τm-HI* N-tele-methylhistamine 1 mM, *PUT* putrescine 1 mM, *πm-HI* pros-methylhistamine 1 mM. *Significantly different from uninhibited samples (p < 0.05). □ Without inhibitors, ▨ +Selegiline 10^{-3} M, ■ +B24 $2 \cdot 10^{-5}$ M

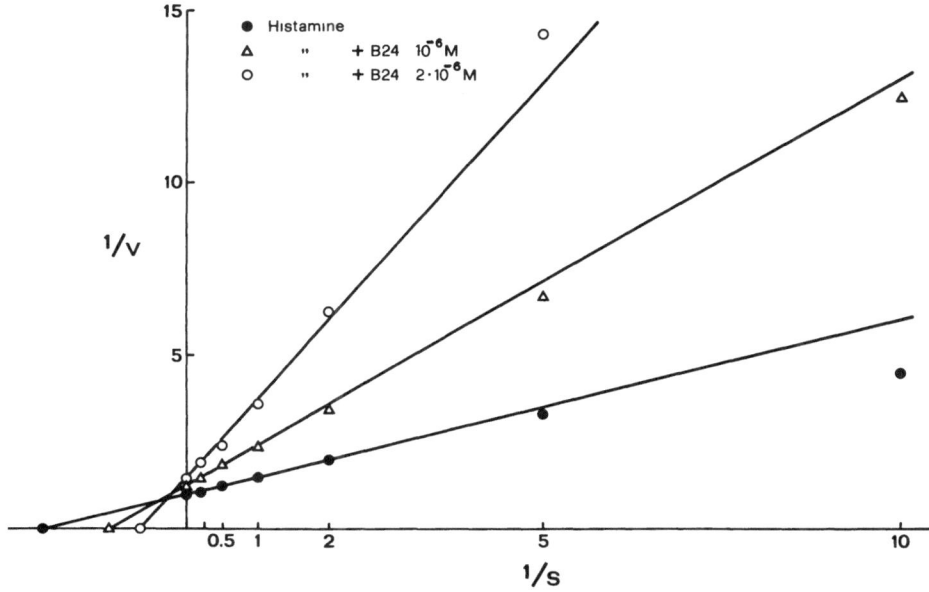

Fig. 2. The metabolism of histamine by homogenates were studied following preincubation with B24 for 30 min at 37°C. The kinetic parameters were computed according to Wilkinson using 6 substrate concentrations ranging from 0.1 to 5 mM

Table 1. Competition between benzylamine and histamine on H_2O_2 production of rat mesenteric artery homogenates

Substrates	nmoles/mg of protein/min		
	Experim.	Theoret.	(T-E)/T%
Benzylamine 25 μM	1.28 ± 0.12		
Histamine 1 mM	1.07 ± 0.12		
Benz. 25 μM + Hist. 1 mM	1.61 ± 0.19*	2.35 ± 0.17	31.5 ± 4.7

Competition between histamine and benzylamine on hydrogen peroxide production further showed that the same enzyme was operative in the oxidation of these two amines (Table 1).

Conclusion

Competition experiments and the use of an inhibitor (B24) which selectively inhibits Bz.SSAO enzymes but does not inhibit DAO (Bertini et al., 1985) or enzymes belonging to the same class (E.C. 1,4,3,6), have shown that, in the mesenteric arteries of the rat, histamine is oxidatively deaminated by the Bz.SSAO. This enzyme also acts on other physiological substrates such as acetylputrescine, 1,4-methylhistamine and methylamine. Bz.SSAO of rat mesenteric arteries therefore, in addition to its direct ability to oxidize histamine, seems to share with the mithocondrial MAO a role in the oxidation of telemethylhistamine and acetylputrescine.

Acknowledgements

We thank the MURST for the financial support and V. Bertini et al. (University of Genova, Italy) for the kind gift of B24.

References

Banchelli G, Buffoni F, Elliott J, Callingham BA (1990) A study of the biochemical pharmacology of 3,5-ethoxy-4-aminomethylpyridine (B24) a novel amine oxidase inhibitor with selectivity for tissue bound semicarbazide-sensitive amine oxidase enzymes. Neurochem Int 17: 215–221

Bertini V, De Munno A, Lucchesini F, Buffoni F, Bertocci B (1985) Italian patent application n.47906-A/85 extended to Europe, U.S.A., Canada, Japan

Coquil JF, Goridis C, Mace G, Neff NH (1973) Monoamine oxidases in rat arteries: evidence for different forms and selective localization. Br J Pharmacol 48: 590–599

Goridis C, Neff NH (1974) Selective localization of monoamine oxidase forms in rat mesenteric artery. Biochem Pharmacol 23 [Suppl] (part I): 106–109

Ignesti G, Banchelli G, Raimondi L, Pirisino R, Buffoni F (1992) Histaminase activity in rat lung and its comparison with intestinal mucosal diamine oxidase. Agents Actions 35: 192–199

Levi R, Rubin L, Gross SS (1991) Histamine in cardiovascular function and disfunction: recent development. In: Born GVR, Cuatrecasas P, Herken H (eds) Handbook of experimental pharmacology, vol 97. Springer, Berlin Heidelberg New York Tokyo, pp 347–383

Lowry HO, Rosebrough NJ, Lewis Farr A, Randal RJ (1951) Protein measurement with the Folin phenol reagent. J Biol Chem 193: 265–275

Maslinski C, Fogel WA (1991) Catabolism of histamine. In: Born GVR, Cuatrecasas P, Herken H (eds) Handbook of experimental pharmacology, vol 97. Springer, Berlin Heidelberg New York Tokyo, pp 165–189

Raimondi L, Pirisino R, Ignesti G, Capecchi S, Banchelli G, Buffoni F (1991) Semicarbazide-sensitive amine oxidase (SSAO) of rat epididymal white adipose tissue. Biochem Pharmacol 41: 467–470

Wilkinson GN (1961) Statistical estimation in enzyme kinetics. Biochem J 80: 324–332

Author's address: Prof. Dr. F. Buffoni, Dipartimento di Farmacologia preclinica e clinica, Università di Firenze, Viale G. B. Morgagni, 65, I-50134 Firenze, Italy

Subject Index